Machine Learning Algorithms for Signal and Image Processing

Machine Learning Algorithms for Signal and Image Processing

Edited by

Deepika Ghai
Lovely Professional University, IN

Suman Lata Tripathi
Lovely Professional University, IN

Sobhit Saxena
Lovely Professional University, IN

Manash Chanda
Meghnad Saha Institute of Technology, IN

Mamoun Alazab
Charles Darwin University, AS

IEEE PRESS
WILEY

Published by John Wiley & Sons, Inc., Hoboken, New Jersey.
Published simultaneously in Canada.

For general information on our other products and services or for technical support, please contact our Customer Care Department within the United States at (800) 762-2974, outside the United States at (317) 572-3993 or fax (317) 572-4002.

Wiley also publishes its books in a variety of electronic formats. Some content that appears in print may not be available in electronic formats. For more information about Wiley products, visit our website at www.wiley.com.

Library of Congress Cataloging-in-Publication Data Applied for:

Hardback ISBN: 9781119861829

Cover Design: Wiley
Cover Image: © Ryzhi/Shutterstock

Set in 9.5/12.5pt STIXTwoText by Straive, Chennai, India

Contents

Editor Biography

Dr. Deepika Ghai received her Ph.D in the area of signal and image processing from Punjab Engineering College, Chandigarh. She received her M.Tech in VLSI Design & CAD from Thapar University, Patiala, and B.Tech in electronics and communications engineering from Rayat Institute of Engineering and Technology, Ropar. She is an Assistant Professor at Lovely Professional University with more than 8 years' academic experience. She received the Dr. C.B. Gupta Award in 2021 at Lovely Professional University. She has published more than 30 research papers in refereed journals and conferences. She has worked as a session chair, conference steering committee member, editorial board member, and reviewer in international/national IEEE journals and conferences. She has also published edited book "Health Informatics and Technological Solutions for Coronavirus (COVID-19)" in CRC Taylor & Francis. She is associated as a life member of the Indian Science Congress. Her area of expertise includes signal and image processing, biomedical signal and image processing, and VLSI signal processing.

Dr. Suman Lata Tripathi received her Ph.D. in the area of microelectronics and VLSI from MNNIT, Allahabad. She received her M.Tech in electronics engineering from UP Technical University, Lucknow, and B.Tech in electrical engineering from Purvanchal University, Jaunpur. In 2022 she has worked as are mote post-doc researcher at Nottingham Trent University, London, UK. She is a Professor at Lovely Professional University and has more than 19 years' academic experience. She has published more than 72 research papers in refereed IEEE, Springer, Elsevier, and IOP science journals and conferences. She has also been awarded 13 Indian patents and 2 copyrights. She has organized several workshops, summer internships, and expert lectures for students. She has worked as a session chair, conference steering committee member, editorial board member, and peer reviewer in international/national IEEE, Springer, Wiley journals and conferences, etc. She received the "Research Excellence Award" in 2019 and "Research Appreciation Award" in 2020, 2021 at Lovely Professional University, India. She received the best paper at IEEE ICICS-2018. She has edited and authored more than 15 books in different areas of electronics and electrical engineering. She has edited works for Elsevier, CRC Taylor and Francis, Wiley-IEEE Press, Nova Science, Apple Academic Press, etc. She is also working as a book series editor for *Smart Engineering Systems* and a conference series editor for *Conference Proceedings Series on Intelligent Systems for Engineering Designs* with CRC Press. She is the guest editor of a special issue in "Current Medical Imaging" Bentham Science. She is a senior member, IEEE, fellow at IETE, and life member at ISC and is continuously involved in different professional activities along with academic work. Her area of expertise includes microelectronics device modeling and characterization, low power VLSI circuit design, VLSI design of testing, and advanced FET design for IoT, embedded system design, reconfigurable architecture with FPGAs, and biomedical applications.

Dr. Sobhit Saxena received his Ph.D. from IIT Roorkee in the area of nanotechnology. He did his M.Tech in VLSI and B.E. in electronics and communication engineering. His area of expertise includes nanomaterial synthesis and characterization, electrochemical analysis and modeling, and simulation of CNT-based interconnects for VLSI circuits. He has designed a new hybrid system of Li-ion batteries and supercapacitors for energy storage applications. He worked as a SEM (scanning electron microscopy) operator for four years against MHRD fellowship. He has a vast teaching experience of more than 14 years in various colleges and universities. Currently, he is working as an Associate Professor in the School of Electronics and Electrical Engineering, Lovely Professional University. He has been awarded the "Perfect Award" four times in consecutive years for achieving 100% result. He has published more than 10 research papers in SCI/Scopus indexed journals and about 20 papers in reputed international conferences/non-indexed journals. He has filed three patents, published an edited book "*Advanced VLSI Design and Testability Issues*" with CRC Press, and two book chapters. He has also published one authored book, *Digital VLSI Design and Simulation with Verilog*, with Wiley. He is an IEEE member and a reviewer at various refereed SCI/Scopus indexed journals and conference proceedings. He also has industrial exposure in two different companies related to manufacturing (PCB) and broadband communication.

Dr. Manash Chanda graduated in electronics and communication engineering from Kalyani Govt. Engineering College in 2005. He obtained his M.Tech degree in VLSI and microelectronics from Jadavpur University. He completed his Ph.D in engineering from ETCE Dept., Jadavpur University, in 2018. At present, he is working as an Assistant Professor in the Department of ECE, Meghnad Saha Institute of Technology, since February 2006. He is a member of IEEE and is currently a member of IEEE Electron Device Society and Solid State Circuit Society. Dr. Chanda is the co-founder of IEEE Student Branch and ED MSIT Student Branch Chapter. At present, he is the Chapter Advisor of ED Meghnad Saha Institute of Technology Student Branch Chapter. Also, he is the Vice Chairman of ED Kolkata Chapter. He served as the Secretary of IEEE ED MSIT SBC from January 2018 to December 2019. He has published more than 65 refereed research papers and conference proceedings. His current research interest spans around the study of analytical modeling of sub 100-nm MOSFETs and nanodevices considering quantum mechanical effects, low-power VLSI designs, SPICE modeling of nanoscale devices, memory designs, etc. He has published papers in refereed international journals of reputed publishers like IEEE, Elsevier, IET, Springer, Wiley, to name a few. He is the reviewer of many reputed international journals and conferences like IEEE TCAS, IEEE TVLSI, IEEE TED, Solid State Circuits (Elsevier), Journal of Computational Electronics (Springer), International Journal of Numerical Modeling: Electronic Networks, Devices and Fields (Wiley), International Journal of Electronics (Taylor and Francis), etc. He is the recipient of University Gold medal in M. Tech from Jadavpur University in 2008. One of his projects was selected in the Top 10 VLSI project design category (including B. Tech and M.Tech) all over INDIA, organized by CADENCE DESIGN CONTEST, BANGALORE, India in 2010.

Dr. Mamoun Alazab is an associate professor at the College of Engineering, IT, and Environment, and the Inaugural Director of the NT Academic Centre for Cyber Security and Innovation (ACCI) at Charles Darwin University, Australia. He is a cyber-security researcher and practitioner with industry and academic experience. His research is multidisciplinary and focuses on cyber security and digital forensics of computer systems with a focus on cybercrime detection and prevention. He has published more than 300 research papers (>90% in Q1 and in the top 10% of journal articles, and more than 100 in IEEE/ACM Transactions) and 15 authored/edited books. He received several awards including the NT Young Tall Poppy (2021) from the Australian Institute of Policy and Science (AIPS), IEEE Outstanding Leadership Award (2020), the CDU College of Engineering, IT and Environment Exceptional Researcher Award in (2020) and (2021), and 4 Best Research Paper Awards. He is ranked in top 2% of world's scientists in the subfield discipline of Artificial Intelligence (AI) and Networking & Telecommunications (Stanford University). He was ranked in the top 10% of 30k cyber security authors of all time. Professor Alazab was named in the 2022 Clarivate Analytics Web of Science list of Highly Cited Researchers, which recognizes him as one of the world's most influential researchers of the past decade through the publication of multiple highly cited papers that rank in the top 1% by citations for field and year in Web of Science. He delivered more than 120 keynote speeches, chaired 56 national events and more than 90 international events; on program committees for 200 conferences. He serves as the Associate Editor of *IEEE Transactions on Computational Social Systems*, *IEEE Transactions on Network and Service Management* (TNSM), *ACM Digital Threats: Research and Practice, Complex & Intelligent Systems*.

List of Contributors

Souid Abdelbaki
Department of Electrical Engineering
MACS Research Laboratory RL16ES22
National Engineering School of Gabes
Gabes University
Gabes
Tunisia

Faris Almalki
Department of Computer Engineering
College of Computers and Information Technology
Taif University
Taif
Kingdom of Saudi Arabia

J. Anil Raj
Department of Electronics and Communication
Muthoot Institute of Technology and Science
Kochi
Kerala
India

and

Department of Computer Science
Cochin University of Science and technology
Kochi
Kerala
India

Sundaram Arun
Department of Electronics and Communication
Engineering
Jerusalem College of Engineering
Chennai
India

Anterpreet K. Bedi
Department of Electrical and Instrumentation
Engineering
Thapar institute of Engineering and Technology
Patiala
Punjab
India

Suman Bera
Department of Computer Science and Engineering
National Institute of Technology
Silchar
Assam
India

Hutashan V. Bhagat
Department of Computer Science and Engineering
Sant Longowal Institute of Engineering and
Technology
Longowal
Sangrur
India

Anupam Biswas
Department of Computer Science and Engineering
National Institute of Technology
Silchar
Assam
India

Angshuman Bora
Department of Computer Science and Engineering
National Institute of Technology
Silchar
Assam
India

Yogini D. Borole
Department of E & TC Engineering
G H Raisoni College of Engineering and Management
SPPU Pune University
Pune
India

Subham Chakraborty
Department of Computer Science and Engineering
National Institute of Technology
Silchar
Assam
India

Koneti Chandra Sekhar
School of Electronics and Electrical Engineering
Lovely Professional University
Phagwara
Punjab
India

Pratik Chattopadhyay
Department of Computer Science and Engineering
Indian Institute of Technology (Banaras Hindu University)
Varanasi
India

Palungbam R. Chanu
Electronics and Communication Engineering
NIT Nagaland
Chumukedima
Nagaland

Kaustav Chaudhury
Electronics and Communication Engineering
Heritage Institute of Technology
Anandapur
Kolkata
India

Aneeta Christopher
Department of Electronics and Communication Engineering
National Institute of Technology Calicut
Calicut
Kerala
India

Debangshu Dey
Department of Electrical Engineering
Jadavpur University
Kolkata
West Bengal
India

Thangaraju Dineshkumar
Department of Electronics and Communication Engineering
Kongunadu College of Engineering and Technology
Trichy
India

Paul Fieguth
Vision Image Processing Lab
Department of Systems Design Engineering
University of Waterloo
Waterloo
Canada

Biswarup Ganguly
Department of Electrical Engineering
Meghnad Saha Institute of Technology
Maulana Abul Kalam Azad University of Technology
West Bengal
India

Deepika Ghai
School of Electronics and Electrical Engineering
Lovely Professional University
Phagwara
Punjab
India

R. Hari Kishan
Department of Electronics and Communication
Engineering
National Institute of Technology Calicut
Calicut
Kerala
India

Samrina Hussain
Department of Drug Design and Pharmacology
University of Copenhagen
Denmark

Sumam M. Idicula
Department of Artificial Intelligence and Data Science
Muthoot Institute of Technology and Science
Kochi
Kerala
India

Hemlata M. Jadhav
Electronics and Telecommunication Department
Marathwada Mitra Mandal's College of Engineering
Pune
India

Makarand M. Jadhav
Electronics and Telecommunication Department
NBN Sinhgad School of Engineering
Pune
India

Neelu Jain
Electronics and Communication Engineering
Department
Punjab Engineering College (Deemed to be University)
Chandigarh
India

Smita Kaloni
Department of Civil Engineering
National Institute of Technology
Uttarakhand
India

Maheshkumar H. Kolekar
Department of Electrical Engineering
Indian Institute of Technology
Patna
Bihar
India

Ranganathan Krishnamoorthy
Centre for nonlinear Systems
Chennai Institute of Technology
Chennai
India

Ashish Kumar
Department of Computer Science and Engineering
Indian Institute of Technology (Banaras Hindu University)
Varanasi
India

Kanak Kumar
Electronics Engineering Department
IEEE Member, Indian Institute of Technology (Banaras Hindu University)
Varanasi
India

Sachin Kumar
Department of Instrumentation and Control
Engineering
Dr B R Ambedkar National Institute of Technology
Jalandhar
India

Sandeep Kumar
Department of Electronics and Communications
Sreyas Institute of Engineering and Technology
Hyderabad
Telangana
India

Sanjeev Kumar
Department of BioMedical Applications (BMA)
Central Scientific Instruments Organisation
(CSIO)-CSIR
Chandigarh
India

P. Leninpugalhanthi
Department of EEE
Sri Krishna College of Technology
Coimbatore
Tamil Nadu
India

Swanirbhar Majumder
Department of Information Technology
Tripura University
Agartala
Tripura
India

Debashish Malakar
Department of Computer Science and Engineering
National Institute of Technology
Silchar
Assam
India

Mudasir Maqbool
Department of Pharmaceutical Sciences
University of Kashmir
Hazratbal
Srinagar
India

Madhusudhan Mishra
Department of ECE
NERIST
Nirjuli
Arunachal Pradesh
India

Indiran Mohan
Department of Computer science and Engineering
Prathyusha Engineering College
Chennai
India

Altaf Mulani
Electronics and Telecommunication Department
SKNSCOE
Pandharpur
India

Sugata Munshi
Department of Electrical Engineering
Jadavpur University
Kolkata
West Bengal
India

Swarup Nandi
Department of Information Technology
Tripura University
Agartala
Tripura
India

Mohamed A. Naiel
Vision Image Processing Lab
Department of Systems Design Engineering
University of Waterloo
Waterloo
Canada

Gadamsetti Narasimha Deva
School of Electronics and Electrical Engineering
Lovely Professional University
Phagwara
Punjab
India

Soufiene B. Othman
Department of Telecom, PRINCE Laboratory
Research, IsitCom, Hammam Sousse, Higher Institute
of Computer Science and Communication Techniques
University of Sousse
Sousse
Tunisia

Shanmugam Padmapriya
Department of Computer Science Engineering
Loyola Institute of Technology
Chennai
India

P. Pandiyan
Department of EEE
KPR Institute of Engineering and Technology
Coimbatore
Tamil Nadu
India

Pooja
Department of Instrumentation and Control
Engineering
Dr B R Ambedkar National Institute of Technology
Jalandhar, India

Dasari L. Prasanna
Department of Electronics and Communication
Engineering
Lovely Professional University
Phagwara
Punjab
India

Vuyyuru Prashanth
School of Electronics and Electrical Engineering
Lovely Professional University
Phagwara
Punjab
India

Himanshu Priyadarshi
Department of Electrical Engineering
Manipal University Jaipur
Jaipur
India

Ranjeet P. Rai
Department of Electronics Engineering
National Institute of Technology
Uttarakhand
India

Zobeir Raisi
Vision Image Processing Lab
Department of Systems Design Engineering
University of Waterloo
Waterloo
Canada

Husam Rajab
Department of Telecommunications and Media
Informatics
Budapest University of Technology and Economics
Budapest
Hungary

Ramya
Department of Electronics and Communication
Engineering
Sri Ramakrishna Engineering College
Anna University (Autonomous)
Coimbatore
India

Roshani Raut
Department of Information Technology
Pimpri Chinchwad College of Engineering
Pune
India

Abhisek Ray
Department of Electrical Engineering
Indian Institute of Technology
Patna
Bihar
India

Rehab A. Rayan
Department of Epidemiology
High Institute of Public Health
Alexandria University
Alexandria
Egypt

Thummala Reddychakradhar Goud
School of Electronics and Electrical Engineering
Lovely Professional University
Phagwara
Punjab
India

Hedi Sakli
Department of Electrical Engineering
MACS Research Laboratory RL16ES22
National Engineering School of Gabes
Gabes University
Gabes
Tunisia

and

EITA Consulting 5 Rue du Chant des oiseaux
Montesson
France

S. Saravanan
Department of EEE
Sri Krishna College of Technology
Coimbatore
Tamil Nadu
India

R. Senthil Kumar
Department of EEE
Sri Krishna College of Technology
Coimbatore
Tamil Nadu
India

Shagun Sharma
Department of Electronics Engineering
National Institute of Technology
Uttarakhand
India

Ashish Shrivastava
Faculty of Engineering and Technology
Shri Vishwakarma Skill University
Gurgaon
India

Ghanapriya Singh
Department of Electronics Engineering
National Institute of Technology
Uttarakhand
India

Kulwant Singh
Department of Electronics and Communication
Engineering
Manipal University Jaipur
Jaipur
India

Manminder Singh
Department of Computer Science and Engineering
Sant Longowal Institute of Engineering and
Technology
Longowal
Sangrur
India

Yeshwant Singh
Department of Computer Science and Engineering
National Institute of Technology
Silchar
Assam
India

Siva Sakthi
Department of Biomedical Engineering
Sri Ramakrishna Engineering College
Anna University (Autonomous)
Coimbatore
India

Gannamani Sriram
School of Electronics and Electrical Engineering
Lovely Professional University
Phagwara
Punjab
India

Sachin Srivastava
Department of Computer Science and Engineering
Indian Institute of Technology (Banaras Hindu
University)
Varanasi
India

P.V. Sudeep
Department of Electronics and Communication
Engineering
National Institute of Technology Calicut
Calicut
Kerala
India

Ramesh K. Sunkaria
Department of Electronics & Communication
Engineering
Dr B R Ambedkar National Institute of Technology
Jalandhar
Punjab
India

K.P. Suresh
Department of EEE
Sri Krishna College of Technology
Coimbatore
Tamil Nadu
India

Ranganathan Thiagarajan
Department of Information Technology
Prathyusha Engineering College
Chennai
India

Suman Lata Tripathi
School of Electronics & Electrical Engineering
Lovely Professional University
Phagwara
Punjab
India

Karan Veer
Department of Instrumentation and Control
Engineering
Dr B R Ambedkar National Institute of Technology
Jalandhar
India

Sidhant Yadav
Department of Electronics Engineering
National Institute of Technology
Uttarakhand
India

Georges Younes
Vision Image Processing Lab
Department of Systems Design Engineering
University of Waterloo
Waterloo
Canada

Imran Zafar
Department of Bioinformatics and Computational
Biology
Virtual University of Pakistan
Lahore
Punjab
Pakistan

John Zelek
Vision Image Processing Lab
Department of Systems Design Engineering
University of Waterloo
Waterloo
Canada

Muhammad Asim M. Zubair
Department of Pharmaceutics
The Islamia University of Bahawalpur
Pakistan

Preface

Machine learning (ML) algorithms for signal and image processing aid the reader in designing and developing real-world applications to answer societal and industrial needs using advances in ML to aid and enhance speech signal processing, image processing, computer vision, biomedical signal processing, text processing, etc. It includes signal processing techniques applied for pre-processing, feature extraction, source separation, or data decompositions to achieve ML tasks. It will advance the current understanding of various ML and deep learning (DL) techniques in terms of their ability to improve upon the existing solutions with accuracy, precision rate, recall rate, processing time, or otherwise. What is most important is that it aims to bridge the gap among the closely related fields of information processing, including ML, DL, digital signal processing (DSP), statistics, kernel theory, and others. It also aims to bridge the gap between academicians, researchers, and industries to provide new technological solutions for healthcare, speech recognition, object detection and classification, etc. It will improve upon the current understanding about data collection and data pre-processing of signals and images for various applications, implementation of suitable ML and DL techniques for a variety of signals and images, as well as possible collaboration to ensure successful design according to industry standards by working in a team. It will be helpful for researchers and designers to find out key parameters for future work in this area. The researchers working on ML and DL techniques can correlate their work with real-life applications of smart sign language recognition system, healthcare, smart blind reader system, text-to-image generation, or vice versa.

The book will be of interest to beginners working in the field of ML and DL used for signal and image analysis, interdisciplinary in its nature. Written by well-qualified authors, with work contributed by a team of experts within the field, the work covers a wide range of important topics as follows:

- Speech recognition, image reconstruction, object detection and classification, and speech and text processing.
- Healthcare monitoring, biomedical systems, and green energy.
- Real applications and examples, including a smart text reader system for blind people, a smart sign language recognition system for deaf people, handwritten script recognition, real-time music transcription, smart agriculture, structural damage prediction from earthquakes, and skin lesion classification from dermoscopic images.
- How various ML and DL techniques can improve the accuracy, precision rate recall rate, and processing time.

This easy-to-understand yet incredibly thorough reference work will be invaluable to professionals in the field of signal and image processing who want to improve their work. It is also a valuable resource for students and researchers in related fields who want to learn more about the history and recent developments in this field.

Acknowledgments

All editors would like to thank the School of Electronics and Electrical Engineering, Lovely Professional University, Phagwara, India; Department of Electronics and Communication Engineering, Meghnad Saha Institute of Technology, Kolkata; and College of Engineering, IT and Environment at Charles Darwin University, Australia; the Ministry of Education of the Republic of Korea and the National Research Foundation of Korea (NRF-2021S1A5A2A03064391); for providing necessary support for completing this book. The authors would also like to thank the researchers who have contributed their chapters to this book.

Section I

Machine Learning and Deep Learning Techniques for Image Processing

1

Image Features in Machine Learning

Anterpreet K. Bedi[1] and Ramesh K. Sunkaria[2]

[1]*Department of Electrical and Instrumentation Engineering, Thapar institute of Engineering and Technology, Patiala, Punjab, India*
[2]*Department of Electronics & Communication Engineering, Dr B R Ambedkar National Institute of Technology, Jalandhar, Punjab, India*

1.1   Introduction

With the speedily inflating visual data accessible in digital format, generating a highly efficient system for different purposes, such as browsing, searching, retrieving, and managing image data from large amount of database has become the high need of the hour. This has made it important for any image-processing application to take into consideration any complex information embedded in the data. An image may consist of both visual as well as semantic content in it. Visual data can be categorized into general and domain-specific. On the other hand, semantic form of data can be extracted using either simple textual annotation or by complicated inference processes depending on certain visual subjects. An image consists of thousands of pixels, each having its individual pixel intensity in several color channels. Images can be characterized by the understanding correlation and relationships between these pixels, thus helping draw a separate identity for each image. In general, it is difficult to integrate the entire information in a reasonable running time for image-processing purposes. Processing an image based on a particular image property like color or image in database may be a tedious task. This is because comparing every image in the database, including all its pixels, gets very difficult. Also, storing the whole image while creating the database might result in a reduction of storage capacity with the user. Hence, in order to overcome these problems, image features are extracted as a representative of each image.

A feature is described as a piece of data that is used to deliver information regarding the content of an image, typically regarding if a particular region of the image consists of some specific characteristics. A feature is any set of information that can be relevant to solving any task related to a particular application. Basically, image features are certain important and salient points on the image that are meaningful and detectable. They consist of relevant information needed to process the images. Features are limited in number and are not affected by irrelevant variations in the input. In cases where input image data is too extensive to be processed, as well as much of it is believed to be redundant, transforming the same into a minimum number of features is of utmost importance.

Features are easier to handle since they remain unchanged even with various image modifications like rotation, scaling, translation, etc. It is important that features of an image should not change with the image transformations. Also, it is required that features have to be insensitive to lighting conditions and color. Working with features using deep-learning techniques is more efficient and beneficial for various image-processing applications. The advantages of features for image-processing applications are described below:

1. Rather than storing an entire image, only important and crucial features can be acquired and saved.
2. It helps in preserving large space for data storage.

Machine Learning Algorithms for Signal and Image Processing, First Edition.
Edited by Deepika Ghai, Suman Lata Tripathi, Sobhit Saxena, Manash Chanda, and Mamoun Alazab.
© 2023 The Institute of Electrical and Electronics Engineers, Inc. Published 2023 by John Wiley & Sons, Inc.

3. It works faster, thus saving large storage and retrieval times.
4. Using features in deep learning image-processing applications help in improving its accuracy.

Thus, by using image features as an index, more efficient and accurate indexing systems can be generated. No single feature is entirely suitable for describing and representing any image. For instance, in image retrieval systems, color features are more suitable for retrieving colored images. Texture feature is best suited for describing and retrieving visual patterns, various surface properties, etc. In fact, various texture features are used in describing and retrieving medical images. Other than these, shape features are most suitable in describing certain specific shapes and boundaries of real-world objects and edges. Thus, no single feature can describe or represent an image completely.

This chapter is arranged as follows: Section 1.2 gives an introduction to feature vectors, followed by a detailed study on various low-level features in Section 1.3. The chapter is concluded in Section 1.4.

1.2 Feature Vector

Image-processing techniques based on deep learning make use of corresponding visual data that can be described in terms of features [1, 2]. Any measurements in an image that are extractable and are able to characterize it are considered as features [3]. Making use of images' contents in order to index the image set is called feature extraction. For computation, each feature that is extracted is further encrypted in multi-dimensional vector, known as feature vector. Feature extraction using deep learning helps in reducing the dimensionality by reducing the raw set of data into more manageable groups for processing. The components of the feature vectors are processed by image-analysis systems, which are further used for comparison among images. Feature vectors for the entire set of images from the database are combined together to result into a feature database.

Feature extraction is a crucial step in construction of the feature characteristics of an image and targets at extraction of pertinent information that can help in any image-processing application. A good set of features consists of unique information, which can distinguish one object from another. Features need to be as robust as possible so as to refrain from generation of different feature sets for a single object. Features can be categorized into global and local. In case of global features, visible features from the complete image are taken in one go, whereas, in case of local features, the image content-ID described using visual descriptors from a region or objects from the image [4]. The main concern for global features is that it might not be able to successfully compare users' regions/objects of interest while searching for images based on a given query image. Local features are extracted from multiple points in an image and hence are more robust to obstructions and clutter. Despite, they might require specific algorithms which might include an inconsistent amount of feature vectors per image. Further, combining the two sets results in errors reduced by a great extent.

Based on the level of abstraction, features can be classified into three broad categories [5]:

1. **Lower-Level Features:** These include visible cues that are obtained directly using raw pixel intensities from an image. These involve color, texture, and shape, etc.
2. **Middle-Level Features:** These consist of blobs or regions acquired as a consequence of image segmentation. Examples in this category include placement of particular kind of objects and scenes.
3. **Higher-Level Features:** These features give us semantic information regarding content of the image and its corresponding component objects along with their specific roles. Examples of higher-level features comprise of impressions, feelings, and meanings of combined perceptual features, viz., objects or roles tending to emotional or religious significance.

Since low-level features are extracted using information available at pixel level only, hence, they are also known as primitive features. Middle-level features demand an additional processing whose end-to-end automatization is still far from current techniques. Further, it is even harder to extract higher-level features from an image without explicit human assistance [6]. Most of the trending image-processing systems rely more on lower-level features for the efficient processing of images [7, 8]. Widely used features at lower-level include color, texture, and shape [9].

1.3 Lower-Level Image Features

Lower-level features refer to the visible contents obtained from the image. As described earlier, these features are extracted at pixel level from the image, without the need of any descriptions. Lower-level features are categorized into general and domain-specific. In case of general low-level features, color, texture, shape, spatial relationship, etc. are included. On the other hand, domain-specific low-level features are application-dependent and require proper domain knowledge. Low-level features form the basis of various techniques of image processing. These features are described below.

1.3.1 Color

Color is a widely exploited visible feature used in the image processing [10–13]. Color feature is used in image processing on the basis of two major reasons [14]: (i) it is considered as an effective feature descriptor that helps in the recognition of an object in a simpler manner, and (ii) it is easier for humans to differentiate between thousands of color intensities, rather than comparing a few dozen of gray shades. Color is a robust feature used for image-processing application. The colors are reflected from an object in query and are perceived by human-visual system. They are defined in different color spaces. Each image pixel is mapped into an element in a three-dimensional color space. Owing to its propinquity to human conceptualization, mostly used spaces for practical applications include RGB, Lab, and HSV.

RGB is a popularly exploited color model in the processing of digital images. The space represents a set of three values at each pixel, viz., red, green, and blue. It is popularly utilized in representation and display of colored images. It is an additive model and produces a color by adding the three components in different proportions. However, RGB color model is not uniform with regard to perception and is device-dependant. Hence, it is not much in use for generic systems. Further, Lab color space consists of one luminance (L) and a pair of chromatic elements (a and b). This space is advantageous since it can make perfect corrections in color balance by changing the chromatic components or by altering the contrast using luminance. The Lab model is more advantageous owing to its device-independent and also perceptually uniform nature.

The HSV color model is an extensively explored color model used in image processing. It comprises of three elements, viz., hue (H), saturation (S), and value (V). The hue indicates what color the component pertains to. The component is represented as an angle and ranging between 0° and 360°, where each degree corresponds to a distinct color. Further, saturation indicates the level of brightness of the corresponding color and may range between 0 and 1. Lastly, the value component describes the intensity of the color that is disintegrated from the color information acquired from the given image. Its value also ranges between 0 and 1. Being closer to human perception and invariant to light, HSV is preferred over other color models. Color-based image-processing applications exploit statistical measures such as histograms, moments, covariance matrix, and correlograms [12, 15].

1.3.1.1 Color Histogram

The feature was presented by Swain and Ballard and is widely used for colored image processing [13]. It constructs a vector by computing frequency of occurrence for various intensities present in different channels. It is an effective method to represent the color content for an image containing a unique pattern of colors when compared to the rest of the database. The histogram vector describes what kind of colors are present in an image, thus giving a distinctive histogram for each image. Also, the feature does not vary with translation or rotation of the plane, thus resulting in good performance [16]. Hence, histograms are well suited for image-processing applications since every histogram is unique and there is no spatial relevance.

1.3.1.2 Color Moments

This feature has been effectively exploited in many image-processing applications where image content comprises mostly of objects [17]. Color moments were introduced by Stricker and Orengo [18], and it defines the three basic central moments of the probability distribution for each color. The first color moment defines the average color of the image. The other two, i.e. second and third moments convey information about the variance and skewness of every channel present in the image respectively. Thus, only three values are used to represent each color channel. Hence, nine values describe the entire color content in the image, thus making the representation more compact as compared to other color features.

1.3.1.3 Color Coherence Vector

The feature was introduced in [19]. It utilizes coherence and incoherence features of image pixels for various image-processing applications. Color coherence vector takes spatial information of color images in account and tags every pixel in the image as either coherent or incoherent, corresponding to a given color. This color feature results in efficient results, since it contains spatial information along with color histogram. It is more applicable to images consisting of mostly uniform regions of color or texture.

1.3.1.4 Color Correlogram

It is a feature used to relate the spatial intensities present in the image [11]. Color correlogram converts color-spatial information of an image into a band of co-occurrence matrices. The 3D histogram comprises of colors of any pair of pixels in the first two dimensions, whereas the remaining dimension is used to mention their spatial distance. Although the feature provides results superior to other color features, it is computationally very expensive owing to its higher dimensionality.

1.3.2 Texture

Texture plays a critical role in various applications of image processing. It is a significant characteristic of appearance of many real-world images. It is a complex pattern comprised of units, known as sub-patterns, which consist of characteristic shape, size, color, brightness, etc. Thus, texture is helpful in similarity grouping in an image [20]. Texture provides the user with all the information related to spatial arrangement of the pixels and their intensities in any particular region in the image. Texture is capable enough to distinguish images from one another [21]. Texture features are described by coarseness, smoothness, and regularity of the image. These texture features can be categorized into signal processing, structural, model-based, and statistical features.

1.3.2.1 Signal Processing-Based Features

Feature extraction techniques using signal processing are widely exploited for extraction of texture features. Features using signal processing-based methods are extracted from spatial domain, frequency domain, and joint spatial/frequency domain. It involves application of various spatial and frequency domain filters and mathematical transformations resulting in coefficients to be later used as features of the corresponding image. A few transformations include Gabor filters [22], discrete wavelet transforms [23], etc. Features from an image, such as edges, lines, and isolated dots, can be extracted using these methods. The signal processing-based features describe the frequency content in spatially localized regions, thus further describing the characteristics of a texture [24]. These features achieve high resolution in both spatial and frequency domains.

1.3.2.2 Structural Features

Structural features correspond to texture primitives or elements, along with the hierarchy of their spatial relationships and arrangements [25]. Structural features have steady patterns large enough to be distinguished and segmented individually [26]. In order to define the structural textures, the primitives, also known as texels, are described along with syntactic rules for their positioning [27]. Structural features provide a satisfactory symbolic representation of an image. Examples of structural features include edges, elongated blobs, bars, and crosses. Mathematical morphology acts as a powerful tool for extraction of these features and is useful in medical-image analysis [28]. Since such regular geometries are rare in most real textures, hence the usage of structural methods is restricted. Also, the features can be ill-defined since there is no clear distinction between the texels and their spatial arrangements in the image.

1.3.2.3 Model-Based Features

Most of the model-based features make use of fractal models [29, 30], autoregressive models [31, 32], random-field models [33], epitome models [34], etc. The texture features in this category are estimated using stochastic and generative models. These methods initially make a model presumption for an input image. Model parameters are further calculated to be exploited as texture features. Model-based features also rely on the local neighborhood of pixels, along with their spatial arrangements. Owing to this fact, these features are extensively used for various medical-image analysis applications [35]. One of the major challenges faced in extracting model-based features involves estimating the model parameters. Hence, these features are generally less used for representing local structures in the image. Markov random field [36] is an example of model-based feature that is popularly used for processing of images.

1.3.2.4 Statistical Features

Statistical texture features extract information related to spatial distribution of different pixel intensities specified at relative positions in an image. Based on the number of pixels in every observation, various statistical features, varying from first to higher-order statistics, have been proposed. First-order statistical features include image histogram, which is easy to extract and is invariant to rotations as well as translations. Statistics of second-order analyze the similarities between a pair of pixels across a given image. A few techniques are co-occurrence matrix, such as Tamura features [37] and Wold features [38]. The idea behind the gray-level co-occurrence matrix (GLCM) is to spatially co-relate image pixels in order to extract texture feature from the image [39]. Further, several other statistical features, such as energy and entropy can also be deduced for texture characterization. Moving further, statistical features of higher-order describe relationships beyond pixel pairs. Features in this category include gray level run length [40], local binary patterns (LBPs) [41], etc.

Gray Level Co-occurrence Matrix (GLCM) Haralick et al. proposed the GLCM as an extensively used statistical method of second-order to extract features from an image [39]. It relates the distribution of co-occurring gray-scaled pixel pairs situated at a specified distance in a particular direction. GLCM features can be described using either a 2D image or a region of image, comprising of N_g quantized gray levels. The count of rows, as well as columns in the matrix, are same as the count of quantized gray levels. GLCM defines the number of times a grayscale occurs based on a specific spatial relationship with respect to some other grayscale in the image. Spatial relationship can be described as a function of inter-pixel distance d, along with direction θ. The co-occurrence matrix, $P(i, j, d, \theta)$, can thus be described in terms of the total number of pixel pairs, with grayscale intensity i for the first pixel, and j for the second pixel, at distance d to each other, in the direction θ. The matrix can be defined in four specific directions, viz., $0°$, $45°$, $90°$, and $135°$. It can be further normalized for subsequent computations.

A total of 22 GLCM texture features are available for a specific value of d and θ. In order to describe these features mathematically, following notions are used.

$P(i,j)$: describes probability of presence of the combination of i and j gray levels in sequence

$p(i,j)$: describes (i,j)th element of the normalized GLCM

$p_x(i)$: describes ith element of the marginal-probability matrix achieved by adding rows of $p(i,j)$

$p_y(j)$: describes jth element of the marginal-probability matrix achieved by adding columns of $p(i,j)$

N_g: number of quantized gray levels in the image

In order to achieve the mean and standard deviation for each available row and column, the following expressions can be used (Eqs. (1.1)–(1.4)):

$$\mu_x = \sum_{i=1}^{N_g}\sum_{j=1}^{N_g} i \cdot p(i,j) \tag{1.1}$$

$$\mu_y = \sum_{i=1}^{N_g}\sum_{j=1}^{N_g} j \cdot p(i,j) \tag{1.2}$$

$$\sigma_x = \sum_{i=1}^{N_g}\sum_{j=1}^{N_g} (i - \mu_x)^2 \cdot p(i,j) \tag{1.3}$$

$$\sigma_y = \sum_{i=1}^{N_g}\sum_{j=1}^{N_g} (j - \mu_y)^2 \cdot p(i,j) \tag{1.4}$$

where the terms μ_x and μ_y denote the mean or average values, whereas σ_x and σ_y refer to the standard deviations of p_x and p_y, respectively. Probabilities related to specified intensity sums and differences, i.e. p_{x+y} and p_{x-y} respectively, are defined by Eqs. (1.5)–(1.6):

$$p_{x+y}(k) = \sum_{\substack{i=1 \\ i+j=k}}^{N_g}\sum_{j=1}^{N_g} p(i,j), \quad k = 2,3,\ldots,2N_g \tag{1.5}$$

$$p_{x-y}(k) = \sum_{\substack{i=1 \\ |i-j|=k}}^{N_g}\sum_{j=1}^{N_g} p(i,j), \quad k = 2,3,\ldots,N_g-1 \tag{1.6}$$

Following features can be calculated using GLCM:

1. **Angular Second Moment:** This measure is also known as energy and is helpful in measuring the uniformity of the texture (Eq. (1.7)).

$$\text{Angular Second Moment} = \sum_{i=1}^{N_g}\sum_{j=1}^{N_g} \{p(i,j)\}^2 \tag{1.7}$$

2. **Contrast:** Contrast measures the amount of variations in pixel intensities between a pixel and its neighborhood (Eq. (1.8)).

$$\text{Contrast} = \sum_{n=0}^{N_g-1} n^2 \left\{ \sum_{\substack{i=1 \\ |i-j|=n}}^{N_g}\sum_{j=1}^{N_g} p(i,j) \right\} \tag{1.8}$$

3. **Correlation:** Correlation calculates the joint probability occurrence of the intended pixel pairs. It measures linear dependence of grayscale between pixels at intended points. Higher correlation value denotes a higher similarity between the specified points (Eq. (1.9)).

$$\text{Correlation} = \frac{\sum_{i=1}^{N_g}\sum_{j=1}^{N_g}(i.j)p(i,j) - \mu_x\mu_y}{\sigma_x\sigma_y} \tag{1.9}$$

4. **Homogeneity:** It is also known as inverse difference moment. It measures how uniformly a given region is structured in terms of gray-level variations, i.e. it measures the distribution of elements in the GLCM (Eq. (1.10)).

$$\text{Homogeneity} = \sum_{i=0}^{N_g-1}\sum_{j=0}^{N_g-1} \frac{1}{1+(i-j)^2}p(i,j) \tag{1.10}$$

5. **Variance:** Corresponding to the second moment, variance is used to compute the heterogeneity of a given image (Eq. (1.11)).

$$\text{Variance} = \sum_{i=1}^{N_g}\sum_{j=1}^{N_g}(i-\mu)^2 p(i,j) \tag{1.11}$$

6. **Sum Average:** It measures the amount of brightness of an image. It computes the average value of grayscale sum distribution of any given image. A brighter image corresponds to a higher value of sum average (Eq. (1.12)).

$$\text{Sum Average} = \sum_{i=2}^{2N_g} i \cdot p_{x+y}(i) \tag{1.12}$$

7. **Sum Entropy:** The sum entropy is used to calculate the amount of disorder corresponding to the grayscale sum distribution of a given image (Eq. (1.13)).

$$\text{Sum Entropy} = -\sum_{i=2}^{2N_g} p_{x+y}(i)\log\{p_{x+y}(i)\} \tag{1.13}$$

8. **Sum Variance:** It computes the amount of dispersion of the sum distribution of gray levels with respect to sum entropy (Eq. (1.14)).

$$\text{Sum Variance} = \sum_{i=2}^{2N_g}(i-\text{Sum Entropy})^2 p_{x+y}(i) \tag{1.14}$$

9. **Entropy:** Entropy is used to quantify the amount of randomness in any given image region (Eq. (1.15)).

$$\text{Entropy} = -\sum_{i=0}^{N_g-1}\sum_{j=0}^{N_g-1} p(i,j)\log\{p(i,j)\} \tag{1.15}$$

10. **Difference Entropy:** It computes the level of disorder in the grayscale difference distribution of a given image (Eq. (1.16)).

$$\text{Difference Entropy} = -\sum_{i=0}^{N_g-1} p_{x-y}(i)\log\{p_{x-y}(i)\} \tag{1.16}$$

11. **Difference Variance:** Difference variance measures the amount of heterogeneity in the difference distribution of gray levels with respect to difference entropy (Eq. (1.17)).

$$\text{Difference Variance} = \sum_{i=0}^{N_g-1}(i-\text{Difference Entropy})^2 p_{x-y}(i) \tag{1.17}$$

12. **Information Measure of Correlation I:** Image with a higher level of homogeneity results in lower information measure of correlation I value (Eqs. (1.18)–(1.23)).

$$\text{Information Measure of Correlation I} = \frac{HXY - HXY1}{\max\{HX, HY\}} \tag{1.18}$$

$$\text{where } HXY = -\sum_{i=1}^{N_g}\sum_{j=1}^{N_g} p(i,j)\log\{p(i,j)\} \tag{1.19}$$

$$HXY1 = -\sum_{i=1}^{N_g}\sum_{j=1}^{N_g} p(i,j)\log\{p_x(i)p_y(j)\} \tag{1.20}$$

$$HXY2 = -\sum_{i=1}^{N_g}\sum_{j=1}^{N_g} p_x(i)p_y(j)\log\{p_x(i)p_y(j)\} \tag{1.21}$$

$$HX = -\sum_{i=1}^{N_g} p_x(i)\log\{p_x(i)\} \tag{1.22}$$

$$HY = -\sum_{j=1}^{N_g} p_y(j)\log\{p_y(j)\} \tag{1.23}$$

13. **Information Measure of Correlation II:** Image with a higher amount of energy results in lower information measure of correlation II value (Eq. (1.14)).

$$\text{Information Measure of Coefficient II} = (1 - \exp[-2.0(HXY2 - HXY)])^{\frac{1}{2}} \tag{1.24}$$

14. **Maximum Correlation Coefficient:** Maximum correlation coefficient can be defined by Eqs. (1.25) and (1.26)):

$$\text{Maximum Correlation Coefficient} = (\text{Second largest eigen value of } Q)^{1/2} \tag{1.25}$$

$$\text{where } Q(i,j) = \sum_{k} \frac{p(i,k)p(j,k)}{p_x(i)p_x(k)} \tag{1.26}$$

15. **Autocorrelation:** Autocorrelation is based on the spatial frequencies and is used for evaluating linear spatial relationships between texture elements. It measures the extent of fineness or coarseness of texture in the image (Eq. (1.27)).

$$\text{Autocorrelation} = \sum_{i=1}^{N_g}\sum_{j=1}^{N_g} (i \cdot j)p(i,j) \tag{1.27}$$

16. **Dissimilarity:** Dissimilarity is the measure of local variability, i.e. variation of gray-level voxel pairs. Higher level of variability results in higher value for dissimilarity (Eq. (1.28)).

$$\text{Dissimilarity} = \sum_{i=1}^{N_g}\sum_{j=1}^{N_g} |i - j|\, p(i,j) \tag{1.28}$$

17. **Cluster Prominence:** This feature is used to evaluate asymmetry of the GLCM. Higher value of cluster prominence indicates that the image is not symmetric. Lower cluster prominence indicates the presence of a peak of distribution, centered around the mean (Eq. (1.29)).

$$\text{Cluster Prominence} = \sum_{i=1}^{N_g}\sum_{j=1}^{N_g} (i - \mu_x + j - \mu_y)^4 p(i,j) \tag{1.29}$$

18. **Cluster Shade:** It measures the level of skewness of the GLCM matrix. Higher value of cluster shade denotes a higher amount of symmetry (Eq. (1.29)).

$$\text{Cluster Shade} = \sum_{i=1}^{N_g}\sum_{j=1}^{N_g} (i - \mu_x + j - \mu_y)^3 p(i,j) \tag{1.30}$$

19. **Maximum Probability:** This feature outputs the most probable gray value intensity between two pixels for a specific distance in a specific direction.

$$\text{Maximum probability} = \max\{P(i,j)\} \tag{1.31}$$

20. **Inverse Difference:** Inverse difference measures the local homogeneity present in a given image. Higher value of the measure indicates that the image is more homogenous (Eq. (1.32)).

$$\text{Inverse Difference} = \sum_{i=1}^{N_g}\sum_{j=1}^{N_g}\frac{1}{1+|i-j|}p(i,j) \tag{1.32}$$

21. **Inverse Difference Normalized:** In this feature measure, the difference between the neighboring gray values is normalized by the total number of discrete gray-level intensities (Eq. (1.33)).

$$\text{Inverse Difference Normalized} = \sum_{i=1}^{N_g}\sum_{j=1}^{N_g}\frac{1}{1+\left(\frac{|i-j|}{N_g}\right)}p(i,j) \tag{1.33}$$

22. **Inverse Difference Moment Normalized:** The feature can be defined by Eq. (1.34)).

$$\text{Inverse Difference Moment Normalized} = \sum_{i=1}^{N_g}\sum_{j=1}^{N_g}\frac{1}{1+\left(\frac{|i-j|}{N_g}\right)^2}p(i,j) \tag{1.34}$$

Gray Level Run Length Matrix (GLRLM) The feature was proposed by Galloway [42]. Gray level run length matrix (GLRLM) gives the length of homogenous runs of each gray level for extraction of texture features from the image in question. GLRLM is a extraction method of higher-order statistical features from the image. For an image, gray level run is explained as a group of consecutive as well as collinear voxels with the identical grayscale intensity. The term run length is defined as the pixel count present in one run, whereas the term run-length value is described as the frequency of occurrence of one such run in the given image. GLRLM is used to represent the coarseness of texture of an image in a specified direction. It is defined by a 2D matrix, where each element $P(i,j|\theta)$ corresponds to the count of homogenous runs of j pixels of grayscale intensity i in the specified direction θ. The count of rows, i.e. the maximum value of i is denoted by the maximum grayscale value, and the count of columns, i.e. the maximum value of j corresponds to the maximum run length N. GLRLM can be calculated in four specified directions, viz., $0°$, $45°$, $90°$, and $135°$.

In all, 11 GLRLM texture features are available for a θ. Considering a GLRLM, the (i,j)th element is defined as $P(i,j)$. Let n_r be defined as the total count of runs, and n_p be defined as the total count of pixels present in a given image. Features that can be extracted using GLRLM are mentioned below:

1. **Short Run Emphasis (SRE):** It defines the distribution of short homogenous runs recorded in a given image (Eq. (1.35)).

$$\text{SRE} = \frac{1}{n_r}\sum_{i=1}^{M}\sum_{j=1}^{N}\frac{P(i,j)}{j^2} \tag{1.35}$$

2. **Long Run Emphasis (LRE):** It defines the distribution of long homogenous runs recorded in a given image (Eq. (1.36)).

$$\text{LRE} = \frac{1}{n_r}\sum_{i=1}^{M}\sum_{j=1}^{N}P(i,j)\cdot j^2 \tag{1.36}$$

3. **Gray-Level Non-uniformity (GLN):** This feature is used to measure the level of non-uniformity detected in the gray levels in a given image (Eq. (1.37)).

$$\text{GLN} = \frac{1}{n_r}\sum_{i=1}^{M}\left(\sum_{j=1}^{N}P(i,j)\right)^2 \tag{1.37}$$

4. **Run-Length Non-uniformity (RLN):** It computes the length of homogenous runs in a given image (Eq. (1.38)).

$$RLN = \frac{1}{n_r} \sum_{i=1}^{N} \left(\sum_{j=1}^{M} P(i,j) \right)^2 \qquad (1.38)$$

5. **Run Percentage (RP):** Run percentage is used to measure the homogeneity of homogenous runs present in the image (Eq. (1.39)).

$$RP = \frac{n_r}{n_p} \qquad (1.39)$$

6. **Low Gray-Level Run Emphasis (LGRE):** This feature is used to represent the distribution of low grayscale runs (Eq. (1.40)).

$$LGRE = \frac{1}{n_r} \sum_{i=1}^{M} \sum_{j=1}^{N} \frac{P(i,j)}{i^2} \qquad (1.40)$$

7. **High Gray-Level Run Emphasis (HGRE):** This feature is used to represent the distribution of high grayscale runs (Eq. (1.41)).

$$HGRE = \frac{1}{n_r} \sum_{i=1}^{M} \sum_{j=1}^{N} P(i,j) \cdot i^2 \qquad (1.41)$$

8. **Short Run Low Gray-Level Emphasis (SRLGE):** It computes the short homogenous run distribution of low grayscale intensities (Eq. (1.42)).

$$SRLGE = \frac{1}{n_r} \sum_{i=1}^{M} \sum_{j=1}^{N} \frac{P(i,j)}{i^2 \cdot j^2} \qquad (1.42)$$

9. **Short Run High Gray-Level Emphasis (SRHGE):** It measures the short homogenous run distribution with high grayscale intensities (Eq. (1.43)).

$$SRHGE = \frac{1}{n_r} \sum_{i=1}^{M} \sum_{j=1}^{N} \frac{P(i,j) \cdot i^2}{j^2} \qquad (1.43)$$

10. **Long Run Low Gray-Level Emphasis (LRLGE):** It measures the long homogenous run distribution with low grayscale intensities (Eq. (1.44)).

$$LRLGE = \frac{1}{n_r} \sum_{i=1}^{M} \sum_{j=1}^{N} \frac{P(i,j) \cdot j^2}{i^2} \qquad (1.44)$$

11. **Long Run High Gray-Level Emphasis (LRHGE):** LRHGE computes the long homogenous run distribution with high grayscale intensities (Eq. (1.45)).

$$LRHGE = \frac{1}{n_r} \sum_{i=1}^{M} \sum_{j=1}^{N} P(i,j) \cdot i^2 \cdot j^2 \qquad (1.45)$$

Laws' Texture Energy Measures (Laws' TEM) Laws proposed a spatial filtering approach for extraction of texture features from a given image [43]. Laws' texture energy measures (TEM) makes use of local masks in order to locate different kinds of textures. In order to measure the amount of variation that occurs in a fixed-size window, a texture-energy approach was developed by constructing a set of nine 5×5 convolutional masks. These masks are obtained by convolving basic 1×5 filters with themselves and one another. These are used in order to measure various local properties in the image, viz., frequency of occurrence of edges, lines, points, and ripples within the texture.

The masks are computed using the following vectors:

Level Detector (L5): 1 4 6 4 1
Edge Detector (E5): −1 −2 0 2 1
Spot Detector (S5): − 1 0 2 0 −1
Ripple Detector (R5): 1 −4 6 −4 1

The 2D convolution of the above-mentioned vertical and horizontal 1D kernels results in 16 filter masks of dimensions 5×5. The filter masks obtained are:

> L5L5 E5L5 S5L5 R5L5
> L5E5 E5E5 S5E5 R5E5
> L5S5 E5S5 S5S5 R5S5
> L5R5 E5R5 S5R5 R5R5

Laws' 5×5 masks are used for extraction of TEMs from an image. An image or a given image region is convolved with each of these 16 masks to result in separate filtered images, respectively. These filtered images are then processed using a moving average non-linear filter using windowing operation. These new filtered images so obtained are known as texture energy measures (TEM) images. Except for L5L5, all other masks are zero-mean. Hence, L5L5 can be considered as normalizing image. Post normalization, the L5L5 image is discarded. Normalizing TEM images with L5L5 image helps in normalizing for contrast. Moreover, similar feature images like L5E5 and E5L5 are averaged in order to remove dimension of directionality. This results in the generation of rotation-invariant features. Thus, in total, nine TEM images are generated. Two first-order statistic features, i.e. mean/average and standard deviation, are computed from each TEM image. Thus, 18 Laws' TEM features are extracted.

First Order Statistics (FOS) Features First-order statistical features (Eq. (1.46)) are computed using normalized histogram of an image. For a grayscale image consisting of L intensity levels, a normalized image histogram can be defined as:

$$p(k) = \frac{\text{number of pixels with intensity } k}{\text{total number of pixels}} \quad \text{where } k = 0, 1, 2, \ldots, L-1 \tag{1.46}$$

First order statistics (FOS) features are measured by computing central tendency, diversity, shape, and entropy of the distribution of pixel intensities in the given region in the image. The FOS features usually extracted are:

1. **Skewness:** It is the computation of level of asymmetry of the probability distribution of a real-valued random variable around its mean, i.e. it measures the lack of symmetry in the histogram. It can be measured using the following equation (Eq. (1.47)):

$$\text{Skewness} = \frac{1}{\sigma^3} \sum_{k=0}^{L-1} (k - \mu)^3 p(k) \tag{1.47}$$

2. **Kurtosis:** This statistical measure describes the distribution of gray-level intensities around the mean. In other words, kurtosis defines the peaked distribution of a real-valued random variable. It can be computed using the following equation (Eq. (1.48)):

$$\text{Kurtosis} = \frac{1}{\sigma^4} \sum_{k=0}^{L-1} (k - \mu)^4 p(k) - 3 \tag{1.48}$$

Local Binary Pattern (LBP) LBP, as suggested by Ojala et al. [41], helps in extracting regional information of a given image by exploiting a pixel block from the neighborhood, with its center pixel considered as the threshold. Every

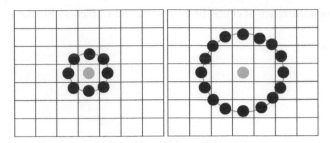

Figure 1.1 Circular (8, 1) and (16, 2) neighborhoods.

voxel in the neighborhood is marked as a one if its value is more compared to the pixel in question, else it is marked as zero. The achieved ones and zeros are further concatenated so as to form a binary number that generates the code for the corresponding pixel in question.

In order to generalize the LBP, let $I(x, y)$ be a given grayscale image. For any pixel from the image, take a local neighborhood of P number of pixels, circular in nature, placed at a length of radius R. The neighborhoods can then be mapped as (P, R). An illustration of one local circular neighborhood around a given pixel is given in Figure 1.1.

Let s_c represent the grayscale value of the center pixel and s_p $(p = 1, 2..., P - 1)$ define the intensity value of P sampling points (Eqs. (1.49)–(1.52)):

$$s_c = I(x, y) \tag{1.49}$$

$$s_p = I(x_p, y_p) \tag{1.50}$$

$$x_p = x + R\cos(2\pi p/P) \tag{1.51}$$

$$y_p = y + R\sin(2\pi p/P) \tag{1.52}$$

Texture T for the image can be defined as the joint distribution of gray levels of P $(P > 1)$ as given by Eq. (1.53)):

$$T = t(s_0, s_1, \dots, s_{P-1}) \tag{1.53}$$

In order to achieve invariability in grayscale, the center pixel intensity s_c is subtracted from the neighboring pixels intensities s_P, thus leading to the following equation (Eq. (1.54)):

$$T = t(s_c, s_0 - s_c, s_1 - s_c, \dots, s_{P-1} - s_c) \tag{1.54}$$

Figure 1.1 shows the figure of circular (8, 1) and (16, 2) neighborhoods.

Taking into assumption the statistical independence of center pixel from difference terms $(s_p - s_c)$, the Eq. (1.55) can be estimated as:

$$T \approx t\left(s_c\right) t\left(s_0 - s_c, s_1 - s_c, \dots, s_{P-1} - s_c\right) \tag{1.55}$$

The factor $t(s_c)$ constitutes the overall distribution of gray levels in the image. To examine the local structures, no information is provided by this term. Thus, Eq. (1.56) can be modeled as:

$$T \approx t(s_0 - s_c, s_1 - s_c, \dots, s_{P-1} - s_c) \tag{1.56}$$

Signed differences $s_p - s_c$ are invariant to any variations in the average grayscale intensity of the image. Hence, to achieve grayscale invariance, just the nature of differences (signs: + or −) in Eqs. (1.57) and (1.58) were considered:

$$T \approx t(q(s_0 - s_c), q(s_1 - s_c), \dots, q(s_{P-1} - s_c)) \tag{1.57}$$

where

$$q(z) = \begin{cases} 1, & z \geq 0 \\ 0, & z < 0 \end{cases} \tag{1.58}$$

defines the thresholding function. For every sign $s(g_P - g_c)$, a weighted factor 2^p is designated and totaled up to achieve a unique $\text{LBP}_{P,R}$ number (Eq. (1.59)) describing the spatial structure of the image.

$$\text{LBP}_{P,R} = \sum_{p=0}^{P-1} q(s_p - s_c)2^p \tag{1.59}$$

The generated LBP map is then turned into histogram using the equations that follow (Eqs. (1.60) and (1.61)):

$$\text{His}(L)|_{\text{LBP}} = \sum_{m=1}^{M}\sum_{n=1}^{N} P_1(\text{LBP}(m,n),L); \quad L \in [0,(2^p - 1)] \tag{1.60}$$

$$P_1(p,q) = \begin{cases} 1 & p = q \\ 0 & \text{else} \end{cases} \tag{1.61}$$

The basic LBP descriptor is unable to capture distinctive information since it uses information only related to the nature of differences present in the local neighborhood. Various LBPs have been designed in order to achieve an improved performance in a variety of applications [44–48].

1.3.3 Shape

Shape is also considered to be a powerful feature representation for image-processing applications. Here, shape feature defines the configuration of a given region being analyzed from an image rather than the shape of the complete image. Shape feature refers to a set of numbers generated for describing a given shape feature. Compared to the previously described features, shape features are generally defined post segmentation, i.e. after an image is divided into regions or objects [49]. Generally used shape features are aspect ratio, circularity, Fourier descriptors, moment invariants, consecutive boundary segments [50], etc. These can be categorized into boundary-based features and region-based features. Boundary-based features, also known as contour-based features extract information from the boundaries of the objects in query. These features are not extractable in cases where a specified boundary of the object is not available. However, region-based features do not depend upon the shape of the boundary, but rather extract information from all the pixels available inside the region or the shape. Since achieving robust and precise segmentation of images is a tedious task, hence extraction of shape features for the purpose of image processing is limited only to specific applications with readily available objects or regions.

1.3.3.1 Shape Features Based on Boundary

Some Simple Descriptors
1. **Perimeter:** It refers to the measure of the smallest exterior contour of the connected region. It is calculated by the total count of pixels that are present on the boundary [14].
2. **Diameter:** It defines the longest measure between two given points present on the boundary contour of the connected region. The measurement of the diameter and orientation of the corresponding line segment which connects the axis of the major boundary are important descriptors of the boundary.
3. **Eccentricity:** It is given as the ratio of major axis to the minor axis. Major axis refers to the length of the diameter of the boundary, whereas the minor axis denotes the line segment placed perpendicular to the major axis.
4. **Curvature:** It can be defined separately for continuous and discrete cases. In case of continuous space, curvature is estimated by the rate at which the slope changes, whereas, in case of discrete space, curvature measures the total count of border pixels to the number of boundary pixels including significant changes in boundary direction.

Fourier Descriptors The Fourier descriptors are established based on the Fourier transform to a shape signature. Fourier transform helps in characterization of a given object by altering sensitivity and representing directly into the frequency domain. The characteristics in frequency domain have an advantage of not being affected easily by minute changes and noise. For describing the border shape quantitatively, discrete Fourier transform of the contour is considered as the fundament of characteristic parameters. These are advantageous since they are insensitive to transitions such as rotation, translation, scale alterations, and the initial point.

Statistical Moments It describes the statistical form of a given image pixel. Quantitative description of the boundary shape can be described by exploiting simple statistical moments, such as mean, variance, and other higher-order moments calculated at pixel level, and insensitive to noise.

1.3.3.2 Shape Features Based on Region
Some Simple Descriptors
1. **Regional Area:** This feature describes the expanse of a given region. It is calculated as the total count of pixels that are present in the given region.
2. **Compactness:** It refers to the roundness of the region. It helps in describing the degrees up to which a given shape is close to circular. A region results in the roundness value equal to 1, when it is circular in shape; else, the value of roundness gets smaller.

Topological Descriptors Topological features identify the image plane region in a complete mode. The most commonly used topological feature is the Euler number, described as the result of subtraction between the total count of connected components and the count of holes. Topological features are simpler to calculate and are insensitive to image transitions, such as translation, rotation, and scale changes.

1.4 Conclusion

In this chapter, various features used for different image-processing systems have been discussed. Features serve the user by bringing down the semantic gaps and also form an important part of an image-processing system. They consist of information related to the images in the dataset in the form of a feature vector. The features are broadly categorized into: lower-level, middle-level, and higher-level features. Low-level features are usually considered for image-processing systems. Color, texture, and shape are the basic lower-level features. Various features can further be extracted from these features for future applications.

References

1 Deng, C., Yang, E., Liu, T. et al. (2019). Unsupervised semantic-preserving adversarial hashing for image search. *IEEE Transactions on Image Processing* 28 (8): 4032–4044.

2 Qayyum, A., Anwar, S.M., Awais, M., and Majid, M. (2017). Medical image retrieval using deep convolutional neural network. *Neurocomputing* 266: 8–20.

3 Fukunaga, K. (2013). *Introduction to Statistical Pattern Recognition*. Elsevier.

4 Chen, Y. and Wang, J.Z. (2002). A region-based fuzzy feature matching approach to content-based image retrieval. *IEEE Transactions on Pattern Analysis and Machine Intelligence* 24 (9): 1252–1267.

5 Sethi, I.K., Coman, I.L., and Stan, D. (2001). Mining association rules between low-level image features and high-level concepts. In: *Data Mining and Knowledge Discovery: Theory, Tools, and Technology III*, vol. 4384, 279–290. SPIE.

6 Cox, I.J., Papathomas, T.V., Ghosn, J. et al. (1997). Hidden annotation in content based image retrieval. *1997 Proceedings IEEE Workshop on Content-Based Access of Image and Video Libraries,* St. Thomas, VI, USA (20–20 June 1997).

7 Brunelli, R. and Mich, O. (2000). Image retrieval by examples. *IEEE Transactions on Multimedia* 2 (3): 164–171.

8 Gevers, T. and Smeulders, A.W. (2000). Pictoseek: combining color and shape invariant features for image retrieval. *IEEE Transactions on Image Processing* 9 (1): 102–119.

9 Smeulders, A.W., Worring, M., Santini, S. et al. (2000). Content-based image retrieval at the end of the early years. *IEEE Transactions on Pattern Analysis and Machine Intelligence* 22 (12): 1349–1380.

10 Huang, J., Kumar, S.R., Mitra, M. et al. (1999). Spatial color indexing and applications. *International Journal of Computer Vision* 35 (3): 245–268.

11 Huang, J., Kumar, S.R., Mitra, M. et al. (1997). Image indexing using color correlograms. *Proceedings of IEEE Computer Society Conference on Computer Vision and Pattern Recognition*, San Juan, USA (17–19 June 1997).

12 Pass, G. and Zabih, R. (1999). Comparing images using joint histograms. *Multimedia Systems* 7 (3): 234–240.

13 Swain, M.J. and Ballard, D.H. (1991). Color indexing. *International Journal of Computer Vision* 7 (1): 11–32.

14 Gonzalez, R.C. and Woods, R.E. (2002). *Digital Image Processing*. Prentice Hall.

15 Pass, G. and Zabih, R. (1996). Histogram refinement for content-based image retrieval. *Proceedings Third IEEE Workshop on Applications of Computer Vision*, Sarasota, USA (02–04 December 1996).

16 Singh, C., Walia, E., and Kaur, K.P. (2018). Color texture description with novel local binary patterns for effective image retrieval. *Pattern Recognition* 76: 50–68.

17 Flickner, M., Sawhney, H., Niblack, W. et al. (1995). Query by image and video content: the QBIC system. *Computer* 28 (9): 23–32.

18 Stricker, M.A. and Orengo, M. (1995). Similarity of color images. *Procedings of the SPIE Storage and Retrieval for Image and Video Databases,* San Jose, CA, United States (23 March 1995).

19 Pass, G., Zabih, R. and Miller, J. (1997). Comparing Images Using Color Coherence Vectors. *Proceedings of the Fourth ACM International Conference on Multimedia*, Boston, USA (February 1997).

20 Rosenfeld, A. and Weszka, J.S. (1980). Picture recognition. In: *Digital Pattern Recognition* (ed. K.S. Fu), 135–166. Berlin, Heidelberg: Springer.

21 He, Z., You, X., and Yuan, Y. (2009). Texture image retrieval based on non-tensor product wavelet filter banks. *Signal Processing* 89 (8): 1501–1510.

22 Ahmadian, A. and Mostafa, A. (2003). An efficient texture classification algorithm using Gabor wavelet. *25th Annual International Conference of Engineering in Medicine and Biology Society*, Cancun, Mexico (17–21 September 2003).

23 Farsi, H. and Mohamadzadeh, S. (2013). Colour and texture feature-based image retrieval by using Hadamard matrix in discrete wavelet transform. *IET Image Processing* 7 (3): 212–218.

24 Chen, C.C. and Chen, D.C. (1996). Multi-resolutional Gabor filter in texture analysis. *Pattern Recognition Letters* 17 (10): 1069–1076.

25 Vilnrotter, F.M., Nevatia, R., and Price, K.E. (1986). Structural analysis of natural textures. *IEEE Transactions on Pattern Analysis and Machine Intelligence* 1: 76–89.

26 Jain, R., Kasturi, R., and Schunck, B.G. (1995). Texture. In: *Machine Vision* (ed. R. Jain, R. Kasturi and B.G. Schunck), 234–248. New York: McGraw-Hill.

27 Fu, K.S. (1982). *Syntactic Pattern Recognition and Applications*. Prentice-Hall Inc.

28 Freeborough, P.A. and Fox, N.C. (1998). MR image texture analysis applied to the diagnosis and tracking of Alzheimer's disease. *IEEE Transactions on Medical Imaging* 17 (3): 475–478.

29 Chaudhuri, B.B. and Sarkar, N. (1995). Texture segmentation using fractal dimension. *IEEE Transactions on Pattern Analysis and Machine Intelligence* 17 (1): 72–77.

30 Chakerian, D. (1984). *The Fractal Geometry of Nature*, 175–177. W. H. Freeman and Co.

31 Mao, J. and Jain, A.K. (1992). Texture classification and segmentation using multiresolution simultaneous autoregressive models. *Pattern Recognition* 25 (2): 173–188.

32 Comer, M.L. and Delp, E.J. (1999). Segmentation of textured images using a multiresolution Gaussian autoregressive model. *IEEE Transactions on Image Processing* 8 (3): 408–420.

33 Li, S.Z. (2009). *Markov Random Field Modeling in Image Analysis*. Springer Science & Business Media.

34 Jojic, N., Frey, B.J., and Kannan, A. (2003). Epitomic analysis of appearance and shape. *IEEE International Conference on Computer Vision,* Nice, France (13–16 October 2003).

35 Sutton, R.N. and Hall, E.L. (1972). Texture measures for automatic classification of pulmonary disease. *IEEE Transactions on Computers* 100 (7): 667–676.

36 Ainhoa, L., Manmatha, R., and Rüger, S. (2010). Image retrieval using Markov random fields and global image features. *Proceedings of ACM International Conference on Image and Video Retrieval*, New York, USA (5–7 July 2010).

37 Tamura, H., Mori, S., and Yamawaki, T. (1978). Textural features corresponding to visual perception. *IEEE Transactions on Systems, Man, and Cybernetics* 8 (6): 460–473.

38 Liu, F. and Picard, R.W. (1996). Periodicity, directionality, and randomness: Wold features for image modeling and retrieval. *IEEE Transactions on Pattern Analysis and Machine Intelligence* 18 (7): 722–733.

39 Haralick, R.M., Shanmugam, K., and Dinstein, I. (1973). Textural features for image classification. *IEEE Transactions on Systems, Man, and Cybernetics* SMC-3 (6): 610–621.

40 Galloway, M.M. (1974). Texture analysis using grey level run lengths. *NASA STI/Recon Technical Report N* 75: 18555.

41 Ojala, T., Pietikäinen, M., and Mäenpää, T. (2002). Multiresolution gray-scale and rotation invariant texture classification with local binary patterns. *IEEE Transactions on Pattern Analysis and Machine Intelligence* 24 (7): 971–987.

42 Galloway, M.M. (1975). Texture analysis using gray level run lengths. *Computer Graphics and Image Processing* 4: 172–179.

43 Laws, K.I. (1980). Rapid texture identification. *Proceedings of Image Processing for Missile Guidance,* San Diego, United States (23 December 1980).

44 Bedi, A.K., Sunkaria, R.K., and Randhawa, S.K. (2018). Local binary pattern variants: a review. *2018 First International Conference on Secure Cyber Computing and Communication (ICSCCC),* Jalandhar, India (15–17 December 2018).

45 Bedi, A.K. and Sunkaria, R.K. (2020). Local tetra-directional pattern–a new texture descriptor for content-based image retrieval. *Pattern Recognition and Image Analysis* 30 (4): 578–593.

46 Bedi, A.K. and Sunkaria, R.K. (2021). Mean distance local binary pattern: a novel technique for color and texture image retrieval for liver ultrasound images. *Multimedia Tools and Applications* 80 (14): 1–30.

47 Subrahmanyam, M., Maheshwari, R.P., and Balasubramanian, R. (2012). Local tetra patterns: a new feature descriptor for content-based image retrieval. *IEEE Transactions on Image Processing* 21 (5): 2874–2886.

48 Verma, M., Raman, B., and Murala, S. (2015). Local extrema co-occurrence pattern for color and texture image retrieval. *Neurocomputing* 165: 255–269.

49 Long, F., Zhang, H., and Feng, D.D. (2003). Fundamentals of content-based image retrieval. In: *Multimedia Information Retrieval and Management*. Berlin, Heidelberg: Springer.

50 Mehrotra, R. and Gary, J.E. (1995). Similar-shape retrieval in shape data management. *Computer* 28 (9): 57–62.

2

Image Segmentation and Classification Using Deep Learning

Abhisek Ray and Maheshkumar H. Kolekar

Department of Electrical Engineering, Indian Institute of Technology, Patna, Bihar, India

2.1   Introduction

Among many, three major concerns surfaced which propel researchers to concentrate more on image processing [1] and video processing [2–4] and especially in the digital domain. These problems can be elaborated as transmission, storage, and printing of images, effective interpretation of an image, and machine vision of an image. Many image-processing techniques had been proposed toward these problems, for example, image coding, digitization, and coding for first; picture restoration, enhancement, and restoration for the second; image description and segmentation for the last one.

A panchromatic image can be interpreted as a two-dimensional light intensity function, $f(x, y)$, in which x and y are the spatial coordinates. The value at this coordinate (x, y) is proportional to the image brightness at that location. For a multispectral image, $f(x, y)$ is a vector with each component specifying the luminance of the scene at point (x, y) in the associated spectral band. A digital picture is a discretized image $f(x, y)$ in both spatial and brightness coordinates. An array of 2D integers, one for each color band, is used to describe it. Gray level refers to a digitized brightness value. Pixel or pel is a term derived from the phrase "picture element" for each element of the array. The dimension of such an array is usually a few hundred pixels by a few hundred pixels, and there are dozens of different gray levels to choose from. The notable contribution of our article can be interpreted as follows:

- We try to cover and discuss in detail all two important domains of image processing, i.e. image segmentation and image classification [5].
- This chapter focuses on various state-of-the-art deep-learning (DL) techniques that should be acknowledged by the research fraternity before diving into image processing.
- Comprised state-of-the-art models are described based on their application through domain-specific research articles. To the best of our knowledge, we only selected the introductory article for a specific application in a particular domain.
- Through this chapter, we try to convey the advantages and limitations of different popular models while applying them in these two domains of image processing.

The organization of the rest of the chapter is based on the above-mentioned four applications in image processing. While Section 2.1 have already discussed the introductory part of image processing, Section 2.2 introduces the domain of image segmentation along with its types, advantages, and applications. In a very similar fashion, Section 2.3 intuitively elaborates the image classification. In the very last part, this chapter is concluded with concluding remarks which are briefed in Section 2.4.

Machine Learning Algorithms for Signal and Image Processing, First Edition.
Edited by Deepika Ghai, Suman Lata Tripathi, Sobhit Saxena, Manash Chanda, and Mamoun Alazab.
© 2023 The Institute of Electrical and Electronics Engineers, Inc. Published 2023 by John Wiley & Sons, Inc.

2.2 Image Segmentation

Image segmentation, also known as pixel-level categorization, is the process of grouping those portions together that correspond to the same object class. Detection is the process of locating and recognizing objects. Image segmentation is a pixel-level predictive process, where each pixel is differentiated into its classification category. Literature can be consulted for more information. Semantic picture segmentation can be used for a variety of tasks, including recognizing road signs, segmenting colon crypts, and classifying land use and land cover. It's also employed in the medical industry for things like detecting brains and tumors, as well as detecting and monitoring medical tools during procedures. Several medical implications of segmentation are listed. Scene parsing is critical in self-driving cars and advanced driver assistance systems (ADASs) because it mainly relies on image segmentation. The accuracy of segmentation has improved dramatically since the re-advent of the deep neural network (DNN) [6]. The methods used before DNN are referred to as traditional approaches. In this sections (Section 2.2), we briefly discuss some popular and widely implemented DL-based segmentation algorithms.

2.2.1 Types of DL-Based Segmentation

2.2.1.1 Instance Segmentation Using Deep Learning

Instance segmentation has surfaced as one of the demanding and complex computer vision research topics. It locates various classes of object instances existing in different photos by anticipating the object class label prediction and the pixel-based object instance-mask anticipation. Instance segmentation application is intended to aid robotics, autonomous vehicles, and surveillance. Many instance segmentation approaches were suggested with the introduction of DL, notably convolutional neural networks (CNNs) [7], for example, in which the segmentation accuracy continued to grow. Mask regional convolutional neural network (R-CNN) is a simple and effective method for segmenting instances. A fully convolutional network (FCN) has been used to predict segmentation masks alongside box-regression and object classification, following the lead of the fast/faster R-CNN. To extract stage-wise network features with high efficiency, a feature pyramid network (FPN) was developed, in which a top-down network path with lateral connections was used to produce semantically strong features. Some relatively recent datasets provide ample opportunity for the proposed methodologies to be improved. The common objects in context (COCO) collection from Microsoft contains 200k pictures. In this dataset's photos, many examples with complicated spatial layouts have been documented. Additionally, the cityscapes dataset and the mapillary vistas dataset (MVD) both offer street scene photos with a huge number of traffic objects per image. These dataset's photos contain blurring, occlusion, and minute examples. Many principles for network architecture for image classification have been offered. Object recognition can benefit greatly from the same. Narrowing the information path, employing dense connections, and improving the versatility and diversity of the information path by creating parallel paths are other examples.

2.2.1.2 Semantic Segmentation Using Deep Learning

The objective of semantic-image segmentation, also known as pixel-level categorization, is to group those sections of a picture that correspond to the very same object class. Modern semantic segmentation algorithms are built on the foundation of fully convolutional networks (FCNs). They frequently forecast outcomes with lower resolution than the input grid and retrieve the additional 8–16 resolution using bilinear upsampling. Dilated/atrous convolutions, which substitute some subsampling layers at the cost of greater memory and processing, may improve outcomes. Encoder–decoder architectures that subsample the grid representation in the encoder and then upsampled it in the decoder, employing skip connections to retrieve filtered features, are an alternative technique. Before performing bilinear interpolation, current techniques combine dilated convolutions with an encoder–decoder architecture to perform operations on four finer grids than the input grids. We present a method for efficiently predicting fine-level details on a grid as dense as the input grid in our chapter.

2.2.2 Advantages and Applications of DL-Based Segmentation

Satellite pictures have been effectively segmented using DL-based segmentation methods in remote sensing applications and mostly in urban planning and precision agriculture. To solve major environmental issues, such as climate change or any meteorological issues, aerial images are taken by drones and airborne equipment and gone under DL-based segmentation algorithms. The larger image size and the scarcity of ground-truth data required to assess the accuracy of the segmentation algorithms are the primary stumbling block. Similarly, in the assessment of building materials, DL-based segmentation approaches confront issues due to a large number of relevant pictures and the lack of reference information. Last but not least, biological imaging is an essential application sector for DL-based segmentation. The potential here is to create standardized image databases that can be used to evaluate novel infectious diseases and track pandemics.

2.2.3 Types and Literature Survey Related to DL-Based Segmentation

DL-based image segmentation techniques can be grouped into nine categories according to their model architecture.

2.2.3.1 Fully Convolution Model

A milestone step toward convolutional segmentation is proposed in [8], as shown in Figure 2.1, comprising a fully connected (FC) network, which includes only convolutional layers to output a segmentation map from the same sized input layer.

This convention was modified and resulted in VGG16, ParseNet, and GoogleNet like frameworks where the final FC layer is cut off from the mainframe to give spatial segmentation map from an arbitrary sized input image instead of a classification score. The skip connection technique came into play where upsampled final layer feature maps got fused with earlier layer resulting in semantic information recombination to achieve a detailed and accurate segmentation technique. The state-of-the-art performance can be achieved by testing on PNYUDv2, PASCAL VOC, and SHIFT Flow in the domain of brain tumor segmentation, iris segmentation, skin lesion segmentation, and instance aware segmentation.

2.2.3.2 CNN with Graphical Model

Due to scene-level semantic context ignorance of FCN, many researchers combine FCN with a probabilistic graphical model like conditional random fields (CRFs) and Markov random fields (MRFs) that can localize image

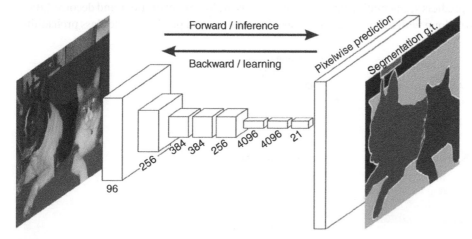

Figure 2.1 A pixel predictive model based on FCN. g.t, means the ground truth. Source: [8], Long et al. (2015), IEEE.

segment boundaries more accurately than previous methods. The invariance features of CNNs make deep CNN (D-CNN) suitable for high-level tasks like classification. And due to this property, the responses from the deeper layers are not sufficiently localized enough for successful object segmentation task. Chen et al. [9] presented a segmentation approach that integrates both CNN and CRF techniques to address this flaw. Other related proposed architectures are

- Joint training method using CNNs and fully connected CRFs proposed by Schwing and Urtasun [10].
- Contextual deep CRFs model explores the patch-background context and patch–patch context in semantic segmentation proposed by Lin et al. [11].
- Parsing network, a CNN model, enables end-to-end analysis in a single go proposed by Liu et al. [12]. Latter this rich information incorporates optimized MRFs to yield satisfactory results.

2.2.3.3 Dilated Convolution Model

The idea behind dilated convolution is to boost the convolutional mask receptive field without an increase in the computational cost. Some real-time segmentation models use this dilated technique e.g. DeepLab family [13], efficient network (Enet) [14], dense atrous spatial pyramid pooling (DenseAPP) [15], dense up-sampling convolution and hybrid dilated convolution (DUC-HDC) [16], and multiscale context aggregation [17].

- Proposed by Chen et al., DeepLabV1 [9] and DeepLabV2 [13] are registered as an efficient and popular image-segmentation technique that has three special features as shown in Figures 2.2 and 2.3 respectively. Dilated convolution mitigates the decreasing resolution due to striding and max pooling, strous spatial pyramid pooling (ASPP) filters the convolution layer at a distinct sampling rate, and combining the structure of DCNN (ResNet or VGGNet) with probabilistic graphical model improves the segmenting mask localization.
- In a very similar fashion, Chen et al. again proposed DeepLabV3 [19] and DeepLabV3+ [18] by combining cascaded and parallel dilated units and by incorporating encoder–decoder architecture with dilated separable convolution unit, respectively.

2.2.3.4 Encoder–Decoder Model

This is the most popular approach in semantic segmentation encompassing DeConvNet, SegNet, HRNet, stacked deconvolutional networks (SDNs), LinkNet, W-Net, U-Net, and V-Net.

- Noh et al. [20] proposed the encoder–decoder DeConvNet model, as shown in Figure 2.4, which recognizes pixel-wise class labels and predicts segmentation layer. This model has multiple convolutions and deconvolution layers adopted from VGG16 along with the pooling and unspooling layer to find pixel-accurate class probability.

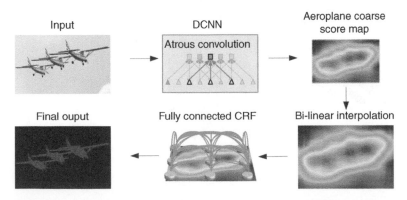

Figure 2.2 The DeepLab architecture. Source: [13], Chen et al. (2017), IEEE.

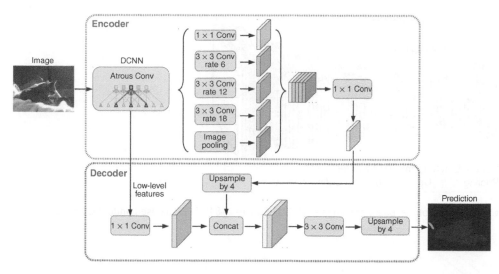

Figure 2.3 The DeepLab-V3+ model architecture. Source: Adapted from [18], Visin et al. (2016), IEEE.

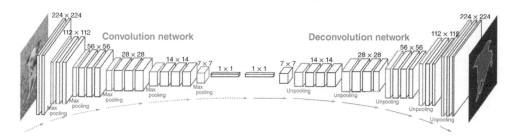

Figure 2.4 Deconvolutional segmentation (semantic). Source: [20], Noh et al. (2015), IEEE.

- Another encoder–decoder architecture derived from VGG16 has a novelty of low-resolution input feature upsampling by max-pooling layer. This model, SegNet as shown in Figure 2.5, is proposed in [21]. It has 13 Conv–Deconv layers and a pixel classification layer.
- Recovering from fine-grained information loss in image resolution during encoding as seen in DeConvNet and SegNet, HRNet not only parallelly connects high-to-low resolution Conv. Stream but also exchange information among themselves.
- For biological microscopy segmenting images, U-Net was proposed by Ronneberger et al. [22] which comprises two paths; one contracting path which is responsible for context capturing and one asymmetric expanding path responsible for localization.
- Milletari et al. [23] proposed V-Net to eradicate the problems that arise due to the voxel count difference between foreground and background by introducing a new loss function taking dice co-efficient.

2.2.3.5 R-CNN Based Model

Initially proposed for object detection, R-CNN uses a region proposal network that extracts region of interest (ROI) and ROI pool layer for coordinate class interference.

- Mask R-CNN, as shown in Figure 2.6, runs on three parallel lines of processing concept which incorporates bounding box computing line, class prediction line, and binary mask computing line. Initially proposed in [24], it shows its efficient capability in object detection and segmentation.

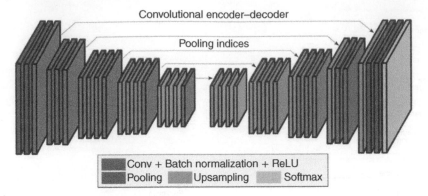

Figure 2.5 The SegNet framework. Source: Badrinarayanan et al. [21] fig 2 (p.2484)/IEEE/Licensed under CC BY 3.0.

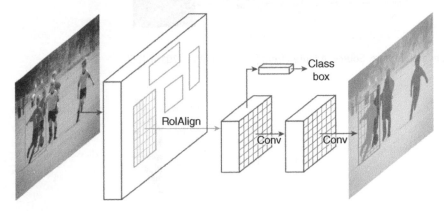

Figure 2.6 Mask R-CNN model architecture. Source: [24], He et al. (2017), IEEE.

- Referring to the concept of FPN and masked R-CNN, Liu et al. [25] proposed path aggregation network (PANet) which uses augmented bottom-up pathway with FPN backbone for lower layer feature improvement.
- Multitask network by Dai et al. [26], MaskLab by Chen et al. [27], DeepMask [28], TensorMask by Chen et al. [29], PolarMask [30], Region-based FCN [31], and CenterMask [32] are among popular R-CNN techniques which do not only yield an efficient detection technique but also produce an accurate segmenting mask.

2.2.3.6 Multiscale Pyramid Based Model

Many neural network (NN) models apply multiscale and pyramid-like structures for segmenting images. Among, FPN [33], dynamic multiscale filter network (DM-Net), pyramid scene parsing network (PSPN) [34], Laplacian pyramid structure [35], adaptive pyramid context network (APC-Net) [36], context contracted network (CCN) [37], salient object segmentation [38], and multiscale context intertwining (MSCI) [39] are most popular.

- FPN was initially proposed for object detection by Lin et al. [33] and later applied for segmentation purposes by merging low and high resolution features through a top-down pathway, a bottom-up pathway, and lateral connections.
- Aiming for global context representation of an image residual network (ResNet) is used as feature extractor in PSPN, as shown in Figure 2.7, which was proposed by Zhao et al. [34]. Extracted feature map first passed through pooling module for distinct (four) pattern study, then a 1×1 Conv layer for dimensionality reduction, and lastly through an upsampling and concatenating unit to yield local as well as global context information. Above mention, three-layer is considered as pyramid pooling modules which later combine with a convolution layer to generate pixel-wise predicted feature maps.

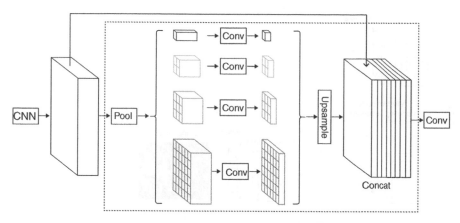

Figure 2.7 The PSPN model architecture. Source: Zhao et al. [34].

2.2.3.7 RNN Based Model

Like CNN, recurrent neural network (RNN) also can improve estimated segmenting maps by exploiting the pixel level short/long term dependencies.

- ReSeg, as shown in Figure 2.8, is a popular RNN based model proposed in [18] which is composed of an image classification model, ReNet [40]. The combination of four RNNs in ReNet is responsible for vertical and horizontal image sweeping, patch/activation encoding, and global information gathering. ReNet is stacked upon the VGG16 convolution layer to extract local descriptions. After that, the initial image resolution is recovered through up-sampling.
- 2D long-short term memory (LSTM) models, proposed in [41], have a greater role in per-pixel image segmentation by learning complex spatial dependencies and textures. This model successfully carries out context integration, segmentation, and classification as well.
- Another approach toward semantic segmentation framework is the graph-LSTM network, as shown in Figure 2.9, proposed in [42] that can yield more structural and global information by augmenting the convolutional layers through graph-LSTM layers. Array structured uniformly arranged data are generalized to graph-structured nonuniform data by taking arbitrary superpixels as consistent nodes and existing relation between superpixels as the edge in an undirected graph.

Apart from these models, data associated recurrent neural network (DA-RNN) proposed by Xiang and Fox [43] in semantic labeling and 3D joint mapping and semantic segmentation algorithm after combining spatial CNN image description and temporal LSTM linguistic description proposed by Hu et al. [44] are popular RNN based algorithm. The sequential nature of RNN does not permit for parallel execution and therefore RNN based models are slower than their CNN counterparts.

2.2.3.8 Generative Adversarial Network (GAN) Based Model

Nowadays generative adversarial networks (GANs) are widely accepted DL techniques in the computer vision domain, not excluding segmentation. Luc et al. [45] proposed a conjoint model by incorporating a combination of

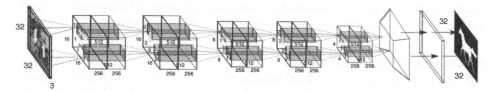

Figure 2.8 The ReSeg framework. Source: [18], Visin et al. (2016), IEEE.

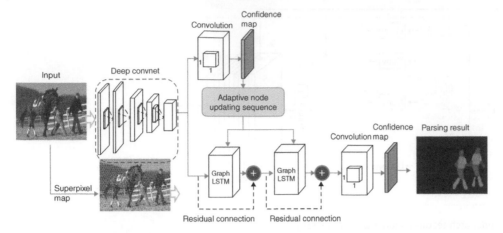

Figure 2.9 The graph-LSTM semantic segmentation model. Source: [42], Liang et al. (2016), Springer Nature.

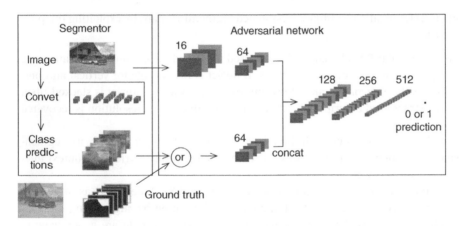

Figure 2.10 The GAN-based semantic segmentation model. Source: [45], Luc et al. (2016), Cornell University.

convolutional segmentation network and adversarial network which distinguishes between generated segmentation map and ground truth segmentation map as shown in Figure 2.10.

- A multiclass classifier or otherwise act as a discriminator using a semi-weekly supervised GAN network was proposed by Souly et al. [46] that either assigns a foreign sample a possible label from trained classes or strike it as an irregular class.
- Another semi-supervised semantic segmentation GAN framework is surfaced by Hung et al. [47] after designing an FCN discriminator that differentiates between predicted segmentation map and ground truth. This spatial resolution-based network has three-loss function terms e.g. adversarial loss, cross-entropy loss, and semi-supervised loss of the segmentation ground truth, discriminator network, and confidence map output respectively.
- Other GAN-based approaches include multiscale adversarial networks [48], cell image GAN segmentation [49], and invisible part segmenting and generating network [50] mostly in the medical image domain.

2.2.3.9 Segmentation Model Based on Attention Mechanism
- Chen et al. [51] weighted multiscale features at every pixel location and proposed a powerful attention-based semantic segmentation model, as shown in Figure 2.11, which assigns distinct weights based on their importance at different scales and positions.

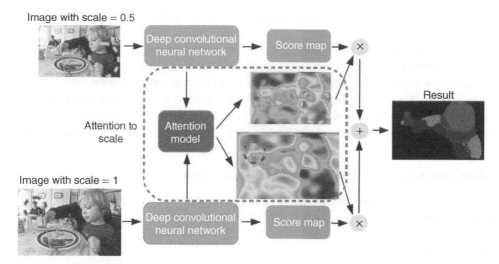

Figure 2.11 Attention-based model for semantic segmentation. Source: [51], Chen et al. (2016), IEEE.

- Li et al. [52] incorporated attention mechanism with spatial pyramid technique, which yields more precise dense features for image-pixel labeling. This model is otherwise known as pyramid attention network (PAN).
- Reverse attention network (RAN) studies the reverse concept features which are not linked with the target class by training the model through a reverse attention mechanism. Proposed by Huang et al. [52], RAN performs both direct and reverse attention to study semantic features.
- Due to persistent exploration, many models like dual attention network (DAN) [53], OCNet [54], ResNest: split attention network [55], criss cross attention network (CCNet) [56], discriminative feature network (DFN) [57], expectation–maximization attention network (EMANet) [58], and height driven attention network (HAN) shows prominent accuracy in attention-based semantic segmentation.

2.3 Image Classification

Image-level classification assigns a particular label to every input image rather than doing pixel-wise prediction. Unsupervised feature learning [74], such as sparse auto-encoders (SAEs) and GANs [75], has been utilized to solve various computer vision problems, and it can be used in medical imaging. The use of neural networks as classifiers, which directly output an individual prediction for one image, is a simple technique to classify images. Alternatively, a network can be employed as feature extractor to build data representations that are supplied to other target classifiers after they have been trained with large-scale data sets.

2.3.1 Types and Schemes in Image Classification

In DL, unsupervised learning and supervised learning have been two intertwined significant themes. Pretraining a DNN, which was subsequently fine-tuned with supervised tasks, was one of the important applications of deep unsupervised learning over the last decade. Stack (denoising) autoencoders, deep belief networks, sparse encoder–decoders, and deep Boltzmann machines are only a few of the deep unsupervised models presented. When the number of available labels was limited, these methods dramatically enhanced the performance of neural networks on supervised tasks.

However, in recent years, supervised learning without any unsupervised pretraining has outperformed unsupervised learning, and it has emerged as the most popular method for training DNNs for practical works like image categorization and object recognition. Purely supervised learning permitted network topologies, such as the inception unit and residual structure, to be more flexible, since they were not constrained by the modeling norms of unsupervised methods. Furthermore, the batch normalization approach has made neural-network learning much easier.

There have been several efforts to combine supervised and unsupervised learning in the same chapter, allowing unsupervised ideas to influence network training after supervised learning. Although these procedures have opened up new possibilities for unsupervised learning, they have yet to be demonstrated to escalate to large counts of labeled and unlabeled data. Many recent papers projected an architecture that is trouble-free to couple with a classification network by outspreading the stacked denoising autoencoder with lateral links, i.e. from the encoder to the same layers of the decoder, and their methods yield promising semi-supervised learning results.

2.3.2 Types and Literature Survey Related to DL-Based Image Classification

Various structures for completing the image-classification problem are enlisted as follows: CNN-based classifier, CNN–RNN-based classifier, auto-encoder-based classifier, and GAN-based classifier.

2.3.2.1 CNN Based Image Classification

- **LeNet:** The CNN was developed by LeCun et al. [59] in 1998 to categorize handwritten digits. LeNet-5 CNN model includes seven weighted (trainable) layers. Three layers of the convolutional block, two layers of average pooling block, an FC layer, and an output layer are among them. Non-linear features from an image are exploited first, followed by the down-sampling operation through the pooling layer. Several consecutive units of Euclidean radial basis function (RBF) are used to categorize 10 digits at the output end of the framework. LeCun et al. [59] used LeNet-5 to train and test the Modified National Institute of Standards and Technology (MNIST) handwritten digits dataset. MNIST dataset comprises 60k and 10k records for training and testing, respectively. The authors used a stochastic gradient descent (SGD) technique to train various variants of LeNet-5 architecture was set with 20 iterations, a momentum of 0.02, and a lower global learning rate for training in each session.

- **AlexNet:** In 2012, Krizhevsky et al. [60] proposed a deep CNN network, AlexNet, for classifying ImageNet data. AlexNet has the same design as LeNet-5, but it is substantially larger. It consists of eight trainable layers. Five Conv layers and three FC layers are included among them. After convolutional and FC layers, they used rectified linear unit (ReLU) non-linearity to assist their model train quicker than equivalent networks with the units. After the first and second convolutional layers, they employed local response normalization (LRN), also known as "brightness normalization," to enhance generalization. The LRN layer and convolutional layer (fifth) are followed by a max-pooling layer that down-sample the dimension of each feature. For the imageNet large scale visual recognition challenge (ILSVRC) – 2010 and ILSVRC – 2012, [60] created AlexNet to classify 1.2 million photos into 1000 classes. It used 256×256 pixels images after down-sampling and centering from ImageNet's variable resolution image. They employed runtime data augmentation and a regularization method termed dropout to minimize the overfitting issue. Both principal component analysis (PCA) and data augmentation techniques are employed for feature reduction and new sample generation purposes, respectively. These translated and horizontally sifted augmented images are retrieved in the form of 10 random patches of dimension 224×224. The authors used SGD to learn AlexNet, which had a batch size of 128, weight decay of 0.0005, and momentum of 0.9.

- **ZFNet:** Zeiler and Fergus [61] presented a CNN architecture called ZFNet in 2014. AlexNet and ZFNet are congruent architectures, except the filter size of the first layer reduce from 11×11 to 7×7. To keep more features, the convolution operation of stride 2 is employed in the first two convolutional layers. It explained the extraordinary performance of deep CNN in the image domain. Another visualization technique, called DeconvNet, is used to reconstruct the activations at higher layers back to the space of the input pixel where it originates. The ImageNet dataset comprising 1.3 million training, 50k validation, and 100k testing images, was used by ZDNet. The hierarchical nature of extracted features in the CNN network can be visualized by projecting each layer in increasing order.

- **VGGNet:** Simonyan and Zisserman proposed VGGNet based on a deeper configuration of AlexNet [62]. This deeper network has kept all the parameters constant except the filter size of the convolution layer. The size of the convolutional filter is fixed to 3×3 throughout the deep network. Initially, they employed four CNN configurations, such as A, A-LRN, B, and C, having four distinct weighted layers of count 11, 22, 13, and 16, respectively. These four CNN configurations are followed by two variants of VGG configuration named D or VGG16 and E or VGG19. To improve non-linearity in model C, three convolutional filters of size 1×1 each is replaced in the 6th, 9th, and 12th position. A pack of three 3×3 convolutional filters has a similar effective receptive field but equipped with more additive non-linearity than a single 7×7. Therefore single 7×7 filter is replaced by a group of three 3×3 filters. Unlike AlexNet, the input image is resized to a dimension of 224×224 in VGGNet.

- **GoogLeNet:** Szegedy et al. [63] developed a novel design for GoogLeNet, which differs from traditional CNN. They used parallel filters termed inception modules of sizes 1×1, 3×3, and 5×5 in each convolution layer to increase the number of units in each convolution layer. The total number of layers in GoogLeNet has become increased to 22 compared to 19 in VGGNet. They kept the computational budget in mind when building this model. To handle several scales, they implemented a sequence of weighted Gabor filters of varied sizes in the inception design. Instead of using the naïve form of the inception module, they employed the inception module with a dimension reduction technique to make the architecture computationally efficient. The disbelief distributed machine learning framework was used to train GoogLeNet. The training process has been accomplished with a reasonable amount of structure and data-processing parallelization. During optimization, the following network has used SGD (asynchronous) as an optimizer with a momentum of 0.9 at a constant learning rate.

- **ResNet:** The vanishing gradient has become a very prominent problem while increasing the number of layers in CNN model. Normalization and intermediate initialization techniques combinedly address these issues very effectively. But, the performance still degrades and this is not due to overfitting. To solve this issue, He et al. [64] introduced a pre-trained shallower model that coupled with an extra layer for identity mapping. The combination of both are sequentially executed to form ResNet architecture. The residual mapping, represented by $H(x) = F(x) + x$, replaces the previous underlying mapping $H(x)$. ResNet model optimization is carried out by SGD as an optimizer with a batch size of 128. The other training parameters, such as weight decay, learning rate, and momentum, are set as 0.0001, 0.1, and 0.9, respectively. At 32k and 48k iterations, the learning rate is manually updated by lowering the initial rate (0.1) and finally stopped at 64k iterations. After each Conv layer, they employed weight initialization and batch normalization. The dropout regularization approach was not used.

- **DenseNet:** Dense Convolutional Networks (DenseNets) were proposed by Huang et al. [65] and included dense blocks in standard CNN. In a dense block, the input of one layer is the concatenation of the outputs of all of the previous levels. The features of earlier layers are reused iteratively to improve the feature quality and lower the vanishing gradient problems. The number of parameters was also minimized by using a modest number of filters. The non-linear transformation functions in a dense block are a combination of batch normalization, ReLU, and the 3×3 convolution operation. To minimize dimensionality, they used the 1×1 bottleneck layer. The DenseNet framework is pre-trained on Canadian Institute for Advanced Research (CIFAR) and ImageNet datasets. This architecture uses SGD as an optimizer with batch sizes of 256 and 64 on the street view house numbers (SVHN) and CIFAR datasets, respectively. The initial learning rate was 0.1 and reduced by 1/10 twice. They employed 0.0001 weight decay, a 0.9 Nesterov momentum, and a 0.2 dropout.

- **CapsNet:** Conventional CNNs has two flaws, as detailed above. The sub-sampling, for starters, eliminates spatial information among higher-level features. Second, it has a hard time generalizing to new viewpoints. It can handle translation but not affine transformations of different dimensions. Geoffrey E. Hinton and coworkers proposed CapsNet [66] in 2017 as a solution to these issues. CapsNet includes capsule components. A capsule is made up of a collection of neurons. CapsNet layers are essentially made up of layered neurons. Sabour et al. introduced the CapsNet, which is made up of three layers: two Conv levels and one FC layer. The first convolutional layer employs ReLU as an activation function and consists of 256 convolutional units with 9×9 kernels of stride 1. This layer identifies local features and delivers them as input to the second layer's major capsules. Each primary capsule holds eight CU, with a stride of 2 kernels. The primary capsule layer comprises a total of 32 6×6 8D capsules. Each digit class has one 16D capsule in the final layer. Between both the primary layer and the DigitCaps layer, the authors employed routing. The MNIST photos are used to train CapsNet. As a regularization strategy, they employed reconstruction loss.

2.3.2.2 CNN–RNN Based Image Classification

Two alternative structures for completing the image-classification problem are: CNN-based classifier and CNN–RNN classifier. Unlike CNN-based generators, CNN–RNN generators can efficiently manipulate the hierarchical labels' dependency, resulting in improved classification results for both fine and coarse classes. As can be observed, this generator outperforms the previous CNN-based generator in terms of fine and coarse grade predictions with and without the help of the data augmentation technique. In particular, this CNN–RNN generator generates out classes the previous CNN-based generator by at least 5.05% for coarse predictions, while by over 6.7% for fine predictions. Figure 2.12 shows the framework of the CNN–RNN pipeline.

- Guo et al. [67] used the dropout version, a bigger mini-batch size (200), and more iterations (7×104 in total, with the learning rate dropping at 2×104, 4×104, and 6×104 iterations) to train the network as shown in Figure 2.12. The following are some more experimental setups. This chapter uses the pre-activation residual block and trains the models for 7×104 iterations with a mini-batch size of 200, a momentum of 0.9, and a weight decay of 0.0005. The learning rate starts at 0.1 and decreases by 0.01 in every 4×104 and 6×104 iterations.
- The proposed CNN–RNN model is a CNN and RNN combined model for image categorization presented by [68]. The convolution computation is a filtering procedure that treats image data as two-dimensional wave data. It can remove non-critical band information from an image while leaving crucial aspects intact. The RNN module is used to calculate the dependency features of intermediate layer output and connect the characteristics of these intermediary tiers to the final full-connection network for classification prediction to improve classification accuracy. At the same time, to satisfy the RNN model's limitation on the size of the input sequence and avoid gradient explosion or disappearance in the network, this research filters the input data using the wavelet transform (WT) method in combination with the Fourier transform. This research will put the suggested CNN–RNN model to the test on the CIFAR-10 dataset, which is extensively utilized.

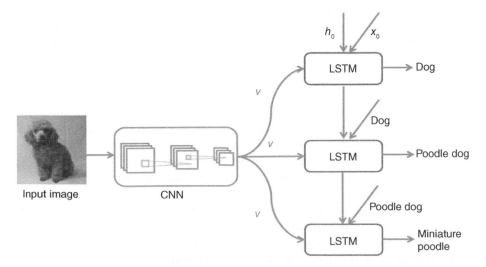

Figure 2.12 The CNN–RNN pipeline framework. Source: [67], Guo et al. (2018), CC BY, Springer Nature.

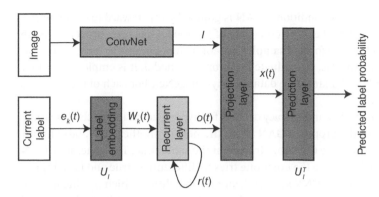

Figure 2.13 The RNN model for multi-label classification. Source: Wang et al. [69].

- Wang et al. [69] used an end-to-end approach to take advantage of semantic redundancy and co-occurrence dependency, which are both necessary for effective multi-label classifications (Figure 2.13). In this challenge, the recurrent neurons model of high-order label co-occurrence dependency is more concise and efficient than other label co-occurrence approaches. The recurrent neurons' implicit attention mechanism adapts picture characteristics to accurately predict small things that require more context.

2.3.2.3 Auto-encoder Based Image Classification

When compared to patch-based sparse coding, convolutional sparse coding (CSC) can simulate local links between visual content and minimize code redundancy. CSC, on the other hand, requires a time-consuming optimization technique to infer the codes (i.e. feature maps).

- Luo et al. [70] introduced a convolutional sparse auto-encoder (CSAE) in this paper, which takes advantage of the convolutional autoencoder's structure and integrates max-pooling to empirically sparsify feature maps for feature learning. This basic sparsifying method, along with competition over feature channels, allows the SGD algorithm to perform quickly for CSAE training; consequently, no complicated optimization process is required. The features learned in the CSAE were used to initialize CNNs for classification, and the results were competitive on benchmark data sets. Lou et al. also offered an approach for constructing local descriptors from the CSAE for categorization by establishing linkages between the CSAE and the CSC. Experiments on Caltech-101 and Caltech-256 proved the efficacy of the suggested strategy and confirmed that the CSAE as a CSC model can investigate connections between nearby image information for classification tasks.
- Liang et al. [71] offered a new approach for image classification remotely built upon stacked denoising autoencoder due to the accuracy bottleneck nature of conventional remote image classification methods. The deep network model is first constructed using denoising autoencoder's stacked layers. The unsupervised greedy layer-wise training algorithm is then used with noised input to train every layer in turn for a more robust expression. Information is retrieved in supervised learning using a back propagation (BP) neural network, and the entire network is optimized using error backpropagation. The total and kappa accuracy are both greater than those of the support vector machine (SVM) and back propagation NNs by a margin of 95.7% and 0.955 respectively processed on Gaofen-1 satellite data. The experimental outcomes recommend that this strategy may effectively increase remote sensing picture categorization accuracy.

2.3.2.4 GAN Based Image Classification

- Kong et al. [72] presented a unique technique for obtaining data labels cost-effectively without querying an oracle. To evaluate the uncertainty level, a new reward for each sample is developed in the model. It is generated through a trained classifier on older labeled data. A conditional GAN is guided by this reward to create useful samples with a greater probability for the specific label. The efficiency of the model has been proven through comprehensive assessments, demonstrating that the formed samples are competent enough to increase performance in common picture classification assignments. To evaluate the proposed model, it is employed on four datasets: CIFAR-10, Fashion-MNIST, MNIST, and a large size dataset Tiny-ImageNet. For each of the 10 broad classes, CIFAR-10 features colorful pictures.

- Zhu et al. [73], for the first time, investigate the utility and efficacy of GAN for hyperspectral image classification. A CNN is utilized to distinguish the inputs in the proposed GAN. With the help of another CNN, false inputs are also generated. The generating and discriminative CNNs are trained at the same time. The generative CNN seeks to create false inputs as realistic as possible, while the other one tries to differentiate true and false inputs. The adversarial training enhances the discriminative CNN's generalization capabilities, which is critical when training samples are restricted. This work proposes two strategies for spectral classifiers: (i) a well-designed 1D-GAN as a spectral classifier; and (ii) a robust 3D-GAN as a spectral–spatial classifier. In addition, the generated adversarial samples are combined with real training data to fine-tune the discriminative CNN, resulting in improved classification performance. Three commonly used hyperspectral data sets are used to test the suggested classifiers: Salinas, Indiana Pines, and Kennedy Space Center. In comparison to state-of-the-art approaches, the collected findings show that the proposed models deliver competitive results. Furthermore, the suggested GANs indicate the enormous potential of GAN-based approaches for the analysis of such complex and intrinsically nonlinear data in the remote sensing community, as well as the immense potential of GAN-based methods for the tough problem of hue-saturation-intensity (HSI) image categorization.

2.4 Conclusion

This chapter went through the numerous DL approaches that are employed in various tasks along the image processing chain in detail. We discuss different models for image segmentation and image classification deeply procured from related articles. Researchers have created a large number of neural architectures in recent decades, and for each model, we explained the methodologies used for certain image-processing applications. In comparison to conventional models, hybrid models have been demonstrated to perform better in case studies. Each task of image segmentation and classification used in the model was detailed in detail, and the model that sets the highest standard was discussed briefly. Despite the excellent accuracy of traditional models, newly designed hybrid models are adapted to a specific use case. Integrating these traditional models according to their use case improves accuracy and boosts the models' performance. As a result, DL algorithms with a highly iterative and learning-based approach perform much better for image processing applications. Through this chapter, we try to convey the advantages and limitations of different popular models while applying them in these two domains of image processing. In the future, we can compare the performance of discussed architecture to conclude a more efficient model for a specific task.

References

1 Chen, Q., Xu, J., and Koltun, V. (2017). Fast image processing with fully-convolutional networks. In: *Proceedings of the IEEE International Conference on Computer Vision*, 2497–2506. IEEE.

2 Kolekar, M.H., Palaniappan, K., Sengupta, S., and Seetharaman, G. (2009). Semantic concept mining based on hierarchical event detection for soccer video indexing. *Journal of Multimedia* 4 (5).

3 Kolekar, M.H. and Sengupta, S. (2010). Semantic concept mining in cricket videos for automated highlight generation. *Multimedia Tools and Applications* 47 (3): 545–579.

4 Kolekar, M.H. and Sengupta, S. (2015). Bayesian network-based customized highlight generation for broadcast soccer videos. *IEEE Transactions on Broadcasting* 61 (2): 195–209.

5 Bhatnagar, S., Ghosal, D., and Kolekar, M.H. (2017). Classification of fashion article images using convolutional neural networks. In: *2017 Fourth International Conference on Image Information Processing (ICIIP)*, 1–6. IEEE.

6 Liu, W., Wang, Z., Liu, X. et al. (2017). A survey of deep neural network architectures and their applications. *Neurocomputing* 234: 11–26.

7 Albawi, S., Mohammed, T.A., and Al-Zawi, S. (2017). Understanding of a convolutional neural network. In: *2017 International Conference on Engineering and Technology (ICET)*, 1–6. Elsevier.

8 Long, J., Shelhamer, E., and Darrell, T. (2015). Fully convolutional networks for semantic segmentation. In: *IEEE Conference on Computer Vision and Pattern Recognition*, 3431–3440. IEEE.

9 Chen, L.C., Papandreou, G., Kokkinos, I., et al. 2014. Semantic image segmentation with deep convolutional nets and fully connected CRFs. arXiv preprint arXiv:1412.7062.

10 Schwing, A.G. and Urtasun, R. 2015. Fully connected deep structured networks. arXiv preprint arXiv:1503.02351.

11 Lin, G., Shen, C., Van Den Hengel, A. et al. (2016). Efficient piecewise training of deep structured models for semantic segmentation. In: *Proceedings of the IEEE Conference on Computer Vision and Pattern Recognition*, 3194–3203. IEEE.

12 Liu, Z., Li, X., Luo, P. et al. (2015). Semantic image segmentation via deep parsing network. In: *Proceedings of the IEEE International Conference on Computer Vision*, 1377–1385. IEEE.

13 Chen, L.C., Papandreou, G., Kokkinos, I. et al. (2017). Deeplab: semantic image segmentation with deep convolutional nets, atrous convolution, and fully connected CRFs. *IEEE Transactions on Pattern Analysis and Machine Intelligence* 40 (4): 834–848.

14 Paszke, A., Chaurasia, A., Kim, S. et al. 2016. Enet: a deep neural network architecture for real-time semantic segmentation. arXiv preprint arXiv:1606.02147.

15 Yang, M., Yu, K., Zhang, C. et al. (2018). Denseaspp for semantic segmentation in street scenes. In: *Proceedings of the IEEE Conference on Computer Vision and Pattern Recognition*, 3684–3692. IEEE.

16 Wang, P., Chen, P., Yuan, Y. et al. (2018). Understanding convolution for semantic segmentation. In: *2018 IEEE Winter Conference on Applications of Computer Vision (WACV)*, 1451–1460. IEEE.

17 Yu, F. and Koltun, V. 2015. Multi-scale context aggregation by dilated convolutions. arXiv preprint arXiv:1511.07122.

18 Visin, F., Ciccone, M., Romero, A. et al. (2016). ReSeg: a recurrent neural network-based model for semantic segmentation. In: *Proceedings of the IEEE Conference on Computer Vision and Pattern Recognition Workshops*, 41–48. IEEE.

19 Chen, L.C., Papandreou, G., Schroff, F. et al. 2017. Rethinking atrous convolution for semantic image segmentation. arXiv preprint arXiv:1706.05587.

20 Noh, H., Hong, S., and Han, B. (2015). Learning deconvolution network for semantic segmentation. In: *Proceedings of the IEEE International Conference on Computer Vision*, 1520–1528. IEEE.

21 Badrinarayanan, V., Kendall, A., and Cipolla, R. (2017). SegNet: a deep convolutional encoder–decoder architecture for image segmentation. *IEEE Transactions on Pattern Analysis and Machine Intelligence* 39 (12): 2481–2495.

22 Ronneberger, O., Fischer, P., and Brox, T. (2015). U-Net: convolutional networks for biomedical image segmentation. In: *International Conference on Medical Image Computing and Computer-Assisted Intervention*, 234–241. IEEE.

23 Milletari, F., Navab, N., and Ahmadi, S.A. (2016). V-Net: fully convolutional neural networks for volumetric medical image segmentation. In: *2016 Fourth International Conference on 3D Vision (3DV)*, 565–571. IEEE.

24 He, K., Gkioxari, G., Dollár, P., and Girshick, R. (2017). Mask R-CNN. In: *Proceedings of the IEEE International Conference on Computer Vision*, 2961–2969. IEEE.

25 Liu, S., Qi, L., Qin, H. et al. (2018). Path aggregation network for instance segmentation. In: *Proceedings of the IEEE Conference on Computer Vision and Pattern Recognition*, 8759–8768. IEEE.

26 Dai, J., He, K., and Sun, J. (2016). Instance-aware semantic segmentation via multi-task network cascades. In: *Proceedings of the IEEE Conference on Computer Vision and Pattern Recognition*, 3150–3158. IEEE.

27 Chen, L.C., Hermans, A., Papandreou, G. et al. (2018). MaskLab: instance segmentation by refining object detection with semantic and direction features. In: *Proceedings of the IEEE Conference on Computer Vision and Pattern Recognition*, 4013–4022. IEEE.

28 Pinheiro, P.O., Collobert, R., and Dollár, P. 2015. Learning to segment object candidates. arXiv preprint arXiv:1506.06204.

29 Chen, X., Girshick, R., He, K., and Dollár, P. (2019). TensorMask: a foundation for dense object segmentation. In: *Proceedings of the IEEE/CVF International Conference on Computer Vision*, 2061–2069. IEEE.

30 Xie, E., Sun, P., Song, X. et al. (2020). Polarmask: single shot instance segmentation with polar representation. In: *Proceedings of the IEEE/CVF Conference on Computer Vision and Pattern Recognition*, 12193–12202. IEEE.

31 Dai, J., Li, Y., He, K., and Sun, J. (2016). R-FCN: object detection via region-based fully convolutional networks. In: *Advances in Neural Information Processing Systems*, 79–387. IEEE.

32 Lee, Y. and Park, J. (2020). CenterMask: real-time anchor-free instance segmentation. In: *Proceedings of the IEEE/CVF Conference on Computer Vision and Pattern Recognition*, 13906–13915. IEEE.

33 Lin, T.Y., Dollár, P., Girshick, R. et al. (2017). Feature pyramid networks for object detection. In: *Proceedings of the IEEE Conference on Computer Vision and Pattern Recognition*, 2117–2125. IEEE.

34 Zhao, H., Shi, J., Qi, X. et al. (2017). Pyramid scene parsing network. In: *Proceedings of the IEEE Conference on Computer Vision and Pattern Recognition*, 2881–2890. IEEE.

35 Ghiasi, G. and Fowlkes, C.C. (2016). Laplacian pyramid reconstruction and refinement for semantic segmentation. In: *European Conference on Computer Vision*, 519–534. Springer.

36 He, J., Deng, Z., Zhou, L. et al. (2019). Adaptive pyramid context network for semantic segmentation. In: *Proceedings of the IEEE/CVF Conference on Computer Vision and Pattern Recognition*, 7519–7528. IEEE.

37 Ding, H., Jiang, X., Shuai, B. et al. (2018). Context contrasted feature and gated multi-scale aggregation for scene segmentation. In: *Proceedings of the IEEE Conference on Computer Vision and Pattern Recognition*, 2393–2402. IEEE.

38 Li, G., Xie, Y., Lin, L., and Yu, Y. (2017). Instance-level salient object segmentation. In: *Proceedings of the IEEE Conference on Computer Vision and Pattern Recognition*, 2386–2395. IEEE.

39 Lin, D., Ji, Y., Lischinski, D. et al. (2018). Multi-scale context intertwining for semantic segmentation. In: *Proceedings of the European Conference on Computer Vision (ECCV)*, 603–619. Springer.

40 Visin, F., Kastner, K., Cho, K. et al. 2015. ReNet: a recurrent neural network based alternative to convolutional networks. arXiv preprint arXiv:1505.00393.

41 Byeon, W., Breuel, T.M., Raue, F., and Liwicki, M. (2015). Scene labeling with LSTM recurrent neural networks. In: *Proceedings of the IEEE Conference on Computer Vision and Pattern Recognition*, 3547–3555. IEEE.

42 Liang, X., Shen, X., Feng, J. et al. (2016). Semantic object parsing with graph LSTM. In: *European Conference on Computer Vision*, 125–143. Springer.

43 Xiang, Y. and Fox, D. 2017. DA-RNN: semantic mapping with data associated recurrent neural networks. arXiv preprint arXiv:1703.03098.

44 Hu, R., Rohrbach, M., and Darrell, T. (2016). Segmentation from natural language expressions. In: *European Conference on Computer Vision*, 108–124. Springer.

45 Luc, P., Couprie, C., Chintala, S. et al. 2016. Semantic segmentation using adversarial networks. arXiv preprint arXiv:1611.08408.

46 Souly, N., Spampinato, C., and Shah, M. (2017). Semi supervised semantic segmentation using generative adversarial network. In: *Proceedings of the IEEE International Conference on Computer Vision*, 5688–5696. IEEE.

47 Hung, W.C., Tsai, Y.H., Liou, Y.T. et al. 2018. Adversarial learning for semi-supervised semantic segmentation. arXiv preprint arXiv:1802.07934.

48 Xue, Y., Xu, T., Zhang, H. et al. (2018). SegAN: adversarial network with multi-scale L_1 loss for medical image segmentation. *Neuroinformatics* 16 (3): 383–392.

49 Majurski, M., Manescu, P., Padi, S. et al. 2019. Cell image segmentation using generative adversarial networks, transfer learning, and augmentations. In *Proceedings of the IEEE/CVF Conference on Computer Vision and Pattern Recognition Workshops,* Long Beach, CA (16–20 June 2019).

50 Ehsani, K., Mottaghi, R., and Farhadi, A. (2018). SegAN: segmenting and generating the invisible. In: *Proceedings of the IEEE Conference on Computer Vision and Pattern Recognition*, 6144–6153. IEEE.

51 Chen, L.C., Yang, Y., Wang, J. et al. (2016). Attention to scale: scale-aware semantic image segmentation. In: *Proceedings of the IEEE Conference on Computer Vision and Pattern Recognition*, 3640–3649. IEEE.

52 Huang, Q., Xia, C., Wu, C., Li, S., Wang, Y., Song, Y., & Kuo, C. C. J. (2017). Semantic segmentation with reverse attention. arXiv preprint arXiv:1707.06426.

53 Fu, J., Liu, J., Tian, H. et al. (2019). Dual attention network for scene segmentation. In: *Proceedings of the IEEE/CVF Conference on Computer Vision and Pattern Recognition*, 3146–3154. IEEE.

54 Yuan, Y., Huang, L., Guo, J. et al. 2018. OCNet: object context network for scene parsing. arXiv preprint arXiv:1809.00916.

55 Zhang, H., Wu, C., Zhang, Z. et al. 2020. ResNeSt: Split-attention networks. arXiv preprint arXiv:2004.08955.

56 Huang, Z., Wang, X., Huang, L. et al. (2019). CCNet: criss-cross attention for semantic segmentation. In: *Proceedings of the IEEE/CVF International Conference on Computer Vision*, 603–612. IEEE.

57 Yu, C., Wang, J., Peng, C. et al. (2018). Learning a discriminative feature network for semantic segmentation. In: *Proceedings of the IEEE Conference on Computer Vision and Pattern Recognition*, 1857–1866. IEEE.

58 Li, X., Zhong, Z., Wu, J. et al. (2019). Expectation–maximization attention networks for semantic segmentation. In: *Proceedings of the IEEE/CVF International Conference on Computer Vision*, 9167–9176. IEEE.

59 LeCun, Y., Bottou, L., Bengio, Y., and Haffner, P. (1998). Gradient-based learning applied to document recognition. *Proceedings of the IEEE* 86 (11): 2278–2324.

60 Krizhevsky, A., Sutskever, I., and Hinton, G.E. (2012). ImageNet classification with deep convolutional neural networks. In: *Advances in Neural Information Processing Systems*, vol. 25, 1097–1105. IEEE.

61 Zeiler, M.D. and Fergus, R. (2014). Visualizing and understanding convolutional networks. In: *European Conference on Computer Vision*, 818–833. Springer.

62 Simonyan, K. and Zisserman, A. 2014. Very deep convolutional networks for large-scale image recognition. arXiv preprint arXiv:1409.1556.

63 Szegedy, C., Liu, W., Jia, Y. et al. (2015). Going deeper with convolutions. In: *Proceedings of the IEEE conference on computer vision and pattern recognition*, 1–9. IEEE.

64 He, K., Zhang, X., Ren, S., and Sun, J. (2016). Deep residual learning for image recognition. In: *Proceedings of the IEEE Conference on Computer Vision and Pattern Recognition*, 770–778. IEEE.

65 Huang, G., Liu, Z., Van Der Maaten, L., and Weinberger, K.Q. (2017). Densely connected convolutional networks. In: *Proceedings of the IEEE Conference on Computer Vision and Pattern Recognition*, 4700–4708. IEEE.

66 Sabour, S., Frosst, N., and Hinton, G.E. 2017. Dynamic routing between capsules. arXiv preprint arXiv:1710.09829.

67 Guo, Y., Liu, Y., Bakker, E.M. et al. (2018). CNN–RNN: a large-scale hierarchical image classification framework. *Multimedia Tools and Applications* 77 (8): 10251–10271.

68 Yin, Q., Zhang, R., and Shao, X. (2019). CNN and RNN mixed model for image classification. In: *MATEC Web of Conferences*, vol. 277, 02001. EDP Sciences.

69 Wang, J., Yang, Y., Mao, J. et al. (2016). CNN–RNN: a unified framework for multi-label image classification. In: *Proceedings of the IEEE Conference on Computer Vision and Pattern Recognition*, 2285–2294. IEEE.

70 Luo, W., Li, J., Yang, J. et al. (2017). Convolutional sparse autoencoders for image classification. *IEEE Transactions on Neural Networks and Learning Systems* 29 (7): 3289–3294.

71 Liang, P., Shi, W., and Zhang, X. (2018). Remote sensing image classification based on stacked denoising autoencoder. *Remote Sensing* 10 (1): 16.

72 Kong, Q., Tong, B., Klinkigt, M. et al. (2019). Active generative adversarial network for image classification. In: *Proceedings of the AAAI Conference on Artificial Intelligence*, vol. 33, no. 01, 4090–4097. IEEE.

73 Zhu, L., Chen, Y., Ghamisi, P., and Benediktsson, J.A. (2018). Generative adversarial networks for hyperspectral image classification. *IEEE Transactions on Geoscience and Remote Sensing* 56 (9): 5046–5063.

74 Aslam, N., and Kolekar, M.H. (2022). Unsupervised anomalous event detection in videos using spatio-temporal inter-fused autoencoder. *Multimedia Tools and Applications*, 1–26.

75 Aslam, N., Rai, P.K., and Kolekar, M.H. (2022). A3N: Attention-based adversarial autoencoder network for detecting anomalies in video sequence. *Journal of Visual Communication and Image Representation*, 87: 103598.

3

Deep Learning Based Synthetic Aperture Radar Image Classification

J. Anil Raj[1,2] *and Sumam M. Idicula*[3]

[1]*Department of Electronics and Communication, Muthoot Institute of Technology and Science, Kochi, Kerala, India*
[2]*Department of Computer Science, Cochin University of Science and technology, Kochi, Kerala, India*
[3]*Department of Artificial Intelligence and Data Science, Muthoot Institute of Technology and Science, Kochi, Kerala, India*

3.1 Introduction

Synthetic aperture radar (SAR) images have many applications in the earth observation (EO) community. SAR remote sensing has many advantages over optical remote sensing. SAR can be operated in extreme weather conditions like cloudy situations, rainy conditions, meteorological effects, or even night. Usually, experimenters use raw SAR images to classify targets as part of the defense and other notable practical interests of the country. So, this research area, SAR automatic target recognition (ATR), is gaining broader attention.

SAR images are usually gray compared with the optical aerial images in the visible region. These grayscale SAR images are generated using the magnitude data of the SAR data. In supplement to this, another similar data, the phase details, is also supplied by the SAR Satellite. Usually, the analysts use the magnitude information alone to classify the SAR image targets. And they discard the phase information. Because it doesn't provide any structural information of the target. Also, the phase data is more difficult to interpret than the magnitude data. But the phase information also contains similar information as magnitudes have. The demanding part of this research is analyzing phase information into meaningful for the object classification task.

This chapter introduces a novel technique by employing the advantages of both magnitude and phase data from SAR data to improve object classification performance. The most difficult part of SAR ATR is the discrimination of small and similar targets [1]. For example, the distinction of military vehicles having a few meters in length. Here the number of pixels representing the target in the SAR image will be a few. So, the deep-learning model should learn the minute features associated with discriminating the target from these pixels. SAR ATR is more complex and challenging than SAR marine target identification (MTI); here, the target information on SAR data is comparably high [2]. So, this chapter focuses on the most challenging part, which is the classification of small targets from SAR data.

In supervised image classification techniques, the images in the datasets are augmented using several image processing techniques to create a balanced dataset. This gives improved performance for the classification of standard RGB color images. But in the case of single channeled SAR images, the usual image processing techniques don't change the key features. So, in this case, the chance of overfitting will be high. In this proposed technique, both magnitude and phase information are logically combined in the pre-processing phase before being in the deep-learning model. This combined data can improve the uniqueness of the whole dataset. And it is used to train the lightweight deep-learning model for classifying the SAR targets.

Machine Learning Algorithms for Signal and Image Processing, First Edition.
Edited by Deepika Ghai, Suman Lata Tripathi, Sobhit Saxena, Manash Chanda, and Mamoun Alazab.
© 2023 The Institute of Electrical and Electronics Engineers, Inc. Published 2023 by John Wiley & Sons, Inc.

3.2 Literature Review

There are many deep learning-based types of research for SAR object detection and classification with reasonable accuracies. The main advantage of using a deep-learning algorithm is automatically learning of features. Deep neural network models with convolutional neural networks (CNNs) have the ability to automatically extract the image features. Also, the availability of a large number of data enriches the model's performance further. Wang et al. [3] developed deep-learning algorithms for classifying the SAR targets with valid accuracy, but the model size is large. Ievgen M. Gorovyi et al. [4] compared the performance of SVM and CNN techniques for SAR target recognition. In this experiment, CNN performed well for almost all classes with the least wrong classifications. But it performed poorly for smaller size targets, the classification accuracy is comparatively less in these cases. Many CNN-based deep learning works are proposed to classify SAR targets [5–7]. They used magnitude information alone for the classification and discarded the phase information for the effective result. Jiao et al. [8] proposed a novel densely connected end-to-end CNN model for SAR-based ship localization. The significant part of this work is the multi-scale SAR ship detection. This work also doesn't consider the application of phase data in the object/pattern detection domain. Cristian Coman and Thaens [9] proposed a phase data technique for deep-learning applications. Recently developed SAR target detection from video SAR using dual faster RCNN has good performance in detecting moving targets [10]. Jiang et al. [11] proposed a ship detection system from SAR using a deep-learning technique with the filtering preprocess method. With a deep-learning technique, this fast non-local means (FNLM) filter could detect ships quickly and accurately. Lan Du et al. [12] proposed a semi-supervised learning technique for SAR target classification. This technique avoids the complicated process of labeling of large SAR data.

3.3 Dataset Description

SAR sensors use microwave electromagnetic radiations with scattering technique for generating images. The satellite transmitter continuously sends the EM signals to the target. And the backscattered signals are collected by the SAR sensor's receiver part. These backscattered signals are signal processed by calculating the time and phase to generate SAR images. The usual grayscale two-dimensional (2D) SAR images, which are normally used by object classification/detection applications, are generated using the backscattered SAR's magnitude information alone.

Along with magnitude, phase information can also be generated from the returned backscattered signal. This phase data depends on the distance between satellite sensor and the target. It also depends on the dielectric property of the material of the target. This phase data can be used to generate a 2D phase image with pixels equal to the sum of all the signals' phase angles reflected from the area corresponding to that pixel. Thus, no structural information is available from phase data.

For the training and testing of the deep-learning model, we used the publicly available standard MSTAR (Moving and Stationary Target Acquisition and Recognition) [13] dataset. This dataset is provided by Sandia National Laboratory project by the Air Force Research Laboratory and the US Defense Advanced Research Projects Agency. The dataset provides aerial SAR images of various defense-based targets. The dataset is more challenging with targets having different shapes, sizes, target types, EM depression angles, image aspect angles, target serial numbers, and articulation. It provides eight classes of targets as described in Table 3.1. The data of MSTAR dataset contains a header part that gives the details of sensor type and the details of the target. The main details, which are the magnitude information and phase information as arrays, are provided after this header section. The magnitude information and phase information of Truck ZIL-131 are visualized as 2D image in Figure 3.1a,b, respectively. From the visual observation of phase data, it is not readily observable.

Table 3.1 Details of MSTAR targets used in this experiment.

Target name	Details	No. of images in 17° depression angle	No. of images in 15° depression angle
2S1_gun	Weapon system	299	274
BRDM-2	Armored carrier	298	274
BTR-60	Armored carrier	256	195
D7	Bulldozer	299	274
ZIL-131	Truck	299	274
ZSU23–4 gun	Air defense unit	299	274
T-62	Tank	299	273
Slicy	Test target	298	274

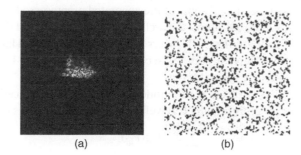

Figure 3.1 Visualization of SAR image of target: Bulldozer-D7. (a) 2D magnitude image and (b) 2D phase image of the target. Source: Keydel et al. [13], from SPIE.

(a) (b)

3.4 Methodology

The SAR data is initially processed to generate the three channeled (channels: A, B, and C) preprocessed images. The proposed methodology of generation of this preprocessed image is as follows:

- First channel (channel A): Lee filtered 2D data of magnitude information
- Second channel (channel B): cosine function of each pixel of phase information
- Third channel (channel C): sine function of each pixel of phase information

The presence of speckle noise reduces the visual quality of the SAR data. In SAR target classification problems, the contribution of speckle for generation of wrong classification is very high. Lee's filter [14] is well known for despeckling and enhancing SAR images. It uses minimum mean square error (MMSE) method for despeckling and improves the visual quality of SAR data. We know that speckle affects both magnitude and phase pixel values. So, Eq. (3.1) describes the Lee filtering action to $r(x, y)$,

$$L(x,y) = r'(x,y) + \gamma(r(x,y) - r'(x,y)) \tag{3.1}$$

Here, $r'(x, y)$ is the mean of $r(x, y)$ and γ is the weighting function and is estimated using Eq. (3.2) as:

$$\gamma = 1 - C_v/C_i \tag{3.2}$$

where C_i is the coefficient of variance of noise-free image and C_v is the coefficient of variance of speckled image.

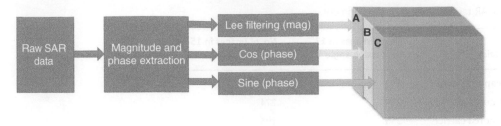

Figure 3.2 Process flow of three channeled preprocessed image generation.

Figure 3.2 shows the visualization three channeled preprocessed image generation processes. Initially, we extracted the magnitude data and phase data from the SAR data files. This 2D magnitude image is passed through Lee filter as described in Eq. (3.1) to generate channel "A." The cos and sine of each pixel of the 2D phase information are calculated and they are the channels "B" and "C," respectively. These channels A, B, and C are combined to form the preprocessed images for this SAR target classification experiment. Figure 3.3 shows the visualization of some of the generated preprocessed images.

Figure 3.4 shows the architecture of the proposed deep-learning model with preprocessed SAR image as input. It contains five CNN layers and two fully connected layers. The last layer is softmax activated to generate the class probability outputs and second last fully connected layer is sigmoid activated. All CNN layers are ReLU activated with (3×3) size kernels.

This problem comes under the multiclass classification problem in the computer vision domain. So categorical cross-entropy, which is well known for similar types of applications is used as the loss function (see Eq. (3.3)).

$$\text{Loss}_i = - \sum T_{i,j} \, \log(P_{i,j}) \tag{3.3}$$

(a)

(b)

(c)

Figure 3.3 Visual comparison of optical images (normal) and generated SAR (aerial) combined images of SAR targets: (a) BRDM-2, (b) ZSU23-4 gun, and (c) T-62 target.

Figure 3.4 Architecture of proposed CNN model with three channel SAR input.

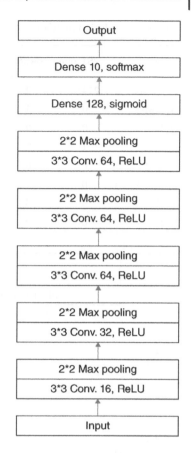

where "T" is SAR target, "p" stands for predictions, and subscripts "i" and "j" denotes data points and class details, respectively. To select the optimizer, we tried with famous stochastic gradient descent (SGD) and Adam optimizer. From these experiments, Adam gives better results because it is appropriate for problems with very noisy gradients.

3.5 Experimental Results and Discussions

The MSTAR data provides SAR data at various depression angles. Here, depression angle is the angle between the radar's line of sight to its horizontal plane, so this can be used for testing the model's generalization ability. In this experiment, SAR data acquired at depression angle 17° are set as train set and those at 15° as the test set. So, we have total 2746 train data and 2425 test data.

The deep-learning model with early discussed parameters is trained with the preprocessed images for 150 epochs, with a fixed batch size of 32 data. By using the callback feature of the Keras technique, the model having weights with improved accuracy at each epoch was saved to hard disk. At the 97th epoch got the best model with the highest validation accuracy of 99.74% with least loss value. Table 3.2 shows the confusion matrix result of SAR target recognition results. Also, Figure 3.5 shows the comparison of precision–recall curves results of different classes for an in-depth analysis. Table 3.3 shows the comparison of various evaluation indexes of models with different types of input images. Here, proposed model with proposed pre-processing technique performed well with the highest precision, recall, and F1-score.

Table 3.2 Confusion matrix of SAR target recognition results.

Category	2S1_gun	BRDM-2	BTR-60	D7	ZIL-131	ZSU23–4 gun	T-62	Slicy	Accuracy (%)
2S1_gun	274	0	0	0	0	0	0	0	100
BRDM-2	0	273	0	0	1	0	0	0	99.64
BTR-60	0	0	195	0	0	0	0	0	100
D7	1	0	0	273	0	0	0	0	99.64
ZIL-131	0	0	0	1	272	1	0	0	99.27
ZSU23-4 gun	0	1	0	0	0	273	0	0	99.64
T-62	0	0	0	0	0	0	273	0	100
Slicy	0	0	0	0	0	0	0	0	100
								Average accuracy	**99.74**

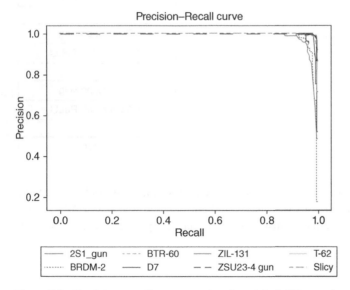

Figure 3.5 Precision–recall curves results of model of different classes.

Table 3.3 Evaluation indexes of models with different types of input images.

Type	Recall (%)	Precision (%)	F1-score (%)
Magnitude alone	98.15	94.16	96.1
Phase alone	47.34	55.65	51.1
Combined	99.74	99.80	99.8

3.6 Conclusion

Deep learning-based SAR target classification with a novel advanced pre-processing technique is implemented. The proposed model was trained and tested using the standard MSTAR dataset. From the results, it is understood that SAR target classification performance is further improved by combining the magnitude and phase data provided by the SAR sensor. These combined results are analyzed and compared with the magnitude alone and phase alone based deep-learning technique. Here combined method yields the highest validation accuracy for the validation dataset. Extracting the features is an important procedure in the SAR target recognition type of problems. By analyzing the result, we can understand that phase information provided by the SAR sensor has the potential to increase the performance and accuracy of SAR target classification. SAR target's phase information along with its magnitude data gives the best classification result in deep learning-based applications.

References

1 Chen, S., Wang, H., Xu, F., and Jin, Y. (2016). Target classification using the deep convolutional networks for SAR images. *IEEE Transactions on Geoscience and Remote Sensing* 54 (8): 4806–4817.

2 Bentes, C., Velotto, D., and Tings, B. (2018). Ship classification in TerraSARX images with convolutional neural networks. *IEEE Journal of Oceanic Engineering* 43 (1): 258–266.

3 Wang, H., Chen, S., Xu, F., and Jin, Y. (2015). Application of deep-learning algorithms to MSTAR data. In: *2015 IEEE International Geoscience and Remote Sensing Symposium (IGARSS)*, 3743–3745. Milan: IEEE.

4 Gorovyi, I.M. and Sharapov, D.S. (2017, 2017). Comparative analysis of convolutional neural networks and support vector machines for automatic target recognition. In: *IEEE Microwaves, Radar and Remote Sensing Symposium (MRRS)*, 63–66. Kiev: IEEE.

5 Deng, S., Du, L., Li, C. et al. (2017). SAR automatic target recognition based on Euclidean distance restricted autoencoder. *IEEE Journal of Selected Topics in Applied Earth Observations and Remote Sensing* 10 (7): 3323–3333.

6 Pei, J., Huang, Y., Huo, W. et al. (2018). Target aspect identification in SAR image: a machine learning approach. In: *IEEE International Geoscience and Remote Sensing Symposium*, 2310–2313. Valencia: IEEE.

7 Liu, Q., Li, S., Mei, S. et al. (2018). Feature learning for SAR images using convolutional neural network. In: *IEEE International Geoscience and Remote Sensing Symposium*, 7003–7006. Valencia: IEEE.

8 Jiao, J., Zhang, Y., Sun, H. et al. (2018). A densely connected end-to-end neural network for multiscale and multiscene SAR ship detection. *IEEE Access* 6: 20881–20892.

9 Coman, C. and Thaens, R. (2018). A deep learning SAR target classification experiment on MSTAR dataset. In: *2018 19th International Radar Symposium (IRS)*, 1–6. Bonn: IEEE.

10 Wen, L., Ding, J., and Loffeld, O. (2021). Video SAR moving target detection using dual faster R-CNN. *IEEE Journal of Selected Topics in Applied Earth Observations and Remote Sensing* 14: 2984–2994. https://doi.org/10.1109/JSTARS.2021.3062176.

11 Jiang, M., Gu, L., and Gong, Z. (2021). Ship target detection method for SAR image of FNLM filtering combined with faster R-CNN. In: *Earth Observing Systems XXVI*, vol. 11829. International Society for Optics and Photonics.

12 Du, L., Wei, D., Li, L., and Guo, Y. (2020). SAR target detection network via semi-supervised learning. *Journal of Electronics and Information Technology* 42 (1): 154–163. https://doi.org/10.11999/JEIT190783.

13 Keydel, E.R., Lee, S.W., and Moore, J.T. (1996). MSTAR extended operating conditions: a tutorial. In: *Proceedings of 3rd SPIE Conference on Algorithms SAR Imagery Text*, vol. 2757, 228–242. SPIE Digital Library.

14 Lee, J.S. (1981). Speckle analysis and smoothing of synthetic aperture radar images. *Computer Graphics and Image Processing* 17 (1): 24–32.

4

Design Perspectives of Multi-task Deep-Learning Models and Applications

Yeshwant Singh, Anupam Biswas, Angshuman Bora, Debashish Malakar, Subham Chakraborty, and Suman Bera

Department of Computer Science and Engineering, National Institute of Technology, Silchar, Assam, India

4.1 Introduction

Machine learning (ML) optimizes a specific measurement metric irrespective of its physical value on a particular standard or a business key performance index. Single-task or multiple single-task models are trained to perform well on new data. The parameters of the models are calibrated to increase the performance metrics until they cannot improve further. We may accomplish satisfactory results by pinpointing a single task, but we may overlook data that could generalize models well in the process. The transformation performed by multi-task neural network (NN) hidden layers on data originates a data representation by overlapping multi-task signals. The model can fully summarize the initial task by distributing information among related tasks. This method is termed multi-task learning (MTL). The difference between single-task learning compared to MTL is given in Figure 4.1.

The motivation behind MTL comes from various directions. For example, humans learn new things in various fields during their life; we regularly employ the information we have accumulated by learning from similar past experiences or tasks. A newborn starts learning how to recognize different people's faces at an early age. The knowledge/information gained is invariant and applicable to other world objects. Moreover, from an instructional point of view, we regularly learn new tasks based on the collective experience gained over other tasks. This makes us perform complex tasks.

Some of the practical applications of MTL are web search, spam filtering, etc. [1]. In web search, MTL with boosted decision trees is incredibly supportive as datasets from multiple countries differ significantly in size due to the expense of editorial judgments. It has been demonstrated that learning from different overlapping tasks can mutually commence considerable improvements in accuracy with quality.

Spam filtering can be treated as particular but related classification tasks over a set of users. In other words, an email considered spam to one user might not appear spam to someone else. The definition of spam emails varies from person to person, yet spam classification has related tasks. Solving every individual user spam classification jointly with the help of MTL can improve all users' performance.

In this chapter, we attempt to summarize MTL from a design perspective. We begin by defining deep learning (DL) in general in Section 4.2 and its relation to MTL. Then we go into some popular classes of neural networks in Section 4.3 to give a base foundation of underlying ideas. Section 4.4 gives an overview of the designs and implementation of various methods in MTL. Section 4.5 presents a few applications of MTL in some domains, followed by evaluation methods used to measure the performances of these models in Section 4.6. Finally, we conclude this chapter in Section 4.7 with challenges and future directions of MTL.

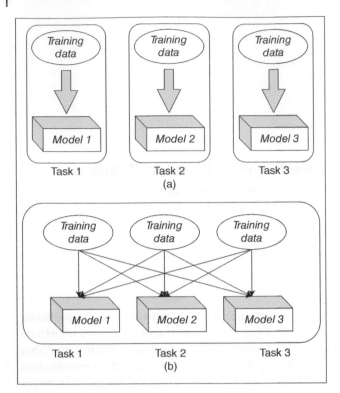

Figure 4.1 Single-task learning versus multi-task learning diagram. (a) Single task learning, (b) Multi-task learning.

4.2 Deep Learning

DL is a sub-domain of ML. It is entirely made up of artificial neural networks aided representation learning. Here, learning is classified into three types: supervised, semi-supervised, and unsupervised [2]. The term "deep" originates from using multiple layers, including hidden layers in the neural networks (NNs). MTL falls in the sub-domain of DL.

In the human brain, there are over 100 billion interconnected neurons. These are the basic working units of the brain. Every neuron is inter-linked through thousands of synapses to neighboring neurons in a giant web-like structure. In DL, neurons are represented by formulating an artificial neural net consisting of nodes (neurons). A neuron takes an input, performs an operation, and gives an output. The operation performed over the input is a linear transformation, making the operation highly parallelable and efficient to compute. Non-linearity is added to the computation of a neuron in terms of the activation function. This non-linearity allows the neural network to learn non-linear patterns from the data. The diagram of a neuron in the brain versus an artificial neural net is shown in Figure 4.2.

Some real-life applications of DL are automatic text generation, here the neural network model first learns from the corpus of text. It learns to spell, form sentences, and the style of writing. After that, the model can generate new texts. Another field where DL is achieving significant research is image recognition; this field deals with how computers can understand and interpret various images or videos. DL-trained models fitted in vehicles can now take input from 360° camera [3] to make automatic driving possible.

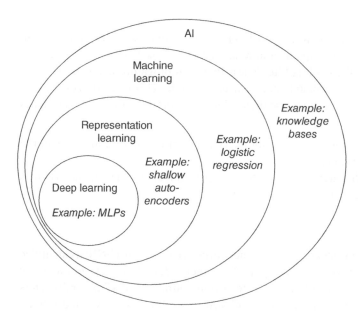

Figure 4.2 AI landscape and sub-emerging fields.

4.2.1 Feed-Forward Neural Network

Generally, the fully connected layer is the final layer in neural network architectures. A feed-forward neural network (FNN) is also called a multi-layer perceptron (MLP). The input to this layer is the output of the previous layer (pooling or convolution), then it is flattened into a one-dimensional input vector. An activation function, for example, the soft-max activation function, is applied to the inputs used as input of the first layer of the fully connected layer. The last layer of this part classifies the images based on the probability of an image that belongs to a particular class. Weights are initialized to the network and multiplied by each input element to determine the probabilities.

4.2.2 Convolution Neural Network

Convolution neural network (CNN) is a specialized type of deep neural network (DNN) model primarily used for working with images (1D, 2D, 3D) to extract the essential features present in the image data using trainable filters. It takes images as inputs to modify the learnable parameters to properly distinguish the salient features and capture their spatial and temporal dependencies. The computation and processing of CNN are much less efficient than most traditional image classification algorithms.

4.2.2.1 Convolution Layer

It is the process of filtering the image with weighted averaging known as the filter kernel. The filter kernel, also known as the convolution matrix, is a small matrix of a given size applied to the image to extract out the desired features. After convolution of the image with the filter kernel, the output is called the feature map. Let I be the image and F be the kernel, and IF, as denoted in Eq. (4.1), be the convolution result of applying F kernel over I image.

$$IF(m, n) = \sum_{i,j} I(m - i, n - j) \cdot F(i, j) \tag{4.1}$$

4.2.2.2 Pooling Layer

CNNs use pooling layers in addition to the convolution layers. This layer is applied in between two convolution layers. This layer down-samples patches of the feature map into a summarized value based on the type of pooling applied over the whole feature map. The size of the pooling filter is less than that of feature maps. If its size is 2×2, it will down-sample the dimension of the feature map to one-quarter of its size. This helps in reducing the dimension of the image data without losing its salient features and thus reduces the computational time. The types of pooling used in CNN are:

(a) **Average Pooling:** This method considers the average value for every patch in the feature map. Then, the down-sampling is done following the average value.
(b) **Max Pooling:** Here the maximum of each patch in the feature map is calculated. Then, the patch is down-sampled with the maximum value calculated.

4.2.3 Recurrent Neural Network

Recurrent neural networks (RNNs) are a significant category of neural networks. It came from a FNN; it can utilize its hidden (internal) to deal with sequences inputs of various lengths. The inner hidden layer acts as a memory bank that helps remember the information already calculated. The "recurrent neural network" generally refers to two broad classes of neural networks with nearly equivalent overall structures: finite impulses and infinite impulses.

In traditional neural networks, every input sample and corresponding output are considered unbiased of each other. However, some cases require anticipating the following words of a sentence or any sequence. The preceding phrases are needed, and there is a need to recollect the previous phrases solely for this reason. Consequently, RNNs came into relevance; they did solve the problem with their hidden layers. The hidden state remembers a few accounts in approximately a sequence.

The significant applications of RNNs are handwriting recognition [4] and speech recognition [5]. Long short-term memory (LSTM) is a prevalent RNN model, primarily due to its feedback connections [6]. It is used in sequence prediction problems.

4.3 Multi-task Deep-Learning Models

MTL is a sub-domain of ML that concentrates on utilizing and learning from many (more than one) tasks concurrently while taking the relevance of the individual tasks into account. The DL model focuses only on one output attribute in single-task learning by defining the problem as a regression, classification, or prediction problem. On the other hand, in MTL, the model focuses on multiple output attributes. Various problems are defined either as the same or a mix of regression, classification, and prediction problems, i.e. multiple relevant classifications or prediction problems can be solved with a single multi-task model considering multiple output attributes for each problem. This section discusses different kinds of multi-task DL models from the types of problems they solve.

4.3.1 Classification Models

Solving different problems as classification problem is quite common in DL. However, unlike single-task learning, the model is designed for multiple-output attributes as several classification tasks in multi-task DL.

4.3.1.1 Multi-attribute Recognition Models Using Joint Learning of Features

Objects located inside an image have a variety of properties. Suppose a picture "P'" containing an item with $T = \{t_1, t_2, t_3, t_4,...\}$ traits. We want to build a model to identify these traits in the given picture. Some traits of various items of the same class are connected. The features of the model need to be shared among the sub-tasks to improve the learning of the model. Therefore, correlated attributes are advantageous for the multi-task DL models. A simple way would be to create a multi-class classification neural network. However, the difference in the

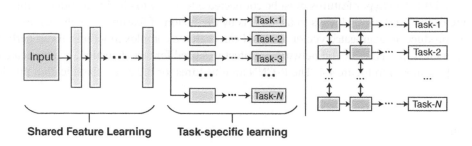

Figure 4.3 Hard parameter (a) and soft parameter (b) sharing MTL using joint feature learning.

feature classes hampers the model's performance. The main idea is to learn shared feature representations and a shared model simultaneously, as shown in Figure 4.3.

Li et al. [7] have developed a joint learning-based multi-task DL model to overcome this issue. The application of MTL is an effective method for addressing this issue. The objects in a picture have various attributes, and some of the properties of distinct objects of the same class correlate. It is possible to accomplish this using feature joint learning in MTL [7]. The two MTL ways to train the model are hard and soft parameter sharing. The designing of the model consists of a series of steps, including image data pre-processing, feature extraction, using a classifier, and dividing into *"n"* number of sub-task classifiers to create a pipeline using MTL and thresholding the classified outputs to identify all the attributes at once [8]. Figure 4.4 shows the flowchart of *region-based convolution neural network (*R-CNN*)* for landmark locations, face detection, gender detection, and pose estimation.

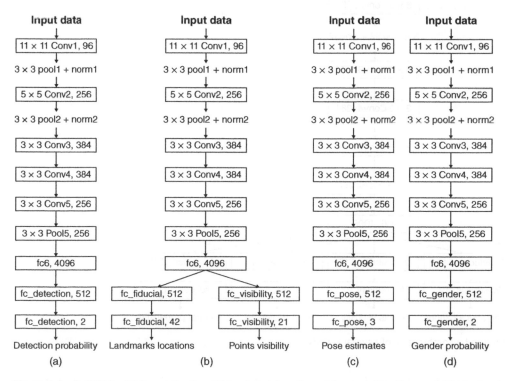

Figure 4.4 R-CNN for (a) face detection, (b) landmark locations, (c) pose estimation, and (d) gender detection.

4.3.1.2 Multi-task Facial Attributes Classification Model Using Feature Fusion

The CNN architecture is made up of numerous layers, each with its own set of activation maps. It makes the hyper features more dimensional [9]. The hyper features must be connected effectively to obtain attributes comparable to numerous tasks. One of the fundamental ways to tackle it is by the fusion of features, which merges vectors representing features to produce an aggregated vector. As CNN can evaluate complex functions, it can perform a fusion of the hyper-parameters. R-CNNs are the network architectures used for face detection, gender detection, and pose estimation [10], as shown in Figure 4.4. The R-CNN architectures for the given classification problems are shown in Figure 4.5.

4.3.2 Prediction Models

4.3.2.1 Multi-tasking on Time-Series Data

The model is built to predict clinical time-series data, including mortality risk, length-of-stay prediction, decomposition, and phenotype classification [11]. Then by using MTL, all these tasks are trained. Each of its tasks is different from the other in terms of output and structure. The use of LSTM neural-network architecture using multi-tasking improves prediction than the linear regression models. The time-series data is being re-sampled to regularly spaced intervals. Channel-wise LSTM, which is a modified version of the LSTM baseline, is used with

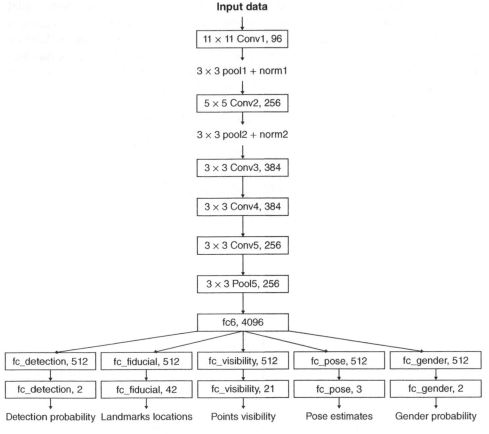

Figure 4.5 MTL architecture by combination of several R-CNN.

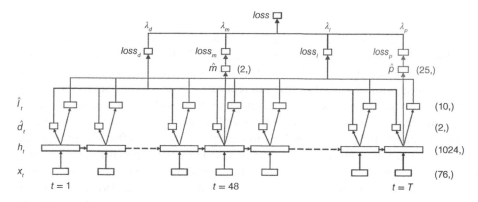

Figure 4.6 LSTM multi-task architecture.

MTL as the regularizer over each task, as shown in Figure 4.6. For prediction purposes, *L*2 regularization is being used. The formula for *L*2 regularization is given below in Eq. (4.2):

$$L(x, y) = \sum_{i=1}^{n} (y_i - h_\theta(xi))^2 + \left[\lambda \sum_{i=1}^{n} \theta_i^2 \right] \tag{4.2}$$

The term in the square bracket is the square magnitude of coefficients summed up in the total loss. The total loss is the summation of all task-specific losses.

4.3.2.2 Multi-step Forecasting on Multivariate Time Series Using Split Layers

In multivariate time series, the tasks are divided mainly into primary and additional tasks. The primary tasks are needed to be formulated for a target series. The additional tasks are needed as the hidden representation improves prediction accuracy. The training of this model follows a split-network architecture, which is expanded into the feature that can be shared and cannot be shared among the primary and supplementary tasks in this multivariate time series. Using split architecture allows extending the number of task-specific representations and shared representations [12]. Split layers properly stabilize the hidden and task-specific representations in the primary and supplementary series. Figure 4.7 shows the representation of the architecture of the split layers. The use of different colors indicates the different task-specific layers. The supplementary targets are weighted less compared to the primary targets so that the model focuses on handling the errors in primary targets more than the supplementary targets.

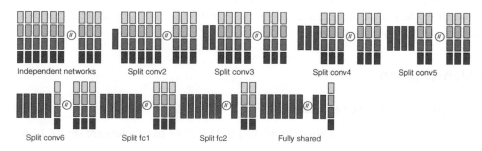

Figure 4.7 Network architecture of the split layers.

4.3.3 Mixed Models

Basic statistics depend mainly on normally distributed data, but these basic statistical methods fail when the datasets fall outside the range. Elementary statistical methods tend to assess only the exact effects shown by the predictor variables, while the problem domain often involves random effects [13]. The most common random effects are visible in the parts of the experimental studies that are replicated spatially. Diversification among individuals is also observed due to random effects. Researchers often ignore the possibility of random effects or use them as fixed factors, which might violate statistical assumptions and reduce the scope of inference. There is an excellent necessity for mixed models in statistical analysis to incorporate random effects in the model. Mixed models are typically used when statistical observations or samples are clustered, especially in ecology and biomedical sciences. Batch effects in biomedical sciences lead to non-independence of statistical observations. We can use a naive linear t to x the linear regression coefficients in the fixed-effects model. However, random effects, i.e. slope and intercept, are no longer fixed but vary around their mean values at individual levels when it comes to the coefficients.

Let us consider a sleep-deprivation study where the sleeping schedule of 18 individuals was restricted, and their reactions were observed. Their reaction, days, and subjects were taken for 10 days. An ordinary least square linear regression was fitted to check the change of response of individuals to sleep deprivation. The reaction versus day's graph, which was made by fitting ordinary least square linear regression, has an increasing trend, but much variation has been observed between days and individuals. It is assumed that all the observations were uncorrelated and hence normally distributed. However, the data points within individuals are not independent instead clustered. When a linear mixed model was fitted with random slopes and intercepts for each individual, two statistics were observed, one for fixed effects and another for random effects. The random effect statistics fixed the error observed between the non-independent samples.

Simpson et al. [14] extended a broad approach to mixed modeling to study system-level properties of the brain across multiple tasks. It allows estimating population network differences between tasks, their relations to outcomes, and evaluating individual variability in network differences. A two-part mixed model is implemented if there is a connection for estimating the strength and probability of the connection [15]. The whole matrix of brain connectivity for every participant comprises the overall model. The endogenous covariates are abstract variables obtained from the network to describe the global topology, whereas the exogenous covariates predict the physiologically related phenotypic traits. The above-described statistical framework provides decent results in group and individual effects. When the formulation of multivariate statistics is provided, the framework accounts for interdependence among the edges present in a network.

The correlation between the covariates and the corresponding tasks of network connectivity can be explained using this statistical framework. It also analyzes network connectivity across groups and tasks and forecasts network connectivity based on nodal network features and task status. This model also provides a more inclusive understanding of topological variability and a medium for measuring goodness-of-t. It provides information related to the changes that occur in the brain during task changes and transitions. Hence, mixed effect models provide a better insight into task analysis in multi-tasking.

4.4 Design and Implementation

4.4.1 Multi-task Learning Methods used for Deep Learning

Soft and hard parameter sharing are the systematic methods employed in MTL. The parameters are shared from the hidden layers.

4.4.1.1 Hard Parameter Sharing
It is one of the frequently used methods to perform MTL in DL [16]. The particular method is applied to each task by sharing hidden layers and maintaining many output layers specifically for each task.

Figure 4.8 Hard parameter sharing.

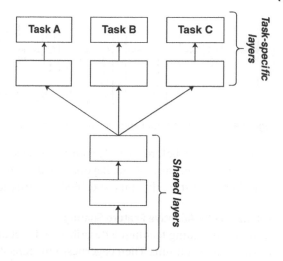

Over-fitting of the model decreases to a large extent by hard parameter sharing. The author [17] has shown the danger of over-fitting is of an order N (total number of tasks). We can also see this intuitively in a model that learns more than one task simultaneously, i.e. the MTL model also has to find a representation that will capture each task. This results in overall less risk concerning over-fitting the actual task. The general architecture of hard parameter sharing is shown in Figure 4.8.

4.4.1.2 Soft Parameter Sharing

While sharing soft parameters, every task has a DL model. Here, the parameters are assumed to be similar to the multi-task model; thereby, the parameters are regularized. Different normalization techniques are used for regularization. Duong et al. [18] used the $L2$ norm, while Yang and Hospedales [19] used the trace norm. Regularization techniques in MTL developed for other DL models have greatly encouraged the constraints of sharing soft parameters, as shown in Figure 4.9.

4.4.2 Various Design of Multi-task Learning

4.4.2.1 Deep Relationship Networks

In MTL, computer vision is often handled by sharing the CNN layers. The model then learns task-specific work utilizing fully connected dense layers. Long et al. [20] provided a way to improve particular models by employing deep relationship networks, shared layers, and individual layers for each MTL job, as illustrated in Figure 4.10.

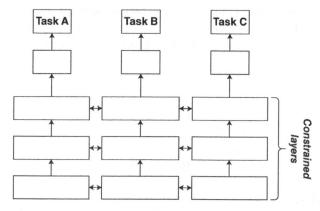

Figure 4.9 Soft parameter sharing.

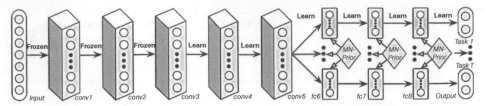

Figure 4.10 Deep relationship network. Source: Adapted from Long et al. [20].

A matrix was put before the fully connected dense layers, allowing the DL model to understand numerous correlations between the tasks. Some Bayesian models are analogous to this technique. However, the suggested technique lacks here as it still relies on a specified structure to exchange diverse parameters.

4.4.2.2 Fully Adaptive Feature Sharing

Another intriguing technique described by Lu et al. [21] is an upside-down strategy that uses a small neural network at the beginning. Then eagerness broadens itself during training using a fundamental notion that supports comparable grouping tasks. The method for widening is to create branches dynamically, as shown in Figure 4.11. However, this greedy approach method might fail to and a globally optimal model. This approach does not permit the model for learning the intricate interactions within different tasks by assigning each branch precisely one task.

4.4.2.3 Cross-stitch Networks

Misra et al. [22] proposed a model with two different architectures, almost the same as in soft parameter sharing. They proposed the model with an architecture that is called cross-stitch units. Then these units are trained from end-to-end by combining the activations from multiple networks. Cross-stitch units support a network to learn a suitable composition of shared and task-specific representations. The generalization across multiple tasks by the model dramatically improves its performance over the other methods with few training examples. This architecture diagram is shown in Figure 4.12. Pooling and fully-connected layers are placed after the cross-stitch units.

4.4.2.4 Weighting Losses with Uncertainty

Kendall and Gal [23] adopt an alternative approach to understanding the structure of sharing by taking the possibility of each task into account. The weight of each task is then managed relative to an error function by calculating a multi-task cost function by improving the maximum likelihood possibility of tasks. They showed that the suggested model outperforms other trained models individually efficiently on their assigned task by learning multi-task weights. The proposed architecture can be seen in Figure 4.13.

4.4.2.5 Tensor Factorization for MTL – Sluice Networks

Some advances were made to generalize existing methods of multi-tasking in DL. The method split the parameters of the model into task-specific and shared for each layer by applying tensor factorization. Yang and Hospedales [24] generalized some of the previously discussed matrix factorization approaches. Sluice networks

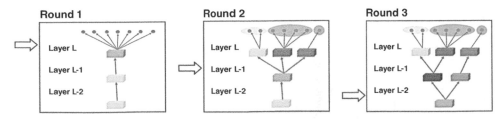

Figure 4.11 Fully adaptive feature sharing. Source: Lu et al. [21].

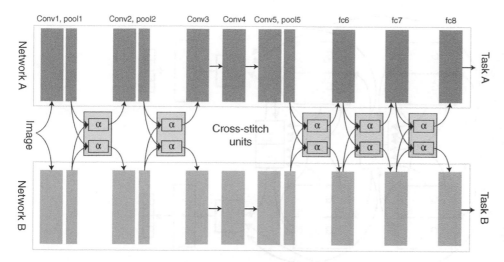

Figure 4.12 Cross-stitch networks. Source: Misra et al. [22].

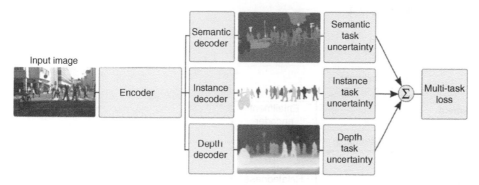

Figure 4.13 Weighting losses with uncertainty. Source: Kendall and Gal [23]/IEEE.

are proposed in [25]. In this model, generalized DL-based MTL approaches such as block-sparse regularization approaches, hard parameter sharing, cross-stitch networks, and recent natural language processing (NLP) methods create task hierarchy. The forwarded model diagram is shown in Figure 4.14.

This proposed model allows it to identify the shareable layers and sub-spaces and share them. This model architecture also determines which layers must have extracted the best description pattern of the supplied input sequences.

4.4.2.6 Joint Many-Task Model

Based on the previous findings, Hashimoto et al. [26] propose a hierarchical design comprising many NLP tasks, as seen in Figure 4.15, as a joint MTL model. The approach used simple regularization terms to allow them to optimize all model weights. This helps in improving one task's loss without seriously affecting the performance of other tasks. This model can provide impressive results using parsing, relatedness, tagging, and entailment tasks.

4.4.3 Common Problems with Design and Implementation

4.4.3.1 Combining Losses

The first challenge of designing the model is to define a single loss for the multi-task model. A single task may have a well-defined loss, but while designing the model for more than one task, selecting a single loss function

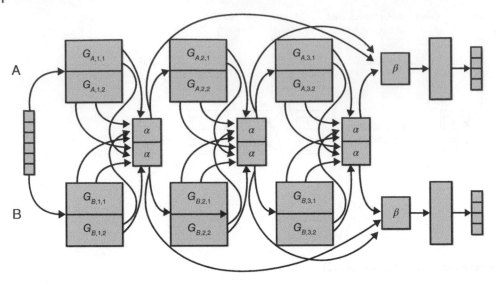

Figure 4.14 Sluice networks. Source: Ruder et al. [25].

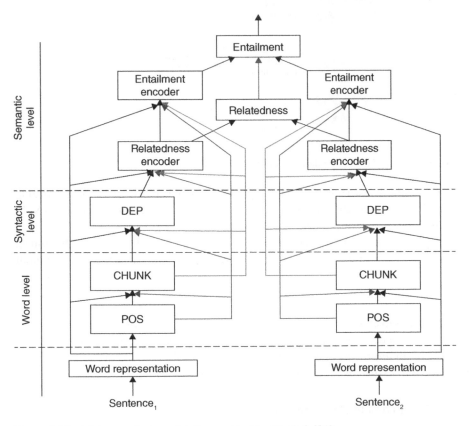

Figure 4.15 Joint many-task model. Source: Hashimoto et al. [26].

might not be efficient. A poor selection of loss functions may lead the model to underperform. Over-fitting here will be less likely since the loss function will prevent the model from performing well even with a training dataset.

As a quick solution, summing the other losses can be tested for the dataset. However, there are specific problems with this approach. The sizes of the losses may be so diverse that one task might dominate the total loss while the other tasks have little chance of influencing the learning process in the common layers. A fast solution to this problem might be to replace the cumulative total of losses with a weighted sum, therefore leveling all losses to approximately the same magnitude. This method, however, includes another hyper-parameter that may need to be tweaked on occasion.

4.4.3.2 Tuning Learning Rates

One of the essential hyper-parameters for tuning neural networks is the learning rate. This is a widespread convention. So, to improve the performance of a MTL model, tuning can be tried to find a learning rate that will look excellent for a task. However, choosing a higher learning rate may kill the performance of activation layers for any undertaken tasks. However, using a low learning rate can also bring a prolonged convergence rate to the model. For solving this problem, a different learning rate for each task-specific subnet can be used along with a single learning rate for the shared subnet.

4.4.3.3 Using Estimates as Features

Following the initial step of developing a DL model capable of predicting several tasks, the next job would be to apply an estimate for one task as a feature to another. In the forward pass, this is simple. Consider the following two tasks: A and B. Assume that the estimate for task A is provided to task B as a feature. However, we already have a label for A, so we are unlikely to propagate the gradients from task B to task A. This approach is helpful for DL models that need to multi-task.

4.5 Applications

MTL has found applications in various fields. Following are some of the essential domains where MTL can be applied.

4.5.1 Image Domain

MTL has proved successful in the image domain. Its applications include semantic scene parsing (SSP), posture estimation, diagnostic medical imaging, dynamic MR image building, etc. Most SSP algorithms require densely labeled data while not ensuring good accuracy. The datasets currently available in this domain are limited, making performance measurements of existing DL models and other CV solutions difficult. The critical issue in the SSP is that colored photos contain very varied data it includes. Fourure et al. [27] proposed a domain adaptation technique based on gradient reversal. A specific loss function uses the train data from several datasets across tasks and significant correlations among them (collected using an auto-context process). Yu and Lane [28] presented an MTL-DL model that can do image classification and image segmentation. The segmentation task increased accuracy twice at the pixel level utilizing the MTL approach, while the classification task's accuracy improved by 2%. Moeskops [29] trained an MTL CNN-based network to segment six MR images of the brain. The CNN learns to recognize brain anatomical features, imaging mode, and tissue classifications. Without any task-specific training, the model learns the various segmentations.

MTL Dynamic Contrast-Enhanced [30] imaging is utilized to objectively assess the new blood vessels of the carotid artery walls. A high connection was discovered between pictures, causing the images to be linearly dependent. Fully supervised DL models often require many pixel-level labels, which are time-consuming to construct manually.

Ke et al. [31] proposed a MTL-based DL model that uses recursive task approximation to decrease the human effort of pixel-level labeling. The authors broke up the image segmentation task into a set of recurrently specified sub-tasks focused on increasing approximation performance. The bulk of the photos used in training were solely labeled using coarse partial masks. During the training phase, the model learns these partial mask statistics. These zones are extended toward the object's boundary, supported by data-driven information.

MTL has also been used to improve images in real-time [32]. A bidirectional guided-up sampling strategy has been presented. The entire concept is built on encoder–decoder architecture, with all components shared by the image-processing technique. This MTL model outperforms DL when using joint up-sampling approaches. MTL has also been used in local seismic image processing [33]. Seismic pictures include three core image processing tasks: recognizing cracks, determining the normal seismic vector to assess local orientation, and reducing noise by smoothing in structures. Mentioned tasks are linked based on examining the same seismic structural elements. Traditional seismic image processing algorithms, on the other hand, see these tasks as separate and encounter significant problems.

Image classification on several dataset like MNIST can benefit by using pooling information learning via MTL on various similar datasets. The exchange of information across concealed layers has the potential to improve learning and hence performance. Due to the scarcity of expertise necessary for data labeling, the medical image processing area suffers from gathering big datasets with clear annotations. Le et al. [34] developed a DL-based MTL model for mammography cancer detection that uses global-level classification annotations and pixel-level segmentation. The given approach is based on FNN, enabling quicker prediction while sharing more features. Furthermore, MTL is also employed for Covid-19 analysis [35] using CT scan images. Instead of classifying and segmentation separately, they designed an architecture for segmentation and reconstruction that included two decoders and one encoder. AFNN is a classification identifier.

4.5.2 Text Domain

DL has achieved many milestones in the domain of NLP, but the main drawback of the existing models is that they have to be trained from scratch, requiring large datasets and high computational costs. Present research on NLP is mainly focused on transductive learning. A new model by Howard and Ruder [36] is based on inductive learning for text classification called universal language model fine-tuning for text classification. It has three stages: the first is used to build a pre-training language model using general domain data. The second stage is used for fine-tuning the language model for the target dataset. The last stage is the fine-tuning classification of the target dataset.

Mrkšić et al. [37] described a multi-tasking RNN model which shares hidden layers among various tasks for training models involving multiple domain dialog state tracking. It learns from the most frequent and general dialog features acquired across various domains. Collobert and Weston [38] proposed a multi-tasking neural network containing time-decay constraints to learn six NLP tasks. The tasks jointly include tagging part-of-speech, recognizing named entity, chunking, labeling of semantic role, language modeling, and related word identification. The model is trained by using unlabeled data. Wu and Huang [39] put forward a multi-tasking model for multi-domain sentiment classification that consists of a common sparse learner that can learn all the tasks and also tasks specific learners.

Multi-tasking is also used for sequence-to-sequence learning in the text domain. Luong et al. [40] used three multi-tasking settings for sequence-to-sequence-based models; the first architecture involves one encoder, which is shared by all tasks, and multiple decoders for specific tasks. The next one involves multiple encoders. Each task having its encoder and one decoder, which is shared among all tasks, and the last one is a more general one that involves task sharing among multiple encoders and decoders.

Above mentioned applications of multi-tasking in the text domain indicates that multi-tasking has been able to give a satisfactory result in this domain. Transfer learning among tasks has been observed as a great advantage to the learning process. It requires less training data and less computational costs. However, this field is quite under-explored, and many new multi-tasking applications are possible in this domain. One can be improving the language model pre-training, augmentation of language model with additional tasks, or tasks where the amount of labeled data is limited.

4.5.3 Others

Apart from the text and image domain, multi-tasking is also used in bioinformatics, health informatics, speech processing, inverse dynamics problem for robotics, stocks, climate prediction, etc. [41]. Multi-tasking is also used for modeling organisms [42]. Another multi-task model has been used to predict cross-platform si-RNA efficacy [43]. A brain–computer interface is constructed [44] based on MTL without using any fixed calibration process for each subject. Xu et al. [45] used multi-task-based approach for predicting location of sub-cellular protein. A multi-task-based approach has been taken for information sharing about known disease genes [46] and prioritization of disease genes by learning from the shared information. Multi-task time series problems can be formulated for the prediction of Alzheimer's disease [47] where at each point of time for parameters organized in the matrix, a learner is associated. Li et al. [48] constructed a multi-tasking classification problem for survival analysis by assuming that an event that occurs once will not take place once again, and MTFS is expanded for solving this problem.

Multi-tasking DL models have also been forwarded for speech synthesis [49]. A stacked DNN consists of numerous neural networks out of which the preceding neural network is used to feed the result of the upper-most hidden layer to the upcoming layer. Each network has two inputs, the first one for the prime task and the other for additional tasks. The multi-task boosting method has been proposed for learning web-search ranking [50] by sharing feature representation between various tasks. In [51] matrix rating in multiple-related domains is done by a multiple domain collaborative filtering method. The MTRL method [52] is forwarded to consider the hierarchical structure and the sparsity in structure for maximum conversion in display advertising. A multi-task neural network is used for predicting a yearly return of stocks [53]. This model shares hidden layers for different prediction tasks. A multi-task model has also been proposed for multi-device localization [54] where learning is acquired from a low-rank transformation. It imposes similar model parameters for different tasks in the transformed space. Multi-tasking is used for solving the inverse dynamics problem in robotics [55]. Hence it can be concluded that MTL has an excellent advantage in every aspect.

4.6 Evaluation of Multi-task Models

For any multi-task model, the primary requirement is to learn whether the model can generalize on unknown data or memorize the data. We normally check our model performance on data other than the training data because the model can easily predict the value for any training data by learning it during the training time, which may lead to over-fitting. Therefore, we use a re-sampling method for the evaluation of the model. Few re-sampling methods to evaluate multi-tasking models are discussed below:

(a) **Single Hold-Out Random Sub-sampling:** In hold-out, we primarily evaluate our model on different data. We divide the whole dataset into two parts: training data comprising about 70% to 90% of the dataset and test data comprising 10% to 30% of data. When both the training and test data are extensive, this is a reliable method for evaluation

(b) **K-fold Cross-validation:** The dataset is divided into k equal disjoint subsets in this technique. Among these k subsets, $k-1$ subsets are taken for training, and the rest is taken for testing purposes. The technique is performed k times, with each subset serving as testing data once. Then the average of the model's performance on these validation data is considered the model's overall performance.

(c) **Leave One Out Cross-validation:** In this technique, one data unit is taken for validation, while the others are training data. The process is repeated for each data unit in the dataset. This technique incurs a substantial computational cost. It can be considered a special case of k-fold cross-validation when $k = n$, n being the number of data points in the dataset.

After considering a suitable method for re-sampling the data, the evaluation is done based on performance metrics. We choose a performance metric based on the multi-tasking task, i.e. classification, clustering, ranking, regression, etc.

Following are some of the powerful metrics used in the chapter we have explored:

(a) **Accuracy:** It is a simple statistic that is used to assess performance. It is calculated as the total number of correct guesses to the total number of forecasts made.

(b) **Precision:** The ratio of the total number of actual positive predictions to overall positive predictions.

(c) **Recall:** It is calculated as the positive correctly predicted observations to the actual number of observations in the current class.

(d) **F-measure:** F-measure is calculated as the harmonic mean of precision and recall. It is used when there is an uneven distribution of class-like segmentation tasks.

(e) **Confusion Matrix:** This metric is normally used for the evaluation of classification-based tasks. Here the number of predictions made for each class is shown in a matrix. The diagonal elements show the values which are correctly predicted (true positive).

(f) **Root Mean Square Error (RMSE):** It is calculated as the standard deviation of the prediction errors. In other words, it is the root of the Euclidian distance between the vector of anticipated values to the vector of values observed.

(g) **The Area Under the Curve (AUC):** It is a technique used by the binary classifier to measure performance by differentiating the true and false classes. A curve is drawn for a square of 1 unit, and the area contained by this curve is the model's accuracy.

Ke et al. [31] used the $F1$ score or dice coefficient to measure the performance of segmentation on the SEM database. Le et al. [34] used the Dice score to evaluate the performance of the segmentation task in cancer mammography, and they used the area under curve metric for measuring the performance of the classifier in predicting cancer. Wang et al. [47] used RMSE between predicted and actual cognitive scores to measure performance. Mrkšić et al. [37] used the geometric mean of the classification accuracy to find the goal accuracy of the shared domains. Ranjan et al. [10] used mean average precision to measure the performance of face detection on AWS and PASCAL datasets, while normalized mean error is used for landmark localization. Das et al. [56] have found the accuracy for three classification tasks: gender, age, and ethnicity on UTK face dataset and BEFA challenge dataset.

4.7 Conclusion and Future Directions

MTL has been used effectively in major ML applications such as computer vision, speech recognition, drug discovery, and NLP. MTL is more successful than single-task learning when he or she has to optimize more than one loss function. These examples will surely aid in explicitly representing the problem in terms of MTL and benefiting from its models. This chapter covers the progression of several types of MTL approaches and briefly summarizes each type. We commonly use DL models to learn latent patterns from the given data to perform better on specific evaluation metrics. Usually, we train models for a single objective function by tweaking the hyper-parameters until the performance does not increase. The model's performance can be further improved by acquiring general representation from several related tasks rather than one.

Acknowledgment

We would like to acknowledge the authors Long et al., Lu et al., Misra et al., Kendall et al., and Ruder et al., for the architecture design diagram of their approach. This research was funded under grant number: ECR/2018/000204 by the Science and Engineering Research Board (SERB), Department of Science and Technology (DST) of the Government of India.

References

1 Chapelle, O., Shivaswamy, P., Vadrevu, S. et al. (2010). Multi-task learning for boosting with application to web search ranking. In: *Proceedings of the 16th ACM SIGKDD International Conference on Knowledge Discovery and Data Mining*, 1189–1198. New York, NY, USA: Association for Computing Machinery.

2 LeCun, Y., Bengio, Y., and Hinton, G. (2015). Deep learning. *Nature* 521 (7553): 436–444.

3 Nelson, F. (2015). Nvidia demos a car computer trained with deep learning. Elizabeth Bramson-Boudreau: Cambridge, Massachusetts

4 Sak, H., Senior, A.W., Beaufays, F. (2014). Long short-term memory recurrent neural network architectures for large scale acoustic modeling. INTERSPEECH. 338-342

5 Li, X. and Wu, X. (2015). Constructing long short-term memory based deep recurrent neural networks for large vocabulary speech recognition. In: *IEEE International Conference on Acoustics, Speech and Signal Processing (ICASSP)*, 4520–4524. IEEE.

6 Hochreiter, S. and Schmidhuber, J. (1997). Long short-term memory. *Neural Computation* 9 (8): 1735–1780.

7 Li, Y., Tian, X., Liu, T., and Tao, D. (2015). Multi-task model and feature joint learning. In: *Proceedings of the 24th International Conference on Artificial Intelligence, IJCAI'15*, 3643–3649. AAAI Press.

8 Wang, X., Zheng, S., Yang, R. et al. (2022). Pedestrian attribute recognition: a survey. *Pattern Recognition* 121: 108220.

9 Agarwal, A. and Triggs, W. (2008). Multilevel image coding with hyper features. *International Journal of Computer Vision* 78: 06.

10 Ranjan, R., Patel, V.M., and Chellappa, R. (2017). Hyperface: a deep multi-task learning framework for face detection, landmark localization, pose estimation, and gender recognition. *IEEE Transactions on Pattern Analysis and Machine Intelligence* 41 (1): 121–135.

11 Harutyunyan, H., Khachatrian, H., Kale, D., and Galstyan, A. (2017). Multitask learning and benchmarking with clinical time series data. *Scientific Data* 6: 1–18.

12 Jawed, S., Rashed, A., and Schmidt-Thieme, L. (2019). Multi-step forecasting via multi-task learning. In: *IEEE International Conference on Big Data (Big Data)*, 790–799.

13 Bolker, B.M., Brooks, M.E., Clark, C.J. et al. (2009). Generalized linear mixed models: a practical guide for ecology and evolution. *Trends in Ecology & Evolution* 24 (3): 127–135.

14 Simpson, S.L., Bahrami, M., and Laurienti, P.J. (2019). A mixed-modeling framework for analyzing multi-task whole-brain network data. *Network Neuroscience* 3 (2): 307–324.

15 Simpson, S.L. and Laurienti, P.J. (2015). A two-part mixed-effects modeling framework for analyzing whole-brain network data. *NeuroImage* 113: 310–319.

16 Caruana, R. (1993). Multi-task learning: a knowledge-based source of inductive bias. In: *Proceedings of the 10th International Conference on Machine Learning*, 41–48.

17 Baxter, J. (1997). A Bayesian/information theoretic model of learning to learn via multiple task sampling. *Machine Learning* 28 (1): 7–39.

18 Duong, L., Cohn, T., Bird, S., and Cook, P. (2015). Low resource dependency parsing: cross-lingual parameter sharing in a neural network parser. In: *Proceedings of the 53rd Annual Meeting of the Association for Computational Linguistics and the 7th International Joint Conference on Natural Language Processing (Volume 2: Short Papers)*, 845–850. Beijing, China: Association for Computational Linguistics.

19 Yang, Y. and Hospedales, T.M. (2016). Trace norm regularised deep multi-task learning. arXiv preprint arXiv: 1606.04038.

20 Long, M., Cao, Z., Wang, J. et al. (2015). Learning multiple tasks with multi-linear relationship network. *Advances in Neural Information Processing Systems* 30.

21 Lu, Y., Kumar, A., Zhai, S. et al. (2017). Fully-adaptive feature sharing in multi-task networks with applications in person attribute classification. In: *Proceedings of the IEEE Conference on Computer Vision and Pattern Recognition*, 5334–5343.

22 Misra, I., Shrivastava, A., Gupta, A., and Hebert, M. (2016). Cross-stitch networks for multi-task learning. In: *Proceedings of the IEEE Conference on Computer Vision and Pattern Recognition*, 3994–4003.

23 Kendall, Y. and Gal, R.C. (2018). Multi-task learning using uncertainty to weigh losses for scene geometry and semantics. In: *Proceedings of the IEEE Conference on Computer Vision and Pattern Recognition*, 7482–7491.

24 Yang, Y. and Hospedales, T. (2016). Deep multi-task representation learning: a tensor factorisation approach. arXiv preprint arXiv: 1605.06391.

25 Ruder, S., Bingel, J., Augenstein, I. (2017). Learning what to share between loosely related tasks. arXiv preprint arXiv:1705.08142.

26 Hashimoto, K., Xiong, C., Tsuruoka, Y. et al. (2016). A joint many-task model: growing a neural network for multiple NLP tasks. arXiv preprintarXiv: 1611.01587.

27 Fourure, D., Emonet, R., Fromont, E. et al. (2017). Multi-task, multi-domain learning: application to semantic segmentation and pose regression. *Neurocomputing* 251: 68–80.

28 Yu, B. and Lane, I. (2014). Multi-task deep learning for image understanding. In: *6th International Conference of Soft Computing and Pattern Recognition (SoCPaR)*, 37–42. IEEE.

29 Moeskops, P., Wolterink, J.M., van der Velden, B.H. et al. (2016). Deep learning for multi-task medical image segmentation in multiple modalities. In: *International Conference on Medical Image Computing and Computer-Assisted Intervention*, 478–486. Springer.

30 Wang, N., Christodoulou, A.G., Xie, Y. et al. (2019). Quantitative 3D dynamic contrast-enhanced (DCE) MR imaging of carotid vessel wall by fast t1 mapping using multi-tasking. *Magnetic Resonance in Medicine* 81: 2302–2314.

31 Ke, R., Bugeau, A., Papadakis, N. et al. (2021). Multi-task deep learning for image segmentation using recursive approximation tasks. *IEEE Transactions on Image Processing* 30: 3555–3567.

32 Kong, K., Lee, J., Song, W. et al. (2019). Multi-task bilateral learning for real-time image enhancement. *Journal of the Society for Information Display* 27 (06): 26.

33 Wu, X., Liang, L., Shi, Y. et al. (2019). Multi-task learning for local seismic image processing: fault detection, structure-oriented smoothing with edge-preserving, and seismic normal estimation by using a single convolutional neural network. *Geophysical Journal International* 219 (3): 2097–2109.

34 Le, T.-L.-T., Thome, N., Bernard, S. et al. (2019). Multi-task classification and segmentation for cancer diagnosis in mammography. arXiv preprint arXiv: 1909.05397.

35 Amyar, A., Modzelewski, R., and Ruan, S. (2020). Multi-task deep learning based CT imaging analysis for covid-19: classification and segmentation. *Computers in Biology and Medicine* 126: 104037.

36 Howard, J. and Ruder, S. (2018). Universal language model fine-tuning for text classification. arXiv preprint arXiv: 1801.06146.

37 Mrkšić, N., Séaghdha, D.O., Thomson, B. et al. (2015). Multi-domain dialog state tracking using recurrent neural networks. arXiv preprint arXiv: 1506.07190.

38 Collobert, R. and Weston, J. (2008). A unified architecture for natural language processing: Deep neural networks with multi-task learning. In: *Proceedings of the 25th International Conference on Machine Learning*, 160–167. New York, NY, USA: Association for Computing Machinery.

39 Wu, F. and Huang, Y. (2015). Collaborative multi-domain sentiment classification. In: *2015 IEEE International Conference on Data Mining*, 459–468. IEEE.

40 Luong, M.T., Le, Q.V., Sutskever, I. et al. (2015). Multi-task sequence to sequence learning. arXiv preprint arXiv: 1511.06114.

41 Zhang, Y. and Yang, Q. (2017). A survey on multi-task learning. IEEE Transactions on Knowledge and Data Engineering.

42 Widmer, C., Leiva, J., Altun, Y., and Rätsch, G. (2010). Leveraging sequence classification by taxonomy-based multi-task learning. In: *Annual International Conference on Research in Computational Molecular Biology*, 522–534. Springer.

43 Liu, Q., Xu, Q., Zheng, V.W. et al. (2010). Multi-task learning for cross-platform siRNA efficacy prediction: an in-silico study. *BMC Bioinformatics* 11 (1): 181.

44 Alamgir, M., Grosse-Wentrup, M., and Altun, Y. (2010). Multi-task learning for brain–computer interfaces. In: *Proceedings of the 13th International Conference on Artificial Intelligence and Statistics, JMLR Workshop and Conference Proceedings*, 17–24.

45 Xu, Q., Pan, S.J., Xue, H.H., and Yang, Q. (2010). Multi-task learning for protein sub cellular location prediction. *IEEE/ACM Transactions on Computational Biology and Bioinformatics* 8 (3): 748–759.

46 Mordelet, F. and Vert, J.P. (2011). Prodige: prioritization of disease genes with multi-task machine learning from positive and unlabeled examples. *BMC Bioinformatics* 12 (1): 389.

47 Wang, H., Nie, F., Huang, H. et al. (2012). High-order multi-task feature learning to identify longitudinal phenotypic markers for Alzheimer's disease progression prediction. *Advances in Neural Information Processing Systems* 1277–1285.

48 Li, Y., Wang, J., Ye, J., and Reddy, C.K. (2016). A multi-task learning formulation for survival analysis. In: *Proceedings of the 22nd ACM SIGKDD International Conference on Knowledge Discovery and Data Mining ((KDD '16))*, 1715–1724. New York, NY, USA: Association for Computing Machinery.

49 Wu, Z., Valentini-Botinhao, C., Watts, O., and King, S. (2015). Deep neural networks employing multi-task learning and stacked bottleneck features for speech synthesis. In: *2015 IEEE International Conference on Acoustics, Speech and Signal Processing (ICASSP)*, 4460–4464. IEEE.

50 Bai, J., Zhou, K., Xue, G. et al. (2009). Multi-task learning for learning to rank in web search. In: *Proceedings of the 18th ACM Conference on Information and Knowledge Management (CIKM '09)*, 1549–1552. New York, NY, USA: Association for Computing Machinery.

51 Zhang, Y., Cao, B., and Yeung, D.-Y. (2012). Multi-domain collaborative filtering. arXiv preprint arXiv: 1203.3535.

52 Ahmed, A., Das, A., and Smola, A.J. (2014). Scalable hierarchical multi-task learning algorithms for conversion optimization in display advertising. In: *Proceedings of the 7th ACM International Conference on Web Search and Data Mining (WSDM '14)*, 153–162. New York, NY, USA: Association for Computing Machinery.

53 Ghosn, J. and Bengio, Y. (1997). Multi-task learning for stock selection. *Advances in Neural Information Processing Systems* 946–952.

54 Zheng, V.W., Pan, S.J., Yang, Q., and Pan, J.J. (2008). Transferring multi-device localization models using latent multi-task learning. *AAAI* 8: 1427–1432.

55 Williams, C., Klanke, S., Vijayakumar, S., and Chai, K.M. (2009). Multi-task Gaussian process learning of robot inverse dynamics. In: *Advances in Neural Information Processing Systems*, 265–272. Curran Associates, Inc.

56 Das, A., Dantcheva, A., and Bremond, F. (2018). Mitigating bias in gender, age and ethnicity classification: a multi-task convolution neural network approach. ECCVW 2018 - European Conference of Computer Vision Workshops, Munich, Germany.

5

Image Reconstruction Using Deep Learning

Aneeta Christopher, R. Hari Kishan, and P.V. Sudeep

Department of Electronics and Communication Engineering, National Institute of Technology Calicut, Calicut, Kerala, India

5.1 Introduction

In imaging systems, the quality of digital images frequently degrades during acquisition. Noise, low resolution (LR), blur, and compression are the main causes of image degradation [1]. It is hard to construct imaging instruments with arbitrarily high resolution (HR) and no degradation in the images acquired. As a result, image reconstruction approaches are required to improve image quality and the interpretability of visual information for human viewers.

To achieve this objective, image restoration (IR) and image enhancement techniques in the spatial or frequency domain can be applied. IR approaches employ prior knowledge of the degradation to reconstruct the underlying image [2], whereas image enhancement does not generally model the degradation [3]. Although it is theoretically possible in many cases to reconstruct the underlying image from poor data acquired with real-world equipment, in practice, image reconstruction from acquired data is a complicated inverse operation due to the absence of information about the actual inverse transform. Thereby the conventional methods of image reconstruction also tend to be inaccurate.

Image reconstruction tasks have seen significant progress in recent years in terms of new modalities, improved temporal and spatial resolution, wider applicability, robustness, speed, and accuracy. The development of rapid, resilient, and accurate reconstruction methods is a desirable but challenging research topic [4].

Modern deep learning (DL) has achieved immense success in computer vision (CV) applications, such as image segmentation, object identification, and image synthesis, outperforming cutting-edge techniques in the presence of large datasets [5]. Besides, DL algorithms have recently emerged as a popular choice for solving low-level vision tasks in image reconstruction, with promising results.

In CV, image reconstruction using DL mainly refers to the techniques for generating high-quality images from damaged or low-quality images. Figure 5.1 shows a generalized depiction of the image reconstruction pipeline applied with different types of image modalities degraded by noise, blurring, compression artifacts, etc. The IR task can be generally perceived as a reconstruction of the image acquired using an imaging device, e.g. a high-definition camera, in case of natural imaging carried out in satellite imaging, underwater imaging, etc.

IR tasks, in particular, denoising, super-resolution (SR), image inpainting (IP), deconvolution, and colorization, offer a wide range of interesting applications [6]. For example, IP attempts to recover the real image from an incomplete damaged image by filling in missing regions. Pixel interpolation, which is related to IP, seeks to replenish noncontiguous removed pixels. Most electronic imaging applications demand HR images because they may provide more information. Using HR satellite photographs, it may be simple to distinguish one item from similar ones, and the efficiency of pattern recognition in CV can be boosted if an HR image is provided. Image reconstruction is capable of creating a 2D or 3D image from partial data, such as radiation results gathered during a medical

Machine Learning Algorithms for Signal and Image Processing, First Edition.
Edited by Deepika Ghai, Suman Lata Tripathi, Sobhit Saxena, Manash Chanda, and Mamoun Alazab.

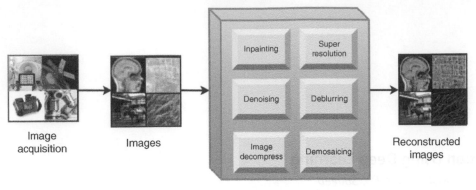

Image reconstruction block

Figure 5.1 Schematic representation of image reconstruction in computer vision (CV). The image reconstruction block depicts different IR tasks.

imaging scan. Image compression is routinely used to compress data to be transmitted, and joint photographic experts group (JPEG) is the most extensively used compression system. JPEG, as a block-based compression strategy, frequently results in blocking artifacts in recovered images. Image deblurring and denoising methods reduce defocus aberration in images and recover images corrupted by sensor noise, respectively.

On the other hand, in the case of medical imaging, the reconstruction process can take place in two formats depending on the algorithm involved as well as the quality of the image acquired. The first format enables us to reconstruct the image from the raw signal acquired, which is modality-specific in nature and is engaged usually when the accuracy being demanded is high. This can be interpreted as the formation of the image from the measurements (see Figure 5.2a). The second format involves the IR task being carried out from the image acquired as in case of the natural imaging as shown in Figure 5.2b. This format is being adhered to when the image acquired is of high quality thus enabling the reconstruction possible from the image itself.

In general, IR task aims at reconstructing a top-quality image from its poor observation that has been degraded by a signal-dependent degrading operator and an additive white Gaussian noise. The nature of degrading operator influences the process of IR. For example, blur removal in images is investigated with the blurring operator and the incoherently under-sampling operator is employed in compressive sensing (CS). Likewise, SR is examined using the blurring and down-sampling operator [6].

Basically, this chapter aims to review a plethora of image reconstruction approaches that rely on DL models such as autoencoders (AEs), convolutional neural networks (CNNs), and generative adversarial networks

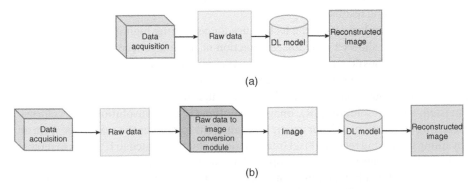

Figure 5.2 (a) Medical image reconstruction (using DL model) from the raw data obtained by measurements in the data acquisition; (b) medical image enhancement using DL models for IR tasks.

(GANs). Also, we discuss different DL models for medical image reconstruction. We disseminated the material in this chapter as follows: Section 5.2 presents DL-based IR methods with various deep neural network (DNN) adaptations embarked for image reconstruction over the past decade. Also, Section 5.3 covers DL-based medical image reconstruction approaches for magnetic resonance (MR), optical coherence tomography (OCT), low-dose computed tomography (LDCT), ultrasound (US), and microscopy images. Finally, we conclude the chapter in Section 5.4.

5.2 DL-IR Methods

This section presents DL-IR methods to generate high signal-to-noise ratio (SNR) images from its poor versions. The minimum mean squared error (MMSE) and the maximum a posteriori (MAP) estimators are widely accepted to produce reliable IR results. DL-MMSE methods, which use supervised learning techniques, are conventionally more popular for image reconstruction due to the simple loss and straightforward training. Although achieving promising results, MMSE methods are commonly corrupted by visual artifacts. MAP methods utilize an explicit image prior or learn an intrinsic prior [7] to execute a variety of IR tasks as a generic framework. Most of MMSE methods are discriminative learning based, whereas MAP methods are blind restoration methods.

5.2.1 DL-MMSE Methods for IR Tasks

The MMSE method is generated by decreasing a distance measure between the result and ground truth (GT). We will discuss different trainable models, which decrease the distance between the results and GT. We examine how different DNN models such as AEs, CNNs, and GANs have been used to improve noisy images contaminated by Gaussian noise.

5.2.1.1 DL-MMSE Methods Using AEs

In 1987, Yann LeCun pioneered the concept of autoencoder [8]. An AE is a self-supervised neural network with an encoder–decoder structure for learning. The encoder provides a parameterized function for extracting features, whereas the decoder tries to reconstruct original data from encoded features. Initially, AEs were used for dimensionality reduction or feature learning. Figure 5.3 depicts an AE-based DL-IR method for image denoising. The encoder and decoder entities are shown separately, and it shows the latent space into which the image is mapped before decoding to the denoised image.

Vincent et al. [9] developed a denoising autoencoder (DAE), which is trained to reconstruct a high-quality image from the corrupted one. Junyuan Xie et al. [10] developed the stacked sparse denoising autoencoder (SSDA) by combining the sparsity-inducing term for regularization and DNN pre-trained with DAE for enhanced performance. When employed for denoising, SSDA, like DAE, is reliant on prior knowledge about the overall nature of the noise.

In [11], the authors used stacked DAE constructed with convolutional layers to denoise medical images. For image denoising, Agostinelli et al. [12] experimented with adaptive multicolumn SSDA and the authors in [13] proposed a nonlocal AE with a nonlocal regularizer in the training objective and a collaborative stabilization method. In 2016, Mao et al. [14] presented an IR method based on AE with skip connections. Skip connection aids in the recovery of clean images and addresses the optimization challenge posed by gradient vanishing, yielding in performance advantages as the network becomes deeper. The authors of [15] examine the performance of convolutional DAE model in the presence of various noises such as Gaussian, impulse, and speckle noise.

It is worth noting that AE-based denoising methods discussed so far have learned the model on a distinct training dataset and have utilized the learned model to reconstruct the test samples. Angshul Majumdar claimed in [16] that such a method fails when the test image (to denoise) is not of the same kind as the models learned with.

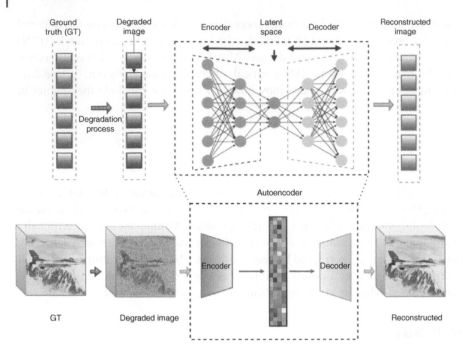

Figure 5.3 Schematic representation of image reconstruction using an autoencoder network. Source: SDASM Archives/Flickr.

In addition, he developed the first AE-based solution in [16] for blind denoising approach, in which the basis is learned from the noisy sample itself during denoising. The authors in [17] demonstrated that the output of an optimum DAE is a local mean of the actual data density, and the AE error is a mean shift vector and built a DAE with natural image prior to pixel domain. In contrast, the method in [18] integrated priors in both the pixel and wavelet domain.

AE-based IR has already been used for a variety of purposes. For example, a cascade of DAEs is built to successfully denoise cryo-electron microscopy images [19]. In [20], the authors used AE to enhance the holographic images. DAE is used as an explicit prior to solving the problem of highly undersampling magnetic resonance imaging (MRI) reconstruction [21], and sparse-view computed tomography (CT) reconstruction [22]. Recently, a convolutional AE with adjustable soft-thresholding units for encoding layers and linear units for decoding layers has been developed for IR [23]. In Table 5.1, we list different AE-based DL methods for different IR tasks such as denoising, IP, SR, decompression, and deblurring.

5.2.1.2 DL-MMSE Methods Using CNNs

As of now, CNNs have clearly outperformed AE and multilayer perceptron [24]-based IR. CNNs outperform conventional sparse representation in terms of image resolution as they have a stronger representation capacity [25]. Jain and Seung pioneered the use of CNN for image denoising and used a five-layer CNN with sigmoid activation [26]. Furthermore, the research showed that CNN works well for both blind and non-blind image denoising, offering equal or even higher performance than wavelet and Markov random field (MRF) techniques. Figure 5.4 shows a collection of CNN-based DL models for image reconstruction tasks such as denoising, IP, SR, decompression, and deblurring.

According to the underlying CNN architecture, the IR models can be categorized into VGGNet based [27–29], ResNet-DenseNet based [30, 31], encoder–decoder based [32], and UNet based [33]. The details on a few popular CNN-based DL-MMSE methods are discussed below.

Table 5.1 List of AE-based DL methods for different IR tasks.

Publisher	Year	IR application				
		Denoising	IP	SR	Compression artifact reduction	Deblurring
Le Cun et al. [8]	2010	✓				
Xie et al. [9]	2012	✓	✓			
Agostinelli et al. [11]	2013	✓				
Gondara et al. [10]	2016	✓				
Wang et al. [12]	2016	✓				
Mao et al. [13]	2016	✓		✓		
Arjomand et al. [16]	2017	✓		✓		✓
Shimobaba et al. [20]	2017	✓				
Majumdar et al. [15]	2018	✓				
Song et al. [14]	2020	✓				
Zhou et al. [17]	2020	✓		✓		✓
Sanqian L et al. [18]	2020	✓			✓	✓
Lei et al. [19]	2020	✓				
Liu et al. [21]	2020	✓		✓		
Zhang et al. [22]	2020	✓				
Fan et al. [23]	2020	✓				✓

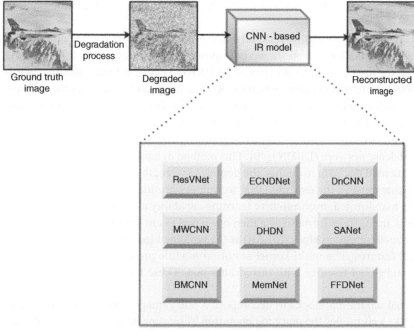

Various CNNs used for IR

Figure 5.4 A generic representation of DL-IR methods using CNNs. The CNN-based IR block represents the architecture of different CNN models used in IR tasks. Source: SDASM Archives/Flickr.

The denoising convolutional neural network (DnCNN) [27] model employs rectifier linear unit (ReLU) [34], residual learning (RL) [35], and batch normalization (BN) [36] to accelerate network training and increase denoising efficiency. The distribution of the data changes after it traverses through the convolution layer of CNN. This is referred to as the internal covariate shift problem. The batch normalization approach can mitigate this problem. Batch normalization initially normalizes the sample data before recovering the training data distribution with scale and shift operations. After the BN of each layer, the activation function is implemented. The ReLU activation function enables quicker and more successful training on big complex datasets. Theoretically, the efficiency of network learning to denoise rises with increase in the network depth. Expanding the network depth, on the other hand, may result in vanishing gradient problems. Applying a residual learning method to CNN is a smart way to solve this problem. It basically combines the information from different feature layers with the input image and feeds it to the next layer to guarantee performance. However, this model did not consider the dilated convolution and used limited receptive fields only. A dilated residual CNN for Gaussian image denoising is proposed in [37]. In [38], a multichannel version of DnCNN is used for enhancing 3D MR images.

Ahn et al. [39] presented a block-matching convolutional neural network (BMCNN) that finds comparable patches and stacks them as a 3D input. Using a set of identical patches as input, the network can learn the non-local self-similarity (NLSS) prior as well as the local prior that conventional CNNs can learn.

The MemNet model in [31] has densely connected blocks that are utilized to produce hierarchical features after a feature extraction unit for generating a basic feature set. A reconstruction unit then restores the clean image. The persistent memory concept was used in densely connected memory blocks made up of a chain of recursive residual blocks. The gate unit in each memory block adaptively determines the long-term memory to be retained and the short-term memory to be stored.

Many plain discriminative learning approaches have achieved highly competitive denoising performance when trained a distinct model for each noise level and necessitate the availability of several models for denoising images with varied noise levels. Also, it lacks the flexibility to cope with non-stationary noise, limiting its usefulness in pragmatic denoising. To overcome such drawbacks, Zhang et al. [28] introduced the fast and flexible denoising CNN (FFDNet) that utilizes noise level map as a part of the input to the model and denoising being done in downsampled sub-image space.

The residual dense network (RDN) model in [30] employs residual dense block (RDB), which are a combination of the residual and dense blocks. The dense block has been shown to improve image processing by fully utilizing multi-scale information. Nevertheless, the residual connection can assist DNNs to learn and reuse features in forward propagation. Furthermore, global feature fusion occurs when the RDN appropriately combines hierarchical features from all RDBs. In comparison to MemNet, the computational cost required in extracting features is lower in RDN since the features are retrieved directly from the LR input, which MemNet does not allow. According to Yuda Song et al., the RDBs in the RDN model are redundant. The residual connection and deep supervision features are used in the dynamic residual dense network (DRDN) [40]. This method dynamically adjusts the number of blocks engaged in denoising by sequential decision to change the denoising strength.

Xinyan Zhang et al. [41] observed that most of the existing DNN techniques are incapable of fully exploiting the model's hierarchical features and have limited discriminative learning capability as they treat all the feature maps equally. To address this situation, the authors introduced a memory-based latent attention (MLA) network. The authors designed memory-based latent attention block (MLAB) as the building block of the network that aids in effective utilization of hierarchical features. The multi-kernel attention module present in the latent branch of MLAB helps in combining local and global features with mixed attention. These features lead to enhanced representational ability of the model.

SANet model [42] uses convolutional separation blocks to decompose a noisy image into several bands with simpler patterns, and deep mapping block simplifies the mapping operation by treating each band independently. After that, the final denoised output is anticipated by concatenating individual mapping results and convolving this feature map with a band aggregate block.

To generalize a denoising model to real-time noise, Shi Guo et al. introduced a CNN model called convolutional blind denoising network (CBDNet) [32]. The authors synthesized noisy images using a realistic noise model and are fed to a noise estimator coupled with a total variation (TV) regularized asymmetric loss to get the estimated noise. Finally, the noisy image and the estimated noise are given as inputs to a non-blind denoiser to produce the clean image.

Guo et al. [43] introduced a noise estimation and removal network (NERNet) for denoising noisy images in the real-world domain. This is an enhanced version of CBDNet with the noise estimator architecture designed to estimate noise levels more accurately. The symmetric dilated convolution ensures extracting more discriminative features and the pyramid feature fusion block enables the utilization of features from different sizes effectively. The noise reduction network included a dilation selective block, which merges features provided by different dilation rate kernels with an attention mechanism to achieve effective denoising.

Although deep CNN-based denoising networks are effective at removing noise, as the network gets deeper, the majority of edge and fine image structure-related features in the final images are lost, resulting in poor image visual quality. To overcome this problem, Yongcun Guo et al. [44] developed a multi-feature extracting CNN with concatenation (McCNN). The McCNN model consists of a multi-feature extraction layer with convolutional kernels to extract various image features and cascade them to propagate into the forward network having skip connections. Skip connection helps to bypass the low-level feature to the output and thereby, reducing image distortion, avoiding the vanishing gradient problem, and increasing the network convergence speed.

It is important in low-level vision to find a balance between receptive field size and efficiency. Typically, plain CNNs improve the receptive field at the penalty of computational cost. To address this issue, dilated filtration was recently implemented. However, in this setting the gridding effect happens, resulting in the receptive field being just a sparse sample of the noisy image having checkerboard patterns. In [45], the authors presented multi-level wavelet-CNN (MWCNN) network made up of a contracting subnetwork with several layers of discrete wavelet transform (DWT) and CNN blocks and an expanding subnetwork made up of multiple levels of inverse discrete wavelet transform (IWT) and CNN blocks. MWCNN is effective at recovering the textures in-depth and well-defined structures from the deteriorated observation due to DWT's invertibility, frequency, and location features. Hence, MWCNN has the ability to broaden the receptive field while balancing efficiency and performance.

In [46], the authors combined the nonsubsampled contourlet transform (NSCT) decomposition with a CNN model to get both frequency and location information of feature maps, which is useful for edge and detail preservation. The deployment of the CNN block following NSCT decomposition is one of the most significant aspects of the proposed countourlet transform based CNN (CTCNN) paradigm. A four-layer fully convolutional network (FCN) with no pooling layer constitutes the CNN block. It is made up of 3×3 filters, batch normalization, and ReLU. In the last CNN block, just the convolution layer is used to predict images. The nonsubsampled shearlet transform (NSST) is used in [47] to decompose the noisy image into sub-bands. Then, to assess whether it belongs to the edge-related class, a CNN is trained using 3D blocks of high-frequency sub-bands along a specified direction. Finally, the NSST coefficients in the edge-related class stay unaltered, and those that are not edge-related are denoised using an adaptive threshold by the shrinkage technique.

The MP-DCNN [48] model is built with adaptive convolutional and residual learning. It incorporates the leaky ReLU for feature extraction from noisy input images. In addition, the first denoised image, produced using MSE loss function, is fed into the SegNet, which can learn the image edge information better, and the final denoised image is obtained using the perceptive loss function. In [49], BM3DNet expanded the computational methods of the BM3D [50] algorithm to be a learnable CNN for transform-domain collaborative filtering.

Tian et al. [51] presented a batch-renormalization denoising network (BRDNet) that is composed of upper and lower networks. The bottom network employs batch renormalization (BRN), residual learning (RL), and dilated convolutions. In the top network, only RL and BRN are used. BRN is utilized to solve the small-batch problem as well as to speed up BRDNet convergence. The RL is being used to clean up latent clean images and minimize noise in noisy images. Widening the receptive field with dilated convolutions allows for more context information to be obtained.

Li et al. presented a detail retaining convolutional neural network (DRCNN) [52] by deducing a mathematical model from a minimization problem to prevent significant structural loss in the restored image. DRCNN requires less parameters and storage space, allowing it to generalize better. GNSCNet was created by Wang et al. [53] by introducing a global non-linear smoothness constraint (GNSC) prior term into a MAP-based cost function. The denoising is performed on a high dimensional transform space to enhance the network outcome. The GNSC prior can fine-tune the structural information in the image leading to the denoising solution. Thus, the important structures like edges and textures are retained with a perceptually fine denoised image.

As the depth of very deep CNNs rises, the impact of shallow layers on deep layers weakens. So, Tian et al. [29] introduced an attention-guided denoising CNN (ADNet), consisting of a sparse block (SB), a feature enhancement block (FEB), an attention block (AB), and a reconstruction block (RB). The SB enlarges the receptive field size by using dilated and standard convolutions. FEB combines the global and local characteristics to improve the model performance. Furthermore, the AB can rapidly capture the significant noisy attributes hidden in the background for real noisy image and blind denoising.

Although the BN can improve network denoising performance, it needs a significant amount of processing. By avoiding BN, Peng et al. developed dilated residual networks with symmetric skip connection (DSNet) [54] by combining symmetric skip connection and dilated convolution. The symmetric skip connections can accomplish the same denoising performance as the BN. Furthermore, dilated residual networks may substantially minimize the number of computations as well as the training and testing time while ensuring performance.

GRADNet [55] computes the image gradient from the denoised network input and concatenates it with shallow layer feature maps to preserve high-frequency textures and edges during denoising. Image gradients have essentially similar characteristics to shallow layer features, indicating that GRADNet's fusion approach is better. In addition, a gradient consistency regularization is used to decrease the gradient disparity between the denoised image and the clean GT.

To leverage self-similarities, a pyramid attention network (PANet) [56] is proposed to collect long-range feature relationships from a multi-scale feature pyramid. The pyramid attention module presented in [56] is a generic building component that can be easily integrated into a wide range of neural architectures. DHDN model [33] employs a U-Net model to make better use of limited memory by reducing the size of the feature maps. The sub-pixel interpolation approach is used to appropriately interpolate the feature maps. The authors were able to induce feature reuse and overcome the vanishing-gradient problem by including dense connectivity and residual learning in the convolution blocks and network. Also, the model employs the model ensemble and self-ensemble methods. In Table 5.2, we listed CNN-based DL-MMSE methods for different IR tasks.

Recently, GAN architectures are used along with CNN-based DL-IR tasks. Two such methods are GCBD [57] and grouped residual dense network (GRDN) [59]. Jingwen Chen et al. [57] introduced a blind denoiser using GAN and CNN that works in the absence of paired training data. The GCBD model synthesizes paired training data from the given noisy images using GANs, and then a CNN denoiser is trained using these images to eliminate the noise from these images. The basic component of a GRDN [59] is a RDN with minor modifications. Cascaded GRDN blocks (GRDB) with attention modules achieve cutting-edge performance. In addition, an enhanced GAN-based real-world noise modeling approach is employed. A generalized representation of image reconstruction using GAN is shown in Figure 5.5.

5.2.2 MAP Based DL-IR Methods

MAP is a Bayesian approach-based probabilistic framework widely useful for problems of IR because it can be used to explicitly model both measurement noise and prior knowledge regarding the image to be estimated. As mentioned earlier, the MAP methods, by learning image prior inherently or not, solve a high dimensional optimization

Table 5.2 List of CNN-based DL-MMSE methods for different IR tasks.

Publication	Year	Denoising	IP	SR	Compression artifact reduction	Deblurring
DnCNN [27]	2017	✓		✓	✓	
MemNet [31]	2017	✓		✓	✓	
BM3DNet [49]	2017	✓				
BMCNN [39]	2018	✓				
MWCNN [45]	2018	✓		✓	✓	
GRADNet [55]	2018	✓				
FFDNet [28]	2018	✓				
GCBD [57]	2018	✓				
DIP [58]	2018	✓	✓	✓	✓	
DRDN [40]	2019	✓				
MP-DCNN [48]	2019	✓				
BRDNet [51]	2019	✓				
SANet [42]	2019	✓				
CBDNet [32]	2019	✓				
DSNet [54]	2019	✓				
NSST-CNN [47]	2019	✓				
GRDN [59]	2019	✓				
PSN [60]	2019	✓		✓		
DHDN [33]	2019	✓				
NERNet [43]	2020	✓				
MLANet [41]	2020	✓		✓	✓	
CTCNN [46]	2020	✓				
DRCNN [52]	2020	✓	✓	✓		✓
GNSCNet [53]	2020	✓				
ADNet [29]	2020	✓				
PANet [56]	2020	✓		✓		
RDN [30]	2021	✓		✓	✓	✓

problem with various optimization techniques such as adaptive moment estimation (Adam) and stochastic gradient Langevin dynamics (SGLD).

A prior is an acronym for prior distribution, which is simply a distribution that encapsulates basic assumptions in the absence of information. In other words, a prior is priori (beforehand) knowledge of how the system under consideration operates. Image prior is prior information on a given set of images that may be used in an image processing problem. For example, a prior distribution over images effectively depicts what we expect the GT images to resemble in an image reconstruction task. It restricts the sample space in which we can find the solution. In the images, the feature-level prior assists in minimizing domain mismatch across degraded images with various noise levels for efficient image denoising, whereas the pixel-level prior drives the denoised image into the photorealistic domain for perceptual enhancement.

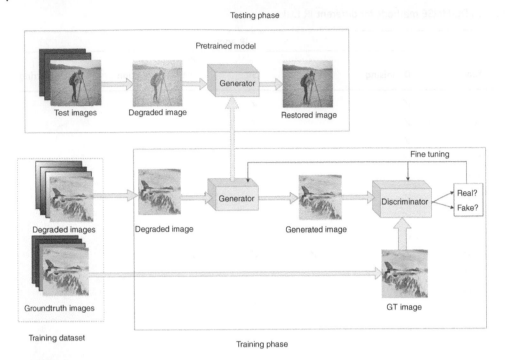

Figure 5.5 Schematic representation of training and testing a GAN network for IR tasks. Source: SDASM Archives/Flickr.

Natural images generally possess smooth areas with sharp edges, resulting in a gradient profile with a heavy tail. To describe heavy-tailed gradient statistics, a generalized Gaussian distribution or a Gaussian mixture can be employed. Prior works by hand-selecting model distribution parameters and fixing them for the whole image, imposing the same image prior everywhere. Unfortunately, even within a single image, various textures have distinct gradient statistics, thus applying a single image prior to the whole image is inefficient.

Several natural image priors have been proposed in recent years, including TV, gradient sparsity priors, models based on image patches, and Gaussian mixtures of local filters, dictionary learning, proximal operators, half quadratic splitting, to mention a few of the more effective ideas. These techniques are based on hand-crafted priors that are approximated through painstaking norm selection. Despite their widespread use, they have a restricted expressive capability of the prior, limiting the quality of the solution. Furthermore, no matter how successful they are, these priors are always inspired by a dedicated purpose. For example, a MRF is often employed to indicate the correlation between neighboring pixels, whereas a dark channel prior and TV are developed for dehazing and denoising.

DL approaches have recently been employed to create generic image priors. In [61], a Bayesian DL approach using a generic image prior is proposed and trained a DAE for finding the gradient of the prior. The prior directly reflects the Gaussian smoothed natural image probability distribution. This approach enables the learning of a single prior and its application to different IR tasks, such as noise-blind deblurring, SR, and image demosaicing.

In [18], a multi-channel and multi-model-based DAE prior is proposed. In this work, based on the fact that an optimum DAE is a local mean of the actual data density, a three-channel DAE prior is developed. The authors used an aggregation function for enhanced network stability and the auxiliary variable method to learn higher-dimensional structural information.

Dmitry Ulyanov et al. [58] demonstrated how a randomly initialized neural network can be utilized as a prior with great performance in common inverse problems and proposed a non-trainable deep image prior. In [62], the authors provide a Bayesian view of the deep image prior and eliminate the requirement of early stopping in [58] by conducting posterior inference with SGLD. Jiaming Liu et al. [63] modified deep image prior (DIP) framework in [58] by combining implicit CNN regularization with an explicit TV penalty. When tested on a variety of conventional IR tasks, the addition of TV yielded considerable performance improvements. Another improvement on [58] using Kullback–Leibler (KL) divergence-based regularization term in the loss function and utilization of Mish activation function is discussed in [64]. Pasquale Cascarano et al. [65] developed a denoising method that combines the DIP framework with a space-variant TV regularizer as well as an automatic estimation of the local regularization parameters. To tackle the resulting optimization problem, the authors also employed a flexible alternating direction method of multipliers (ADMMs) algorithm.

A CNN denoiser can be used to leverage the multi-scale redundancies of natural images [66]. Both the CNN denoisers and other network parameters may be simultaneously improved through end-to-end training. Regardless of the type of deterioration, plug-and-play denoisers can be utilized to execute generic IR tasks [6]. In [6], the authors have devised a framework with ADMM optimization strategy to enhance the efficiency of solving the MAP objective with an explicit image prior.

Unlike the existing methods that use a fixed and fully instantiated prior for IR problem, Aljadaany et al. [67] introduced a PSN framework for solving IR using proximal operator that has been modeled as a CNN. This in effect resulted in a function class image prior during training, allowing it to be adaptable to data.

Apart, SR, deblurring, IP, low light image enhancement, and compression artifact reduction are the common restoration tasks. SR refers to the task of producing a HR image from a LR input. Image deblurring is the act of recovering the latent clear image with crisp details from a blurred observation. Low light image enhancement is the technique of predicting the well light image from a low light image. IP refers to the process of rebuilding damaged areas of an image by computing the gray values and gradients of damaged parts such that they extrapolate the surrounding region's values.

5.2.3 Other DL-SR Methods

The use of DL to image SR applications has attracted the attention of many researchers in recent years. A significant number of DL methods for SR applications have been found supervised, which means they have been trained on pairs of LR and HR images. Supervised SR, which depends on defined degradation to generate LR images from HR images, is insufficient for many real-world applications. As a result, several academics are now focusing on DL-based unsupervised SR models. Here, we review a few methods exclusively developed for DL-SR applications.

5.2.3.1 Supervised SR Techniques

Zhen Cui et al. [60] proposed a DL-based SR approach termed deep network cascade (DNC) to overcome the constraints of small upscale factors with the conventional SR techniques. The DNC network is made up of a series of stacked layers, each with a NLSS search unit and a collaborative local autoencoder (CLA). This setup performs gradual upscaling layer by layer with small scaling factors. The NLSS search unit provides high-frequency texture details and the CLA removes noise introduced by the NLSS unit and collaborates compatibility among overlapping patches. Cascading of multiple such stacked CLAs gradually upscale the LR input to the final desired HR image. Nonetheless, the method is computationally expensive and requires layer-wise optimization.

Chao Dong et al. [68] proposed SRCNN method for an end-to-end CNN based single image super-resolution (SISR). SRCNN has a patch extraction layer, a layer for nonlinear mapping onto HR patches, and a layer for HR image reconstruction. In [69], the authors proposed a very deep SISR (VDSR) method by increasing network depth. Hence, it overcomes the limitations of SRCNN, such as slow convergence, applicability to a single scale only, and reliance on the context of small image regions. In [70], the authors presented sparse coding based networks (SCNs) by combining the capabilities of sparse coding and DL. The CSCN model is a cascaded version of SCN that enables SR for arbitrary factors.

Jiwon Kim et al. [71] developed a deeply recursive convolutional network (DRCN) by utilizing a single convolution layer to produce a deep network recursively without adding a new weight layer. The authors adopted recursive supervision and high learning rates to ease the training. In [71], the authors attempt to overcome the drawbacks of using very deep networks such as over-fitting and significant storage requirements, as well as making network training difficulties caused by vanishing and exploding gradients. Ying Tai et al. [72] introduced a deep recursive residual network (DRRN) for SISR. In DRRN, the use of a recursive block by stacking a set of residual units that share a weight set reduces the number of parameters to be learned by the network.

The aforementioned approaches use bicubic operator to pre-upscale the input image and, hence, have an extra computational expense and a long run time. To solve this, Dong et al. [73] introduced a fast SR CNN (FSR-CNN) approach by adding a deconvolution layer at the network's end to aid in learning the upscaling filters rather than pre-upscaling and using smaller filter sizes with more mapping layers. Tong et al. [74] investigated the use of dense nets [75] with skip connections to improve the training and reconstruction performance of deep networks for SR and proposed the SRDenseNet. In [76], Namhyuk Ahn et al. introduced a deep SISR method with the goal of minimizing the number of operations, as opposed to earlier methods that focused solely on reducing the number of parameters. CARN model uses residual architecture with a block- and layer-wise cascading mechanism that provides an efficient flow of information and gradients.

Many of the existing deep recursive SR techniques operate at a fixed spatial resolution and can be reformulated as a single state recurrent neural network (RNN) with finite unfolding. Wei Han et al. [77] investigated new structures for SR that exploits both LR and HR signals jointly and proposed dual state residual network (DSRN). Recurrent signals are transmitted between these states via delayed feedback in both directions (LR to HR and HR to LR). Muhammad Haris et al. [78] proposed a framework that produce SR image by exploiting the relation between the LR and HR images by introducing an error feedback mechanism and named the resulting network the deep back-projection network (DBPN). This network employs iterative up and down sampling layers, and projection errors from these layers are used to guide the network. The method produced superior results even on large scaling rates of 8×. The SR feedback network (SRFBN), proposed by Zhen Li et al. [79], is a RNN with a feedback block (FB) that aids in the efficient handling of feedback information flow and feature reuse. Additionally, a curriculum learning technique enables the model to perform effectively in complex IR tasks.

Existing CNN-based SR methods treat all the channel features equally and lag in producing more representative features. To improve the representational power of CNNs to generate more informative features, channel attention (CA) mechanism was proposed in [80]. Based on this mechanism, Yulun Zhang et al. [81] introduced a CA module in their SR framework called a very deep residual channel attention network (RCAN). The CA module provided discriminative representation ability by adaptively rescaling channel-wise features based on the channel's interdependencies and aided in achieving visual improvements in HR image obtained. Similar to this, Tao Dai et al. introduced a second order CA (SOCA) module in their proposed second order channel attention network (SAN) [82]. The SOCA module rescales channel features based on feature statistics of second-order and further improved HR image quality. Yulun Zhang et al. [83] introduced a nonlocal attention mechanism in their SR network RNAN. The proposed network used local and nonlocal attention modules that aid in learning local attention and predicting scaling weights for different feature maps, respectively. This further enhanced the representational ability.

The majority of existing SR approaches optimized the network using MSE loss as a criterion. Despite the high peak signal-to-noise ratio (PSNR) values obtained, the MSE loss is observed to result in perceptually low-quality HR images. To solve this, Christian Ledig et al. [84] introduced SRGAN, a GAN-based SR technique. The authors used a GAN to train a perceptual loss function that included MSE-based content loss, a loss based on feature mappings produced using a VGG network, and an adversarial loss. Even with significant upscaling factors, SRGAN ensured photorealistic SR results (4×). Wang et al. introduced an upgraded version of SRGAN, dubbed enhanced SRGAN (ESRGAN) [85], to improve visual quality even further. The authors reconfigured the generator module of existing SRGAN by removing batch normalization and replacing the basic blocks with residual-in-residual dense block. In addition, the authors replaced the discriminator in GAN with a relativistic discriminator, which attempts to differentiate inputs based on their relative realistic nature. Finally, they included a more effective perceptual loss that constrains features prior to activation rather than after activation, which helped them avoid the inadequate supervision and uneven reconstructed brightness that SRGAN had.

Bee Lim et al. proposed an improved deep SR network (EDSR) that outperformed existing SR techniques. The proposed technique was created using SRResNet. The large performance improvement of the proposed model can be attributed to optimization by deleting unnecessary modules like normalizing layers in standard residual networks. Expanding the model size while stabilizing the training technique improved performance even more. Similarly, Lai et al. proposed the deep Laplacian pyramid SR network (LapSRN) [86] to overcome the limitations of earlier DL-based SR methods by introducing progressive reconstruction of HR images over pyramid levels with deep supervision for better HR reconstruction, replaced the MSE loss function with the robust Charbonnier loss functions, and provided a multi-scale model of the network that learns inter-scale correlations and is referred to as MS-LapSRN.

5.2.3.2 Unsupervised SR Techniques

Despite their great performance, supervised SR algorithms are generally constrained to known and ideal image acquisition circumstances, failing to conduct SR in real-time scenarios with uncertain acquisition processes. Patch-based unsupervised learning was developed as a way to exploit image-specific information using a single image; however, it fails to generalize well to patches that don't exist in the LR image and is incapable of adopting nonuniform recurring structures within the image. Assaf Shocher et al. [87] proposed a CNN-based unsupervised SR technique called "Zero-Shot" SR to overcome these issues. The proposed method trains an image-specific CNN to capture the cross-scale internal recurrence of image-specific information by relating LR image and multiple downscaled versions. This learned relationship is then used to produce desired HR from the LR input image.

Unlike existing SR techniques that rely on supervised learning with HR and LR image pairs and the assumption of known downscaling, Yuan et al. [88] proposed unsupervised SR method with cycle-in-cycle generative adversarial network (CinCGAN) model. After translating noisy and blurred input into a noise-free low-resolution space using a cycle GAN, a pre-trained deep model that consists of another cycle GAN is used to upsample the image.

5.2.4 Other DL-IR Tasks

Based on the knowledge on the blur kernel, image deblurring can be classified as blind or non-blind methods. Also, these methods can be noisy or noiseless depending on the noise characteristics. Yuesong Nan et al. [89] presented a noise blind image deblurring technique capable of estimating noise level using CNN-based learnable image prior construction and measuring image prior uncertainty using MLP-based predictions. In [90], Jian Sun et al. proposed a CNN to suppress nonuniform motion blur by estimating the probabilistic distribution of motion blur at the patch level. In [91], Sameer Malik and Soundararajan trained CNNs on noisy lowlight and clean well-lit image pairs. The authors achieved lowlight IR via denoising and contrast enhancement using separate CNNs.

In [92], depth IP method developed by Zun Li and Wu turned the depth IP problem into the related denoising problem. The denoising problem that results is then handled by a deep CNN-based network that learns the denoiser prior.

Lukas Cavigelli et al. [93] presented 12-layer CNN architecture with hierarchical skip connections and a multi-scale loss function for removing compression artifacts in JPEG compressed image data. The authors adjusted the network architecture to reduce average path length and avoid the training challenges associated with very deep networks. This feature allows the proposed method to converge to levels of accuracy that were previously unattainable.

5.3 DL-Based Medical Image Reconstruction

Medical imaging modalities such as MRI, CT, US, and OCT have low SNR and contrast-to-noise ratio (CNR) owing to deterioration and visual artifacts. As a result, the diagnoses of many diseases by visual examination and quantification from these images, as well as treatment planning by the physician in cancer care, are difficult.

In CT, short acquisitions deteriorate the quality of scans and long acquisitions cause motion artifacts, patient discomfort, or even physical harm to the patient. Hence, there is a need for signal processing algorithms to reduce radiation dose in CT. David Boublil et al. [94] introduced a strategy for increasing the quality of a reconstructed CT image that enables proper LDCT reconstruction. A feed-forward ANN is used to realize a regression function and performs a local nonlinear fusion of the resulting image estimates corresponding to different control parameter values. The local fusion has improved the variance resolution trade-off of the reconstruction algorithm and improved the perceptual quality of the CT images. In [95], Kyong Hwan Jin et al. utilized a CNN-based algorithm for sparse view CT reconstruction. The authors explored the relationship between CNNs and iterative optimization methods. The authors employed filtered back projection (FBP) and a CNN that uses residual learning and multilevel learning.

In [96], the authors introduced LDCT image reconstruction by relying on a directional wavelet transform to recognize the directional components of noise. The authors used a deep CNN in the contourlet transformed domain. The limitation of this method is its lengthy training time as this is a very deep network with 26 layers. In contrast, a lightweight CNN network with three layers, which directly operated in the image domain has been proposed [97]. This network is referred to as residual encoder–decoder CNN (RED-CNN). The layers of the network correspond to the patch encoding, nonlinear filtering, and reconstruction steps, respectively. The network is trained to estimate the parameters using a dataset consisting of pairs of low-dose and corresponding normal-dose image patches. The encoder part of the net enabled it to suppress noise and reduce artifacts while the decoder part helped in recovering structural details. Finally, residual learning eased the training of deep networks.

The parameters of the aforementioned network were optimized based on the MSE loss function. Since the regular dose CT image may also contain noise and the predicted value is the mean of these values results in smoothed images, it lacks the texture of a typical normal-dose CT image. To overcome this, Wolterink et al. [98] proposed a GAN-based pix2pix framework for noise reduction in LDCT images. This network consists of a generator and discriminator CNNs are trained to translate the LDCT image into an approximation of the regular dose CT image and to discriminate the generator output from the routine-dose CT images, respectively. Despite their high performance, the original GANs had training issues caused by the vanishing gradients problem during the generator loss minimization phase. To address this, a LDCT denoising technique based on the Wasserstein generative adversarial network (WGAN) has been proposed [99]. The features extracted enable the network to retain vital image information when reducing the noise in it. To obtain better performance, a structurally sensitive multi-scale GAN (SMGAN-3D) has been proposed [100]. The authors formulated a structurally sensitive loss by combining adversarial loss, perceptually favorable structural loss, and pixel-wise L_1 loss. A noise learning least squares generator adversarial network (LSGAN) is proposed in [101] with least squares, structural similarity, and L_1 losses. The model learns the noise distribution in the LDCT image and then, subtracted it from the input to produce the denoised output. Structural similarity index (SSIM) loss term added importance to the preservation of structural details, and the L_1 loss ensures the sharpness preservation in the output.

In [102], Jonas Adler and Oktem presented a DL-CT image reconstruction based on unrolling the iterating scheme of primal dual hybrid gradient (PDHG) algorithm. Each proximal operator in the subproblem of PDHG is approximated by a neural network. This model has a significant performance boost compared with FBP and some handcrafted reconstruction models. Haimiao Zhang et al. [103] proposed JSRNet to jointly reconstruct CT images and their associated Radon domain projections using CNN. The network is optimized using a hybrid loss function as a combination of SSIM loss, MSE, and semantic loss.

In clinical MRI, accelerated MR image acquisition is highly desirable to reduce scan cost and less patient burden. To achieve this, CS-MRI reconstructs an MR image with partially completed k-space. In CS-MRI, regularization related to the data prior is significant, and ADMM is a widely accepted iterative algorithm for optimizing it. Yan Yang et al. [104] proposed a CS-MR image reconstruction system that incorporates both standard model-based CS and data-driven DL. The authors proposed a deep ADMM-Net as a solver of the ADMM iterative algorithm for the reconstruction problem. The suggested network is made up of a deep architecture with multiple stages, each of which corresponds to one ADMM iteration involving reconstruction, denoising, and Lagrangian multiplier update steps. The deep network's flexibility in learning parameters aided in obtaining high reconstruction accuracy while maintaining the ADMM algorithm's computational efficiency.

DL-based undersampled MR image reconstruction method [105] consists of a CNN trained to learn an end-to-end mapping between zero-filled MR images obtained by direct inversion of observed undersampled k-space data and fully-sampled MR images as GT. This trained network can be used to integrate with classic CS-MRI methods either as a two-phase reconstruction to initialize CS-MRI with these learned parameters or directly integrating the network output as an additional regularization term. This network can not only restore the features and fine structures of MR images, but it can also be used with any online reconstruction algorithm for more efficient and effective imaging.

Jo Schlemper et al. [106] proposed CS cardiac MR image reconstruction method consists of two CNNs, the first CNN being a de-aliasing CNN and the second CNN learning to reconstruct from the output of the first CNN. In effect, two CNNs cascaded to iterate between intermediate de-aliasing and the data consistency reconstruction in this cascading network. In DAGAN [107] for CS-MRI reconstruction, a U-net generator is implemented with skip connections, and the discriminator is used to classify the de-aliased reconstruction and the fully sampled GT reconstruction. The authors introduced content loss that is coupled to adversarial loss considering both pixel-wise mean square error (MSE) and perceptual loss defined by pre-trained deep convolutional networks from VGG. The method also introduced refinement learning to stabilize the training and reduce the complexity of learning as it is made to generate only missing information. The automated transform by manifold approximation (AUTOMAP) framework [108] employs fully connected (FC) layers followed by a sparse convolutional AE. It achieves unified image reconstruction by learning a direct mapping from sensor domain data to an underlying image. The FC layers approximate the between manifold projection from sensor domain to image domain. In addition, the sparse AEs collect high-level features from the input and force the image to be represented sparsely in the convolutional feature space.

To despeckle US images, many despeckling techniques have been proposed in frequency and spatial domains. However, DL-based despeckling methods have gained prominence in recent decades due to their applicability in real-time denoising and their improved effectiveness in despeckling as it can automatically learn the intrinsic features from the training data. In [109], the authors proposed DL framework for real-time denoising of US images through the learning of various smoothing algorithms. The method generalizes on the input data (such as noise type, image dimension [2D, 3D], and resolution) and brings out a specialization of the denoising algorithms based on algorithm parameter tuning for generic to US images. This method enables preserving the main features of the underlying image (e.g. edges, grayscale) while respecting the industrial and production requirements (e.g. real-time computation, memory overhead, hardware configurations). Lan et al. [110] proposed a mixed attention mechanism-based residual UNet (MARU) for real-time despeckling. It has a lightweight mixed-AB for maintaining both channel and spatial attention using separation and refusion method. For that, it needs only a small additional memory and time.

Recently, Qiu et al. [111] proposed an OCT denoising technique consisting of a DNN trained with the loss function that is based on MS-SSIM metric to ensure better perceptual quality of the denoised OCT images. This method, which proved to be very efficient specifically for the OCT denoising, was outperformed by unsupervised learning-based nonlocal deep image prior proposed by Fan et al. [112], which entails the popular deep image prior [58] model. Microscopy is another smart visualization tool widely used in the field of biology, dealing with the imaging at cellular level catering to the innovative and exploratory studies in this field. These images are typically disturbed by blur and background noise. The blur element is caused by the resolution limit of the microscope, while the background noise pertains to the limitation of the imaging system, particularly the acquisition sensor. DL-based denoisers in the supervised data-driven fashion have put forth excellent performance with the CNN architecture [113]. In structured illumination microscopy (SIM), Chatton [114] proposed the DL-based restoration algorithm handling SR and denoising parallelly for the wide-field images. The model is based on ESRGAN [85] and has reduced the imaging time required for HR image acquisition and improved the lifetime of the cells. In two-photon fluorescence microscopy (2PM) imaging, which has led to the 3D neural imaging of deep cortical regions, Lee et al. [115] introduced MuNet that consists of multiple U-Nets with 3D convolutions such that each U-Net removes the noises at different scales and then provides a performance improvement based on a coarse-to-fine strategy. This model refrains from pre/post-processing and facilitates end-to-end learning. Regularization is accommodated by a GAN which is involved in the training phase that enables "near-real" clear image generation. MuNet can preserve the image details well even in very noisy conditions.

Microscopy image denoising has also seen adaptation of filter augmented algorithms for regularization in DL-based denoising approach. Pronina et al. [116] have proposed a unifying framework of algorithms for Gaussian image deblurring and denoising. These algorithms are based on DL techniques for the design of learnable regularizers integrated into the Wiener–Kolmogorov filter, leading to a lower computational complexity. This reconstruction pipeline augmented with variance stabilizing transformation can be successfully applied for Poisson image deblurring.

Single-photon emission computerized tomography (SPECT) is a nuclear imaging scan that integrates computed tomography (CT) and a radioactive tracer that enables the medical experts to evaluate the blood flow patterns in tissues and organs. SPECT reconstruction methods are broadly categorized as analytic and iterative methods. The analytic methods include FBP, Fourier rebinning (FORE), and three-dimensional reprojection (3DRP) algorithms. Iterative approaches consist of methods like ordered subsets expectation maximization (OSEM) [117] based on maximum likelihood estimation, which tends to increase the image noise level as the number of iterations is increased. Post reconstruction techniques involve filters such as the three-dimensional (3D) Gaussian filter [118], but it affects the spatial resolution of the reconstructed images as the noise level varies. With the advancements in image processing with DL approaches, denoising techniques adapted to them outperforming the conventional smoothing filters.

Liu et al. presented a DL approach that involves Noise2Noise (N2N) [119] model for denoising cardiac SPECT-MPI (single-photon emission computerized tomography-myocardial perfusion imaging) images. In this approach, both the input and target images are noisy in nature and the network is trained to separate the common signal content in the noisy image. The author used coupled-Unet (CUNet) [120] network structure as it improves the training efficiency. The DL approach greatly reduced the noise level in the reconstructed images improving the perfusion defect accuracy.

positron emission tomography (PET) is a functional imaging technique using radiotracers to visualize and quantify the changes of metabolic processes as well as physiological activities such as blood flow, regional chemical composition, and absorption. Low-dose PET (LD PET) imaging has been encouraged to bring down the radiation exposure due to the increased radiotracer dosage. Methods based on prior information have improved the reconstruction of full-dose PET (FD PET), but it enables to directly incorporate the imaging physics information into the reconstruction process. The drawback lies in its computational complexity and physics model accessibility.

An iterative PET image reconstruction framework has been developed by Gong et al. [121] which involved a residual convolutional AE framework. D. Nie et al. [122] proposed a cycle Wasserstein regression adversarial training framework (CycleWGANs) for amplifying the LD PET image quality. The Wasserstein distance replaces the conventional Jensen–Shannon (JS) divergence to nullify the training problem in GANs. Unsupervised PET denoising has also been explored by researchers because of the unavailability of large datasets. Hashimoto et al. [123]

proposed a 4D DIP framework for dynamic PET image denoising without using any anatomical information. However, this method is easy to overfit. H. Sun et al. [124] experimented with the combination of DIP [58, 125] and RED by considering the prior images as the network input.

5.4 Conclusion

DL-based image reconstruction approaches have grown rapidly in recent years and outperform traditional image reconstruction techniques, particularly in the context of noisy and restricted data representation. In this chapter, we reviewed the application of DL models on various image reconstruction tasks such as IP, deblurring, SR, compression artifact removal, and denoising. The different DL models include CNN, AE, and GAN-based methods. Also, we investigated the significance of DL models in connection with medical image reconstruction in the field of LDCT, MRI, OCT, US, and microscopy images. DL has aided in accelerating the reconstruction process with improved accuracy and enabled to reduce the effective imaging time. The generalization and robustness of DL methods with respect to different imaging modalities and machine specifications and the reliability independent of the supervision from a radiologist are major concerns. The DL techniques are computationally expensive and require a large amount of training datasets in case of non-blind methods.

Acknowledgment

This work was supported by the Science and Engineering Research Board (Department of Science and Technology, India) through project funding SRG/2019/001586.

References

1 Al-Shaykh, O. and Mersereau, R. (1998). Lossy compression of noisy images. *IEEE Transactions on Image Processing* 7: 1641–1652.

2 Banham, M. and Katsaggelos, A. (1997). Digital image restoration. *IEEE Signal Processing Magazine* 14: 24–41.

3 Zamperoni, P. (1995). Image Enhancement. In: *Advances in Imaging and Electron Physics* (ed. P.W. Hawkes), 1–77. Elsevier.

4 Antun, V., Renna, F., Poon, C. et al. (2020). On instabilities of deep learning in image reconstruction and the potential costs of AI. *Proceedings of the National Academy of Sciences of the United States of America* 117: 30088–30095.

5 Shukla, P.K., Sandhu, J.K., Ahirwar, A. et al. (2021). Multi-objective genetic algorithm and convolutional neural networks based COVID-19 identification in chest X-ray images. *Mathematical Problems in Engineering* 2021: 1–9.

6 Zhang, K., Li, Y., Zuo, W. et al. (2021). Plug-and-play image restoration with deep denoiser prior. *IEEE Transactions on Pattern Analysis and Machine Intelligence* 1.

7 Bigdeli, S., Honzátko, D., Süsstrunk, S. et al. (2020). Image restoration using plug-and-play CNN map denoisers. *Proceedings of the 15th International Joint Conference on Computer Vision, Imaging and Computer Graphics Theory and Applications*, Malta.

8 Le Cun, Y. and Fogelman-Soulié, F. (1987). Intellectica. *Revue De L'Association Pour La Recherche Cognitive* 2: 114.

9 Vincent, P., Larochelle, H., Lajoie, I. et al. (2010). *Journal of Machine Learning Research* 11 (12): 3371–3408.

10 Xie, J., Xu, L., and Chen, E. (2012). Image denoising and inpainting with deep neural networks. In: *Advances in Neural Information Processing Systems*, 341–349. United States of America: Morgan Kaufmann Publishers, Inc.

11 Gondara, L. (2016). Medical image denoising using convolutional denoising autoencoders. *2016 IEEE 16th International Conference on Data Mining Workshops (ICDMW)* (12–15 December 2016), Spain.

12 Agostinelli, F., Anderson, M.R., and Lee, H. (2013). Adaptive multi-column deep neural networks with application to robust image denoising. In: *Advances in Neural Information Processing Systems* (ed. C.J. Burges, L. Bottou, M. Welling, et al.), 1493–1501.

13 Wang, R. and Tao, D. (2016). Non-local auto-encoder with collaborative stabilization for image restoration. *IEEE Transactions on Image Processing* 25: 2117–2129.

14 Mao, X.J., Shen, C., and Yang, Y.B. 2016). Image restoration using convolutional auto-encoders with symmetric skip connections. arXiv preprint arXiv:1606.08921.

15 Song, J.H., Kim, J.H., and Lim, D.H. (2020). Image restoration using convolutional denoising autoencoder in images. *Journal of the Korean Data and Information Science Society* 31: 25–40.

16 Majumdar, A. (2019). Blind denoising autoencoder. *IEEE Transactions on Neural Networks and Learning Systems* 30: 312–317.

17 Arjomand Bigdeli, S. and Zwicker, M. (2018). Image restoration using autoencoding priors. *Proceedings of the 13th International Joint Conference on Computer Vision, Imaging and Computer Graphics Theory and Applications* (5–8 December 2013), Portugal.

18 Zhou, J., He, Z., Liu, X. et al. (2020). Transformed denoising autoencoder prior for image restoration. *Journal of Visual Communication and Image Representation* 72: 102927.

19 Lei, H. and Yang, Y. (2021). CDAE: a cascade of denoising autoencoders for noise reduction in the clustering of single-particle cryo-EM images. *Frontiers in Genetics* 11: 627746.

20 Shimobaba, T., Endo, Y., Hirayama, R. et al. (2017). Autoencoder-based holographic image restoration. *Applied Optics* 56: –F27, F30.

21 Liu, Q., Yang, Q., Cheng, H. et al. (2019). Highly undersampled magnetic resonance imaging reconstruction using autoencoding priors. *Magnetic Resonance in Medicine* 83: 322–336.

22 Zhang, F., Zhang, M., Qin, B. et al. (2021). REDAEP: robust and enhanced denoising autoencoding prior for sparse-view CT reconstruction. *IEEE Transactions on Radiation and Plasma Medical Sciences* 5: 108–119.

23 Fan, F., Li, M., Teng, Y., and Wang, G. (2020). Soft autoencoder and its wavelet adaptation interpretation. *IEEE Transactions on Computational Imaging* 6: 1245–1257.

24 Burger, H.C., Schuler, C.J., and Harmeling, S. (2012). Image denoising: Can plain neural networks compete with BM3D? *2012 IEEE Conference on Computer Vision and Pattern Recognition*, Rhode Island, 2392-2399.

25 Dong, C., Loy, C.C., He, K., and Tang, X. (2016). Image super-resolution using deep convolutional networks. *IEEE Transactions on Pattern Analysis and Machine Intelligence* 38: 295–307.

26 Jain, V. and Seung, H.S. (2008). Natural image denoising with convolutional networks. In: *Advances in Neural Information Processing Systems 21 (NIPS 2008)*, vol. 8, 769–776.

27 Zhang, K., Zuo, W., Chen, Y. et al. (2017). Beyond a Gaussian denoiser: residual learning of deep CNN for image denoising. *IEEE Transactions on Image Processing* 26 (7): 3142–3155.

28 Zhang, K., Zuo, W., and Zhang, L. (2018). FFDNet: toward a fast and flexible solution for CNN-based image denoising. *IEEE Transactions on Image Processing* 27 (9): 4608–4622.

29 Tian, C., Xu, Y., Li, Z. et al. (2020). Attention-guided CNN for image denoising. *Neural Networks* 124: 117–129.

30 Zhang, Y., Tian, Y., Kong, Y. et al. (2021). Residual dense network for image restoration. *IEEE Transactions on Pattern Analysis and Machine Intelligence* 43 (7): 2480–2495.

31 Tai, Y., Yang, J., Liu, X. et al. (2016). MemNet: a persistent memory network for image restoration. *IEEE Transactions on Image Processing*, Italy.

32 Guo, S., Yan, Z., Zhang, K. et al. (2019). Toward Convolutional Blind Denoising of Real Photographs. *2019 IEEE/CVF Conference on Computer Vision and Pattern Recognition (CVPR)* (15–20 June 2019), USA.

33 Park, B., Yu, S., and Jeong, J. (2019). Densely connected hierarchical network for image denoising. *2019 IEEE/CVF Conference on Computer Vision and Pattern Recognition Workshops (CVPRW)* (16–20 June 2019), USA.

34 Krizhevsky, A., Sutskever, I., and Hinton, G.E. (2017). ImageNet classification with deep convolutional neural networks. *Communications of the ACM* 60: 84–90.

35 He, K., Zhang, X., Ren, S. et al. (2016). Deep residual learning for image recognition. *2016 IEEE Conference on Computer Vision and Pattern Recognition (CVPR)* (27–30 June 2016), USA.

36 Ioffe, S. and Szegedy, C. (2015). Batch normalization: accelerating deep network training by reducing internal covariate shift. In: *International Conference on Machine Learning*, 448–456. France: JMLR.org.

37 Wang, T., Sun, M., and Hu, K. (2017). Dilated deep residual network for image denoising. *2017 IEEE 29th International Conference on Tools with Artificial Intelligence (ICTAI)* (6–8 Nov 2017), USA.

38 Jiang, D., Dou, W., Vosters, L. et al. (2018). Denoising of 3D magnetic resonance images with multi-channel residual learning of convolutional neural network. *Japanese Journal of Radiology* 36 (9): 566–574.

39 Ahn, B., Kim, Y., Park, G. et al. (2018). Block-matching convolutional neural network (BMCNN): improving CNN-based denoising by block-matched inputs. *2018 Asia-Pacific Signal and Information Processing Association Annual Summit and Conference (APSIPA ASC)* (12–15 November 2018), Honolulu, Hawaii, USA.

40 Song, Y., Zhu, Y., and Du, X. (2019). Dynamic residual dense network for image denoising. *Sensors* 19 (17): 3809.

41 Zhang, X., Gao, P., Zhao, K. et al. (2020). Image restoration via deep memory-based latent attention network. *IEEE Access* 8: 104728–104739.

42 Zhang, L., Li, Y., Wang, P. et al. (2019). A separation–aggregation network for image denoising. *Applied Soft Computing* 83: 105603.

43 Guo, B., Song, K., Dong, H. et al. (2020). NERNet: noise estimation and removal network for image denoising. *Journal of Visual Communication and Image Representation* 71: 102851.

44 Guo, Y., Jia, X., Zhao, B. et al. (2020). Multifeature extracting CNN with concatenation for image denoising. *Signal Processing: Image Communication* 81: 115690.

45 Liu, P., Zhang, H., Zhang, K. et al. (2018). Multi-level wavelet-CNN for image restoration. *2018 IEEE/CVF Conference on Computer Vision and Pattern Recognition Workshops (CVPRW)*, USA.

46 Lyu, Z., Zhang, C., and Han, M. (2020). A nonsubsampled countourlet transform based CNN for real image denoising. *Signal Processing: Image Communication* 82: 115727.

47 Shahdoosti, H.R. and Rahemi, Z. (2019). Edge-preserving image denoising using a deep convolutional neural network. *Signal Processing* 159: 20–32.

48 Gai, S. and Bao, Z. (2019). New image denoising algorithm via improved deep convolutional neural network with perceptive loss. *Expert Systems with Applications* 138: 112815.

49 Yang, D. and Sun, J. (2018). BM3D-net: a convolutional neural network for transform-domain collaborative filtering. *IEEE Signal Processing Letters* 25 (1): 55–59.

50 Dabov, K., Foi, A., Katkovnik, V., and Egiazarian, K. (2007). Image denoising by sparse 3-D transform-domain collaborative filtering. *IEEE Transactions on Image Processing* 16 (8): 2080–2095.

51 Tian, C., Xu, Y., and Zuo, W. (2020). Image denoising using deep CNN with batch renormalization. *Neural Networks* 121: 461–473.

52 Li, X., Xiao, J., Zhou, Y. et al. (2020). Detail retaining convolutional neural network for image denoising. *Journal of Visual Communication and Image Representation* 71: 102774.

53 Wang, C., Ren, C., He, X., and Qing, L. (2021). Deep recursive network for image denoising with global non-linear smoothness constraint prior. *Neurocomputing* 426: 147–161.

54 Peng, Y., Zhang, L., Liu, S. et al. (2019). Dilated residual networks with symmetric skip connection for image denoising. *Neurocomputing* 345: 67–76.

55 Liu, Y., Anwar, S., Zheng, L. et al. (2020). GradNet image denoising. *2020 IEEE/CVF Conference on Computer Vision and Pattern Recognition Workshops (CVPRW)* (14–19 June 2020), USA.

56 Mei, Y., Fan, Y., Zhang, Y. et al. (2020). Pyramid attention networks for image restoration. arXiv preprint arXiv:2004.13824.

57 Chen, J., Chen, J., Chao, H. et al. (2018). Image blind denoising with generative adversarial network based noise modeling. *2018 IEEE/CVF Conference on Computer Vision and Pattern Recognition* (18–22 June 2018), USA.

58 Ulyanov, D., Vedaldi, A., and Lempitsky, V. (2020). Deep image prior. *International Journal of Computer Vision* 128: 1867–1888.

59 Kim, D.W., Chung, J.R., and Jung, S.-W. (2019). GRDN: grouped residual dense network for real image denoising and GAN-based real-world noise modeling. *2019 IEEE/CVF Conference on Computer Vision and Pattern Recognition Workshops (CVPRW)* (16–17 June 2019), USA.

60 Cui, Z., Chang, H., Shan, S. et al. (2014). Deep network cascade for image super-resolution. In: *European conference on Computer Vision (ECCV)*, 49–64.

61 Bigdeli, S.A., Jin, M., Favaro, P. et al. (2017). Deep mean-shift priors for image restoration. arXiv preprint arXiv:1709.03749.

62 Cheng, Z., Gadelha, M., Maji, S.et al. (2019). A Bayesian perspective on the deep image prior. *2019 IEEE/CVF Conference on Computer Vision and Pattern Recognition (CVPR)* (15–20 June 2019), USA.

63 Liu, J., Sun, Y., Xu, X., and Kamilov, U.S. (2019). Image restoration using total variation regularized deep image prior. *2019 IEEE International Conference on Acoustics, Speech and Signal Processing (ICASSP)*, UK.

64 Hari Kishan, R., Aneeta, C., and Sudeep, P.V. (2021). Blind image restoration with CNN denoiser prior. *Proceedings of Second International Conference on Data Science and Applications 2021* (10–11 April 2021), India.

65 Cascarano, P., Sebastiani, A., Comes, M.C. et al. (2020). Combining weighted total variation and deep image prior for natural and medical image restoration via ADMM. arXiv preprint arXiv:2009.11380.

66 Dong, W., Wang, P., Yin, W. et al. (2019). Denoising prior driven deep neural network for image restoration. *IEEE Transactions on Pattern Analysis and Machine Intelligence* 41 (10): 2305–2318.

67 Aljadaany, R., Pal, D.K., and Savvides, M. (2019). Proximal splitting networks for image restoration. *Lecture Notes in Computer Science* 11662: 3–17.

68 Dong, C., Loy, C.C., He, K., and Tang, X. (2014). Learning a deep convolutional network for image super-resolution. In: *European Conference on Computer Vision (ECCV)*, 184–199.

69 Kim, J., Lee, J.K., and Lee, K.M. (2016). Accurate image super-resolution using very deep convolutional networks. *2016 IEEE Conference on Computer Vision and Pattern Recognition (CVPR)* (23–26 June 2016), USA.

70 Wang, Z., Liu, D., Yang, J. et al. (2015). Deep networks for image super-resolution with sparse prior. *2015 IEEE International Conference on Computer Vision (ICCV)* (11–18 December 2015), Chile.

71 Kim, J., Lee, J.K., and Lee, K.M. (2016). Deeply-recursive convolutional network for image super-resolution. *2016 IEEE Conference on Computer Vision and Pattern Recognition (CVPR)* (27–30 June 2016), USA.

72 Tai, Y., Yang, J., and Liu, X. (2017). Image super-resolution via deep recursive residual network. *2017 IEEE Conference on Computer Vision and Pattern Recognition (CVPR)* (21–26 July 2017), USA.

73 Dong, C., Loy, C.C., and Tang, X. (2016). Accelerating the super-resolution convolutional neural network. In: *European Conference on Computer Vision (ECCV)*, 391–407. The Netherlands, Springer.

74 Tong, T., Li, G., Liu, X., and Gao, Q. (2017). Image super-resolution using dense skip connections. *2017 IEEE International Conference on Computer Vision (ICCV)* (22–29 October 2017), Italy.

75 Huang, G., Liu, Z., Van Der Maaten, L. et al. (2017). Densely connected convolutional networks. *2017 IEEE Conference on Computer Vision and Pattern Recognition (CVPR)*, USA.

76 Ahn, N., Kang, B., and Sohn, K.-A. (2018). Fast, accurate, and lightweight super-resolution with cascading residual network. In: *European conference on computer vision (ECCV)*, 256–272. Germany: Springer.

77 Han, W., Chang, S., and Liu, D. et al. (2018). Image super-resolution via dual-state recurrent networks. *2018 IEEE/CVF Conference on Computer Vision and Pattern Recognition* (18–23 June 2018), USA.

78 Haris, M., Shakhnarovich, G., and Ukita, N. (2018). Deep back-projection networks for super-resolution. *2018 IEEE/CVF Conference on Computer Vision and Pattern Recognition* (18–23 June 2018), USA.

79 Li, Z., Yang, J., Liu, Z. et al. (2019). Feedback network for image super-resolution. *2019 IEEE/CVF Conference on Computer Vision and Pattern Recognition (CVPR)* (15–20 June 2019), USA.

80 Hu, J., Shen, L., Albanie, S. et al. (2020). Squeeze-and-excitation networks. *IEEE Transactions on Pattern Analysis and Machine Intelligence* 42 (8): 2011–2023.

81 Zhang, Y., Li, K., Li, K. et al. (2018). Image super-resolution using very deep residual channel attention networks. In: *European Conference on Computer Vision (ECCV). ECCV 2018*, 294–310. Germany: Springer.

82 Dai, T., Cai, J., Zhang, Y. et al. (2019). Second-order attention network for single image super-resolution. *2019 IEEE/CVF Conference on Computer Vision and Pattern Recognition (CVPR)* (15–20 June 2019), USA.

83 Zhang, Y., Li, K., Li, K. et al. (2019). Residual non-local attention networks for image restoration. arXiv preprint arXiv:1903.10082.

84 Ledig, C., Theis, L., Huszar, F. et al. (2017). Photo-realistic single image super-resolution using a generative adversarial network. *2017 IEEE Conference on Computer Vision and Pattern Recognition (CVPR)* (21–26 July 2017). HI, USA.

85 Wang, X., Yu, K., Wu, S. et al. (2019). ESRGAN: enhanced super-resolution generative adversarial networks. *Lecture Notes in Computer Science* 11133: 63–79.

86 Lai, W.S., Huang, J.-B., Ahuja, N., and Yang, M.-H. (2019). Fast and accurate image super-resolution with deep Laplacian pyramid networks. *IEEE Transactions on Pattern Analysis and Machine Intelligence* 41 (11): 2599–2613.

87 Shocher, A., Cohen, N., and Irani, M. (2018). Zero-shot super-resolution using deep internal learning. *2018 IEEE/CVF Conference on Computer Vision and Pattern Recognition* (18–23 June 2018), USA.

88 Yuan, Y., Liu, S., Zhang, J. et al. (2018). Unsupervised image super-resolution using cycle-in-cycle generative adversarial networks. *2018 IEEE/CVF Conference on Computer Vision and Pattern Recognition Workshops (CVPRW)* (18–23 June 2018), USA.

89 Nan, Y., Quan, Y., and Ji, H. (2020). Variational-EM-based deep learning for noise-blind image deblurring. *2020 IEEE/CVF Conference on Computer Vision and Pattern Recognition (CVPR)* (14–19 June 2018), USA.

90 Sun, J., Cao, W., Zongben, X., and Ponce, J. (2015). Learning a convolutional neural network for non-uniform motion blur removal. In: *2015 IEEE Conference on Computer Vision and Pattern Recognition (CVPR)*, 769–777. USA: Institute of Electrical and Electronics Engineers (IEEE).

91 Malik, S. and Soundararajan, R. (2020). A model learning approach for low light image restoration. *2020 IEEE International Conference on Image Processing (ICIP)* (19–22 June 2020), USA.

92 Li, Z. and Wu, J. (2019). Learning deep CNN denoiser priors for depth image inpainting. *Applied Sciences* 9: 1103.

93 Cavigelli, L., Hager, P., and Benini, L. (2017). CAS-CNN: a deep convolutional neural network for image compression artifact suppression. *2017 International Joint Conference on Neural Networks (IJCNN)* (14–19 May 2017), USA.

94 Boublil, D., Elad, M., Shtok, J., and Zibulevsky, M. (2015). Spatially-adaptive reconstruction in computed tomography using neural networks. *IEEE Transactions on Medical Imaging* 34 (7): 1474–1485.

95 Jin, K.H., McCann, M.T., Froustey, E., and Unser, M. (2017). Deep convolutional neural network for inverse problems in imaging. *IEEE Transactions on Image Processing* 26 (9): 4509–4522.

96 Kang, E., Min, J., and Ye, J.C. (2017). A deep convolutional neural network using directional wavelets for low-dose X-ray CT reconstruction. *Medical Physics* 44 (10): e360–e375.

97 Chen, H., Zhang, Y., Kalra, M.K. et al. (2017). Low-dose CT with a residual encoder–decoder convolutional neural network. *IEEE Transactions on Medical Imaging* 36 (12): 2524–2535.

98 Wolterink, J.M., Leiner, T., Viergever, M.A., and Isgum, I. (2017). Generative adversarial networks for noise reduction in low-dose CT. *IEEE Transactions on Medical Imaging* 36 (12): 2536–2545.

99 Yang, Q., Yan, P., Zhang, Y. et al. (2018). Low-dose CT image denoising using a generative adversarial network with Wasserstein distance and perceptual loss. *IEEE Transactions on Medical Imaging* 37 (6): 1348–1357.

100 You, C., Cong, W., Wang, G. et al. (2018). Structurally-sensitive multi-scale deep neural network for low-dose CT denoising. *IEEE Access* 6: 41839–41855.

101 Ma, Y., Wei, B., Feng, P. et al. (2020). Low-dose CT image denoising using a generative adversarial network with a hybrid loss function for noise learning. *IEEE Access* 8: 67519–67529.

102 Adler, J. and Oktem, O. (2018). Learned primal-dual reconstruction. *IEEE Transactions on Medical Imaging* 37 (6): 1322–1332.

103 Zhang, H., Dong, B., and Liu, B. (2019). JSR-Net: a deep network for joint spatial-radon domain CT reconstruction from incomplete data. In: *2019 IEEE International Conference on Acoustics, Speech and Signal Processing (ICASSP)*, 3657–3661. UK.

104 Yang, Y., Sun, J., Li, H. et al. (2017). ADMM-Net: a deep learning approach for compressive sensing MRI. arXiv preprint arXiv:1705.06869.

105 Wang, S., Su, Z., Ying, L. et al. (2016). Accelerating magnetic resonance imaging via deep learning. In: *Proceedings of the IEEE International Symposium of Biomedical Imaging*, 514–517.

106 Schlemper, J., Caballero, J., Hajnal, J.V. et al. (2018). A deep cascade of convolutional neural networks for dynamic MR image reconstruction. *IEEE Transactions on Medical Imaging* 37 (2): 491–503.

107 Yang, G., Yu, S., Dong, H. et al. (2018). DAGAN: deep de-aliasing generative adversarial networks for fast compressed sensing MRI reconstruction. *IEEE Transactions on Medical Imaging* 37 (6): 1310–1321.

108 Zhu, B., Liu, J.Z., Cauley, S.F. et al. (2018). Image reconstruction by domain-transform manifold learning. *Nature* 555 (7697): 487–492.

109 Cammarasana, S., Nicolardi, P., and Patanè, G. (2021). A universal deep learning framework for real-time denoising of ultrasound images. arXiv preprint arXiv:2101.09122.

110 Lan, Y. and Zhang, X. (2020). Real-time ultrasound image despeckling using mixed-attention mechanism based residual unet. *IEEE Access* 8: 195327–195340.

111 Qiu, B., Huang, Z., Liu, X. et al. (2020). Noise reduction in optical coherence tomography images using a deep neural network with perceptually-sensitive loss function. *Biomedical Optics Express* 11 (2): 817–830.

112 Fan, W., Yu, H., Chen, T., and Ji, S. (2020). OCT image restoration using non-local deep image prior. *Electronics* 9 (5): 784.

113 Weigert, M., Schmidt, U., Boothe, T. et al. (2018). Content-aware image restoration: pushing the limits of fluorescence microscopy. *Nature Methods* 15 (12): 1090–1097.

114 Chatton M. (2020). Microscopy image restoration using deep learning on W2S. arXiv preprint arXiv:2004.10884.

115 Lee, S., Negishi, M., Urakubo, H. et al. (2020). Mu-net: multi-scale U-net for two-photon microscopy image denoising and restoration. *Neural Networks* 125: 92–103.

116 Pronina, V., Kokkinos, F., Dylov, D.V., and Lefkimmiatis, S. (2020). Microscopy image restoration with deep Wiener–Kolmogorov filters. In: *16th European Conference on Computer Vision (ECCV)*, 202–221. Glasgow, UK: Springer.

117 Lalush, D.S. and Tsui, B.M. (2000). Performance of ordered-subset reconstruction algorithms under conditions of extreme attenuation and truncation in myocardial SPECT. *Journal of Nuclear Medicine* 41 (4): 737–744.

118 Kim, H.S., Cho, S.G., Kim, J.H. et al. (2014). Effect of post-reconstruction Gaussian filtering on image quality and myocardial blood flow measurement with N-13 ammonia PET. *Asia Oceania Journal of Nuclear Medicine and Biology* 2 (2): 104–110.

119 Liu, J., Yang, Y., Wernick, M.N., et al. (2021). Improving diagnostic accuracy of reduced-dose studies with full-dose noise-to-noise learning in cardiac SPECT. *2021 IEEE 18th International Symposium on Biomedical Imaging (ISBI)*, 1173–1176 (13–16 April 2021).

120 Tang, Z., Peng, X., Li, K., and Metaxas, D.N. (2020). Towards efficient U-Nets: a coupled and quantized approach. *IEEE Transactions on Pattern Analysis and Machine Intelligence* 42 (8): 2038–2050.

121 Gong, K., Guan, J., Kim, K. et al. (2019). Iterative pet image reconstruction using convolutional network representation. *IEEE Transactions on Medical Imaging* 38 (3): 675–685.

122 Nie, D., Trullo, R., Lian, J. et al. (2018). Medical image synthesis with deep convolutional adversarial networks. *IEEE Transactions on Biomedical Engineering* 65 (12): 2720–2730.

123 Hashimoto, F., Ohba, H., Ote, K. et al. (2021). 4D deep image prior: dynamic pet image denoising using an unsupervised four-dimensional branch convolutional neural network. *Physics in Medicine and Biology* 66 (1): 015006.

124 Sun, H., Peng, L., Zhang, H. et al. (2021). Dynamic pet image denoising using deep image prior combined with regularization by denoising. *IEEE Access* 9: 52378–52392.

125 Mataev, G., Milanfar, P., and Elad, M. (2019). DeepRED: deep image prior powered by RED. *Proceedings of the IEEE/CVF International Conference on Computer Vision Workshops* (11–17 October 2021), Seoul, Korea.

6

Machine and Deep-Learning Techniques for Image Super-Resolution

Ashish Kumar, Sachin Srivastava, and Pratik Chattopadhyay

Department of Computer Science and Engineering, Indian Institute of Technology (Banaras Hindu University), Varanasi, India

6.1 Introduction

Image super-resolution (SR) is a class of image processing methods that aims at upscaling an image, thereby improving its resolution. The key idea of super-resolution is to produce a high-quality image from its equivalent low-quality image. To date, researchers have developed several methods for image super-resolution, from using simple bicubic interpolation (BI) to using highly sophisticated generative adversarial network (GAN) models. The image super-resolution problem has broad applicability in the areas of medical imaging, forensics, security, surveillance, imaging by satellite, etc. Due to the advancement of deep-learning [1–3], new and more effective approaches toward solving this problem have emerged. In super-resolution imaging, a high resolution (HR) image is generated from an input low resolution (LR) counterpart. In general, the LR image I_l can be written as the degradation of the HR image I_h, as shown in Eq. (6.1):

$$I_l = H(I_h; \eta) \tag{6.1}$$

where H refers to the function used for degradation, and η refers to the parameters used for degrading the HR image.

Figure 6.1 shows a schematic diagrammatic representation of the single image super-resolution framework. Given I_l as input, the process of super-resolution aims to obtain an approximation I'_h of the ground truth I_h. Mathematically, this is stated in Eq. (6.2).

$$I'_h = M(I_l; \theta) \tag{6.2}$$

where the super-resolution model is denoted by M, and the model parameters are denoted by θ. Then, the objective of the super-resolution model M is given by Eq. (6.3)

$$\theta' = \mathrm{argmin}_\theta \, L(I'_h, I_h) + \lambda \psi(\theta) \tag{6.3}$$

where $L(I'_h, I_h)$ refers to the loss function, λ denotes the parameter controlling relative importance of the $\psi(\theta)$ term, and $\psi(\theta)$ denotes regularization. However, it may be noted that the task of super-resolving an image is not a straightforward problem since there can be many HR images for a single LR image.

The goals of this chapter are enlisted as follows:

- Learn about single image super-resolution and its practical applications
- Understand various traditional, machine-learning, and deep-learning approaches to image super-resolution
- Understand how deep-learning techniques produce better results than traditional techniques
- Evaluate the performance of various existing techniques

Machine Learning Algorithms for Signal and Image Processing, First Edition.
Edited by Deepika Ghai, Suman Lata Tripathi, Sobhit Saxena, Manash Chanda, and Mamoun Alazab.

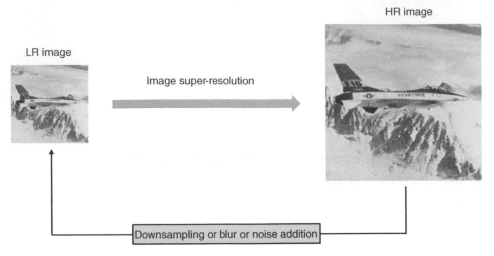

Figure 6.1 Single image super-resolution framework. Source: SDASM Archives/Flickr.

6.1.1 Motivation

Image super-resolution is nowadays a standard feature that can be found in various smartphones and cameras. The intrinsic details of an LR image may not be visible clearly when zoomed in. However, super-resolution enables us to produce an image with more pixels within a given height and width. As a result, the LR image gets transformed into an HR image, and the details can be viewed clearly even when it is significantly zoomed in. The desired resolution of an image depends on the dimensions in which the image has to be displayed.

One of the earliest smartphones with this feature was the Oppo Find 7, which has an ultra-HD multi-shot feature that is capable of producing 50-megapixel images from a 13-megapixel camera. Some smartphones like the Google pixel series have a super res zoom feature [4], using which one can get more details in the picture by zooming in before pressing the shutter button. Latest smartphones such as Redmi 8 pro, OnePlus series phones, etc., have UltraShot super-resolution feature that extracts key features from multiple images and combines them to get a super-resolution image.

The photo-editing software like Adobe Photoshop, GNU Image Manipulation Program, etc., also employs traditional image upsampling techniques like the nearest neighbor, bilinear, or bicubic interpolation to create better resolution images. Figure 6.2 shows a snapshot of the upsampling methods available in GIMP, an open-source photo editing software. These perform image super-resolution by combining the sub-pixel details from multiple images captured simultaneously with a slight shift in the position, similar to the technique, which is used by digital cameras and smartphones. This method of combining multiple images is useful only for landscape photography. In the case of portrait images or moving images, the moving entity would become hazy or blurred after super-resolution. In those cases, more advanced deep-learning techniques can be used to get better resolution images from a single still image instead of combining multiple images. Hence, the super-resolution techniques seem to have a lot of importance in the commercial retouching software or other available products.

6.1.2 Applications of Image Super-Resolution

Super-resolution has many real-world uses in a variety of areas. Some of its applications are described in Sections 6.1.2.1–6.1.2.4.

Figure 6.2 Traditional upsampling methods available in GIMP, photo-editing software. Source: GIMP.

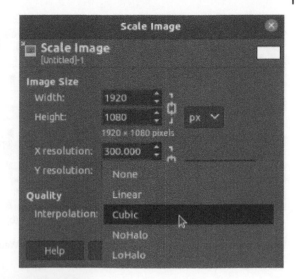

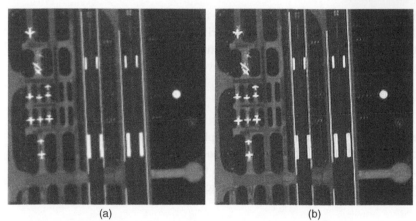

(a)   (b)

Figure 6.3 Application of SR on remote sensing data. (a) Bilinearly interpolated image and (b) super-resolved image. Source: Zhang et al. [5], MDPI,CC BY 3.0.

6.1.2.1 Satellite Imaging

Super-resolution imaging has been used to enhance the image resolution captured by satellites. The earliest example of this method is the enhancement of images obtained by the Landsat 4 satellite. Over the last few decades, the technique of super-resolution imaging has been extensively used in astronomical observations or in obtaining information about an area or an object remotely. One such application of remote sensing data is shown in Figure 6.3, in which the intrinsic details of the remotely sensed image become clear in the super-resolved image. Super-resolution can aid scientific organizations in space exploration by increasing the resolution of the images of astronomical objects or celestial events.

6.1.2.2 Medical Diagnosis

The medical imaging technique is used to acquire anatomical information by obtaining images of the interior of a body. Super-resolution is used in medical image processing, such as ultrasound imaging, positron emission

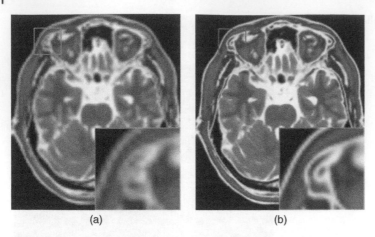

(a)   (b)

Figure 6.4   Application of SR on brain MRI scan (BrainWeb database). (a) Bicubic image and (b) super-resolved image. Source: Du and He [6], MDPI,CC BY 4.0.

tomography (PET), X-rays image enhancement, and magnetic resonance imaging (MRI). Figure 6.4 shows the applications of super-resolution on MRI scan, again showing the superiority of image super-resolution over the traditional approaches. In general, the images obtained by the image capturing equipment are degraded due to resolution limitations. So, super-resolution imaging can be applied to improve the clarity of these degraded medical images.

6.1.2.3   Surveillance

The super-resolution techniques are used in surveillance and security applications to obtain fine details from a scene, e.g. through sign or number plate reading. These techniques are also employed in biometric identification methods, including iris recognition, fingerprint image enhancement, and facial image enhancement. Although there have been progressive developments in these techniques, their practical application is still a challenge.

6.1.2.4   Video Enhancement

The super-resolution technique can also be employed to obtain a high-quality video from a standard quality video. Many researchers have developed traditional and deep-learning models to achieve this goal. Certain telecommunications companies make use of this approach to provide the users an HD view of a standard definition television. The application of SR techniques in video enhancement has been very effective in recent years.

6.1.3   Major Contributions and Organization of the Chapter

The main contributions of this chapter are as follows:

- We have given a thorough overview of the various machine and deep-learning approaches to image super-resolution as well as the trend of research work in this domain.
- We have also discussed a few practical applications of image super-resolution so that interested readers can work on any such application and extend the research work done in that area.
- We have made a high-level categorization of the existing approaches, made a comparative study among some recent approaches, and also pointed out future scopes of research.
- We have described various datasets and the performance metrics used in research on image super-resolution that will also help the researchers interested in working on this topic.

The rest of the chapter is organized as follows. In Section 6.2, we describe the traditional upsampling approaches and the primitive machine-learning-based approaches to image super-resolution. Next, in Section 6.3 we present an overview of the primitive machine- learning-based approaches to image super-resolution. Following this, the recently developed deep-learning-based approaches have been detailed in Section 6.4, with a focus on the recently proposed convolutional neural network (CNN) and GAN-based approaches. Thereafter, in Section 6.5 we present a thorough evaluation of these methods using the objective and subjective image quality assessment metrics and perform a comparative analysis of these techniques. Finally, in Section 6.6 we summarize the chapter and present the future research scopes in the domain of image super-resolution.

6.2 Traditional Upsampling Approaches

In a particular stage of super-resolution, the dimensions of the LR image are increased to get an upsampled LR image. Depending on the model of the SR framework, this upsampled image can be fed to the CNN for reconstruction. The upsampling of an LR image can be done either during the initial stage of the SR process, or the final stage, or progressively, or even iteratively. This section describes briefly the various traditional approaches used for super-resolution.

6.2.1 Nearest Neighbor Interpolation

Nearest neighbor interpolation is one of the straightforward and intuitive interpolation algorithms. This algorithm selects the intensity value of only that pixel that is nearest to a given unknown pixel, ignoring the other pixels. As a result, although this is a fast form of upsampling, it creates photos that look pixelated due to preserving the information by just duplicating the pixels. Deliberate pixelation can be used to obscure a person's identity, otherwise, for image super-resolution tasks; it produces a low-quality image. The advantage of this interpolation method is that owing to a low computational cost, it can be used in real-time situations, where the quickness of the super-resolution is given more importance than the performance of the method.

6.2.2 Bilinear Interpolation

Bilinear interpolation [7] is a step forward from nearest-neighbor interpolation that takes into consideration more pixels in addition to the single nearest neighbor, during the upsampling operation. This method assumes that the unknown pixel value is not a duplicate of the nearest pixel, rather it is a linear interpolation of them. It first interpolates on one of the image's axes linearly and then extends it to a 2D image by performing interpolation on the other axis. Equivalently, the value of an unknown pixel is determined by averaging the pixel values of the closest 2×2 pixels. This results in quadratic interpolation with four pixels instead of one and usually performs better than the nearest neighbor interpolation without increasing the computational cost significantly. Figure 6.5 shows the comparison between the nearest neighbor and bilinear interpolation methods for upsampling a 3×3 image.

6.2.3 Bicubic Interpolation

The bicubic interpolation [8] method is a level up from bilinear interpolation. Rather than linearly interpolating, it assumes that the intensity of the unknown pixel lies on a smooth curve passing through the nearby pixels. Thus, it conducts a cubic interpolation on one dimension and then does a cubic interpolation on the other, thereby taking into consideration 4×4 pixels. As a consequence, this interpolation performs better than bilinear interpolation. However, the difference between them is not noticeable for small magnifications. It removes the artifacts created

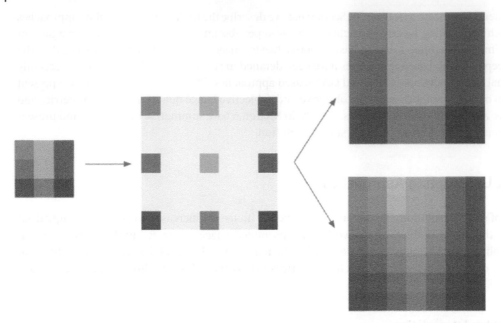

Figure 6.5 Upsampling a 3×3 image using nearest-neighbor interpolation (top), and bilinear interpolation (bottom).

by the nearest neighbor or bilinear interpolation, but it is much slower than those approaches. When converting HR images to LR images, bicubic interpolation and anti-aliasing (smoothing of edges and colors) are often used. It is also often used in the super-resolution process to upsample the image. Readers may refer to the survey paper [9] for further details on the above techniques.

6.3 Primitive Machine-Learning-Based Approaches

Since the last decade, many researchers have been working on different approaches to image super-resolution. Most of the earlier contributions toward the task were based on primitive machine learning. Figure 6.6 shows a high-level representation of this machine- learning framework. The machine-learning-based approaches can be broadly divided into frequency and spatial domains, which are described in Sections 6.3.1 and 6.3.2.

6.3.1 Frequency Domain

For super-resolution tasks, several researchers employ the frequency domain information. In this approach, the LR image is first converted into frequency domain with the help of suitable transformations (typically, wavelet transform and fast Fourier transform (FT)) and then the HR image is constructed. The HR image is eventually converted

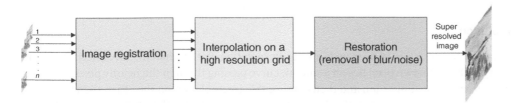

Figure 6.6 Block diagram of the machine-learning framework for image super-resolution. Source: SDASM Archives/Flickr.

back into the spatial domain after construction to produce the final output. There are two major algorithms in the frequency domain approach depending on the method employed for transformation. These are described in Sections 6.3.1.1 and 6.3.1.2.

6.3.1.1 Fast Fourier Transform

The reconstruction method is based on the following properties of the Fourier transform (FT):

(i) Shifting property,
(ii) Band-limitation of the HR image, and
(iii) Anti-aliasing relation of the discrete Fourier transform (DFT) and the continuous Fourier transform (CFT).

Borman and Stevenson [10] propose that a relationship can be formed between the DFT coefficients of the aliased images and the CFT of the unknown scene using the above properties. By using this relation, frequency-domain coefficients of the original image can be obtained which can be recovered by the application of inverse DFT.

This approach was used on the photos collected by the Landsat 4 satellite. From a sequence of K globally translated photographs f_k of the same area of the planet, the satellite produced a continuous scene f. This yields $f_k(x, y) = f(x + \Delta x_k, y + \Delta y_k)$, where $k = 1, 2, ..., K$. The shifting property of the Fourier transform can be employed to account for these shifts, as shown in Eq. (6.4).

$$F_k(u, v) = e^{i2\pi(u\Delta x_k + v\Delta y_k)} \cdot F(u, v) \tag{6.4}$$

where $F(u, v)$ and $F_k(u, v)$ denote the CFT of $f(x, y)$ and $f_k(x, y)$, respectively. The discrete samples of the continuous LR image f can be sampled with periods T_x and T_y to obtain the sampled images $g_k(m, n) = f(mT_x + \Delta x_k, nT_y + \Delta y_k)$ where $m = 0, 1, ..., M - 1$ and $n = 1, 2, ..., N - 1$. Let $G_k(k, l)$ denote the DFT of the image $g_k(m, n)$. Then, the DFT G_k of the shifted and the sampled image, and the CFT F_k of the image can be related by Eq. (6.5).

$$G_k(k, l) = \frac{1}{T_x} \frac{1}{T_y} \sum_{p=-\infty}^{\infty} \sum_{q=-\infty}^{\infty} F_k \left(\frac{k}{MT_x} + p\frac{1}{T_x}, \frac{1}{NT_y} + q\frac{1}{T_y} \right) \tag{6.5}$$

Assuming that $f(x, y)$ is band-limited, Eq. (6.5) is written in the matrix form as Eq. (6.6).

$$G = \phi F \tag{6.6}$$

where ϕ is a matrix that connects the DFT of LR images, denoted by G with the CFT of an HR scene, denoted by F. Hence, the super-resolution process can be summarized as obtaining the DFTs of K LR images, constructing the relation matrix ϕ, then using the Eq. (6.6) to obtain the CFT F, and finally producing the super-resolved image using inverse DFT.

6.3.1.2 Wavelet Transform

The wavelet transform is another super-resolution method based on the frequency domain approach, and it finds many applications in the field of super-resolution. Several tiny patches in the image have been observed to repeat themselves on the same scale as well as across scales. This observation has been used widely for various image enhancement techniques like image denoising, edge detection, etc., and is referred to as self-similarity. It may also be used to break down an image into structurally correlated sub-images in a learning-based image super-resolution task.

Demirel and Anbarjafari [11] propose a method where the input image is first broken down into different sub-bands, such as the low low, low high, high low, and high high by using the discrete wavelet transform (DWT). Let us denote these sub-bands as LL, LH, HL, and HH, respectively. Then, the last three sub-bands with high frequency, namely LH, HL, and HH, along with the input image are interpolated using the bicubic interpolation. The interpolated sub-bands are then improved by adding the result of the stationary wavelet transform (SWT) of their corresponding high-frequency sub-bands. Finally, the HR image is generated by combining the interpolated input image and the high-frequency sub-bands using an inverse discrete wavelet transform (IDWT).

Since the interpolation of high-frequency sub-band images and the insertion of SWT of their corresponding sub-band images retain much of the high-frequency elements, the HR image produced using this method has sharper edges than the direct interpolation of the input signal. Although the wavelet-based approach is theoretically simple, it has been observed that it is difficult to efficiently implement the degraded convolution filters, which is easier in the Fourier transform. Therefore, both approaches are often combined to obtain better results.

6.3.2 Spatial Domain

The spatial domain of an image defines the actual spatial coordinates of the pixels of the image. Hence, the image super-resolution in the spatial domain is based on the direct manipulation of pixels. The spatial domain of an RGB image is represented by a three-dimensional vector consisting of two-dimensional matrices, each of which denotes the color intensity values for red, green, and blue color respectively. There are some problems with the frequency domain since Fourier transform assumes uniformly-spaced samples, so the inter-frame motion is constrained to be translational. Also, the prior knowledge cannot be easily incorporated into the frequency domain, which could have been used to regularize the super-resolution task. These problems of expressing the prior knowledge and nonrestricted motion between the frames are solved by the spatial domain. The spatial domain includes algorithms such as interpolation-based, iterative back projection, and maximum a posteriori (MAP) which are described as follows.

6.3.2.1 Iterative Back Projection

Iterative back projection is one of the earliest methods for spatial super-resolution. It starts with an input LR image, which generates the guess of an initial HR image using some method. One of the most straightforward approaches for doing this is to map the pixels of the LR image on the HR image matrix and then take the average of the nearby pixels. The initial HR image is then refined over several iterations to get the final HR image. To refine the initial HR image f_0, some imaging model is utilized to compress and reduce the dimension of the HR image to generate the simulated LR images, $g_i^{(k)}$, $k = 1, \ldots, K$, where the ith iteration is denoted by the symbol i. The error between the actual and the simulated LR images is determined, and then it is back projected to the previous HR image to refine it further. The procedure is performed over a certain amount of iterations before the HR image shows no further change.

Mathematically, the HR image in the $(i+1)$th iteration can be obtained by Eq. (6.7).

$$f_{i+1} = f_i + c \sum_{k=1}^{K} \left(g^{(k)} - g_i^{(k)} \right) \times h^{\text{BP}} \tag{6.7}$$

where f_i is the HR image generated in the ith iteration $g_i^{(k)}$ is the kth LR image generated in ith iteration, $g^{(k)}$ is the kth original LR image, h^{BP} is the back projection kernel, and c is a constant that normalizes the effect of projection. In Eq. (6.7), replacing the back-projection error with median instead of mean can speed up the algorithm. One of the demerits of this approach is that the number of iterations may not converge, but oscillate between the possible solutions. This issue can be solved if we have an a priori knowledge of the solution. In this case, an extra term $\lambda \parallel \rho(f) \parallel^2$ can be added while calculating the error, where ρ and λ are the constraint and the regularization parameters, respectively. Moreover, replacing the $L_2 - \text{norm}$ by the $L_1 - \text{norm}$ in Eq. (6.7) can speed up the algorithm and make it robust against the outliers.

6.3.2.2 Maximum Likelihood Estimation

We estimate the parameters of the SR image using maximum likelihood estimation by optimizing a likelihood function in such a way that the probability of the observed data is maximized. If we suppose that the image has

Gaussian noise with a mean of 0 and a variance of σ^2, we can calculate the likelihood of the observed LR image g_k given an approximation of the SR image f' by Eq. (6.8).

$$P(g_k \mid f') = \prod_{\forall m,n} \frac{1}{\sigma\sqrt{2\pi}} \exp\left(\frac{g_k'(m,n) - g_k(m,n)}{2\sigma^2} \right) \tag{6.8}$$

and the corresponding log-likelihood function is given by Eq. (6.9).

$$L(g_k) = -\sum_{\forall m,n} \left(g_k'(m,n) - g_k(m,n) \right)^2 \tag{6.9}$$

We maximize this function over all the observed images, represented by Eq. (6.10) to get the maximum likelihood estimate f_{ML} of the SR image as shown in Eq. (6.11).

$$f_{\mathrm{ML}} = \operatorname{argmax}_f \left(\sum_{\forall m,n} L(g_k(m,n)) \right) \tag{6.10}$$

$$f_{\mathrm{ML}} = \operatorname{argmin}_f \left(\| g_k(m,n)' - g_k(m,n) \|_2^2 \right) \tag{6.11}$$

For the super-resolution problem, the resulting maximum likelihood solution can then be obtained by representing Eq. (6.11) in matrix form, and then solving it, to get Eq. (6.12).

$$f_{\mathrm{ML}}' = (A^T A)^{-1} A^T g \tag{6.12}$$

The maximum likelihood solution obtained is sensitive to noises, and in certain cases, it might not give a unique solution. If we have additional constraints, then these issues can be resolved. One of the additional constraints can be the prior knowledge of the desired image, which can then transform this to the MAP estimation problem.

6.3.2.3 Maximum A Posteriori (MAP) Estimation

If we have the prior probability distribution of the SR image, then the MAP optimizer can be used to regularize the estimation of the super-resolution image. Suppose $L(f)$ is the measure of the likelihood of a particular estimate f, then the MAP estimator is given by Eq. (6.13), which can be simplified further to get Eq. (6.14).

$$f_{\mathrm{MAP}} = \operatorname{argmax}_f \left(\sum_{\forall m,n} L(g_k(m,n)) + \lambda L(f) \right) \tag{6.13}$$

$$f_{\mathrm{MAP}} = \operatorname{argmax}_f \left(\sum_{\forall m,n} L(g_k(m,n)) + \lambda \sum_{\forall m,n} \rho(\nabla f(m,n)) \right) \tag{6.14}$$

Various regularization parameters can be used as the prior. Here, we have used the Huber–Markov prior that applies a penalty to the gradient of image ∇f, and the penalty function $\rho(x)$ is the Huber function, given by Eq. (6.15).

$$\rho(x) = \begin{cases} x^2, & x \le \alpha|x| \\ 2\alpha|x| - \alpha^2, & \text{otherwise} \end{cases} \tag{6.15}$$

Since sharp edges carry important information, the penalty function should be lenient toward those edges. The penalty function used above is appropriate as it encourages a piecewise constant solution by having a local smoothness, and it is also more lenient toward sharp edges than the function $\rho(x) = x^2$.

6.3.2.4 Self-Similarity-Based Approach

The wavelet transform is used in this approach to achieve image super-resolution. This approach also aims to exploit the self-similarity of image patches as described in the wavelet transform section. In the self-similarity-based super-resolution technique, similar image patches are scanned across image pyramid scales: within the same image scale at sub-pixel misalignments and across the different image scales. Then the resolution is increased by choosing the best patch within and across the scales for each pixel. Thus, the self-similarity-based super-resolution methods produce good results without having to train on large extensive datasets.

6.3.2.5 Learning-Based Approach

The learning-based approach uses a pair of LR and HR patches to analyze the correlation between a LR image and its HR equivalent. Then, the training data are used to estimate the HR image. In the method proposed by Yang and Yang [12], a significant number of LR patches are obtained from their corresponding HR patches. Using this method, the feature space is split into numerous subspaces and the mapping can be learned by employing simple functions. Since each cluster represents a subspace, it must be ensured that there are sufficient data for each cluster centroid. To learn the mapping function, for each cluster centroid, a large number of exemplar patches are collected. When the test image is fed to the model, it crops each LR patch to extract the features and finds the nearest cluster centroid to predict the corresponding HR patch. The method is fairly effective and computationally efficient.

6.3.2.6 Sparse-Based Approach

This method employs a faster technique to perform the super-resolution task without compromising on the quality. It relies on a glossary of patch-based elements, called atoms, which are used to efficiently construct a representation of image patches. Through this, the super-resolved image can be quickly constructed from the LR patches and the corresponding dictionary. Further, the sparse coding approach by Yang et al. [13] is employed to reduce the size of dictionaries for bigger training samples. Sparse coding uses sparse dictionaries to train the LR and HR dictionaries used to describe LR patches and their HR equivalents. Representing the LR patch sparsely with respect to its LR dictionary helps in explicitly retrieving the equivalent HR patch from the HR dictionary. The method is employed on a large set of input patches with the use of a sparsity constraint. The results obtained from the sparse-based approach can be further improved by the use of back projection.

6.4 Modern Deep-Learning-Based Approaches

Recently, deep learning has been seen to perform superior to machine learning in terms of accuracy when trained extensively on large datasets. Various modern deep- learning-based approaches for super-resolution have also been proposed in recent times. These techniques seem very promising and are capable of attaining state-of-the-art results. Some of them are discussed in Sections 6.4.1 and 6.4.2.

6.4.1 Upsampling-Based Classification

Any super-resolution model framework based on deep learning consists of the phases of upsampling, feature extraction, and image reconstruction. In the upsampling phase, the input LR image is transformed into an interpolated LR image using the upsampling approaches. We described some traditional upsampling approaches in a previous section: nearest neighbor, bilinear interpolation, or bicubic interpolation. Some other upsampling approaches include deconvolution, sub-pixel, or meta upscale. The essential features from the image are extracted in the feature extraction process, and the image restoration phase aids in the reconstruction of the HR image based on the extracted features.

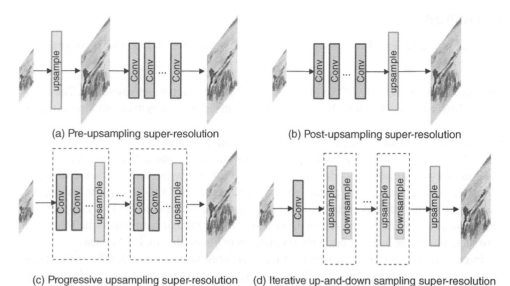

(a) Pre-upsampling super-resolution   (b) Post-upsampling super-resolution

(c) Progressive upsampling super-resolution   (d) Iterative up-and-down sampling super-resolution

Figure 6.7 Upsampling-based classification of the super-resolution frameworks. (a) The image is first upsampled and then refined by the convolutional layers; (b) the image in the LR space is first refined by convolutional layers, and then upsampled; (c) the image is refined and upsampled in various stages to produce the final image; and (d) the image is refined through back projection in stages to learn the mutual dependence, and finally upsampled to get the final image. Source: SDASM Archives/Flickr.

In a super-resolution model, the upsampling process occurs during the initial, final, incremental, or iterative stages, resulting in various SR model structures. Thus, the super-resolution frameworks can be classified into four categories based on the upsampling methodology used. Figure 6.7 shows the diagrammatic representation of the four upsampling-based classification schemes. Each classification has its own set of advantages and drawbacks, as described below:

- **Pre-upsampling Super-Resolution:** Traditional models use a simple approach in which the input LR image is first upsampled to an interpolated LR image, and then refined by the CNN. The intuition for this method is that it is easier to upsample the image to get a high-dimensional image, and then refine that image in the high-dimensional space. However, the pre-upsampling of the image amplifies the noise and may lead to blurring effects in the final image.
- **Post-upsampling Super-Resolution:** In this approach, the feature extraction and other computation are performed in the LR image, and thus it helps in reducing the computational complexity. In the last stage, the upsampling of the image is done by end-to-end learnable layers.
- **Progressive Upsampling Super-Resolution:** The above two approaches cannot produce good results with large scaling factors as it is difficult to learn the mapping for large factors. Also, for different scaling factors, different models are needed to be trained. Hence, the progressive upsampling upsamples the image in stages so that it can easily adapt to large-scale factors and the intermediate images produced could be used as HR images for different intermediary scaling factors. Thus, this approach helps in reducing the learning difficulty by splitting a task into different smaller tasks.
- **Iterative Up and Down Sampling:** This approach helped in refining the image through the back projection at each stage. Iteratively upsampling the image and then down-sampling it helps to deduce the mutual dependence between the HR and LR image. The back-projection stage feeds the reconstruction error back to the image so that the final HR image can be greatly refined.

6.4.2 Network-Based Classification

In this section, we describe various CNN-based image super-resolution approaches developed in recent years.

6.4.2.1 Linear Networks

The linear networks were the earliest and classical CNN-based approaches for image super-resolution. They consist of simple layer-based architecture, as shown in Figure 6.8, and are efficient and lightweight structures for super-resolution tasks.

Super-resolution convolutional neural network (SRCNN) by Dong et al. [14] and fast super-resolution convolutional neural networks (FSRCNNs) by Dong et al. [15] are the popular ones among the initial CNN-based approaches. The SRCNN consists of a simple network with three convolutional layers that learns the mapping between the images. It consists of three parts: (i) the patch extraction and representation, (ii) nonlinear mapping, and finally, (iii) image reconstruction. First, the LR image is pre-upsampled to the interpolated LR image. Then, the first layer extracts the features of the image, the second layer learns the nonlinear mapping to the HR patch, and the third layer reconstructs the image by combining the predictions to obtain the final HR image.

The SRCNN model is improved by the FSRCNN, which introduces more layers in SRCNN for better mapping. Also, it employs a deconvolution layer i.e. an upscaling block at the end of the network to substitute the pre-upsampling process with post-upsampling, allowing the mapping between the images to be learned directly. It uses a smaller filter size with more mapping layers and shrinks the input feature dimension before mapping to reduce the computational cost.

6.4.2.2 Residual Networks

Residual learning can be used to construct very deep networks by making use of skip connections to avoid the vanishing gradient problem. The networks learn the residue, i.e. high frequencies between LR image and its HR counterpart. Figure 6.9 shows the diagrammatic representation of the architecture of residual networks. Some of the recent residual learning-based networks used to perform super-resolution are the enhanced deep super-resolution (EDSR) and the multi-scale deep super-resolution (MDSR).

The EDSR proposed by Lim et al. [16] employs residual scaling techniques to stably train the large network. It is built on the super-resolution residual network (SRResNet) model proposed by Ledig et al. in [17]. It simplifies the residual blocks in SRResNet architecture by eliminating the batch normalization layers and the outside blocks by removing ReLu activation. This method is further expanded to operate on different scales from a single scale.

The MDSR architecture, also proposed by Lim et al. [16], reduces the training time of the process and the model size. The multi-scale paradigm can easily cope up with different scales of super-resolution in a single context, thanks to the scale-based modules and a shared main network. l_1 loss is employed to train both the architectures. The models also employ a geometric self-ensembling technique that involves applying data augmentation

LR image
(input)

SR image
(output)

Figure 6.8 Linear networks for super-resolution. Source: SDASM Archives/Flickr.

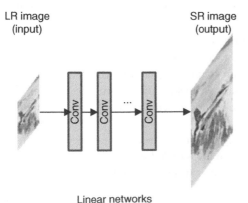

Linear networks

Figure 6.9 Residual networks used to perform super-resolution. Source: SDASM Archives/Flickr.

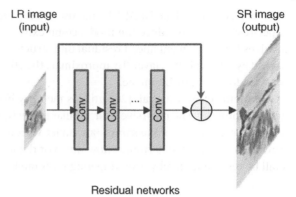

Residual networks

(flips and rotations) to the input images to create augmented images that can be merged to come up with the super-resolved image. The advantage of using a self-ensembling strategy is that it allows us to get results that are comparable to conventional ensemble-based models without having to train on numerous separate models. Compared to the older similar architectures, these models achieve better performance in terms of quantitative estimation.

6.4.2.3 Recursive Networks

Recursive learning is used to learn higher-level features by employing recursively connected convolution layers in the model. The purpose is to minimize the number of parameters by breaking down the more difficult super-resolution problem into smaller, easier-to-solve problems. Adding more weight layers introduces more parameters and can cause overfitting. To overcome this issue, a recursive layer is used in the network such that more recursion can be performed in the recursive layer without increasing the model parameters. Figure 6.10 shows the representation of recursive networks used for image super-resolution. The deep recursive convolutional network (DRCN), the deep recursive residual network (DRRN), and the memory network (MemNet) are some of the recursive networks used in image super-resolution.

The DRCN proposed by Kim et al. [18] employs the recursively connected CNNs in the architecture. Three subnetworks make up this network: (i) the embedding net, (ii) the inference net, and (iii) the reconstruction net. The feature maps are created by the embedding net using the input image. The inference net, the next subnetwork, generates HR feature maps. It inspects the input by adding a single layer of CNN and ReLu recursively. After each recursion, the size of the receptive field increases. Finally, the reconstruction net converts the HR feature maps to super-resolved images.

Figure 6.10 Recursive networks for super-resolution. Source: SDASM Archives/Flickr.

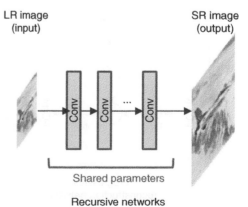

Recursive networks

Tai et al. [19] developed a DRRN that uses a deep architecture with many convolutional layers to perform image super-resolution. To reduce the model complexity, residual learning is combined with recursive learning, which involves learning an enhanced residual unit structure in a recursive block, which are next placed together accompanied by a convolution layer. To approximate the HR image, the residual image from a global identification branch is added to the input LR image.

The MemNet by Tai et al. [20] utilizes memory blocks composed of recursive and gate units for the adaptive learning process. The recursive unit, similar to ResNet, consists of two convolutional layers and it learns different level representations of the same state under various receptive fields. The gate unit, which is made up of a 1×1 convolutional sheet, determines how much of the former state shall be saved and how much of the current state shall be reserved, thereby reserving long-term memory and saving short-term memory.

6.4.2.4 Progressive Reconstruction Networks

In any CNN-based image translation task, we expect the final output to be generated after a single pass through the trained model. However, scaling an LR image directly to large-scale factors, such as 8× or 16× using a CNN is likely to produce an HR image with certain noise and irregularities. Instead, the progressive reconstruction networks progressively build the final image in several steps, first with a scaling factor of 2×, then 4×, and so on, until the desired scaling factor is achieved. In addition to this, the intermediate images of smaller scaling factors are also generated using the same network. Unlike traditional CNN, in which we require different models to generate different upscaled images, the same model with progressive reconstruction design can generate upscaled images of various scaling factors in the intermediate steps. This can be better understood from the visual representation of the progressive reconstruction networks shown in Figure 6.11.

The Laplacian pyramid super-resolution network (LapSRN) by Lai et al. [21] is one of the networks that progressively reconstruct the residuals of the HR image using the Laplacian pyramid at each stage. It has a feature extraction branch for extracting features at each stage and an image reconstruction branch that combines the upsampled picture with the residual image produced from the function extraction branch to generate the HR image. Similarly, the image restoration branch of the next stage receives the output HR image from the previous

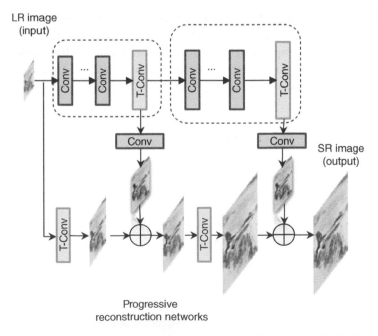

Figure 6.11 Progressive reconstruction networks for super-resolution. Source: SDASM Archives/Flickr.

stage to progressively build the resultant super-resolution image. Finally, it uses a Charbonnier loss function, a differentiable variant of the L_1 loss, to handle the outliers.

6.4.2.5 Densely Connected Networks

In densely connected networks, the idea is to combine the feature maps available along the network depth to obtain richer feature representation, thereby boosting the quality of the reconstructed image. Figure 6.12 shows the architecture of the densely connected networks. Some of the recently proposed densely connected networks are the super resolution dense network (SRDenseNet) and the residual dense network (RDN).

The SRDenseNet by Tong et al. [22] makes use of the dense skip connections between the layers. This allows the output of all previous layers to operate directly on the subsequent layers. The network consists of a convolution layer, blocks of the DenseNet architecture, deconvolution layer, and the reconstruction layer. The dense connections improve the transfer of information to the high-level feature layers and help in alleviating the vanishing gradient problem. Each layer's feature maps are propagated into subsequent layers, eliminating unnecessary features and allowing high-level and low-level features to be combined effectively. Three variations of SRDenseNet are proposed by the authors. The first model uses a dense block arrangement accompanied by deconvolution layers to generate the HR image utilizing only the high-level elements. The second model combines the high-level features with low-level features obtained from the initial layers by using skip connections before constructing the HR image. In the third model, numerous skip associations between the dense blocks and the low-level features are used to merge all the features. This helps in improving the information flow and yields the best results among the other variants.

The RDN by Zhang et al. [23] learns local patterns by collecting hierarchical features from the LR image using residual dense blocks (RDBs). It combines residual skip connections from the SRResNet and the dense connections from the SRDenseNet. At both the local and global levels, residual connections are utilized. At the local level, the input of each block is added to the block's output and also forwarded to all the layers within the block. This approach allows each block to focus more on residual patterns. Moreover, in each RDB, a local feature fusion method with 1×1 convolution is employed to decrease the dimensions. This is because the dense connections soon lead to high dimensional outputs. Finally, at the global level, the features from multiple RDBs are fused in a global feature fusion method.

6.4.2.6 Attention-Based Networks

In general, for the super-resolution image, the spatial positions and channels are of varying significance. Attention-based networks provide the flexibility to selectively choose the features at any given layer. Figure 6.13 shows the diagrammatic representation of the attention-based networks used for image super-resolution.

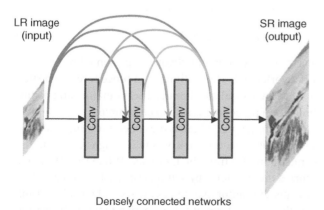

Figure 6.12 Densely connected networks used in image super-resolution. Source: SDASM Archives/Flickr.

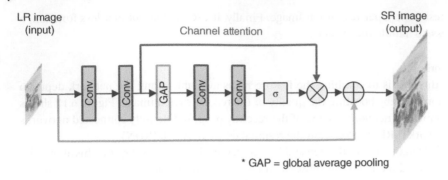

Figure 6.13 Attention-based networks for super-resolution. Source: SDASM Archives/Flickr.

The residual channel attention network (RCAN) by Zhang et al. [24] uses an attention-based technique for super-resolution. RCAN builds a deep network of multiple residual groups linked together by skip connections by making use of the residual in residual (RIR) framework. The idea is to allow multiple pathways for the flow of information between initial and final layers. Thus, inside every block of a global residual network, the residual connections are present. The RIR discards the low-frequency data so that the main network can concentrate on the high-frequency data. Moreover, a channel attention function is often used by each local residual block. By considering interdependencies across channels, the channel attention mechanism can be used to adaptively rescale channel-wise features. This enables the model to concentrate on the most relevant function maps. It is observed that the RCAN achieves superior results in comparison to the other contemporary approaches.

6.4.2.7 GAN-Based Networks

GANs are introduced by Goodfellow et al. [25], which have become a major success in the deep-learning field since then. GANs consist of two networks: the generator and the discriminator, that compete or interact with each other in a min–max game. As an analogy, the generator is like a fraudster that generates fake images, and a discriminator is like a police that tries to differentiate between the real and the fake images. In the super-resolution task, the generator and the discriminator play the same min–max game, in which the role of the generator is to create a super-resolution image such that a discriminator cannot distinguish between this image and the real image. Figure 6.14 shows an image super-resolution framework using GAN-based networks. In the game theory scenario, the objective function $V(G_\theta, D_\phi)$ given by Eq. (6.16) can be used to train a GAN.

$$\min_{G_\theta} \max_{D_\phi} V(G_\theta, D_\phi) = E_{x \sim p_{\text{data}}(x)}[\log D_\phi(x)] + E_{z \sim p_G(z)}[\log(1 - D_\phi(G_\theta(z)))] \tag{6.16}$$

Here, x denotes the HR image, and z denotes the LR image. The two neural networks represented by D and G update alternatively, keeping the parameters of the other network fixed. This optimization procedure is repeated until a state arises when the generator can generate images that are so realistic that the discriminator can't distinguish them from the actual images. At this stage, the optimization procedure is said to be complete.

The discriminator maximizes this objective function V w.r.t parameter ϕ, whereas the generator minimizes this function w.r.t parameter θ. The discriminator performs a binary classification task, where it assigns class 1 to the HR image from the dataset, and assigns class 0 to the generated samples from the LR image. When the HR image is taken from the training dataset, i.e. $x \sim p_{\text{data}}(x)$, it maximizes the first term by setting $D_\phi(x) \approx 1$, whereas when the SR image is generated by the generator, i.e. $z \sim p_G(z)$, the discriminator maximizes the second term by setting $D_\phi(G_\theta(z)) \approx 0$. On the other hand, the generator minimizes the second term of this objective function by setting

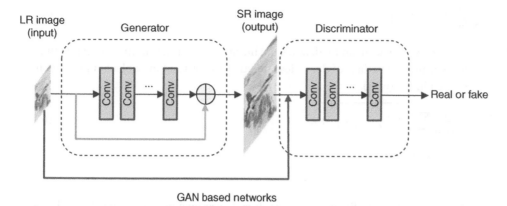

LR image (input)   Generator   SR image (output)   Discriminator   Real or fake

GAN based networks

Figure 6.14 GAN-based networks for super-resolution. Source: SDASM Archives/Flickr.

$D_\phi(G_\theta(z)) \approx 1$, and thus fools the discriminator into predicting the generated SR image as the real HR image from the dataset.

Some of the recent GAN-based networks used for image super-resolution are the super-resolution generative adversarial networks (SRGANs) by Ledig et al. [17], the EnhanceNet by Sajjadi et al. [26], the enhanced super-resolution generative adversarial networks (ESRGANs) by Wang et al. [27]. The SRGAN uses the same adversarial objective function described above and uses multitask loss formulation, namely, the perceptual loss that consists of adversarial loss (GAN objective) and content loss (Euclidean distance between SR and HR image), and the mean squared error (MSE) loss for computing pixel-wise dissimilarity. Mean opinion score (MOS) is used for evaluating the effectiveness of this network in test scenarios. The EnhanceNet introduces perceptual loss on the pre-trained network's intermediate function representation, and the texture transfer loss to balance the texture of the HR and LR images. The ESRGAN enhances the SRGAN by using an RIR dense block instead of batch normalization layers, and by using features before activation to reduce perceptual loss.

6.4.3 Discussion on Different Types of Loss Functions

Loss functions are used to measure the error in reconstruction when the model generates an HR image. A model gets trained in multiple epochs by optimizing certain predefined loss functions. Previously, researchers usually employed L_2 loss function while training the models. Further study revealed that L_2 loss alone is not sufficient to compute a difference between the real image and the generated images. Later, several other loss functions have been developed for training deep neural network models more effectively. A few important such loss functions are described next.

6.4.3.1 Pixel Loss

Pixel loss is one of the standard loss functions used by most super-resolution models. As the name suggests, this loss function measures the pixel-by-pixel difference between the two images: the original HR image and the produced super-resolved image. The pixel-wise difference can be computed broadly in two ways giving rise to the two pixel-loss functions, namely the L_1 loss and the L_2 loss. The L_1 loss provides the mean absolute error, whereas the L_2 loss provides the MSE between the pixel intensities of the two images. Let I' be the reconstructed image and I be the ground truth, and then mathematically, these can be represented by Eqs. (6.17) and (6.18).

$$L_{\text{pixel}-L_1}(I', I) = \frac{1}{hwn} \sum_{i,j,k} |I'_{i,j,k} - I_{i,j,k}| \tag{6.17}$$

$$L_{\text{pixel}-L_2}(I', I) = \frac{1}{hwn} \sum_{i,j,k} \left(I'_{i,j,k} - I_{i,j,k} \right)^2 \tag{6.18}$$

where h denotes the height, w denotes the width, and n denotes the number of channels in the ground truth I.

The Charbonnier loss function is another variant of the L_1 loss function that is used in Laplacian pyramid networks. It is mathematically defined in Eq. (6.19).

$$L_{\text{pixel}-\text{Charbonnier}}(I', I) = \frac{1}{hwn} \sum_{i,j,k} \sqrt{\left(I'_{i,j,k} - I_{i,j,k} \right)^2 + \varepsilon^2} \tag{6.19}$$

where ε is a small nonzero hyper parameter.

6.4.3.2 Content Loss

Content loss is used to assess the perceptual quality of the image. It is determined by using a pre-trained image classification model to compute the semantic dissimilarity between the original and the generated super-resolved image. The model extracts the features and forms high-level representations of both images, and the content loss is determined by taking the Euclidean distance between these high-level representations. Let f denote the image classification model, I' be the reconstructed image and I be the ground truth, then this can be mathematically represented in Eq. (6.20).

$$L_{\text{content}}(I', I; f, t) = \frac{1}{h_t w_t n_t} \sqrt{\sum_{i,j,k} \left(f^{(t)}_{i,j,k}(I') - f^{(t)}_{i,j,k}(I) \right)^2} \tag{6.20}$$

where $f^{(t)}(I)$ is the high-level representation on tth layer, and h, w, and n are height, width, and the number of channels in I. Unlike pixel loss, content loss relies on visual similarities between the output image and the ground reality rather than pixel-wise similarity. Therefore, it finds wide use in the field of super-resolution.

6.4.3.3 Texture Loss

The texture of an image provides knowledge about the spatial arrangement of colors, intensities, or contrast around a region within the image. Texture loss is employed to measure the quality of texture reconstruction by the model. As described in Wang et al. [28], the image texture can be quantified using the Gramian matrix $G^{(t)} \varepsilon R^{n_t \times n_t}$, where n_t denotes the number of feature maps in the tth layer, and $G^{(t)}_{ij}$ denotes the inner product between the feature maps i and j on tth layer, as shown in Eq. (6.21).

$$G^{(t)}_{ij}(I) = \text{dot} \left(\text{vec} \left(f^{(t)}_i(I) \right), \text{vec} \left(f^{(t)}_j(I) \right) \right) \tag{6.21}$$

where vec(.) is the vectorization operation, $f^{(t)}_i(I)$ is the ith channel of the feature maps on the tth layer of the image I, and dot(.) refers to the dot product. Mathematically, the texture loss can be represented by Eq. (6.22).

$$L(I', I; f, k) = \frac{1}{n_t^2} \sqrt{\sum_{i,j} \left(G^{(t)}_{ij}(I') - G^{(t)}_{ij}(I) \right)^2} \tag{6.22}$$

6.4.3.4 Adversarial Loss

The adversarial loss function is used for adversarial training in GANs. Adversarial learning is a relatively new training method that has been extensively used in several image translation tasks including image super-resolution. Here, the SR model is used as the generator G, which is trained to generate HR images. Similarly, a discriminator D distinguishes between the real image sampled from dataset I and the generated image I'. These two steps are alternately done during the training phase: first, the generator is trained to generate real-looking images and the discriminator is kept fixed, and then, the generator is fixed, and the discriminator is used to distinguish the images. This results in a min–max game in which the generator tries to reduce its loss L_G and the discriminator minimizes the discriminator loss L_D. The loss functions for the generator and discriminator are formulated in Eqs. (6.23) and (6.24).

$$L_G(I'; D) = -\log D(I') \tag{6.23}$$

$$L_G(I', I; D) = -\log D(I) - \log(1 - D(I')) \tag{6.24}$$

where I' is given as $G(z)$ where G is the generator and z is the input to the generator.

Generally, it is observed that the peak signal-to-noise ratio (PSNR) score of super-resolution images trained using the content loss and adversarial loss is lower than that of images trained using the pixel loss. Yet, the perceptual quality of the images trained using the adversarial loss is found to be significantly better. Therefore, it is a suitable loss function to be used when the perceptual quality of the images is considered important.

6.5 Performance Metrics and Comparative Study of Existing Techniques

6.5.1 Objective Evaluation

Several assessment criteria have been introduced to measure image quality. These metrics can be used to objectively assess the image quality based on a predefined function. Some of the evaluation metrics are as follows.

6.5.1.1 Peak Signal-to-Noise Ratio (PSNR)

MSE is one of the trivial measures of image quality, represented in Eq. (6.25).

$$\text{MSE} = \frac{1}{mn} \sum_{i=0}^{m-1} \sum_{j=0}^{n-1} [I'(i,j) - I(i,j)]^2 \tag{6.25}$$

where m and n are the image dimensions. However, a limitation of MSE is that it is strongly dependent on the scale of the image intensity. E.g.: For an 8-bit image, MSE of 100 would be quite high, whereas the same MSE value for a 16-bit image would be low enough. As an alternative, the PSNR metric can be used that scales the MSE value according to the image range.

PSNR is defined as the ratio of the maximum potential power of a signal to the power of the noise and is expressed in decibels. In calculating the PSNR value for image quality assessment, the maximum possible power is given by L^2 ($L = 255$ for an 8-bit image), and the noise strength is determined by the MSE. Thus, the PSNR can be calculated by Eqs. (6.26) and (6.27).

$$\text{PSNR} = 10 \times \log_{10} \left(\frac{MAX_I^2}{\text{MSE}} \right) \tag{6.26}$$

$$\text{PSNR} = 10 \times \log_{10} \left(\frac{L^2}{\frac{1}{mn} \sum_{i=0}^{m-1} \sum_{j=0}^{n-1} [I'(i,j) - I(i,j)]^2} \right) \tag{6.27}$$

where I is an $m \times n$ image, L is the maximum possible pixel intensity value in I, and I' is the reconstructed image.

PSNR is widely used by researchers for estimation. It measures the reconstruction quality of a lossy transformation. However, it is not an ideal metric as it can only compare the restoration results of the same image. But comparing the PSNR value between two different images is completely meaningless. Also, PSNR is more concerned with the pixel-level similarity rather than the perceptual quality. Hence, despite its popularity, it might not always be suitable and can fail in some cases.

6.5.1.2 Structural Similarity Index Measure (SSIM)

The visual discrepancy between the restored image and the ground reality is measured using structural similarity index measure (SSIM). It compares three parameters, viz. luminance (l), contrast (c), and structure (s), to determine structural similarity. Let I be an image containing n pixels, and I' be the reconstructed image. Then, the luminance of the image is given by Eq. (6.28), contrast is given by Eq. (6.29), and structure is given by Eq. (6.30).

$$l(I', I) = \frac{2\mu_{I'}\mu_I + c_1}{\mu_{I'}^2 + \mu_I^2 + c_1} \tag{6.28}$$

$$c(I', I) = \frac{2\sigma_{I'}\sigma_I + c_2}{\sigma_{I'}^2 + \sigma_I^2 + c_2} \tag{6.29}$$

$$s(I', I) = \frac{\sigma_{I'I} + c_3}{\sigma_{I'}^2 \sigma_I^2 + c_3} \tag{6.30}$$

where,

- $\mu_I = \frac{1}{n}\sum_{i=1}^{n} I(i)$ is the mean of image intensity.
- $\sigma_I = \sqrt{\frac{1}{n-1}\sum_{i=1}^{n}(I(i) - \mu_I)^2}$ is the standard deviation of image intensity.
- $\sigma_{I'I} = \frac{1}{n-1}\sum_{i=1}^{n}(I'(i) - \mu_{I'})(I(i) - \mu_I)$ is the covariance between images I' and I.
- $c_1 = (k_1 L)^2$, $c_2 = (k_2 L)^2$, $c_3 = \frac{c_2}{2}$ are the nonzero stability constants.
- L is the range of the pixel values.

SSIM is defined as the weighted combination of the above functions and is given by Eq. (6.31).

$$\text{SSIM}(I', I) = l(I', I)^\alpha \times c(I', I)^\beta \times s(I', I)^\gamma \tag{6.31}$$

where the parameters α, β, γ can be adjusted according to the relative importance of the terms.

Following that, the human vision is receptive to image structures, SSIM performs much better in the perceptual assessment of the reconstructed image than PSNR. As a result, SSIM is extensively used in the area of super-resolution.

6.5.2 Subjective Evaluation

In certain situations, objective assessment metrics might be unable to measure the restored image's perceptual quality. Hence, it may be useful to use subjective evaluation to assess the image quality.

The MOS is a subjective assessment technique that asks people to assess the restored image on a scale of 1 to 5, with 1 being the lowest and a user-defined number (usually, 5 or 10) being the highest. An average of the ratings given by several users provides the final MOS score. Despite being effective, MOS takes a longer time and is more expensive than objective measurement metrics, owing to the involvement of the people.

6.5.3 Datasets

The selection of the dataset is a crucial step in the training and testing of the super-resolution algorithms. Sections 6.5.3.1 and 6.5.3.2 list the various training and testing datasets usually employed for the super-resolution task.

6.5.3.1 Training Dataset

The following datasets are widely used for the training of super-resolution models:

(i) **DIV2K [29]**: It contains 1000 diversified 2K resolution images containing images of people, animals, scenery, objects, etc. Out of the 1000 images, 800 images are present in the training dataset, and it has both LR and HR images for 2×, 3×, and 4× downscaling factors.

(ii) **ImageNet** [30]: It is a large database for object recognition that is structured according to the WordNet hierarchy. It contains over 14 million images with around 20 000 categories, in which the images reflect each node of the hierarchy. For the image super-resolution task, some random samples of 350 000 images are generally used.

(iii) **T91** [31]: An image dataset with 91 images, in which the images are divided into image patches for training purposes.

(iv) **BSD200** [32]: It contains 200 professional photographic images taken from the Berkeley Segmentation Dataset.

(v) **General100** [15]: It contains 100 BMP-format images, without any compression, suited for the image super-resolution task.

Generally, the datasets are used with data augmentation, such as padding, flipping, rotation, and random cropping, so that the models are trained with diversified data without adding any data from other sources. The recent models widely use the DIV2K dataset for training purposes as it contains a large number of diverse images. The initial models for image super-resolution used the T91, BSD200, and General100 datasets.

6.5.3.2 Testing Dataset

The results are generally evaluated on the following five benchmark datasets:

(i) **Set5** [33]: It contains five images of butterfly, baby, bird, woman, and head, usually used for visualization purposes.

(ii) **Set14** [34]: It contains 14 images of animals, vegetables, flowers, comics, etc., therefore has more categories than Set5, so it is sometimes used for manual visualization of the super-resolution task.

(iii) **B100** [32]: It consists of 100 images from the Berkeley Segmentation Dataset. It is generally used for performance evaluation on natural scene images, and it also contains object-specific images like plants, food, etc. Collectively, the dataset is denoted as BSD300, consisting of 200 training images and 100 testing images. It is now expanded to BSD500 with 200 newly added images.

(iv) **Manga109** [35]: It consists of 109 images of Manga volume from Japanese comic books and is generally used for evaluating the performance of the super-resolution task on drawing images.

(v) **Urban100** [36]: It contains 100 HR images of real-world scenes such as city, architecture, and urban and is generally used to investigate self-similarity in the image.

Set5 and Set14 are the widely used datasets as they contain common images, and also, the results for all the images can be visualized manually owing to the small count of images.

6.5.4 Evaluation Results

The simulation results for the models SRCNN, FSRCNN, RDN, SRDenseNet, DRRN, LapSRN, EDSR, MemNet, and SRGAN are mentioned in Table 6.1. The SR results were evaluated using PSNR and SSIM metrics on the five benchmark datasets: Set5, Set14, B100, Urban100, and Manga109 using bicubic down-sampling using bicubic interpolation (BI) as the degradation model. To maintain consistency with the models, the results are evaluated only on luminance (Y) channels with 2×, 3×, and 4× scaling factors.

The results show that the recursive residual networks like DRRN achieve an overall good performance, with very good PSNR and SSIM values. However, good performance generally comes with a slightly high computational cost. RDN achieves the highest PSNR value among the densely connected networks, followed by SRDenseNet, which performs well on both SSIM and PSNR. EDSR also has good performance; however, it is generally not preferred, owing to the high computational cost, large model size, and a large number of parameters. Lastly, even though SRGAN has a low PSNR/SSIM value, it has been found to achieve a higher MOS score than others after sufficient training.

Table 6.1 Average PSNR/SSIM values for the scale factors 2×, 3×, and 4×, evaluated on the five benchmark datasets. The text in bold shows the best performance for each scaling factor, and the text in italics shows the second-best performance for each scaling factor.

Method	Scale	Set5	Set14	B100	Urban100	Manga109
Bicubic	2×	33.71/0.9306	30.25/0.8718	30.98/0.8751	31.31/0.9409	26.90/0.8418
SRCNN	2×	36.65/0.9540	32.34/0.9070	33.05/0.9144	36.14/0.9705	29.39/0.8951
FSRCNN	2×	37.12/0.9568	32.68/0.9108	33.35/0.9177	37.39/0.9740	29.84/0.9021
RDN	2×	**38.18**/0.8817	**33.90**/0.8048	**34.19**/0.8411	**39.50**/0.8715	**32.46**/0.8292
DRRN	2×	*37.57*/**0.9590**	*33.09*/**0.9147**	*33.73*/**0.9226**	*37.93*/**0.9770**	*30.88*/**0.9158**
LapSRN	2×	37.32/*0.9580*	32.92/0.9126	33.53/0.9197	37.19/*0.9753*	30.38/0.9102
MemNet	2×	36.72/0.9542	32.54/*0.9110*	33.10/*0.9184*	36.16/0.9699	30.69/*0.9153*
SRGAN	2×	35.51/0.9428	32.13/0.9009	32.71/0.9102	34.90/0.9620	30.15/0.9055
Bicubic	3×	30.92/0.8828	27.90/0.7938	27.66/0.7554	27.28/0.8648	25.23/0.7612
SRCNN	3×	33.29/0.9211	29.55/0.8369	28.86/0.7999	30.61/0.9163	27.05/0.8216
FSRCNN	3×	33.22/0.9182	29.57/0.8349	28.81/0.7980	30.45/0.9100	26.99/0.8178
RDN	3×	**34.73**/0.7896	**30.81**/0.6671	**29.49**/0.6401	**33.12**/0.6995	**28.63**/0.6784
DRRN	3×	*34.41*/**0.9347**	*30.34*/**0.8509**	*29.44*/**0.8143**	*32.62*/**0.9398**	*28.37*/**0.8549**
MemNet	3×	33.64/*0.9286*	29.66/*0.8468*	29.02/*0.8139*	31.08/*0.9312*	27.66/*0.8508*
Bicubic	4×	28.47/0.8106	25.99/0.7092	26.71/0.6976	25.31/0.8012	23.17/0.6612
SRCNN	4×	30.25/0.8577	27.25/0.7526	27.66/0.7371	27.64/0.8573	24.32/0.7168
FSRCNN	4×	30.50/0.8577	27.42/0.7537	27.76/0.7382	27.99/0.8497	24.48/0.7206
RDN	4×	**32.40**/0.7065	**28.91**/0.5654	**28.62**/0.5506	**31.42**/0.6569	**26.40**/0.5810
SRDenseNet	4×	31.72/*0.8880*	28.27/*0.7812*	*28.39*/**0.7605**	*30.19*/**0.9071**	*25.89*/**0.7786**
DRRN	4×	31.40/0.8829	28.04/0.7740	28.22/0.7552	29.51/0.8975	25.28/0.7574
LapSRN	4×	31.55/0.8857	28.09/0.7773	28.25/**0.7574**	29.40/0.8996	25.31/0.7622
EDSR	4×	*31.87*/**0.8896**	*28.10*/0.7779	27.25/0.7526	29.99/*0.9057*	25.49/*0.7691*
MemNet	4×	30.73/0.8742	27.41/0.7648	27.79/0.7489	28.28/0.8857	24.74/0.7500
SRGAN	4×	30.63/0.8623	27.57/0.7577	27.85/0.7403	28.94/0.8784	25.06/0.7435

6.6 Summary and Discussions

Image super-resolution is an image processing technique that aims at improving image resolution. With the advent of deep-learning tools and techniques, it has matured into an active research problem and several interesting approaches to image super-resolution have been proposed to date. Nowadays, various smartphones and camera devices use super-resolution techniques to produce high-quality pictures by exploiting the powerful generalization capability of deep learning and achieve much better results than previous photo-editing software that employs traditional image upsampling such as the nearest neighbor, bilinear, or bicubic interpolation to enhance the image resolution. Super-resolution techniques have a large number of applications in aerial or satellite photography, medical imaging, and forensics, security, as well as various other common computer vision tasks. Upsampling of an image is performed at a specific stage in the super-resolution process to increase the dimensions of the LR image. Nearest neighbor, bilinear, and bicubic interpolation techniques are some examples of common upsampling techniques.

Some of the earlier contributions in the field of super-resolution were based on traditional machine-learning algorithms, which are broadly classified into two categories: (i) frequency domain and (ii) spatial domain. In the frequency domain approach, the LR image is first converted into the frequency domain, following which the HR image is constructed. The HR image is eventually transformed back into the spatial domain to produce the final output. Depending on the type of transformation employed, the frequency-domain approach is divided into fast Fourier transform, and wavelet transform. Unlike the frequency domain, in the spatial domain, we work directly on the manipulation of pixels. The spatial domain approach aims to overcome some of the problems concerning the frequency domain approach, such as the restricted motion between the frames and exclusion of prior knowledge. Some of the algorithms based on the spatial domain approach are maximum likelihood estimation, MAP estimation, iterative back projection.

Recently, deep learning is gaining immense popularity owing to its supremacy in performance when trained extensively on large datasets. Researchers have developed numerous deep-learning models which are promising and have shown effectiveness in the image super-resolution task. A super-resolution deep-learning model framework, in general, consists of the phases of upsampling, feature extraction, and image reconstruction. In the upsampling phase, the LR image is transformed to an interpolated LR image using the upsampling approaches. In the feature extraction phase, the key features are extracted from the image, and in the image reconstruction phase, the HR image is reconstructed based on the extracted features. Linear networks consisting of simple layer-based architecture were the earliest deep-learning models used in super-resolution. Because of the signal decomposition that makes the learning process simpler, residual networks are commonly used in super-resolution models for improving efficiency. GAN-based deep-learning models have been seen to produce visually appealing, high-quality outputs. Popular evaluation metrics for image quality assessment are PSNR and SSIM. However, these objective evaluation metrics may fail to assess the perceptual quality of the reconstructed image, which can be overcome by using more expensive subjective evaluation techniques such as the MOS.

To obtain a deeper understanding and comparison between the recent and past super-resolution frameworks, the Survey Papers by Wang et al. [28], Anwar et al. [37], or the Review Paper by Yang et al. [38] can be studied. Overall, with the development of more and more sophisticated deep-learning methodologies, the effectiveness of the super-resolution task has increased dramatically in recent years. However, deep networks suffer from an inherent problem of high network complexity and a lot of parameter tuning, due to which executing the models on low-end devices such as smartphones often becomes a challenge. Also, these methods have shortcomings such as the inability to manage real-life degradations, high complexity, and poor metrics, which hinder their use in real-world scenarios. We anticipate that new methods to solve these critical problems in the area of super-resolution will be established in the near future.

References

1 Deep learning applications for cyber security. https://link.springer.com/book/10.1007/978-3-030-13057-2.

2 Malicious spam emails developments and authorship attribution, https://ieeexplore.ieee.org/abstract/document/6754642.

3 A review on deep learning for future smart cities. https://onlinelibrary.wiley.com/doi/full/10.1002/itl2.187.

4 Wronski, B. and Milanfar, P. (2018). See better and further with Super Res Zoom on the Pixel 3. Google AI Blog [Online]. Available: https://ai.googleblog.com/2018/10/see-better-and-further-with-super-res.html (accessed 5 October 2021).

5 H. Zhang, Z. Yang, L. Zhang, H. Shen, "Super-resolution reconstruction for multi-angle remote sensing images considering resolution differences," *Remote Sensing*, vol. 6, no. 1, pp. 637–657, Jan. 2014.

6 X. Du, Y. He, "Gradient-guided convolutional neural network for MRI image super-resolution," *Applied Sciences*, vol. 9, no. 22, pp. 4874, Nov. 2019.

7 Smith, P. (1981). Bilinear interpolation of digital images. *Ultramicroscopy* 6 (2): 201–204.

8 Keys, R. (1981). Cubic convolution interpolation for digital image processing. *IEEE Transactions on Acoustics, Speech, and Signal Processing* 29 (6): 1153–1160.

9 Lehmann, T., Gonner, C., and Spitzer, K. (1999). Survey: interpolation methods in medical image processing. *IEEE Transactions on Medical Imaging* 18 (11): 1049–1075.

10 Borman, S. and Stevenson, R.L. (1998). Super-resolution from image sequences – a review. In: *Proceedings of the IEEE Midwest Symposium on Circuits and System*, 374–378. Notre Dame, Indiana.

11 Demirel, H. and Anbarjafari, G. (2011). Image resolution enhancement by using discrete and stationary wavelet decomposition. *IEEE Transactions on Image Processing* 20 (5): 1458–1460.

12 Yang, C.Y. and Yang, M.H. (2013). Fast direct super-resolution by simple functions. In: *Proceedings of the IEEE International Conference on Computer Vision*, 561–568. Sydney, NSW, Australia.

13 Yang, J., Wright, J., Huang, T., and Ma, Y. (2010). Image super-resolution via sparse representation. *IEEE Transactions on Image Processing* 19 (11): 2861–2873.

14 Dong, C., Loy, C., He, K., and Tang, X. (2014). Learning a deep convolutional network for image super-resolution. In: *Proceedings of European Conference on Computer Vision*, 184–199. Cham, Zurich, Switzerland: Springer.

15 Dong, C., Loy, C.C., and Tang, X. (2016). Accelerating the super-resolution convolutional neural network. In: *Proceedings of European Conference on Computer Vision*, 391–407. Cham, Amsterdam, The Netherlands: Springer.

16 Lim, B., Son, S., Kim, H. et al. (2017). Enhanced deep residual networks for single image super-resolution. In: *Proceedings of the IEEE Conference on Computer Vision and Pattern Recognition Workshops*, 136–144. Honolulu, Hawaii.

17 Ledig, C., Theis, L., Huszar, F. et al. (2017). Photo-realistic single image super-resolution using a generative adversarial network. In: *Proceedings of the IEEE Conference on Computer Vision and Pattern Recognition*, 105–114. Honolulu, Hawaii.

18 Kim, J., Lee, J.K., and Lee, K.M. (2016). Deeply-recursive convolutional network for image super-resolution. In: *Proceedings of the IEEE Conference on Computer Vision and Pattern Recognition Workshops*, 1637–1645. Las Vegas, Nevada.

19 Tai, Y., Yang, J., and Liu, X. (2017). Image super-resolution via deep recursive residual network. In: *Proceedings of the IEEE Conference on Computer Vision and Pattern Recognition Workshops*, 2790–2798. Honolulu, Hawaii.

20 Tai, Y., Yang, J., Liu, X., and Xu, C. (2017). MemNet: a persistent memory network for image restoration. In: *Proceedings of the IEEE Conference on Computer Vision and Pattern Recognition Workshops*, 4539–4547. Honolulu, Hawaii.

21 Lai, W.S., Huang, J.B., Ahuja, N., and Yang, M.H. (2017). Deep laplacian pyramid networks for fast and accurate super-resolution. In: *Proceedings of the IEEE Conference on Computer Vision and Pattern Recognition Workshops*, 624–632. Honolulu, Hawaii.

22 Tong, T., Li, G., Liu, X., and Gao, Q. (2017). Image super-resolution using dense skip connections. In: *Proceedings of the IEEE Conference on Computer Vision and Pattern Recognition Workshops*, 4799–4807. Honolulu, Hawaii.

23 Zhang, Y., Tian, Y., Kong, Y. et al. (2018). Residual dense network for image super-resolution. In: *Proceedings of the IEEE Conference on Computer Vision and Pattern Recognition Workshops*, 2472–2481. Salt Lake City, Utah.

24 Zhang, Y., Li, K., Li, K. et al. (2018). Image super-resolution using very deep residual channel attention networks. In: *Proceedings of the European Conference on Computer Vision*, 286–301. Munich, Germany: Springer.

25 Goodfellow, I.J., Abadie, J.P., Mirza, M. et al. (2014). Generative adversarial nets. In: *Proceedings of Neural Information Processing Systems*, 2672–2680. Montreal, Canada.

26 Sajjadi, M.S.M., Scholkopf, B., and Hirsch, M. (2017). Enhancement: single image super-resolution through automated texture synthesis. In: *Proceedings of IEEE International Conference on Computer Vision*, 4501–4510. Venice, Italy.

27 Wang, X., Yu, K., Wu, S. et al. (2018). ESRGAN: enhanced super-resolution generative adversarial networks. In: *Proceedings of European Conference on Computer Vision Workshops*, 63–79. Munich, Germany: Springer.

28 Wang, Z., Chen, J., and Hoi, S. (2021). Deep learning for image super-resolution: a survey. *IEEE Transactions on Pattern Analysis and Machine Intelligence* 43 (10): 3365–3387.

29 Agustsson, E. and Timofte, R. (2017). Ntire 2017 challenge on single image super-resolution: dataset and study. In: *Proceedings of the IEEE Conference on Computer Vision and Pattern Recognition Workshops*, 1122–1131. Honolulu, Hawaii.

30 Russakovsky, O., Deng, J., Su, H. et al. (2015). ImageNet large scale visual recognition challenge. *International Journal of Computer Vision* 115 (3): 211–252.

31 Yang, J., Wright, J., Huang, T., and Ma, Y. (2008). Image super-resolution as sparse representation of raw image patches. In: *Proceedings of the IEEE Conference on Computer Vision and Pattern Recognition Workshops*, 1–8. Anchorage, AK.

32 Martin, D., Fowlkes, C., Tal, D., and Malik, J. (2001). A database of human segmented natural images and its application to evaluating segmentation algorithms and measuring ecological statistics. In: *Proceedings of the IEEE Conference on Computer Vision and Pattern Recognition Workshops*, 416–423. Vancouver, Canada.

33 Bevilacqua, M., Roumy, A., Guillemot, C., and Morel, M. (2012). Low-complexity single-image super-resolution based on nonnegative neighbor embedding. In: *Proceedings of the British Machine Vision Conference*, 1–10. Surrey: BMVA Press.

34 Zeyde, R., Elad, M., and Protter, M. (2010). On single image scale-up using sparse-representations. In: *Proceedings of International Conference on Curves and Surfaces*, 711–730. Arcachon, France, Springer, Berlin, Heidelberg.

35 Fujimoto, A., Ogawa, T., Yamamoto, K. et al. (2016). Manga109 dataset and creation of metadata. In: *Proceedings of the International Workshop on Comics Analysis, Processing and Understanding*, 1–5. New York, USA: ACM.

36 Huang, J., Singh, A., and Ahuja, N. (2015). Single image super-resolution from transformed self-exemplars. In: *Proceedings of the IEEE Conference on Computer Vision and Pattern Recognition Workshops*, 5197–5206. Boston, MA, USA.

37 Anwar, S., Khan, S., and Barnes, N. (2020). A deep journey into super-resolution. *ACM Computing Surveys* 53 (3): 1–34.

38 Yang, W., Zhang, X., Tian, Y. et al. (2019). Deep learning for single image super-resolution: a brief review. *IEEE Transactions on Multimedia* 21 (12): 3106–3121.

Section II

Machine Learning and Deep Learning Techniques for Text and Speech Processing

7

Machine and Deep-Learning Techniques for Text and Speech Processing

Dasari L. Prasanna[1] and Suman Lata Tripathi[2]

[1]*Department of Electronics and Communication Engineering, Lovely Professional University, Phagwara, Punjab, India*
[2]*School of Electronics & Electrical Engineering, Lovely Professional University, Phagwara, Punjab, India*

7.1 Text Processing

The rapid growth in e-documents on online databases such as news articles, blogs, journals, and digital libraries has created a mass of unstructured data. Huge amount of data are very challenging to organize and search manually. Text categorization is an automatic method of text processing and organizing such large quantities of textual data, which maps the text to predefined classes.

7.1.1 Automatic Text to Image Generation or Vice-Versa Using Machine and Deep Learning

Vast amounts of data have been pilled-up in scientific and academic research fields, due to the expeditious growth in information technology (IT). Different observation methods can be used to obtain the same information. The goal of this machine and deep learning is to find the information that relates to the same concept or belongs to the same field.

Increasing interest in exploring the relationship between language and images leads to various studies of conversion or classifications of concept. In the field of computer vision, numerous experiments have been conducted to improve the learning of text to image (TTI) and conversion of image to text (ITT) connections. In this, automatic image captioning and image retrieval from the text, tasks are performed. The complexity of TTI filtering techniques and their dependence on many algorithms has revealed the excessive work involved in these systems.

The steps involved in TTI are keyword extraction, image selection, image captioning, sentence similarity, and image evaluation. An approach was introduced to extract the image. The basic flow diagram (Figure 7.1) of TTI uses machine-learning algorithms for text processing and deep-learning model for retrieval and captioning of image search from online [1]; an automatic captioning process was used to rank the top-10 images. To rank the best image from the list of thousand pictures from Google, deep-learning model was used. It can automatically convert simple Arabic stories of children to high-quality related images from Google search. Machine learning was used to translate the Arabic text to English, and convolution neural network (CNN) was used to caption the images, which will be included in the search, automatically.

The author [2] proposed a TTI heterogeneous transfer learning using a nonnegative and vigorous collective matrix factorization (CMF) design in noisy domains. A basis factor was identified by removing noise and adding $l_{1,2}$ (factorization norm) regularizations to the co-occurrence data. The classification method (support vector machine [SVM] classifier) was applied to the training image and predicted the class's testing image information. The CMF model transfers the accurate data from TTI domain and handles the noise in both domains. Experimental results

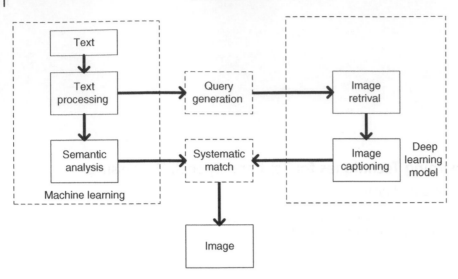

Figure 7.1 The architecture of the automatic text to image system.

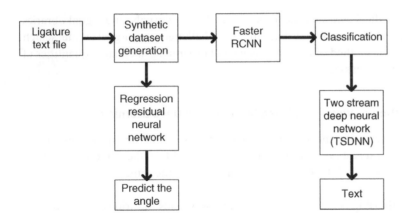

Figure 7.2 The methodology of ligature detection, orientation, and recognition of urdu text.

showed that an efficient iterative method was used to solve the CMF design and had better iterative method convergence and accuracy or performance. A deep-learning model was implemented [3] to detect Urdu text from an image using various sub-architectures. Four modules are involved in detecting the text from image, which are synthetic dataset, faster region-based convolution neural network (RCNN) builder, regression residual neural network (RRNN), and two stream deep neural network (TSDNN). Figure 7.2 shows the methodology of ITT block diagram for the recognition of text from the ligature text file. The text was detected from outdoor images by including different phases, such as text detection, orientation determination, and recognition. The five sub-datasets were generated for evaluating photo-optical character recognition (OCR) phases. Experimental results showed that when using 4k images 79% and for 51k images 99% of accuracy was achieved. RCNN was found to be best for detecting text rotation.

7.1.2 Automatic Image Caption Generation Using Machine and Deep Learning

The contents of an image describing or automatically captioning is a very challenging task; it will greatly impact blind people when understanding the contents of images on Google or documents. Image captioning tells about the objects and includes the relation between the objects, characteristics, and the activities involved in the image.

The language model is also needed to express in the natural language. Bilingual evaluation understudy (BLEU) score compares the score of text translation from one natural language to another reference. A higher score of BLEU shows that the caption generated was more similar to the actual caption on the image presented.

A generative model was presented [4], and trained to generate the sentence or caption. An end-to-end neural networks (NNs) system was used to view the image automatically and generate an appropriate caption or description for an image. CNN followed by recurrent neural network (RNN) was used in the system. CNN encodes the image to a compacted representation, and RNN generates the sentence corresponding to the image. Both qualitative and quantitative evaluations verified the model. The state-of-the-art (STA) of BLEU metrics were used to calculate the generated caption quality. The results of the experiment show that the BLEU-1 score was higher than other approaches: 59, 66, and 28 for Pascal, Flickr30k, and SBU datasets. For the COCO dataset, BLEU-4 of 27.7 was achieved.

Implemented a CNN–RNN model [5] for generating the caption for images. By tuning the design or model with different hyper-parameters, caption generation can be improved. The model was recognized and converted into audio and text using long short-term memory (LSTM) and Google text-to-speech (GTTS). The OpenCV was used to identify the object in the image correctly. The block diagram of captioning the images (Figure 7.3) uses the CNN model following LSTM for captioning the data. By using LSTM, we can process the sequence of data because of its feedback connection. This method was designed for blind people to achieve their potential. The model was implemented on Keras framework, which runs wholly on the system. Keras offers strong support for numerous graphical processing units (GPUs), back-end engines, distributed training, and a wide range of deployment of production options support.

The recognition of automatic image captions for different language text, uses different methods and datasets. Mentioned methods, datasets and findings of automatic image captions in Table 7.1. For automatic image captioning, we have to consider large data to train the model. Using CNN based network gives better performance.

7.1.3 Manipuri Handwritten Script Recognition Using Machine and Deep Learning

In pattern recognition, extracting the data from the printed or handwritten text from the image or scanned document is called optical character recognition (OCR), which converts the data into a machine-readable form for editing the data. In pattern recognition and computer vision, OCR is one of interesting and stimulating topic, which recognizes the text in the image and converts to electronic text from physical paper or image. It can extract the script, even printed or handwritten text from the scanned image offline or online. Extracting or recognizing text from the printed script is easy compared to handwritten text because it has a standard style and size. In the handwritten script, different individuals have different writing styles, and the script's variation in size makes it challenging to extract text. To recognize the text from the data using OCR algorithm, the image acquisition, pre-processing module, and extraction of features were considered [11]. The applications of handwritten script recognition are numerous, such as bank data processing, document digitization, image recognition, restoration,

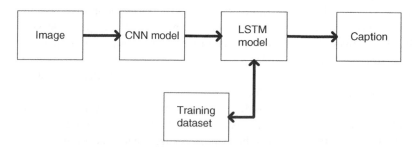

Figure 7.3 The block diagram of captioning the image.

Table 7.1 Automatic image caption generation using machine- and deep-learning method.

Proposed method	Dataset	Findings
Deep-learning model was proposed [6]	MS-COCO	Better performance and optimization of the generator in real-time and high-quality captions were produced by using TensorFlow.
Model used attributes and objects information of images, and semantic ontology was used to reconstruct the sentence [7]	MS-COCO	The model used visual and semantic attention for caption generation of the image. Domain-specific words replaced specific words in the caption.
CNN and NLP [8]	MS-COCO	To generate the descriptive captions, LSTM was used, and it was paired with CNN. The model does not need a large dataset for generating the caption of the image.
RCNN with ResNet101 [9]	MS-COCO	The result of quantitative and qualitative shows the importance of geometric attention for image captioning, improvements on all common captioning metrics on the dataset.
CNN and LSTM are involved to detect the object and captioning [10]	Flicker 8k and ResNet	Tested around 300 images of different classes and 178 images got perfect caption by observing. These pictures have only a few objects, and images with multiple and moving objects are not recognized properly. The precision of the dataset was 63%.

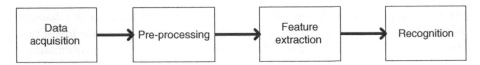

Figure 7.4 Block diagram of MM handwritten recognition model.

reading postal addresses and forms. Figure 7.4 shows the recognition of handwritten Manipuri scripts done in four steps – data acquisition to retrieve the data from image, preprocessing the data image, extracting the features of the data using extraction methods and classifiers to recognition, the block diagram of Manipuri script recognition.

The Manipuri language, also called Meitei Mayek (MM), is an official language of Manipur (north–east state of India); recognizing the Manipuri script is a bit hard because of the structural complexity. Various handwritten text or script extraction or recognition systems have been developed for different languages like English, Chinese, Urdu, and Arabic. Nevertheless, few papers exist for Manipuri handwritten script for recognition. Manipuri handwritten script recognition using artificial neural networks (ANNs) was proposed by authors [12] in 2010. The character pattern was segmented from the converted binarized image using connected-component analysis. Fuzzy, probabilistic, and combination of both features used to perform the system to recognize the script.

Figure 7.5 shows the Manipuri language alphabets and numerals, lot of researches are done to identify the script from handwritten images. The authors [13] explained the modeling and simulation of Handwritten Meitei Mayek Digit Recognition (HMMDR) using NNs method. In this experiment, researcher used thousands samples of hand written Meitei Mayek digits from numerous people. Before passing the data from NN, different pre-processed techniques or steps involved in collecting the samples. After completing the preprocessing, collected samples are send to trained NN and tested on these samples. NN solves problems by self-learning and self-organization. A systematic method of training multilayer ANN is called back propagation. This method has a better application potential and using high level mathematical base for backpropagation.

Table 7.2 discusses about the existing methodologies and their findings. A lot of training data was needed for recognition of handwritten text because different people use different text, using multiple networks for extracting the features and considering large dataset for training the image will helps to recognize the data accurately.

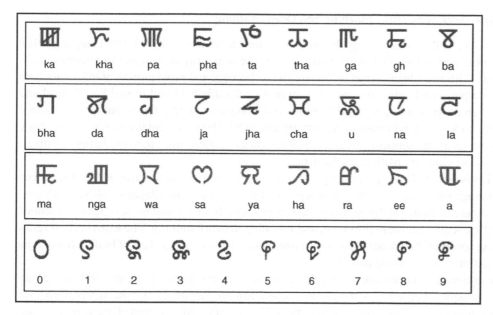

Figure 7.5 Meitei alphabets and numerals.

Table 7.2 Recognition of MM handwritten script methods.

Proposed method	Dataset	Findings
SVM machine learning classifier was used to recognition of handwritten numerals of MM. Different techniques of feature techniques such as projection histogram (PH), background directional distribution (BDD), zone-based diagonal, histogram oriented gradient (HOG), and feature vectors (FVs) are used to extract the data [13].	Proposed techniques test 2000 data samples. Twenty different age groups of different professional people were written the dataset.	Experimental results showed that the HOG features were more appropriate than others for finding numerals of MM Script. For the values of the parameters $h = 0.4$ and cost $C = 500$, accuracy was calculated. For FV10 and HOG (FV8), 95.16% and 94.24% accuracy was obtained respectively.
A methodology was proposed for the recognition of the 27 alphabets of the MM script. Four steps of the model – image acquisition, pre-processing, feature extraction, and script recognition are involved in recognizing the handwritten characters. SVM learning classifier was used for recognition [14].	Twenty-seven alphabets of MM were collected as the database in four sets. The first set consists of 35 samples for each alphabet and a total of 945 samples. Out of 945 samples, 540 were used for training and 405 for testing.	The recognition rate (RR) are 90.62%, 92.59%, 93.33%, and 91.36% of the four sets, respectively. The third set achieved the highest accuracy. Alphabets attained 100% RR, and the accuracy of the two classes is 53.33%, and 20% are the only less.
The author proposed a robust classifier of handwritten manipuri meetei-mayek (HMMM) characters for 56 different classes. Two different systems are compared, extracting the features from training samples [15].	The author collected different handwritten scripts of 5600 samples from 56 people.	By extracting the feature of each letter from the group, the noise was removed using the preprocessed technique. The accuracy of HOG–CNN with KNN classifier is 98.714% when $k = 3$.
Proposed a CNN-based HMM recognition of three image channels – gradient direction, gradient magnitude, and gray channel. Each pixel value indicates the intensity of gray color in the gray channel. Each pixel value represents the largest possible intensity and magnitude of the change in the second and third channels [16].	The model had experimented with a dataset of 14 700 sample images collected from around 240 people of different age groups and educational backgrounds. Before the operation, all the images are normalized to 32×32 volumes.	An accuracy of 98.70% was obtained to recognize HMM when gradient and gray image channels are considered together. When experiments are done separately, the accuracy of gradient direction, gradient magnitude, and grayscale images are 94.44%, 95.34%, and 97.87%, respectively.

7.1.4 Natural Language Processing Using Machine and Deep Learning

Natural language processing (NLP) is a technology that aids computers to understand the human language. NLP is a branch of artificial intelligence (AI) that trains computers in the same way as humans to understand the text and spoken language. With the advancement in science and technology and rapid growth of information on the internet, different data types and structures are piled up in the scientific field and in real life. To read and understand humans' spoken or written language, NLP segments the language, so that the meaning and sentence of words are analyzed and understood as humans. The primary task of NLP is to segment the words, and the basic units are deep-level grammatical and semantic analysis. NLP has many applications – auto prediction, chatbots, smart assistance, speech processing, translation, medical, etc.

A multi-modal neural network (NN) was proposed [17] to extract and analyze the features of NLP using deep learning for the English language. Each mode has an independent structure of a multilayer sub-NN, which converts the similarities of different modes to the same modal features. Conditional random field (CRF) annotated the sentence text semantics, sequence, prediction time, and lessening network training. The experimental result shows that the processing speed of the prediction was 1.94 times improved efficiency than of bi-directional long short-term memory conditional random field model.

In paper [18], author proposed a framework for assessing hospital readmissions for chronic obstructive pulmonary disease (COPD) patients. The subsystems selection, classification, feature extraction, and performance evaluation are used in this framework. To test the system 1248 COPD patient's clinical record of five years was used as test data. Extracting the features of data by clinical text analysis and knowledge extraction system (cTAKES) and bag-of-words. Feature selection lessens the feature numbers by removing unwanted or less information. Four classifiers are used and compared performance. Naïve Bayes (NB), K-nearest neighbors (KNN), and SVM, random forest (RF), classifiers are used and calculated the area under the ROC (receiver operating characteristic) curve (AUC) and time. The experimental results show that the NB classifier with chi-squared (CS) feature extraction was effective as compared to other classifiers.

The author designed [19] a privacy-preserving system of NLP. Many applications use NLP, and models collect the data from different sources, privacy effects when collecting such massive data. Two protocols using secure multi-party computation (MPC) were designed. The performance of SecureNLP considered the metrics of accuracy and efficiency. Private sigmoid and tanh activation function is 59% and 40%. NLP applications for computational phenotyping of electronic health records (EHRs) were reviewed. It concludes that supervised machine learning, rule-based, and keyword search NLPs are widely used methods; keywords search and rule-based NLP showed high accuracy [20]. NLP using deep learning was discussed [21]. Summarized a lot of relevant contributions and processing issues of linguistics were discussed and provided recommendations for future research work in linguistics.

7.2 Speech Processing

Audio data or speech signal has significant part in computer applications. The database of computer multimedia contains a large amount of audio clips of animal sounds, ambient sounds, music, desk noise, speech sounds, and several non-speech vocal sounds.

7.2.1 Smart Sign Language Recognition System for Deaf People Using Deep Learning

Sign language (SL) is commonly used by people who are deaf and dumb for conversation among themselves. It is a medium for communication and is composed of various gestures, shapes of hand, movements of the hand, and facial expressions. Due to its complexity, people with limited knowledge often find it hard to understand. Different parts of the world have different sign languages. In India, for instance, the Indian Sign Language (ISL) is used to represent the English alphabet. The representation of words may be single or two-handed gestures. ISL follows

Figure 7.6 Block diagram of smart sign language recognition system.

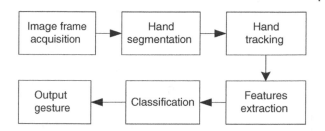

two hands to represent the words. American Sign Language (ASL) has more research compared to other sign languages for sign language recognition or gesture recognition. It is a well-researched topic for ASL. Less research on ISL and varies significantly from ASL. ASL uses a single hand, whereas ISL uses two hands for communicating. Using both hands for sign gesturing often leads to occlusion, so it is hard to detect the correct output because of overlapping of hands, variance in the locality, and lack of dataset, the existence of multiple gestures for the same sign character results to restrain the ISL gesture detection.

The basic block diagram of recognition of the smart sign language is given in Figure 7.6, which includes the methods of frame acquisition of images, skin segmentation, tracking the moment, extracting the features from the hand moments, and classifying the features and recognizing the moment of hand. The first stage in recognizing sign language is to segment the skin part from the image; the remaining part can be considered as noise. The second stage is extracting the relevant features from the segmented skin image for the next stage. In the third stage, learning and classification models use these extracted or collected features from the second stage are used as input. Various supervised methods are used for learning models to train the data, and the final step is classification, done by trained models [22].

Extracting the features from the image when both hands are used is difficult because of the overlapping hands. Deep learning CNN was chosen to avoid this problem. CNN trains two models for top and bottom view. The same CNN was used for both top and bottom view models, and the only difference was output units. Sign language recognition (SLR) system will be difficult if all the possible combinations of gestures are considered. The system was divided into simpler problems to avoid this problem. The dropout technique was used to improve the model capability of generalization between the fully connected layers. Figure 7.7 shows the signs of English alphabets using single hand (https://towardsdatascience.com/sign-language-recognition-using-deep-learning-6549268c60bd).

An automatic ISL recognition for number signs was designed; a standard digital camera was used to collect the sign images. The system converted the isolated digit sign images to text. Naive Bayes and KNN classifiers were used to classify the data and compared the performance of classifiers [23]. Numeric signs of 5000 static signs were captured from 100 singers – the collected database was of different age groups and gender of the singers. The experimental results of classifiers are compared. The various performance characters of classifiers – false positive and false negative (FP and FN) and true positive and true negative (TP and TN), of confusion matrix, $F1$ score, accuracy, sensitivity, specificity, and precision of Naive Bayes and KNN classifiers were calculated and compared for ISL digit recognition. The better results were shown when the system compared to previous existing research papers. The recognition model gets confused with several signs because of some similarities of gestures. The systems could be improved by collecting more quality data, CNN architectures, and redesigning the vision systems.

7.2.2 Smart Text Reader for Blind People Using Machine and Deep Learning

Due to digitalization, articles, journals, academic and research books are converted to online and offline digital or paper documents. People with vision impairment normally use Braille for reading and writing. Reading a text from a digital document helps blind people to learn or understand the contents. Text documents are converted to voice documents using machine- and deep-learning models.

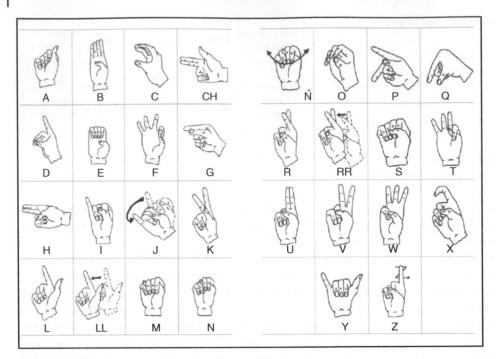

Figure 7.7   Signs for English alphabets can be used as dataset for the system deep-learning model.

The author implemented [24] a smart reader for blind and visually impaired (BVI) people; the design consists of three main blocks – input, processing unit, and output modules. The camera captures the text, and the microphone was used for queries, and speakers or headphones were used as outputs. By OCR, text images are captured, and OpenCV recognizes the characters. Text-to-speech (TTS) was used to convert the audio from text images. Figure 7.8 shows the methodology of a smart reader block diagram, which consists of an image pre-processing converter and classification.

The detection and recognition system [25] was designed for text spotting. For text recognition, the system used CNN. The authors evaluated a huge dictionary of 90k words. The model was trained on synthetic data. A combination of a sliding window detector and an object agnostic region was used to spot text. High individual word recall coverage of 98% for street view text and the ICDAR 2003 (IC03) dataset was obtained. RF classifier was used for calculating FN from the confusion matrix, and CNN was used to train the data. The pipeline technique was used for searching the large-scale text. A novel framework was proposed in [26] for data points identification on images using generative adversarial networks (GANs), which will train the CNN by creating stains on images. MNIST dataset was used for validating the method's effectiveness. The recognition of text from CNN accuracy was 97.38%, and trained data with 97.3% accuracy.

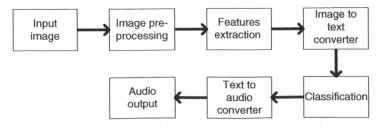

Figure 7.8   The methodology stages of smart reader for blind people.

The system tries to collect the characters from the picture or image data in the first step [27]. For recognition of character, it zooms in and retakes the resolution image. Four-character extraction methods are used and calculated the performance for the IC03 dataset. The characters are extracted from the edge of the image, reverse edge, color-based character, and connected component selection rule. The character of big size was detected effectively. In [28], the system was designed in MATLAB using OCR, Rasberrypi, and a webcam to convert the text into voice signal. To extract or detect the data, the system was divided into two parts, one for to detect the text image and the other to train the image. The classifier was used to detect the data.

7.2.3 The Role of Deep Learning Paradigm in Building the Acoustic Components of an Automatic Speech Recognition System

The natural communication way for all the human beings is speech. Age and educational background does not count when communicating in one language. The automatic speech recognition (ASR) is useful to convert the speech to words sequence by computer program. The applications of ASR converts to the text, identify the sound of animals or instruments, and can be helpful to convert the language.

The acoustic model was used in the ASR system to extract the information from the signals. The recognition units are acoustically modeled based on vocabulary description. The model was designed using Gaussian mixture model (GMM) and hidden Markov model (HMM) for speech signal acoustic (SSA) [29]. Figure 7.9 shows the block diagram of ASR system using acoustic and language model. The improvements of ASR are made in preprocessing, extraction of features, and language model domains of ASR. The sequence of words (W) are considered classes and input speech signal as X. Equation (7.1) is useful to find the word sequence (W), which maximize the quantity. The word sequence defined by classes from closed or similar vocabulary and X is input speech/voice signal of parametric representation. The classification of speech signal is to find the W, which maximizes the quantity.

$$P(W|X) = \frac{P(X|W)\,P(W)}{P(X)} \tag{7.1}$$

where $P(X|W)$ known as quantity of acoustic model, describes the sequence of acoustic observation statics.

Different types of acoustic models are used in ASR. Acoustic models are noise-independent, distortion-independent, noise-dependent, noise mismatch, joint, and clean acoustic models [30]. The clean speech was used in clean acoustic model to train the data, which refers to speech without noise. Using only one type of noise and testing on the same type of noise signal trains a noise-dependent acoustic model. If a single type of noise is used but tested on different noises that signal trains a mismatched acoustic model. For the joint acoustic model, both training and testing noises are the same. Six types of acoustic models were examined. Among the six models of acoustics distortion-independent acoustic model was performed best. The independent distortion model is also able to work with different DNN based speech enhancements.

In [31], front-end uses speech processing, and back-end uses acoustic model, which is called an end-to-end ASR paradigm. Here the input acoustic features are converted directly into sequence of words. Deep neural networks (DNNs) used for speech processing led to high accuracy and enhanced speech quality, which can handle a wide range of reverberation time (RT). Both clean and multi-condition speech can consider by using DNN. Performance of the ASR will increase by using end-to-end joint learning DNN. The ASR system uses acoustic and myoelectric signals [32]; the results show that an increase in acoustic noise affects the performance of an acoustic expert; the noise of acoustic reduces with an increase in myoelectric signals (MES) ASR expert. A system for automatic speech recognition and understanding (ASRU) using deep learning in an acoustic modeling frame has

Figure 7.9 The stages of recognition of ASR system.

been designed [33]. Different acoustics scene classification (ASC) models are compared [34], the proposed model using two dataset has better performance than other models. This baseline model is suitable for ASC development. The author proposed a computational notebook paradigm (CNP) having entities and the process. The experimental results enhanced by every element and also discussed about multi-paradigm modeling and CNP interaction and improvement [35].

The paradigm model was highlighted and relied on HMM. The experimental results show that the accuracy obtained for continuous speech recognition of very large vocabulary in a single layer was 68% and 70% for hidden layers. The acoustic model is also known as pronunciation modeling. The role of the acoustic model is to represent the audio signal and linguistic relation in ASR. The process of statistical representation for a sequence of the feature vector (FV) for a particular sound is called the acoustic component, which will be used in the classifier of the ASR system. The accuracy of features obtained by many layers was high compare to the features extracted by single layer; these layers are trained deeply to extract the features. This experiment is done using continuous voice recognition vocabulary. The average sentence accuracy of deep-learning method using multiple hidden layers is comparatively higher than single layer. In order to outperform the STA, paradigm was highlighted and relied on HMMs [36, 37].

The strategy of extracting the particular or useful patterns or data from the raw data, using deep learning is easier to extract and more efficient. The accuracy is also high for recognition of speech. The basic recognition of speech uses some lexical description to modeled acoustically, which maps acoustic measurement unit and phoneme. These mappings are learned by using a finite training set of utterances. An acoustic model was used in ASR systems to collect or extract the information from acoustic signals and emotions recognition and analysis of sentiment [38, 39].

7.3 Conclusion

The development of artificial intelligence can make life easy without depending on manually hard work. Machine vision, NLP, and speech recognition applications have been developing for the past three decades. This chapter discussed the applications of deep and machine learning for speech and text processing and methods for recognition or generation of text or image, captioning the image, handwritten MM script, sign language, text reader, and ASR. To obtain better accuracy, consider the large dataset for the training model and multiple classes for the classification model. Using CNN can help design multiple-layer classes or networks, and KNN or SVM learning classifiers followed by CNN increase the regression rate (RR). The better quality data input also increases the performance of accuracy for OCR-based models.

References

1 Zakraoui, J., Elloumi, S., Alja'am, J.M., and Yahia, S.B. (2019). Improving Arabic text to image mapping using a robust machine learning technique. *IEEE Access* 7: 18772–18782. https://doi.org/10.1109/ACCESS.2019.2896713.

2 Yang, L., Jing, L., and Ng, M.K. (2015). Robust and non-negative collective matrix factorization for text-to-image transfer learning. *IEEE Transactions on Image Processing* 24 (12): 4701–4714. https://doi.org/10.1109/TIP.2015.2465157.

3 Arafat, S.Y. and Iqbal, M.J. (2020). Urdu-text detection and recognition in natural scene images using deep learning. *IEEE Access* 8: 96787–96803. https://doi.org/10.1109/ACCESS.2020.2994214.

4 Vinyals, O., Toshev, A., Bengio, S., and Erhan, D. (2015). Show and tell: a neural image caption generator. In: *IEEE Conference on Computer Vision and Pattern Recognition (CVPR), Boston, MA, USA*, 3156–3164. https://doi.org/10.1109/CVPR.2015.7298935.

5 Krishnakumar, B., Kousalya, K., Gokul, S. et al. (2020). Image caption generator using deep learning. *International Journal of Advanced Science and Technology* 29 (3): 975–980.

6 P. Mathur, A. Gill, A. Yadav, A. Mishra and N. K. Bansode, "Camera2Caption: a real-time image caption generator," *2017 International Conference on Computational Intelligence in Data Science(ICCIDS), Chennai, India*, 2017, pp. 1–6, https://doi.org/10.1109/ICCIDS.2017.8272660.

7 Han, S.-H. and Choi, H.-J. (2020). Domain-specific image caption generator with semantic ontology. In: *2020 IEEE International Conference on Big Data and Smart Computing (BigComp), Busan, Korea. S.*, 526–530. https://doi.org/10.1109/BigComp48618.2020.00-12.

8 Subash, R., Jebakumar, R., Yash, K., and Nishit, B. (2019). Automatic image captioning using convolution neural networks and LSTM. *Journal of Physics: Conference Series* 1362: 012096. https://doi.org/10.1088/1742-6596/1362/1/012096.

9 Herdade, S., Kappeler Armin, Boakye Kofi, Soares Joao (2019).Image captioning: transforming objects into words. *33rd Conference on Neural Information Processing Systems (NeurIPS 2019)*, Vancouver, Canada.

10 Mandal, S., Lele, N., and Kunawar, C. (2021). Automatic image caption generation system. *International Journal of Innovative Science and Research Technology* 6 (6): 1034–1037.

11 Laishram, R., Singh, A.U., Singh, N.C. (2012). Simulation and modeling of handwritten Meitei Mayek digits using neural network approach. *Proceedings of the International Conference on Advances in Electronics, Electrical and Computer Science Engineering-EEC*, Tula's Institute of Engineering, Dehradun, Uttarakhand, India.

12 Tangkeshwar, T., Bansal, P.K., Vig, R., and Bawa, S. (2010). Recognition of handwritten character of Manipuri script. *Journal of Computers* 5: 1570–1574. https://doi.org/10.4304/jcp.5.10.1570-1574.

13 Kumar, C.J. and Kalita, S.K. (2013). Recognition of handwritten numerals of Manipuri script. *International Journal of Computer Applications* 84 (17): https://doi.org/10.5120/14674-2835.

14 Inunganbi, S. and Choudhary, P. (2018). Recognition of handwritten Meitei Mayek script based on texture feature. *International Journal on Natural Language Computing (IJNLC)* 7 (5): https://doi.org/10.5121/IJNLC.2018.7510.

15 Nongmeikapam, K., Wahengbam, K., Meetei, O.N., and Tuithung, T. (2019). Handwritten Manipuri Meetei-Mayek classification using convolutional neural n. *ACM Transactions on Asian and Low-Resource Language Information Processing* 18 (4), Article 35: 1–23. https://doi.org/10.1145/3309497.

16 Inunganbi, S., Choudhary, P., and Manglem, K. (2020). Handwritten Meitei Mayek recognition using three-channel convolution neural network of gradients and gray. *Computational Intelligence* 37 (6): https://doi.org/10.1111/coin.12392.

17 Wang, D., Su, J., and Yu, H. (2020). Feature extraction and analysis of natural language processing for deep learning English language. *IEEE Access* 8: 46335–46345. https://doi.org/10.1109/ACCESS.2020.2974101.

18 Agarwal, A., Baechle, C., Behara, R., and Zhu, X. (2018). A natural language processing framework for assessing hospital readmissions for patients with COPD. *IEEE Journal of Biomedical and Health Informatics* 22 (2): 588–596. https://doi.org/10.1109/JBHI.2017.2684121.

19 Feng, Q., He, D., Liu, Z. et al. (2020). SecureNLP: a system for multi-party privacy-preserving natural language processing. *IEEE Transactions on Information Forensics and Security* 15: 3709–3721. https://doi.org/10.1109/TIFS.2020.2997134.

20 Zeng, Z., Deng, Y., Li, X. et al. (2019). Natural language processing for EHR-based computational phenotyping. *IEEE/ACM Transactions on Computational Biology and Bioinformatics* 16 (1): 139–153: https://doi.org/10.1109/TCBB.2018.2849968.

21 Otter, D.W., Medina, J.R., and Kalita, J.K. (2021). A survey of the usages of deep learning for natural language processing. *IEEE Transactions on Neural Networks and Learning Systems* 32 (2): 604–624. https://doi.org/10.1109/TNNLS.2020.2979670.

22 Shirbhate, R.S., Shinde, V.D., Metkari, S.A. et al. (2020). Sign language recognition using machine learning algorithm. *International Research Journal of Engineering and Technology (IRJET)* 7 (3): 2122–2125.

23 Sahoo, A.K. (2021). Indian sign language recognition using machine learning techniques. In: *Macromolecular Symposia*, vol. 397, no. 1, 2000241. https://doi.org/10.1002/masy.202000241.

24 Geetha, M.N., Sheethal, H.V., Sindhu, S. et al. (2019). Survey on smart reader for Blind and Visually Impaired (BVI). *Indian Journal of Science and Technology* 12 (48): https://doi.org/10.17485/ijst/2019/v12i48/149408.

25 Jaderberg, M., Simonyan, K., Vedaldi, A. et al. (2014). Reading text in the wild with convolutional neural networks. arXiv: 1412.1842.

26 Brewer, M.B., Catalano, M., Leung, Y., and Stroud, D. (2020). Reading PDFs using Adversarially trained Convolutional Neural Network based optical character recognition. *SMU Data Science* 3 (3): 1–23.

27 Ezaki, N., Bulacu, M., and Schomaker, L. (2004). Text detection from natural scene images: towards a system for visually impaired persons. In: *Proceedings of the 17th International Conference on Pattern Recognition, 2004. ICPR 2004., Cambridge, UK*, vol. 2, 683–686. https://doi.org/10.1109/ICPR.2004.1334351.

28 Musale, S. and Ghiye, V. (2018). Smart reader for visually impaired. In: *2018 2nd International Conference on Inventive Systems and Control (ICISC)*, 339–342. https://doi.org/10.1109/ICISC.2018.8399091.

29 Sarma, K. and Sarma, M. (2015). Acoustic modeling of speech signal using Artificial Neural Network: A review of techniques and current trends. In: *Intelligent Applications for Heterogeneous System Modeling and Design* (ed. K.K. Sarma et al.), 282–299. IGI Global https://doi.org/10.4018/978-1-4666-8493-5.ch012.

30 Wang, P., Tan, K., and Wang, D.L. (2020). Bridging the gap between monaural speech enhancement and recognition with distortion-independent acoustic modeling. *IEEE/ACM Transactions on Audio, Speech and Language Processing* 28: 39–48. https://doi.org/10.1109/TASLP.2019.2946789.

31 Wu, B., Li, K., Ge, F. et al. (2017). An end-to-end deep learning approach to simultaneous speech dereverberation and acoustic modeling for robust speech recognition. *IEEE Journal of Selected Topics in Signal Processing* 11 (8): 1289–1300. https://doi.org/10.1109/JSTSP.2017.2756439.

32 Chan, A.D.C., Englehart, K.B., Hudgins, B., and Lovely, D.F. (2006). Multiexpert automatic speech recognition using acoustic and myoelectric signals. *IEEE Transactions on Biomedical Engineering* 53 (4): 676–685. https://doi.org/10.1109/TBME.2006.870224.

33 Gavat, I. and Militaru, D. (2015). Deep learning in acoustic modeling for automatic speech recognition and understanding - an overview. In: *2015 International Conference on Speech Technology and Human-Computer Dialogue (SpeD)*, 1–8. https://doi.org/10.1109/SPED.2015.7343074.

34 Kek, X.Y., Siong Chin, C., and Li, Y. (2019). Acoustic scene classification using bilinear pooling on time-liked and frequency-liked convolution neural network. In: *2019 IEEE Symposium Series on Computational Intelligence (SSCI)*, 3189–3194. https://doi.org/10.1109/SSCI44817.2019.9003150.

35 Oakes, B.J., Franceschini, R., Van Mierlo, S., and Vangheluwe, H. (2019). The computational notebook paradigm for multi-paradigm modeling. In: *2019 ACM/IEEE 22nd International Conference on Model Driven Engineering Languages and Systems Companion (MODELS-C)*, 449–454. https://doi.org/10.1109/MODELS-C.2019.00072.

36 Zhang, P. and Li, H. (2009). Hybrid model of continuous hidden Markov model and multi-layer perceptron in speech recognition. In: *2009 Second International Conference on Intelligent Computation Technology and Automation*, 62–65. https://doi.org/10.1109/ICICTA.2009.252.

37 Najkar, N., Razzazi, F., and Sameti, H. (2010). A novel approach to HMM-based speech recognition systems using particle swarm optimization. *Mathematical and Computer Modelling* 52 (11–12): 1910–1920. https://doi.org/10.1016/j.mcm.2010.03.041.

38 Rudnicky, A., Baumeister, L., DeGraaf, K., and Lehmann, E. (1987). The lexical access component of the CMU continuous speech recognition system. In: *ICASSP '87. IEEE International Conference on Acoustics, Speech, and Signal Processing*, 376–379. https://doi.org/10.1109/ICASSP.1987.1169705.

39 Bayerl, S.P., Tammewar, A., Riedhammer, K., and Riccardi, G. (2021). Detecting emotion carriers by combining acoustic and lexical representations. In: *2021 IEEE Automatic Speech Recognition and Understanding Workshop (ASRU)*, 31–38. https://doi.org/10.1109/ASRU51503.2021.9687893.

8

Manipuri Handwritten Script Recognition Using Machine and Deep Learning

Palungbam R. Chanu

Electronics and Communication Engineering, NIT Nagaland, Chumukedima, Nagaland

8.1 Introduction

Optical character recognition (OCR) of handwritten text is an exciting research field nowadays. OCR reads documents that have been sensed optically so that human-readable texts can be translated into codes that are readable by machines. With the development of sophisticated gadgets like smartphones and tablets, OCR has become a medium for interaction between computers and humans. Handwritten documents are stored to record many events in terms of culture, literature, history, etc. for future reference. Due to variations in different individuals' writing styles, it is challenging to recognize handwritten texts efficiently.

Manipur is a northeastern state of India, and Manipuri Meitei Mayek (MMM) is its original script. "Meiteilol" is commonly spoken by the ethnic "Meitei" group that resided mainly in the valley region. "Mayek" refers to script in the Manipuri language. Therefore, the official language of the Manipuri people is known as Manipuri Meitei Mayek. It is also spoken by Manipuri people habituated in different parts of other Indian Northeastern states such as Tripura and Assam, and neighboring countries such as Myanmar and Bangladesh. In [1–3], detailed research works are presented on the origin and evolution of MMM in history. All the characters of Meitei Mayek are inspired from different body parts of human anatomy and are shown in Figure 8.1. It has 56 characters, which are categorized into: (a) Eyek Eepee: it represents the original letters and has 18 alphabets; (b) Lom Eeyek: it is additional letters consisting of 9 letters, which are derived from the original 18 alphabets; (c) Lonsum Eeyek: 8 letters with abrupt endings; (d) Cheitap Eeyek: 8 vowel signs; (e) Cheising Eeyek: 10 digits; and (f) Khudam Eeyek: 3 punctuation marks.

Manipur was an independent princely kingdom in ancient times, and Meitei Mayek was widely used. Meitei forefathers used to write about Manipuri traditions, folks, cultures, and legends in a narrative form and are known as "PuYa" (https://en.wikipedia.org/wiki/History_of_Manipur). This PuYa was used to be handed by ancestral to future generations. But, with introduction of Vaisnavism in the beginning of eighteenth century, all the PuYas were burnt. This incident is known as "PuYa Meithaba," or the burning of PuYas. After this incident, the original Meitei Mayek was replaced by the Bengali script. It depicts the necessity of protecting written texts and documents in a secured form such as the digital form to save them from destruction. Manipuri put a lot of effort into reviving the lost Meitei Mayek, and it was officially approved in 1980. After 25 years of its approval in 2005, the Manipuri government included Meitei Mayek in schools' academic curriculum. Since then, research on MMM becomes an essential topic of interest. Many machine-learning techniques such as artificial neural networks (ANNs) and convolutional neural networks (CNNs) are used for the purpose of classifying and recognizing handwritten scripts due to their good learning capabilities.

In this work, survey on literature of handwritten character recognition is given briefly. Many neural-network techniques are explained which are used for the recognition of handwritten Manipuri script. An algorithm is

Machine Learning Algorithms for Signal and Image Processing, First Edition.
Edited by Deepika Ghai, Suman Lata Tripathi, Sobhit Saxena, Manash Chanda, and Mamoun Alazab.
© 2023 The Institute of Electrical and Electronics Engineers, Inc. Published 2023 by John Wiley & Sons, Inc.

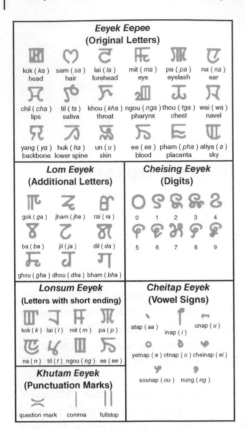

Figure 8.1   Meitei Mayek script.

designed for the classification of MMM numerals based on histogram of oriented gradient (HOG). In HOG, the gradient is used for computing the magnitude and direction with the largest variation in pixel's intensity confined in a local neighborhood. Multilayer perceptron (MLP) is used for further training the features. This process is called stacking, and it aims at obtaining a more robust and discriminative feature, which is again given to a SoftMax classifier for MMM recognition. The proposed algorithm is compared with some existing algorithms of Manipuri OCR to show the effectiveness of the work.

This chapter is organized as follows: literature review is described in Section 8.2. Section 8.3 gives the proposed work. The experimental results are discussed in Section 8.4 followed by conclusions in Section 8.5.

8.2   Literature Survey

In literature, many studies are carried out on handwritten character recognition of some common scripts of India. These scripts are Bangla, Tamil, Telugu, Devanagari, Oriya, Gujarati, Malayalam, Urdu, etc. The commonly used classifier is based on neural network and its types such as CNN, *k*-nearest neighbor (*k*-NN), and other methods based on support vector machine (SVM), HOG, hidden Markov model (HMM), and MLP. The classification and recognition of handwritten MMM are becoming a new research area. Bangla script is used in Manipur as an official language for many years, but in recent years, efforts are made to replace Bangla script with MMM in academic and official places. Therefore, developing new methods for MMM classification will be very beneficial for general public and government. In [4], an algorithm is developed for the recognition of handwritten MMM. In this work, first, the image is binarized and by using the heuristic segmentation technique, the character pattern is segmented.

For extracting features and segmented patterns, KL technique of divergence is used along with a simulator called noncognition. Another work is proposed in [5] which are based on backpropagation neural network. It starts by thresholding the gray images into binary image. Then, the character patterns are segmented by using component analysis which is connected from binarized image. The probabilistic and fuzzy features are extracted from the resized character matrix. Recognition of handwritten MMM numerals is discussed by Laishram et al. in [6] based on multilayer nearest neighbor (NN) with a feed-forward nature learned by backpropagation. Also, a handwritten character recognition for MMM alphabets technique is developed by Laishram et al. in [7]. Their work is focused on segmenting character from whole scanned document. An algorithm based on histogram is used for segmenting lines and words. Connecting component method is used for segmenting characters. Based on Gabor filter and SVM, Maring and Dhir [8] proposed an algorithm for the recognition of handwritten and printed MMM numerals. Kumar and Kalita [9] designed an algorithm for the recognition of MMM numerals by using various feature extraction techniques. These include HOG, zone-based diagonal, background directional distribution (BDD), and projection histogram with SVM classifier. Nongmeikapam et al. [10] designed a recognition method by using HOG features with k-NN classifier and achieved an overall accuracy of 94.29%. In their other work [11], another technique for classifying 56 different characters of MMM is proposed. They exploit HOG feature vectors for training a CNN for producing more efficient features, which is again trained by k-NNs classifier. In [12], a public dataset of MMM character is developed. For character recognition, CNN-based architecture is proposed in which the effect of data augmentation is analyzed for avoiding overfitting of deep network. In [13], a recognition model is developed for recognition of offline handwritten MMM by using texture features and local binary patterns (LBPs). Another recognition system of handwritten MMM based on CNN is found in [14]. Recognition of characters is done using the grayscale of the image of each character. In this work, the corresponding gradient direction and gradient magnitude images are considered for creating three-channel image of every character, which helps in efficient recognition. In [15], an algorithm is developed for recognition of MMM by using the local texture descriptor and projection histogram feature. This algorithm uses certain types of LBP, such as uniform local binary pattern (ULBP), center-symmetric local binary pattern (CS-LBP), and improved local binary pattern (ILBP). These features and the projection histogram are combined with various machine-learning techniques, random forest (RF), SVM, and k-NN for characters' classifications.

Numerical system recognition is designed for many languages worldwide. For Indian languages, a Gujarati digit recognition algorithm is found in [16], and it exploits a NN approach. For analysis purposes, three different sizes of images are used and achieve an accurate rate of 87.29%, 88.76%, and 88.52%. In [17], a neural network-based classification method is developed for Devanagari numerals. It achieves 97.87% accuracy by using multiple stages of classifiers. In [18] a hybrid method of different features is designed for the classification of handwritten Gurumukhi digits. For recognition of Assamese digits, a mathematical morphology-based algorithm is designed in [19], which can attain recognition of 80% accuracy. However, very limited research is performed for recognition of MMM script because of its underlying complexity. In [6], a recognition method based on multilayer feedforward NN is developed for MMM classification. It achieves 85% accuracy by using the backpropagation learning approach. In [8], a recognition algorithm of MMM is found, which utilizes Gabor filter, in feature extraction and SVM in classification. It is 89.58% accurate. In [9], another work on SVM based recognition system of MMM is observed achieving an accuracy rate of 95%.

8.3 Proposed Work

The characters of MMM are very much similar in a complex way and this complicates the recognition process. Thus, it is required to design an efficient feature detector that helps in discriminating every texture or shape of the script clearly. We have proposed a classification algorithm for handwritten MMM digits with the help of HOG methods and MLP. HOG is used to extract the features and these features are trained by MLP. All the script images

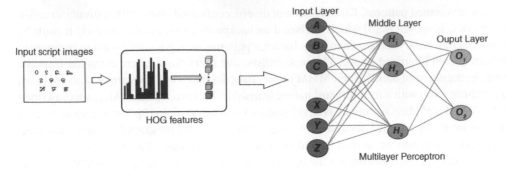

Figure 8.2 Proposed model.

are pre-processed for removing noise. They are given to HOG feature descriptors. At the last, the features extracted from HOG are used for training and recognition purposes by MLP. MLP can be defined as a supplement of ANN and has a feedforward form. In MLP, there exists a network consisting of connected neurons. It has three layers of connected nodes at the minimum level. These layers are (i) input layer, (ii) hidden layer, and (iii) output layer. Every neuron in a layer is connected to other neurons in the preceding layer with a weighted link. A non-activation function is used by every node except the input node. To train the features, MLP utilizes the backpropagation method of supervised learning. In [20], the brief explanation of MLP is given.

Figure 8.2 illustrates the process of MMM digit recognition using HOG and MLP. In first stage, feature vectors of the script input images are extracted using HOG. HOG is a commonly used feature descriptor in image detection. It was first used by Dalal and Triggs [21]. In this work, HOG's feature vectors are used for assisting in deep feature extraction. This aids in discriminating the shapes or textures of script images so that every character can be differentiated in a peculiar way. HOG is associated with gradients of image. Gradients of an image are observed by a directional change with respect to color or intensity of image's pixel. For computing HOG, we divide the input script image into smaller regions, also known as cells. Then, for every cell, we find the corresponding gradients of each pixel in the horizontal and vertical direction by using Eqs. (8.1) and (8.2).

$$G_h(r, s) = I(r + 1, s) - I(r - 1, s) \tag{8.1}$$

$$G_v(r, s) = I(r, s + 1) - I(r, s - 1) \tag{8.2}$$

where $I(r, s)$ represents the pixel's value at (r, s) coordinate of input image, $G_h(r, s)$ and $G_v(r, s)$ are gradients of image at horizontal and vertical direction. Finally, the magnitude of gradient $G(r, s)$ and its orientation $\alpha(r, s)$ are calculated by using Eqs.(8.3) and (8.4)

$$G(u, v) = \sqrt{|G_h(r, s)^2| + |G_v(r, s)|^2} \tag{8.3}$$

$$\alpha(r, s) = \arctan \frac{G_v(r, s)}{G_h(r, s)} \tag{8.4}$$

Referring to Figure 8.2, we have divided the input script image into size of 80×80 which is again sub-divided into smaller sub-blocks of 25, each of size being 60×60. In order to increase the correlation of pixels in the localized region, we have utilized an overlapping process with a 5-pixels stride. Again, every block is again sub-divided into smaller cells of 36 with pixel size of 10×10.

8.4 Experimental Results and Discussions

We collected a dataset of 20 000 samples for testing the proposed algorithm. These samples are collected from persons of varying age group as well as profession. Figure 8.3 is a copy of script image from the collected datasets used for MMM digits recognition.

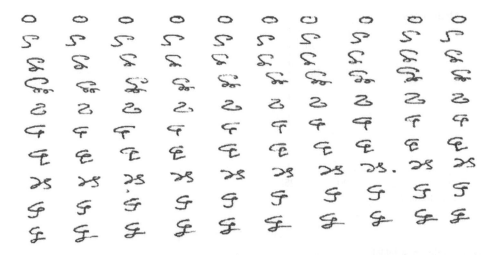

Figure 8.3 An image script sample from MMM numerals dataset.

(a) **Training Set:** It comprises of collected script images utilize for learning. About 2000 samples of handwritten images for every 10 numerals are collected. From this dataset, 1900 samples are reserved for training the NN while the remaining are utilized in testing. The testing dataset is exploited in evaluating a generalization approach for the proposed method toward unseen data.

Two charts are obtained after training the system:

1. The plot depicting accuracy of the training along with the validation datasets over training iterations (also known as epochs) as shown in Figure 8.4.
2. The plot depicting loss of the training along with the validation datasets over training iterations (also known as epochs) as shown in Figure 8.5.

In case of the validation dataset, the word "test" is labeled since it belongs to the datasets used for testing the proposed model. It is observed from Figure 8.4 that the training model is not over-learning on both the training and validation datasets. At around epoch 18, we can achieve the best validation accuracy and beyond this, there is no increase in accuracy.

In Figure 8.5, it is seen that the model is having a comparable performance for both the training and validation datasets. Hence, we can train the proposed model further but since no increase in accuracy is observed after 18th epoch, we have stopped training. Table 8.1 gives the confusion matrix.

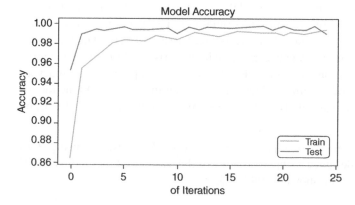

Figure 8.4 Accuracy of model.

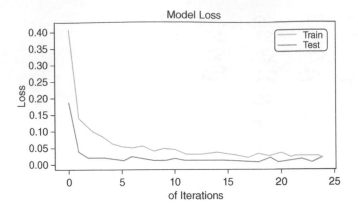

Figure 8.5 Model loss.

Table 8.1 Confusion matrix of 10 digits of MMM.

	0	1	2	3	4	5	6	7	8	9
0	100	0	0	0	0	0	0	0	0	0
1	0	99	0	0	0	0	0	1	0	0
2	0	0	100	0	0	0	0	0	0	0
3	0	0	1	99	0	0	0	0	0	0
4	0	0	0	0	100	0	0	0	0	0
5	0	0	0	0	0	99	1	0	0	0
6	0	0	0	0	0	0	100	0	0	0
7	0	1	0	0	0	0	0	99	0	0
8	0	0	0	0	0	0	0	0	100	0
9	0	0	0	2	0	0	0	0	0	98

(b) **Testing Set:** For testing purposes, 1000 sample images are used and each numeral has 100 sample images. We develop a matrix from this experiment known as the confusion matrix. The confusion matrix gives the information about correct and false recognition of a particular digit. Diagonal elements represent the number of digits recognized correctly while the remaining rows give false recognition number for digits in the first column. For the case of digit "9," it is correctly recognized 98 times out of 100, and recognized incorrectly twice as "3." The digits "0", "2", "3", "4", "6," and "8" are recognized 100% correctly. The remaining digits have some incorrect classifications. The similarity between the digits gives rise to the confusion matrix. True prediction is 995 out of 1000 whereas false prediction is 5. Finally, we also computed the accuracy as, accuracy = 995/1000 = 0.995 = 99.5%. The overall accuracy of each digit is given in Table 8.2. As mentioned above, 100 images are used for every digit for recognition. Here, the correct recognition, false recognition, and accuracy are given.

To prove the superiority of the proposed algorithm, we compare the proposed method designed using HOG features stacked by MLP, and proposed method designed using MLP only. In this case, the first method (HOG stacked with MLP) shows better results. The results are shown in Table 8.3.

Table 8.2 Percentage accuracy for recognition.

Digits	No. of attempts	Correct recognition	False recognition	Accuracy (%)
0	100	100	0	100
1	100	99	1	99
2	100	100	0	100
3	100	100	0	100
4	100	100	0	100
5	100	99	1	99
6	100	100	0	100
7	100	99	1	99
8	100	100	0	100
9	100	98	2	98

Table 8.3 Comparison of performance.

Test	HOG integrated MLP (%)	MLP only (%)
Loss	15.2	17.9
Accuracy	99.3	98.3

Table 8.4 Comparison between the proposed model with existing models.

Models	No. of digits used	Types of digits	Accuracy (%)
Laishram et al. [6]	10	Manipuri	85
Dhir and Maring [8]	10	Manipuri	89.58
Kalita and Kumar [9]	10	Manipuri	95
Proposed method	10	Manipuri	99.3

It is found in the literature that many methods are designed for recognition of numerals for different languages. For Indian languages, recognition systems on Gujarati, Assamese, Gurumukhi as well as Devanagari are found. However, for Manipuri digits recognition only limited works exist. In Table 8.4 we compare the proposed technique with some existing techniques used for MMM numeral recognition. Laishram et al. [6] used a neural network method for recognition achieving 85% accuracy. In [8], SVM is used for classification achieving 89%. The proposed method develop using HOG stacked by MLP shows better performance with an accuracy of 99.3%.

The challenge in the work is finding an optimal size of the feature block of the HOG. Different cell sizes such as 6×6, 7×7, and 8×8 are used for finding an optimized accuracy and cell size contributing to the largest accuracy is used. Due to difference in cell size, timing for training task also varies.

8.5 Conclusion

We have designed a novel technique for MMM numeral recognition using HOG stacked with MLP. The handwritten script image samples are collected from varying age group of persons with different professions. A maximum accuracy of 99.3% is observed as compared to existing methods. Main novelty of the work belongs to the use of HOG stacked by deep learned features of MLP. In future, our work can be used for classification and recognition of degraded texts.

References

1 Wangkhemcha, C. (2007). *A Short History of Kangleipak (Manipur) Part-II*. Sagolband Thangjam Leirak, Imphal, India: Kangleipak Historical and Cultural Research Centre.

2 Kangjia, M. (2003). *Revival of a Closed Account, a Brief History of Kanglei Script and the Birth of Phoon (Zero) in the World of Arithmetic and Astrology*. Lamshang, Imphal: Sanmahi Laining Amasung Punshiron Khupham (SalaiPunship-Ham).

3 Hodson, T.C. (1989). *The Meitheis*. Delhi: Low Price Publications.

4 Tangkeshwar, T. and Bonsai, R. (2005). A novel approach to off-line handwritten character recognition of Manipuri script. In: *Soft Computing* (ed. D. Garg and A. Singh), 365–371. New Delhi: Allied Publishers Pvt. Ltd.

5 Tangkeshwar, T., Bansal, P.K., Vig, R., and Bawa, S. (2010). Recognition of handwritten character of Manipuri script. *Journal of Computers* 5: 1570–1574.

6 Laishram, R., Singh, A.U., Singh, N.C. et al. (2012). Simulation and modeling of handwritten Meitei Mayek digits using neural network approach. In: *Proceedings of the International Conference on Advances in Electronics*, 355–358. Uttarakhand, India (7–9 July): Electrical and Computer Science Engineering—EEC.

7 Laishram, R., Singh, P.B., Singh, T.S.D. et al. (2014). A neural network based handwritten Meitei Mayek alphabet optical character recognition system. In: *2014 IEEE International Conference on Computational Intelligence and Computing Research*, 1–5. Coimbatore, India (18–20 December).

8 Maring, K.A. and Dhir, R. (2014). Recognition of Cheising İyek/Eeyek-Manipuri digits using support vector machines. *International Journal of Computer Science and Information Technologies* 1: 1694–2345.

9 Kumar, C.J. and Kalita, S.K. (2013). Recognition of handwritten numerals of Manipuri script. *International Journal of Computer Applications* 84: 1–5.

10 Nongmeikapam, K., Kumar, W., and Singh, M.P. (2017). Exploring an efficient handwritten Manipuri Meetei-Mayek character recognition using gradient feature extractor and cosine distance based multiclass k-nearest neighbor classifier. In: *Proceedings of the 14th International Conference on Natural Language Processing*, 328–337. Kolkata, India (December. 2017).

11 Nongmeikapam, K., Wahengbam, K., Meetei, O.N., and Tuithung, T. (2019). Handwritten Manipuri Meetei-Mayek classification using convolutional neural network. *ACM Transactions on Asian and Low-Resource Language Information Processing* 18: 1–23.

12 Hijam, D. and Saharia, S. (2018). Convolutional neural network based Meitei Mayek handwritten character recognition. In: *Intelligent Human Computer Interaction. IHCI 2018*, Lecture Notes in Computer Science, vol. 11278 (ed. U. Tiwary). Cham: Springer.

13 Inunganbi, S. and Choudhary, P. (2018). Recognition of handwritten Meitei Mayek script based on texture feature. *International Journal on Natural Language Computing* 7: 99–108.

14 Inunganbi, S., Choudhary, P., and Manglem, K. (2020). Handwritten Meitei Mayek recognition using three-channel convolution neural network of gradients and gray. *Computational Intelligence* 37: 70–86.

15 Inunganbi, S., Choudhary, P., and Singh, K.M. (2020). Local texture descriptors and projection histogram based handwritten Meitei Mayek character recognition. *Multimedia Tools and Applications* 79: 2813–2836.

16 Avani, R.V., Sandeep, R.V., and Kulkarani, G.R. (2012). *Performance Evaluation of Different image Sizes for Recognizing Offline Handwritten Gujarati Digits Using Neural Network ApproachInternational Conference on Communication Systems and Network Technologies*, 270–273. Rajkot, Gujarat, India (11–13 May).

17 Rajiv, K. and Kiran, K.R. (2014). Handwritten Devnagari digit recognition: benchmarking on new dataset. *Journal of Theoretical and Applied Information Technology* 60: 543–555.

18 Gita, S., Rajneesh, R., and Renu, D. (2012). Handwritten Gurmukhi numeral recognition using zone based hybrid feature extraction techniques. *International Journal of Computer Applications* 47: 24–29.

19 Medhi, K. and Kalita, S.K. (2014). Recognition of Assamese handwritten numerals using mathematical morphology. In: *IEEE International Advance Computing Conference*, 1076–1080. Gurgaon, India (21–22 February).

20 Wilson, E. and Tufts, D.W. (1994). Multilayer perceptron design algorithm. In: *Proceedings of IEEE Workshop on Neural Networks for Signal Processing*, 61–68. Ermioni, Greece (6–8 September).

21 Dalal, N. and Triggs, B. (2005). Histograms of oriented gradients for human detection. In: *IEEE Computer Society Conference on Computer Vision and Pattern Recognition*, 886–893. San Diego, CA, USA (20–25 June).

9

Comparison of Different Text Extraction Techniques for Complex Color Images

Deepika Ghai[1] and Neelu Jain[2]

[1] *School of Electronics and Electrical Engineering, Lovely Professional University, Phagwara, Punjab, India*
[2] *Electronics and Communication Engineering Department, Punjab Engineering College (Deemed to be University), Chandigarh, India*

9.1 Introduction

Text embedded in a multimedia image involves high domain knowledge that may be utilized to determine the image's contents. The expression contained in the image sometimes cannot explain its identity but the text expresses its meaning. Mixed text, image, and graphical areas may be found in many television shows, pamphlets, posters, book covers, journals, newspapers, magazines, advertisements, and educational and training programs [1, 2]. As the number of multimedia documents grows, so does the demand for annotation, indexing, and retrieval. As a result, much work on the text extraction system is required. It has been widely used in the real world, for example, in content analysis, vehicle number plate detection, factory automation part identification, street signs, libraries for automated book storage, postal code from the address on the envelope, text translation for foreigners, and bank cheque processing forms. Variations in text alignment, slant, font, font size, languages, contrast, color, lighting, and complex backdrop in photographs have a significant impact on the effectiveness of text extraction systems [3].

In recent years, the topic of text extraction from images has received more attention. Recent International Conference on Document Analysis and Recognition (ICDAR) Robust Reading Competitions in the years 2003, 2005, 2011, 2013, 2015, 2017, 2019, 2021, and 2023, as well as bi-annual International Workshops on Camera-Based Document Analysis and Recognition (CBDAR) held from 2005 to 2022, all point to this. Text extraction is highly difficult due to the large number of images collected from multiple datasets (ICDAR, Street View Text [SVT], KAIST, MSRA-TD500, and so on) [4–8]. These datasets provide a broad range of photos with varying background complexity, text color, font, font size, orientation, alignment, and uneven lighting.

The increasing availability of mobile phones and digital cameras creates an opportunity for image acquisition. So, images available on internet or captured by mobile phones/digital cameras and in databases are growing every day. However, as the quantity of available photographs grows at an exponential rate, users are finding it increasingly difficult to find specific images of interest. It is easy to get a word in an image from an electronic dictionary and transform it into another language if the text region is extracted efficiently from photographs. By indexing the picture with the extracted text information as a classification label, it becomes simple to create an image database or retrieve an image from one [9].

Document texts, scene texts, and caption texts are the three forms of text that may be found in images. The pages of a journal, newspaper, book cover, printed, and handwritten document is scanned to create document text images [10, 11]. Natural pictures with graphic text that appear in the actual world are known as scene text images. People use digital cameras and mobile phones to capture these kinds of photographs in their daily lives

Machine Learning Algorithms for Signal and Image Processing, First Edition.
Edited by Deepika Ghai, Suman Lata Tripathi, Sobhit Saxena, Manash Chanda, and Mamoun Alazab.
© 2023 The Institute of Electrical and Electronics Engineers, Inc. Published 2023 by John Wiley & Sons, Inc.

and when traveling. Text is the most natural method to obtain information. Text present in images such as in advertisements, street names and signs, shops names, road signs, traffic information, banners, vehicle nameplates, text on T-shirts, etc. provides semantic information. This helps us in understanding the various aspects of the captured image [9, 12, 13]. Caption texts are also known as superimposed text or artificial text and are achieved by artificially inserting or overlaying text in the image or video frames during the editing process. The text in these images generally provides information about when, where, and who the activities are taking place [14, 15]. Scene texts are the most difficult to extract since they often have a complex background, and the text in these photos may be in a variety of styles, sizes, colors, orientations, and alignments. As a result, several researchers are working on natural scene text extraction.

Text extraction techniques may be divided into two types: region-based and texture-based. The attributes of color or grayscale in a text region, or their variance with the equivalent properties of the background, are used in region-based algorithms. It is further broken down into two types: edge-based and connected component (CC)-based. The significant contrast between text and background is emphasized via edge-based approaches [16–21]. Text edges are found, combined, and subsequently, non-text portions are filtered out using various procedures. Shadow, highlight, and a complicated background are all susceptible to this method. CC-based approaches work from the bottom up (grouping tiny components into larger components) until all of the text areas in the image are found. This technique uses various approaches such as split-and-merge [22–25], adaptive local connectivity map (ALCM) [26, 27], conditional random field (CRF) analysis [28], etc. Its efficiency is inhibited by noise, a complicated background, multicolored text, low resolution, and small font size. Texture-based approaches are based on the fact that the text in imagery has specific textural features that distinguish it from the background. Even in the presence of noisy, complicated backgrounds and poor resolution photos, it performs well. Using Gabor filters [29, 30], wavelet transform (WT) [31], fast Fourier transform (FFT) [32], spatial variance [33], and discrete cosine transform (DCT) [34], these approaches decomposes an image into blocks and extract features of text and non-text regions. To detect the text portions, these attributes are fed into one of three classifiers: neural network (NN), support vector machine (SVM), or clustering method. The processing of NN and SVM classifiers takes time since they must be trained on a dataset with a variety of images before being tested. As a result, k-means clustering is a superior alternative than NN and SVM classifiers for classifying the image into text and non-text regions.

One of the most significant technologies for text segmentation is optical character recognition (OCR), but it can only recover characters from clean backgrounds (white text on a black background) and cannot extract text from complicated backgrounds [35]. In our daily life, we are mostly encountered with images having complex background on which OCR fails. As a result, an effective text extraction technique that works well on complicated background images is required so that OCR can function properly.

The rest of the chapter is presented as follows. Related work is described in Section 9.2. Section 9.3 discusses edge-based method introduced by Liu and Samarabandu [17] and CC-based method formulated by Gllavata et al. [36]. The proposed text extraction technique is explained in Section 9.4. Section 9.5 contains the results and discussion, whereas Section 9.6 has the conclusions.

9.2 Related Work

The goal of this chapter is to show how to use the edge-based, CC-based, and texture-based approaches and compare them. Many researchers worked to extract text from photos and video frames. Jung et al. [37] proposed SVM-based technique for reliable text identification in pictures. It has three modules: (i) edge-CC analysis for text detection, (ii) for text verification, a learner merging of normalized gray level and constant gradient variance was used, and (iii) SVM for text line refining. If there is distortion and skewing in text lines, this technique will not work. Phan et al. [18] proposed a Laplacian operator-based technique for video text detection. Firstly, a 3×3

Laplacian mask was applied to the input grayscale image to detect edges in four directions: horizontal, vertical, up-left, and up-right. Because the filtered image had both positive and negative values, the maximum gradient difference (MGD) method was employed to find a correlation between them. Text regions have higher MGD values than non-text regions, owing to their vast number of positive and negative values. The k-means clustering was utilized to partition the image into two clusters: text and non-text, relying on Euclidean distance between MGD values. Next, Sobel edge map has been applied only to text regions undergoes projection profile analysis to obtain refined text blocks. On images with non-horizontal text, this technique does not perform efficiently. For video text detection, Shivakumara et al. [20] presented a gradient difference based approach. Firstly, the input grayscale image has been convolved with horizontal mask $[-1\ 1]$ to obtain a gradient image (G). Then, for each pixel in G, the gradient difference was calculated by taking the difference between the maximum and minimum gradient values inside a local window of size $1 \times n$ centered at the pixel, where n is a value that depends on the character stroke width. Because of the strong contrast of text in the photos, text regions exhibit higher positive and negative gradient values than non-text regions. To detect significant gradient values, global thresholding was used, which is based on the gradient difference's mean value. After that, instead of employing a projection profile-based method, which fails when there isn't enough space between text lines, the zero crossing technique was used to determine detected text lines in photos. This method is inefficient for images having staggered and skewed text lines. Yu et al. [38] suggested a text detection and recognition approach based on edge analysis in natural scene photos. Firstly, canny edge detector was used to extract edges. After that, the edge input image was divided to separate the text edges from the background. Many candidate borders are formed by combining adjacent edge segments with comparable stroke width and color. Candidate boundaries were concatenated into text chains in the boundary classification stage, followed by chain classification using character-based and chain-based characteristics. Histogram of gradient features and random forest (RF) classifier has been used to recognize characters. Shivakumara et al. [21] developed a video text detection approach based on the study of edge and texture features in low and high contrast images. For categorization of high contrast (amount of Sobel edge components exceeds amount of canny edge components) and low contrast video pictures, two heuristic rules, R1 and R2, were adopted. On the input image, the Sobel edge operator was used to generate four directional edge maps: horizontal, vertical, up-right, and up-left. To capture textural property, statistical characteristics such as mean, standard deviation, energy, entropy, inertia, and local homogeneity are obtained from each edge map. The feature vector was then classified into text and background clusters using the k-means technique. Morphological opening as well as dilation operations have been used to remove too small background objects and connect the character edges respectively. Using average gradient edge map, two threshold values have been selected automatically for high and low contrast images to detect final text blocks. When the same threshold value is applied for both low and high contrast video images, this approach fails to detect the text lines. Shivakumara et al. [39] developed a Laplacian and Sobel operations-based video scene text identification approach, preceded by Bayesian classification and boundary expanding. The low accuracy rate and F-measure of this method are limitations due to the increased number of false alarms. Wei and Lin [40] suggested a three-step technique for identifying video text: (i) using pyramidal gradients and k-means clustering to detect text regions; (ii) employing projection profile analysis to refine text; (iii) text classification using geometrical, texture, and SVM features. This method has a poor detection rate (DR) and results in a higher false alarm rates. To extract text from historical document images, Shi et al. [26] introduced the ALCM technique. ALCM has been implemented to a grayscale document image. The average foreground pixel density for the original document image is represented by an ALCM image. After that, thresholding was used to ALCM to produce text line patterns in terms of CCs. CCs have been further grouped into location masks by using grouping algorithm. To extract the text line components, position masks in ALCM were mapped back onto the binary image to extract text lines. Finally, a splitting technique was used to solve the issue of components touching multiple lines. Using CC clustering and non-text filtering, Koo and Kim [41] demonstrated a scene text detection system. CCs in image have been extracted and partitioned into clusters by using maximally stable extremal region (MSER) method so as to generate candidate regions. AdaBoost classifier has been trained to determine adjacency pairwise relationship between cluster CCs. Finally, a multilayer perceptrons-based text/non-text classifier determines if the region is text or non-text. Oriented text line images do not give better results. Yan and Gao [42] presented a layered-based method for detecting and recognizing overlay text in images. Color clustering method has been applied to input color image, which divides it into many layered images. To find candidate text areas, each layered image was subjected to CC analysis. After that, a cascade AdaBoost classifier was trained to extract actual text portions. Next, an OCR was utilized to identify real text sections. Processing performance needs to be improved as each image gets divided into many layered images and each such image requires processing. Moreover, small text regions cannot be detected accurately. Using convolutional neural networks (CNNs), Xu and Su [43] suggested a hierarchical text detection technique for scene text. The boosted CNN filtering is used to determine candidate character components using MSER. For finer filtering, a random forest (RF) classifier was used to feature sets of text lines. Zhang et al. [44] introduced a symmetry-based technique for text region segmentation from natural scene images. This approach fails to recognize all characters in photos with extremely low contrast, a brightly lit background, or a significant change in character size. For text localization in born-digital images, Chen et al. [45] presented an efficient local contrast-based segmentation algorithm. In the event of curved text lines, this approach fails to locate text. From low-resolution natural scene photographs, Angadi and Kodabagi [34] developed a texture-based method for text extraction. The input image was broken into 8×8 blocks, with DCT applied to each one. Horizontal frequencies grow from left to right and vertical frequencies rise from top to bottom in the respective DCT block. The low frequency components in the top left corner contain the majority of the energy, but the high frequency components in the bottom right corner are essentially blank (contain zero values). A high pass filter was used to remove the continuous background. The DC component is attenuated by the high pass filter, which stores a zero value in every top left co-ordinate of the 8×8 DCT block. The inverse DCT has been performed on every 8×8 block to obtain processed image. At 0°, 45°, 90°, and 135° orientations, texture properties such as homogeneity and contrast were retrieved from each 50×50 block of the processed output image. Based on texture properties and discriminant functions, these blocks were divided between text and non-text areas. To extract text regions, the detected text blocks were combined and improved. This method is not suitable well with text lines from different languages. Zhao et al. [31] proposed a sparse representation with discriminative dictionaries word identification approach for pictures. Wavelet transform was used to detect the edges of the input image, which were then divided into blocks using a sliding window. Then, two learned discriminative dictionaries (i.e. text and background) have been used to classify candidate text areas. Finally, the adaptive run-length smoothing algorithm (ARLSA) and projection profile analysis were used to refine candidate text areas. Aradhya et al. [46] proposed a wavelet transform and entropy-based multilingual text detection technique. After wavelet transform, Gabor filter has been used to extract texture information. Gaussian smoothing was used to remove the background information from the resulting Gabor image. The smoothed image provides texture feature energy information. The image was broken into smaller blocks and wavelet entropy was applied to each grid to extract energy information. The threshold has been set at the average energy of all grids. It is text area if grid entropy is greater than threshold. It is otherwise a non-text area. Zhang and Chong [47] suggested a discrete Shearlet transform-based text localization method. The use of the discrete Shearlet transform yielded a series of directional sub-bands with textural features. Each of the Shearlet analysis elements produces a set of well-localized pulses at different scales, orientations, and positions. Binarization with dynamic thresholding has been applied to directional sub-bands to preliminary remove non-text edges. Morphological dilation and logical AND operator have been used to obtain candidate text regions. Finally, voting decision mechanism has been used to refine text regions.

According to the literature review, despite the fact that much research has been done in text extraction, a more efficient and effective technique that works well on images, including text in various fonts, font sizes, colors, contrasts, skewed and languages, light text on a dark background, background complexity, and text of poor quality, is still needed.

9.3 Edge-Based and CC-Based Methods

The edge-based and CC-based approaches have been addressed in this section.

9.3.1 Edge-Based Method Introduced by Liu and Samarabandu [17]

For complicated images, Liu and Samarabandu [17] proposed a multi-scale edge-based text extraction technique. Text contained in images can be detected using three major characteristics: edge strength, density, and orientation variance. The main phases of the edge-based text extraction method are outlined below and shown in Figure 9.1.

(a) Convoluting the input image with a line detection mask to identify edges in the horizontal (0°), vertical (90°), and diagonal directions (45° and 135°) yields four oriented edge intensity images.
(b) Gaussian pyramids construct multi-scale images by convolving the input image with a Gaussian kernel and successively down-sampling each direction by half.
(c) Each directional filter is convolved with every image in the Gaussian pyramid.
(d) All images are resized to original image size and combined to build feature map.
(e) A global thresholding is applied on feature map to generate a binary image.
(f) Close edges are connected and far away edges are isolated by using morphological operation.
(g) Finally, non-text parts are filtered out using several constraints such as area and width-to-height ratio, and the resultant image is ready to be presented to an OCR engine for character recognition.

Figure 9.2 shows the results of each step involved in edge-based method suggested by Liu and Samarabandu [17] for an input complex color image.

9.3.2 CC-Based Method Formulated by Gllavata et al. [36]

Gllavata et al. [36] proposed a method for detecting, localizing, and extracting text from complicated background images. The important stages of the CC-based text extraction method are described below and illustrated in Figure 9.3.

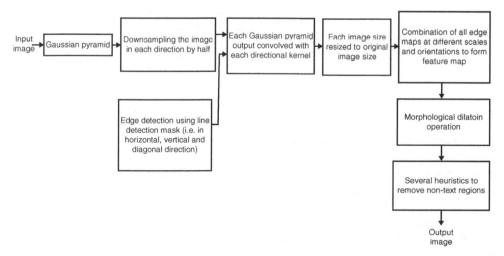

Figure 9.1 Flowchart of edge-based method. Source: Adapted from Liu and Samarabandu [17].

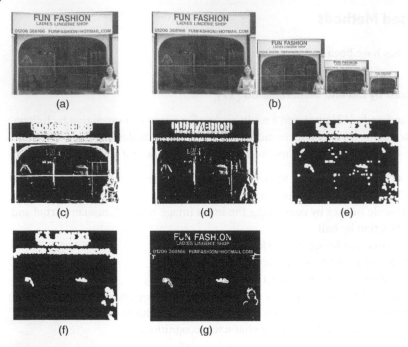

Figure 9.2 Steps involved in edge-based method. Source: Liu and Samarabandu [17], from IEEE. (a) Input image; (b) Gaussian pyramid with four levels of resolution; (c) combination of all edge maps; (d) morphological dilation operation; (e) generated feature edge map by using closing, thinning and dilation operation; (f) geometrical properties of CC for removing non-text regions; and (g) final text extraction results.

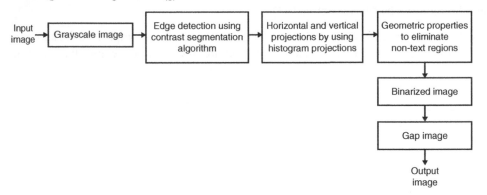

Figure 9.3 Flowchart of CC-based method. Source: Adapted from Gllavata et al. [36].

(a) Only the luminance (Y-channel) is used for further processing, once the input color image is converted to a YUV color space image (luminance + chrominance). The output Y-channel is a grayscale image. The image's brightness or intensity is described by the Y-channel, while the actual color data is described by the U- and V-channels. Because text in an image has a higher contrast with its background when only the Y-channel is used.

(b) After that, the contrast segmentation technique is used to create an edge image, whereby each pixel value in the main image is updated by the main difference between it and its neighbors (in horizontal, vertical, and diagonal direction).

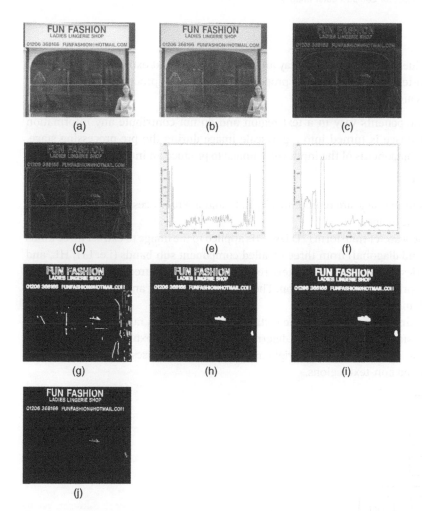

Figure 9.4 Steps involved in CC-based method formulated. Source: Gllavata et al. [36], IEEE. (a) Input image; (b) pre-processing; (c) edge image; (d) sharpened edge image; (e) vertical projection profile; (f) horizontal projection profile; (g) total edge image; (h) heuristics filtering; (i) gap filling operation; and (j) text region extraction output.

(c) The edge image's horizontal and vertical projection profiles are evaluated using a histogram with a suitable threshold value to locate text regions and construct bounding boxes. Text lines and words are separated using Y- (or horizontal) histogram projection and X- (or vertical) histogram projection in histogram projection.

(d) To filter out non-text regions, several limitations of text characters are imposed, including height, width, and the width-to-height ratio for every bounding box.

(e) Thresholding is then applied to generate the binary image.

(f) Finally, detected text regions are refined by the gap filling process and the output image is ready to be passed to an OCR engine.

Figure 9.4 shows the results obtained from each step of CC-based technique formulated by Gllavata et al. [36] for complex color image.

9.4 Proposed Methodology

The proposed method is based on the idea that image edges play an essential part in text extraction because text edges have a greater intensity than non-text edges [48–50]. The proposed methodology's process flow is presented in Figure 9.5, and the following phases are listed:

(a) **Pre-processing:** Because color components vary in a text region and do not contribute any information in text extraction, the input color image is turned into a grayscale image during the pre-processing stage. Equation (9.1) combines the RGB components of the input color image to produce an intensity image (Y).

$$Y = 0.299\,R + 0.587\,G + 0.114\,B \tag{9.1}$$

where red, green, and blue regions in an image are represented by R, G, and B. Pre-processing is not necessary if the input image is really a grayscale image.

(b) **2D DWT:** Two dimensional discrete wavelet transform (2D DWT) is applied to the image to extract three kinds of edges (i.e. horizontal, vertical, and diagonal) from three detailed component sub-bands (i.e. LH, HL, and HH). Because the textural properties of the edges of the text region are better captured by high-frequency wavelet coefficients, they provide a priority for text extraction. The horizontal, vertical, and diagonal edges are detected by LH, HL, and HH sub-bands, respectively.

(c) **Feature Extraction Through Sliding Window:** Since the texture of text in image is rather irregular, statistical parameters, such as mean and standard deviation, are determined to reflect the texture attribute. As four pixels overlapping, a sliding window is shifted from left to right and top to bottom. It uses wavelet coefficients to extract characteristics from text and non-text regions.

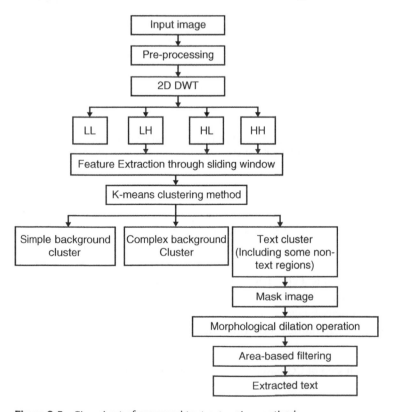

Figure 9.5 Flowchart of proposed text extraction method.

(i) **Sliding Window:** Each high-frequency sub-band (i.e. LH, HL, and HH) of size $(M \times N)$ is scanned by the sliding step size $(s_1 \times s_2)$, using a tiny size overlapped sliding window of size $(w \times h)$. The 4×4 sliding step size is utilized in this study because it covers most area of the preceding window, ensuring that no text is missing.

Equations (9.2) and (9.3) calculate two variables p_1 and p_2 for zero padding.

$$p_1 = \text{Mod}(M - w, s_1) \tag{9.2}$$

$$p_2 = \text{Mod}(N - h, s_2) \tag{9.3}$$

where w and h represent the number of rows and columns in a sliding window.

If both $p_1 = 0$ and $p_2 = 0$ are correct, the sliding window should scan all sub-band areas.

If $p_1 \neq 0$ or $p_2 \neq 0$ is case, the sliding window won't be able to scan every area. As a result, zero padding is required at the end of each row and column to cover the entire area of each sub-band.

The rows (w_1) and columns (h_1) to be filled with zeros are given by Eqs. (9.4) and (9.5), respectively.

$$w_1 = s_1 - p_1 \tag{9.4}$$

$$h_1 = s_2 - p_2 \tag{9.5}$$

The size of the sliding window is determined by the criteria listed in Table 9.1.

The majority of the text in images is aligned horizontally; hence window widths (i.e. $w = h$ and $w < h$) are typically employed. For high font sizes and low contrast texts, the window size should be adjusted to guarantee that characters are not misread.

(ii) **Feature Extraction:** It is focused on the selection of the most suitable set of fundamental statistical features in order to reduce computing complexity and increase the accuracy of text segmentation against a complicated background. The mean (μ) (Eq. (9.6)) and standard deviation (σ) (Eq. (9.7)) is determined from each sliding window for each sub-band:

$$\mu = \frac{1}{w \times h} \sum_{i=1}^{w} \sum_{j=1}^{h} I(i,j) \tag{9.6}$$

$$\sigma = \sqrt{\frac{1}{w \times h} \sum_{i=1}^{w} \sum_{j=1}^{h} [I(i,j) - \mu]^2} \tag{9.7}$$

where $I(i,j)$ is the high-frequency wavelet coefficients.

The mean and standard deviation values for the background region are probably to be zero, however for the text areas; these values are substantially greater due to abrupt changes in pixel values.

Every detail component sub-band yields two features, which are concatenated to generate a single feature space X (Eq. (9.8)) of size $(n \times 6)$.

$$X = [\mu_{\text{LH}}, \sigma_{\text{LH}}, \mu_{\text{HL}}, \sigma_{\text{HL}}, \mu_{\text{HH}}, \sigma_{\text{HH}}]_{n \times 6} \tag{9.8}$$

The range of sliding windows from which attributes are obtained is n in each detail component sub-band.

In comparison to non-text regions, text regions have a higher mean and standard deviation values. On the basis of these features, k-means clustering is used to construct text and non-text groupings.

(d) **k-Means Clustering:** The feature vector is classified between text, simple background, and complex background regions using the k-means clustering algorithm. The value of k should be selected on the type of background images (i.e. simple and complex background) as shown in Table 9.2.

The steps involved in k-means clustering algorithm are shown in Figure 9.6.

Table 9.1 Size of sliding window.

Condition	Text alignment
$w < h$	Horizontal direction
$w > h$	Vertical direction
$w = h$	Both horizontal and vertical direction

Table 9.2 Selection of k values.

No. of clusters (k value)	Type of background present in image	Clusters
$k = 2$	Simple background images	Background and text clusters
$k = 3$	Complex background images	Simple background, complex background, and text clusters

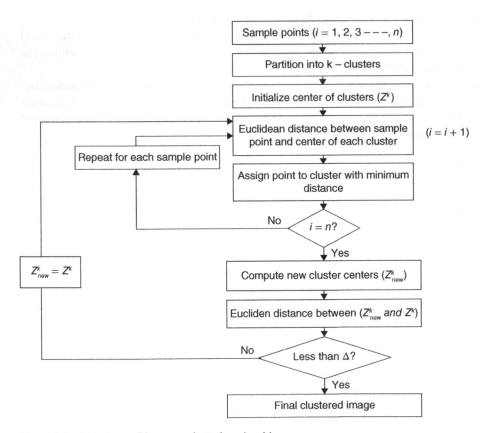

Figure 9.6 Flowchart of k-means clustering algorithm.

(e) **Masking of Image:** The masked image is created via a voting decision procedure. The voting image (V) and masked image (M) are created using the procedures below (M).

  (i) If the image's sliding window is categorized as a text cluster, the voting image's related pixel gets voted once.

  (ii) A masked image is formed when all sliding windows have been voted on. If a pixel in the voting picture has a vote higher than 1, its value in the masking image is set to 1, else it is set to 0. This was chosen to eliminate noisy pixels around the text, leaving only the text region in the masked image. This approach removes small background blobs that were mistakenly recognized as text regions in complicated images, and area-based filtering removes the remaining non-text regions.

(f) **Morphological Dilation Operation:** The dilation technique adds pixels to the borders of text sections, preserving text information around the segmented edge areas. It is used to precisely detect the text region on the masked image (M). In this chapter, structuring elements of size $[4 \times 4]$ and $[4 \times 9]$ are mostly used.

(g) **Area-Based Filtering:** To filter out the remaining non-text areas in the image, several heuristics are employed, including area and inverse of extent. The retrieved text region is further refined using global thresholding.

Figure 9.7 depicts the stages necessary for extracting text from a complex color image.

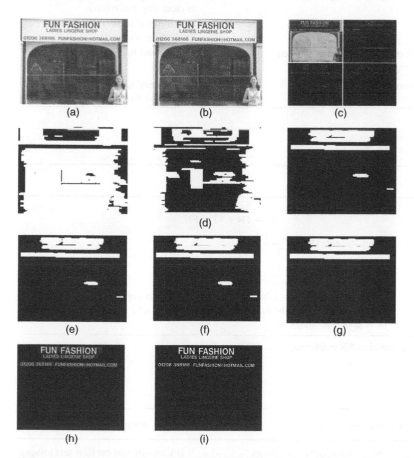

Figure 9.7 Text extraction steps for a complicated color image: (a) input image; (b) pre-processing; (c) 2D DWT; (d) *k*-means clustering algorithm (simple background, complex background, and text clusters); (e) masked image; (f) morphological dilation operation; (g) area-based filtering; (h) extracted text; and (i) global thresholding.

9.5 Experimental Results and Discussion

The proposed approach is run on an Intel (R) Core (TM) i5 CPU running at 1.80 GHz with 4 GB of RAM. Experiments are conducted on ICDAR 2013, KAIST, and own created dataset and as shown in Table 9.3.

The performance of the proposed technique is examined using a number of evaluation criteria, which are mentioned in Table 9.4.

In k-means clustering, the input parameters (k) for the clusters are determined empirically on several sets of photos. The number of clusters chosen is determined by the image complexity rather than the image size and is shown in Table 9.5.

Table 9.3 Datasets description used for experimentation.

Datasets	Images present in various datasets	Features
ICDAR 2013	Challenge 1 (reading text in born-digital images)	• Horizontal text
	Challenge 2 (reading text in scene images)	• English text
KAIST	Indoor and outdoor scene images	• Different lighting conditions (such as clear day and night)
		• Multilingual texts (i.e. Korean, English, and mixed)
Own created dataset	Images of book covers, magazines, printed materials, and sign-boards	• Skewed and handwritten text images
	Caption and scene text images	• Multilingual texts (i.e. English, Hindi, and Punjabi)

Table 9.4 List of evaluation parameters and their formula.

Sr. No.	Parameters name	Formula
1.	Detection rate (DR)	$\text{Detection rate (DR)} = \dfrac{\text{TP} + \text{TN}}{\text{TP} + \text{TN} + \text{FP} + \text{FN}}$
2.	Precision rate (PR)	$\text{Precision rate (PR)} = \dfrac{\text{TP}}{\text{TP} + \text{FP}}$
3.	Recall rate (RR)	$\text{Recall rate (RR)} = \dfrac{\text{TP}}{\text{TP} + \text{FN}}$
4.	Processing time (PT)	It refers to how long it takes to process an image. It depends on the approach, how much text is in the foreground, how complicated the background is, and how good the image is.

TP = true positive, TN = true negative, FP = false positive, FN = false negative.

Table 9.5 Number of clusters chosen.

No. of clusters (k)	Type of background present in image	Type of images
$k = 2$	Simple and less complex background images	Document and caption text images
$k = 3$	More complex background images	Scene text images

9.5.1 Sample Test Results

Two existing methods i.e. edge-based method (Liu and Samarabandu [17]) and CC-based method (i.e. Gllavata et al. [36]) have also been implemented along with the proposed texture-based method for the performance comparison. The proposed method's text extraction results are shown in Figure 9.8c, as well as the results of the two current approaches in Figure 9.8a,b. The edge-based and CC-based methods can work well with images having high contrast, simpler background, and bigger font size. But, these methods are unable to detect small text characters and low-contrast text. When opposed to texture-based approaches, edge-based and CC-based methods take longer to process. The proposed method correctly detects all text lines and extracts very low contrast text against a background with no false positives. It can work well for different languages as well as aligned text as shown in Table 9.6. The proposed method is superior w.r.t. all the performance evaluation metrics as shown in Table 9.7.

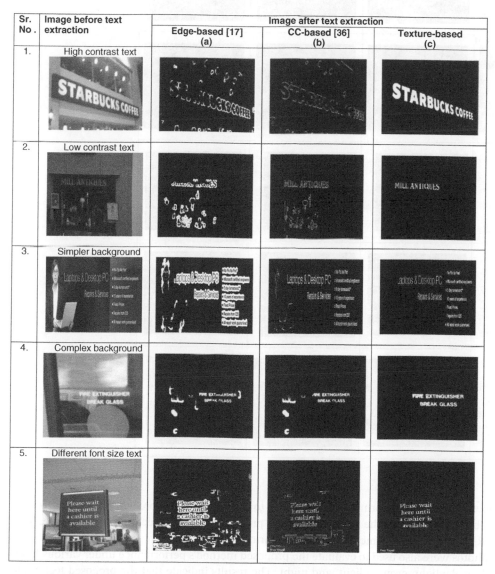

Figure 9.8 Images from complex color images before and after text extraction. Source: (a) from KAIST public dataset; (b) from ICDAR public dataset; (c) own created dataset: Sarab Sukh Charitable Trust Moga.

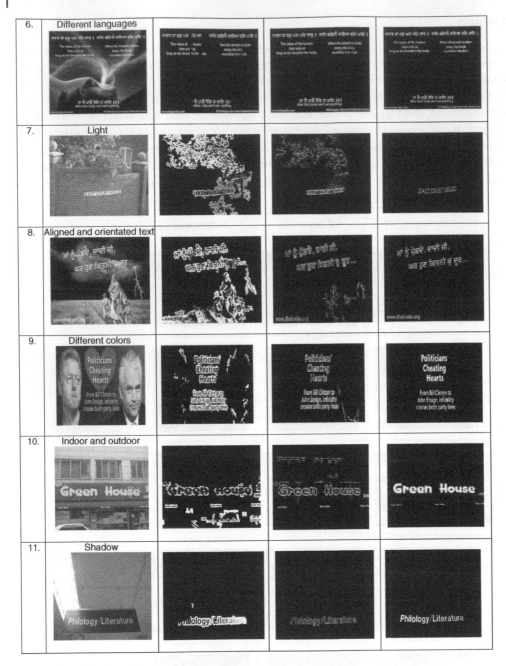

Figure 9.8 *(Continued)*

In Table 9.8, the proposed method's performance is compared to that of existing approaches. In comparison to the edge-based and CC-based techniques, the proposed method provides the best detection rate (DR), precision rate (PR), recall rate (RR), and processing time (PT) values for various test datasets.

Indoor and outdoor photos were also used to evaluate the proposed method's performance under various lighting circumstances, such as daylight, evening light, and night. The results indicate that the proposed technique achieves a higher average detection, precision, and recall rates than the edge-based and CC-based algorithms. The above-mentioned drawbacks of edge-based and CC-based approaches (Table 9.6) restrict text extraction efficiency,

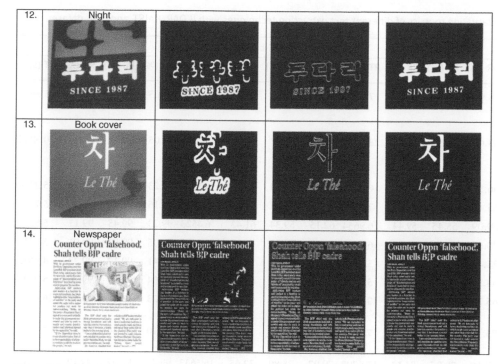

Figure 9.8 (*Continued*)

Table 9.6 Comparison between different text extraction techniques (i.e. edge-based, CC-based, and texture-based).

Existing methods		Proposed method
Edge-based method	**CC-based method**	**Texture-based method**
• Not suitable for dense and sparse text lines	• Not suitable for dense and sparse text lines	• Suitable for dense and sparse text lines
• Misses some of the text characters when text and background are of low contrast	• Misses some of the text characters when text and background are of low contrast	• Works well for low-contrast images
• Not suitable for aligned text	• Not suitable for aligned text	• Suitable for aligned text
• Not suitable for multi-colored text in background	• Not suitable for multi-colored text in background	• Works well when multi-colored text present in images
• More false positives in complex background and shadow images	• Less false positives in complex background and shadow images as compared to edge-based method	• Works well for complex background and shadow images

resulting in lower detection, precision, and recall rates in edge-based and CC-based methods (Table 9.7). In comparison to edge-based and CC-based methods for text region extraction from sample-tested photos, the proposed method is more reliable and insensitive to scale, lightning, and orientation.

9.5.2 Comparison of Proposed Method with Existing State-of-the-Art Methods

Table 9.8 summarizes the performance of the proposed technique in comparison to existing strategies. The following are primary benefits of the proposed algorithm are shown in Table 9.9.

Table 9.7 Performance evaluation metrics of different text extraction techniques with ICDAR, KAIST, and own dataset.

Dataset	Image type	Edge-based				CC-based				Texture-based			
		DR (%)	PR (%)	RR (%)	PT (s)	DR (%)	PR (%)	RR (%)	PT (s)	DR (%)	PR (%)	RR (%)	PT (s)
ICDAR 2013	Challenge 1 (born-digital images)	68.90	67.02	68.24	7.3412	69.99	68.34	69.67	6.9783	99.65	98.87	99.12	5.5978
	Challenge 2 (scene text images)	63.67	59.54	63.20	23.8878	65.97	61.02	64.23	22.9087	99.05	98.15	99.34	20.3845
KAIST	Indoor	66.56	62.14	64.17	6.908	67.34	63.29	64.45	6.5676	99.23	99.13	99.29	5.2230
	Outdoor	65.45	61.21	63.75	7.6912	65.67	64.15	65.56	6.7244	99.16	99.02	99.24	5.7896
	Light	60.09	55.45	59.44	6.9654	62.15	59.34	61.43	6.4523	99.23	99.45	99.11	5.8267
	Shadow	70.67	68.11	69.66	6.2349	72.87	70.22	71.97	5.2450	99.05	99.01	98.01	5.0314
	Night	71.45	69.91	70.56	5.2379	72.76	70.09	71.68	4.9627	99.41	99.12	99.23	4.9672
	Others	64.87	62.76	63.89	5.7949	66.56	64.94	65.72	5.2355	99.04	99.10	99.26	5.2345
Own	Scene and caption text	65.23	63.90	64.29	19.6878	66.45	65.89	66.24	17.1342	99.23	98.67	99.44	16.675
	Book-cover images	75.12	72.99	74.87	7.3456	76.78	74.34	75.16	6.453	99.10	99.34	99.56	5.3299
	Newspaper images	77.16	70.87	74.13	17.6797	79.97	75.89	72.65	16.5643	99.45	99.22	99.12	14.9876
	Skewed text images	60.19	56.20	59.38	22.5676	64.90	61.97	58.09	20.7654	99.58	98.15	98.34	19.2432

Table 9.8 Comparison of proposed method with other techniques for complex color images.

Author (year)	Method	Dataset	Source No.	DR (%)	PR (%)	RR (%)	Average PT (s)
Liu and Samarabandu [17] (2005)	Edge-based (edge detection using directional filters and Gaussian pyramid)	Own dataset (book covers, object labels, indoor lab name plates, and outdoor information signs)	75	—	91.80	96.60	14.5 (for 480 × 640 sized image)
Phan et al. [18] (2009)	Edge-based (3 × 3 Laplacian mask, MGD, k-means clustering, Sobel edge map, projection profile analysis)	Own dataset (English, Chinese, and Korean text video images from news, sport, and movie)	101	93.30	—	—	—
Shivakumara et al. [20] (2009)	Edge-based (gradient difference-based approach, zero crossing technique)	Own dataset (English, Chinese, and Korean text video images from news, sport, movie, music, and web images)	488	95.40	—	—	4 (for 352 × 288 sized image)
Shivakumara et al. [21] (2010)	Edge-based (Sobel and Canny edge operators, feature extraction through sliding window, k-means clustering, morphological operations)	Own dataset (video images taken from movies, news, sports, music)	2580	85.60	—	—	21

(continued)

Table 9.8 (Continued)

Author (year)	Method	Dataset	Source No.	DR (%)	PR (%)	RR (%)	Average PT (s)
Shivakumara et al. [39] (2012)	Edge-based (Sobel and Laplacian operations, Bayesian classifier, boundary growing method)	ICDAR 2003 scene images	251	—	72	87	7.9
		Own dataset (video and camera text data images)	1020	—	74	87	9.3
Wei and Lin [40] (2012)	Edge-based (bilinear interpolation, gradient difference, k-means clustering, projection profile analysis, DWT, SVM)	Own dataset (TV news, web images, and movie clips)	200	95.10	89.60	—	8.7 (for 288×352 sized image)
Gllavata et al. [36] (2003)	CC-based (contrast segmentation algorithm, projection profile analysis)	Own dataset (advertisements, video frames from films, and news)	326	—	83.90	88.70	—
Xu and Su [43] (2015)	CC-based (MSER, CNN, RF classifier)	ICDAR 2003	—	—	77	75	—
		ICDAR 2011	—	—	80	77	5.3
		ICDAR 2013	—	—	84	78	—
Zhang et al. [44] (2015)	CC-based (text line identification method based on symmetry)	ICDAR 2013	—	97.70	88	74	—
Chen et al. [45] (2015)	CC-based (local contrast based segmentation method, CCs generation, and analysis)	ICDAR 2013 Challenge 1: reading text in born-digital images (web and email)	—	—	91.57	85.80	—
Angadi and Kodabagi [34] (2010)	Texture-based (text region extraction using DCT)	Own dataset (indoor and outdoor low-resolution natural scene display board images)	100	96.60	—	—	6 10 (for 240×320 sized image)
Zhao et al. [31] (2010)	Texture-based (wavelet transform, discriminative dictionaries, ARLSA and projection profile analysis)	Microsoft common test set	45	—	98.80	94.20	—
Proposed method	Texture-based (DWT, feature extraction through sliding window, k-means clustering, voting decision process, and area-based filtering)	ICDAR 2013	424	99.35	98.51	99.23	5.5978–20.3845
		KAIST	496	99.19	99.14	99.02	5.3454 (for 640×480 sized image)
		Own — Scene and caption text	364	99.23	98.67	99.44	3.2699–16.6354
		Book-cover	65	99.10	99.34	99.56	5.3299
		Newspapers	40	99.45	99.22	99.12	14.9876
		Skewed text	35	99.58	98.15	98.34	19.2432

Table 9.9 Benefits of the proposed method.

Sr. No.	Benefits	Existing methods		Proposed method	Remarks
		Edge-based	CC-based	Texture-based	
1.	Fast and accurate edge detection	Not works well in case of background complexity [17], noisy [39], and skewed text [18, 20]	Not works well in case of background complexity [36], low contrast [44], strong illumination [44], skewed text [45], and multi-script text lines [34]	Works well for background complexity, noisy, low contrast, skewed text, strong illumination, and multi-script text lines	For complicated backgrounds, 2D DWT edge detection is more precise than canny-based and Sobel-based methods [17, 21, 39] and contrast segmentation algorithm [36]. Hence, 2D DWT works well in all the conditions
		Lower PR due to more false positives	Improves PR as compared to edge-based method	PR is further improved with proposed method as compared to edge- and CC-based method	Reduce false positives with the help of 2D DWT
		Not works well for small font sized text or significant difference in font size of text [21, 39]	Not works well for small font sized text or significant difference in font size of text [44]	Works well for small font sized text or significant difference in font size of text	For high contrast and caption text images, Canny and Sobel edge-based algorithms work well, but not for low contrast and scene text images. As a result, conventional approaches have a low DR. So, 2D DWT works well for low and high contrast, and different types of images
		Not works well due to the limits of uniform colored pixels between text and background [21]	Not works well due to the limits of uniform colored pixels between text and background [44]	Works well when there is non-uniformity of color pixels between text and background	Due to differing colored pixels in the same text lines w.r.t. background, this causes misdetection of text characters, resulting in a decrease in recall rate [21, 44]. So, 2D DWT detects differing colored pixels in the same text lines w.r.t. background
		Sobel [18, 21, 39] and Canny [21] edge operator	—	2D DWT	It detects three different types of edges simultaneously. So, processing time of 2D DWT is lesser than traditional edge detection filter
2.	Fast detection	Supervised techniques (SVM [40])	Supervised techniques (CNN [43])	Unsupervised techniques (k-means clustering)	The unsupervised k-means clustering algorithm is faster than SVM and CNN in classifying photos into text and background groups because these classifiers must be trained on a dataset with a variety of images before being tested
3.	Reduced number of texture features	Six statistical features (i.e. mean, standard deviation, energy, entropy, inertia, and local homogeneity) are used [21]		Only two statistical features (mean and standard deviation) are employed in this study	Enhance segmentation accuracy while minimizing computational complexity
4.	Less mis-detection of text	No sliding window is used		Sliding window is used	Overlapping sliding window is used as it covers the maximum area of the preceding window which ensures that no text is missed
5.	Robust text extraction	Sensitive to background complexity [17, 36], noisy [39], low contrast [44], skewed text [18, 20, 45], strong illumination [44], multi-script text lines [34] and small font sized text or significant difference in font size of text [21, 39, 44]		Insensitive to background complexity, noisy, low contrast, skewed text, strong illumination, multi-script text lines and small font sized text or significant difference in font size of text	Better performance evaluation metrics in all the conditions as compared to edge-based and CC-based methods

9.6 Conclusions

This chapter proposes a new method for extracting text regions from color images. Due to variances in font size, color, and background complexity, extracting text regions from pictures is extremely difficult. DWT is firstly used to detect the edges of an input image. The texture features are being retrieved using a small overlapping sliding window to examine high-frequency component sub-bands. The image is classified into text, simple background, and complicated background groups using k-means clustering based on these characteristics. The false positives are further removed by using the voting decision process and geometrical properties of CCs. Results show that DR, PR, RR, and processing time are best as compared to edge-based and CC-based techniques. Experiments show that our system is capable of extracting text sections in a variety of font sizes, colors, languages, orientations, and backgrounds. When the photos are strongly illuminated and the watermark text is large, the proposed approach performs poorly. Our future efforts will be directed at finding a solution to this issue.

Acknowledgment

The authors would like to thank ECE Department, Punjab Engineering College (Deemed to be University), Chandigarh and SEEE Department, Lovely Professional University, Phagwara for providing necessary facilities and CSIR for providing funds (grant file No: 08/423(0001)/2015-EMR-1) required for carrying out this research work.

References

1 Zhang, H., Zhao, K., Song, Y.Z., and Guo, J. (2013). Text extraction from natural scene image: a survey. *Neurocomputing* 122: 310–323.

2 Jung, K., Kim, K.I., and Jain, A.K. (2004). Text information extraction in images and video: a survey. *Pattern Recognition* 37: 977–997.

3 Zhu, Y., Yao, C., and Bai, X. (2016). Scene text detection and recognition: recent advances and future trends. *Frontiers of Computer Science* 10: 19–36.

4 Lucas, S.M., Panaretos, A., Sosa, L. et al. (2003). ICDAR 2003 robust reading competitions. In: *Proceedings of 7th International Conference on Document Analysis and Recognition*, Edinburgh, UK (6 August)., 682–687.

5 Lucas, S.M. (2005). ICDAR 2005 text locating competition results. In: *Proceedings of 8th International Conference on Document Analysis and Recognition*, Olympic Parktel, Seoul, Korea (29 August–1 September)., 80–84.

6 Shahab, A., Shafait, F., and Dengel, A. (2011). ICDAR 2011 robust reading competition challenge 2: reading text in scene images. In: *Proceedings of 2011 IEEE International Conference on Document Analysis and Recognition*, Beijing, China (18–21 September)., 1491–1496.

7 Karatzas, D., Shafait, F., Uchida, S. et al. (2013). ICDAR 2013 robust reading competition. In: *12th International Conference on Document Analysis and Recognition* (23–28 August)., 1484–1493. Washington, DC: IEEE.

8 Karatzas, D., Bigorda, L.G., Nicolaou, A. et al. (2015). ICDAR 2015 competition on robust reading. In: *13th International Conference on Document Analysis and Recognition* (23–26 August)., 1156–1160. Tunis, Tunisia: IEEE.

9 Sumathi, C.P., Santhanam, T., and Devi, G.G. (2012). A survey on various approaches of text extraction in images. *International Journal of Computer Science and Engineering Survey* 3: 27–42.

10 Ghai, D. and Jain, N. (2013). Text extraction from document images – a review. *International Journal of Computer Applications* 84: 40–48.

11 Nagabhushan, P. and Nirmala, S. (2010). Text extraction in complex color document images for enhanced readability. *Intelligent Information Management* 2: 120–133.

12 Crandall, D. and Kasturi, R. (2001). Robust detection of stylized text events in digital video. In: *Proceedings of the 6th International Conference on Document Analysis and Recognition*, 865–869. Seattle, WA, USA: IEEE.

13 Chen, D., Luettin, J., and Shearer, K. (2000). A survey of text detection and recognition in images and videos. Institut Dalle Molle d'Intelligence Artificielle perceptive (IDIAP) research report, IDIAP-RR 00-38, August.

14 Zhong, Y., Zhang, H., and Jain, A.K. (2000). Automatic caption localization in compressed video. *IEEE Transactions on Pattern Analysis and Machine Intelligence* 22: 385–392.

15 Su, Y.M. and Hsieh, C.H. (2006). A novel model-based segmentation approach to extract caption contents on sports videos. In: *2006 IEEE International Conference on Multimedia and Expo* (9–12 July)., 1829–1832. Toronto, ON: IEEE.

16 Liu, X. and Samarabandu, J. (2005). An edge-based text region extraction algorithm for indoor mobile robot navigation. In: *2005 IEEE International Conference on Mechatronics & Automation* (29 July–1 August)., 701–706. Niagara Falls, Canada: IEEE.

17 Liu, X. and Samarabandu, J. (2006). Multiscale edge-based text extraction from complex images. In: *2006 IEEE International Conference on Multimedia and Expo* (9–12 July)., 1721–1724. Toronto, ON: IEEE.

18 Phan, T.Q., Shivakumara, P., and Tan, C.L. (2009). A Laplacian method for video text detection. In: *10th International Conference on Document Analysis and Recognition* (26–29 July)., 66–70. Barcelona, Spain: IEEE Computer Society.

19 Shivakumara, P., Phan, T.Q., and Tan, C.L. (2009). Video text detection based on filters and edge features. In: *International Conference on Multimedia and Expo* (28 June–3 July)., 514–517. New York, USA: IEEE.

20 Shivakumara, P., Phan, T.Q., and Tan, C.L. (2009). A gradient difference based technique for video text detection. In: *10th International Conference on Document Analysis and Recognition* (26–29 July)., 156–160. Barcelona, Spain: IEEE Computer Society.

21 Shivakumara, P., Huang, W., Phan, T.Q., and Tan, C.L. (2010). Accurate video text detection through classification of low and high contrast images. *Pattern Recognition* 43: 2165–2185.

22 Lienhart, R. and Stuber, F. (1996). Automatic text recognition in digital videos. In: *Proceedings of SPIE, Image and Video Processing IV, San Jose, CA, United States* (13 March)., vol. 2666, 180–188.

23 Lienhart, R. and Effelsberg, W. (1998). Automatic text segmentation and text recognition for video indexing. Technical Report TR-98-009, Praktische Informatik IV, University of Mannheim.

24 Lienhart, R. and Effelsberg, W. (2000). Automatic text segmentation and text recognition for video indexing. *Multimedia Systems* 8: 69–81.

25 Zhan, Y., Wang, W., and Gao, W. (2006). A robust split-and-merge text segmentation approach for images. In: *18th International Conference on Pattern Recognition* (20–24 August)., 1002–1005. Hong Kong, China: IEEE Computer Society.

26 Shi, Z., Setlur, S., and Govindaraju, V. (2005). Text extraction from gray-scale historical document images using adaptive local connectivity map. In: *Proceedings of the 8th International Conference on Document Analysis and Recognition* (29 August–1 September)., 794–798. Olympic Parktel, Seoul, Korea: IEEE Computer Society.

27 Shi, Z., Setlur, S., and Govindaraju, V. (2009). A steerable directional local profile technique for extraction of handwritten Arabic text lines. In: *10th International Conference on Document Analysis and Recognition* (26–29 July)., 176–180. Barcelona, Spain: IEEE Computer Society.

28 Pan, Y.F., Hou, X., and Liu, C.L. (2009). Text localization in natural scene images based on conditional random field. In: *10th International Conference on Document Analysis and Recognition* (26–29 July)., 6–10. Barcelona, Spain: IEEE Computer Society.

29 Raju, S., Pati, P.B., and Ramakrishnan, A.G. (2004). Gabor filter based block energy analysis for text extraction from digital document images. In: *Proceedings of the 1st International Workshop on Document Image Analysis for Libraries* (23–24 January)., 233–243. Palo Alto, CA, USA: IEEE Computer Society.

30 Qiao, Y.L., Li, M., Lu, Z.M., and Sun, S.H. (2006). Gabor filter based text extraction from digital document images. In: *Proceedings of the 2006 International Conference on Intelligent Information Hiding and Multimedia Signal Processing* (December)., 297–300. Pasadena, CA, USA: IEEE Computer Society.

31 Zhao, M., Li, S., and Kwok, J. (2010). Text detection in images using sparse representation with discriminative dictionaries. *Image and Vision Computing* 28: 1590–1599.

32 Chun, B.T., Bae, Y., and Kim, T.Y. (1999). Automatic text extraction in digital videos using FFT and neural network. In: *Proceedings of the IEEE International Fuzzy Systems Conference* (22–25 August)., 1112–1115. Seoul, South Korea: IEEE.

33 Zhong, Y., Karu, K., and Jain, A.K. (1995). Locating text in complex color images. *Pattern Recognition* 28: 1523–1535.

34 Angadi, S.A. and Kodabagi, M.M. (2010). A texture based methodology for text region extraction from low resolution natural scene images. *International Journal of Image Processing* 3: 229–245.

35 Jianyong, S., Xiling, L., and Jun, Z. (2009). An edge-based approach for video text extraction. In: *2009 International Conference on Computer Technology and Development* (13–15 November)., 331–335. Kota Kinabalu: IEEE Computer Society.

36 Gllavata, J., Ewerth, R., and Freisleben, B. (2003). A robust algorithm for text detection in images. In: *Proceedings of the 3rd International Symposium on Image and Signal Processing and Analysis* (18–20 September)., 611–616. IEEE.

37 Jung, C., Liu, Q., and Kim, J. (2009). Accurate text localization in images based on SVM output scores. *Image and Vision Computing* 27: 1295–1301.

38 Yu, C., Song, Y., Meng, Q. et al. (2015). Text detection and recognition in natural scene with edge analysis. *IET Computer Vision* 9: 603–613.

39 Shivakumara, P., Sreedhar, R.P., Phan, T.Q. et al. (2012). Multioriented video scene text detection through Bayesian classification and boundary growing. *IEEE Transactions on Circuits and Systems for Video Technology* 22: 1227–1235.

40 Wei, Y.C. and Lin, C.H. (2012). A robust video text detection approach using SVM. *Expert Systems with Applications* 39: 10832–10840.

41 Koo, H.I. and Kim, D.H. (2013). Scene text detection via connected component clustering and nontext filtering. *IEEE Transactions on Image Processing* 22: 2296–2305.

42 Yan, J. and Gao, X. (2014). Detection and recognition of text superimposed in images base on layered method. *Neurocomputing* 134: 3–14.

43 Xu, H. and Su, F. (2015). A robust hierarchical detection method for scene text based on convolutional neural networks. In: *IEEE International Conference on Multimedia and Expo* (29 June–3 July)., 1–6. Turin, Italy: IEEE.

44 Zhang, Z., Shen, W., Yao, C., and Bai, X. (2015). Symmetry-based text line detection in natural scenes. In: *IEEE Conference on Computer Vision and Pattern Recognition* (7–12 June)., 2558–2567. Boston, MA, USA: IEEE.

45 Chen, K., Yin, F., Hussain, A., and Liu, C.L. (2015). Efficient text localization in born-digital images by local contrast-based segmentation. In: *13th International Conference on Document Analysis and Recognition* (23–26 August)., 291–295. Tunis, Tunisia: IEEE.

46 Aradhya, V.N.M., Pavithra, M.S., and Naveena, C. (2012). A robust multilingual text detection approach based on transforms and wavelet entropy. *Procedia Technology* 4: 232–237.

47 Zhang, J. and Chong, Y. (2013). Text localization based on the discrete Shearlet transform. In: *2013 4th IEEE International Conference on Software Engineering and Service Science* (23–25 May)., 262–266. Beijing: IEEE.

48 Ghai, D. and Jain, N. (2013). Comparison of various text extraction techniques for images – a review. *International Journal of Graphics and Image Processing* 3: 210–218.

49 Ghai, D. and Jain, N. (2016). A new approach to extract text from images based on DWT and *K*-means clustering. *International Journal of Computational Intelligence Systems* 9: 900–916.

50 Ghai, D. and Jain, N. (2019). Comparative analysis of multi-scale wavelet decomposition and *k*-means clustering based text extraction. *Wireless Personal Communications* 109: 455–490.

10

Smart Text Reader System for People who are Blind Using Machine and Deep Learning

Zobeir Raisi, Mohamed A. Naiel, Georges Younes, Paul Fieguth, and John Zelek

Vision Image Processing Lab, Department of Systems Design Engineering, University of Waterloo, Waterloo, Canada

10.1 Introduction

The World Health Organization (WHO) estimates that at least 2.2 billion people have a near or distance vision impairment worldwide [1, 2]. As of 2020, there are around 43.3 million blind people and about 285 million suffering from moderate to severe vision impairment [3–6]. Unfortunately, the statistics reported by WHO over the years show an increasing trend in the number of people with visual impairments that may suffer from several difficulties while performing day-to-day tasks like navigation, reading, communicating, etc. Coming up with solutions that can positively impact their lives is then of utmost importance.

With the recent advancement of several technologies like machine learning, computer vision, and the internet of things, many attempts were made to improve the quality of life for people who are visually impaired [7, 8]. For example, wearable assistive devices have been developed that can be categorized into (i) sensorial network systems that use electronic devices to get information from the environment, or (ii) video-based systems that rely on cameras accompanied with computer vision and machine-learning algorithms to help the blind go about their day [6].

One of the most challenging tasks for blind people in their daily life is reading text from documents or images in indoor or outdoor environments [9]. Text in the form of printed fonts on a signage board in the wild images is specifically prepared for people with normal vision, and it is difficult for blind people to acquire this information. If we could extract all the text in the scene, this would be very informative, particularly for sign detection and recognition, storefronts, traffic signs, indoor navigation with labels on doors, such as washrooms and offices. In this chapter, we study the state-of-the-art advancement in the text detection and recognition field and their capability to parse scene texts into strings that a speech engine can use to relay audible or other means of feedback. We also study the challenging scenarios in the wild images that may affect the performance of intelligent text readers.

Text is a vital-communication tool and plays an important role in our daily lives. As a result, it is frequently embedded into a wide variety of documents or scenes as a means of conveying information [10–12]. Identifying text can be considered as a main building block for a variety of computer vision-based applications, such as robotics [13, 14], industrial automation [15], image search [16, 17], instant translation [18, 19], automotive assistance [20] and sports videos [21]. Generally, text identification can be organized into two categories:

1. Identifying text in scanned printed documents;
2. Text in the wild or scene text, how text appears in daily scenes (e.g. text captured on highways, buildings, and subject to geometric, illumination, and environmental distortions).

Figure 10.1 illustrates examples of both categories. For text in scanned documents, optical character recognition (OCR) methods are widely used [10, 25–27], however, these methods face complex challenges when attempted to

Machine Learning Algorithms for Signal and Image Processing, First Edition.
Edited by Deepika Ghai, Suman Lata Tripathi, Sobhit Saxena, Manash Chanda, and Mamoun Alazab.
© 2023 The Institute of Electrical and Electronics Engineers, Inc. Published 2023 by John Wiley & Sons, Inc.

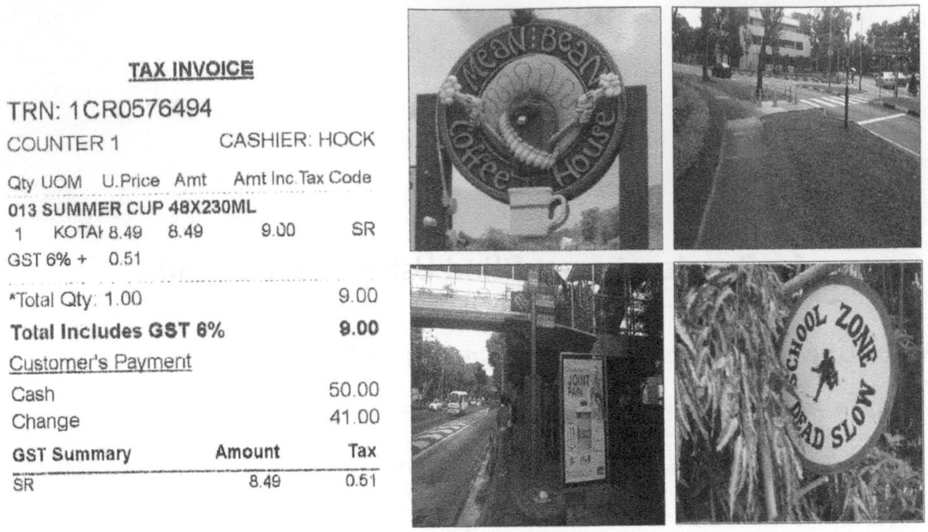

Figure 10.1 Examples for two main classes of text: text in a printed document (a) and text captured in the wild (b). Source: [22–24].

be used to detect and recognize text in images captured in the wild, causing them to fail in most cases [10, 11, 28]. These challenges include as follows:

- **Text Diversity:** text can exist in a wide variety of colors, fonts, orientations, and languages.
- **Scene Complexity:** images may include elements having an appearance similar to text, such as signs, bricks, and symbols.
- **Distortion Factors:** the effects due to a wide variety of contributing factors, including blur, limited resolution, field of view, and occlusion [10–12].

In the literature, many techniques have been proposed to address the challenges of scene text detection and/or recognition, often divided into those based on *classical machine learning-based* [29–41], and those based on *deep learning-based* [42–68]. Classical approaches often combine feature extraction with a machine-learning model [32, 69, 70]; although some of these methods [69, 70] achieve good performance in detecting or recognizing horizontal text [10, 12], these methods typically fail on images containing multi-oriented (MO) or curved text (CT) [11, 12]. On the other hand, for text captured under adverse situations deep-learning-based methods have shown effectiveness in detecting text [45–52, 54, 55, 58, 59, 62, 71], recognizing text [44, 64–68, 72–83], and end-to-end detection, and recognition [60–63].

Earlier surveys [10, 84] offered comprehensive reviews of classical methods, whereas more recent surveys [11, 12, 85] focused on deep learning, however, all these surveys concentrated on summarizing and comparing the results reported in the surveyed papers.

This chapter aims to address the gap in the literature by not only reviewing the recent advances in scene text detection and recognition, with a focus on the deep learning-based-methods, but also using the same evaluation methodology to assess the performance of some of the best state-of-the-art methods on challenging benchmark datasets. Further, this chapter studies the shortcomings of existing techniques by conducting an extensive set of experiments. Finally, the chapter proposes potential future research directions and best practices, which potentially would lead to designing better models that are able to handle scene text detection and recognition under adverse situations. Figure 10.1 shows the examples for two main classes of text: text in a printed document (a) and text captured in the wild (b), images taken from [22–24].

10.2 Literature Review

10.2.1 Smart Text Reader System for Blind People

In the last decades, many capable assistive tools [8, 86–90] have been proposed for aiding the visually impaired in moving and exploring unfamiliar environments. Some methods [86–89] used camera sensors and machine-learning techniques to help people who are blind find their way in indoor/outdoor environments in order to avoid obstacles and dangers. For example, Kang et al. [86] introduced an obstacle avoidance method that uses a deformable grid in order to accurately measure the collision risk by taking perspective projection geometry into account. Afif et al. [89] proposed an indoor signage recognition system that uses a deep convolutional neural network (CNN) that classifies indoor signage into four sign classes to help visually impaired people. Other methods [91–93] proposed walking assistance using visual assistance technology to reduce the daily difficulties of blind people. These methods find the location information [92], shortest path [93], and floor level in front of users [91]. They have been designed with audio-signal feedback for helping the visually impaired [5].

Extracting text information from pre-trained scene text detection and recognition models are practical in visual question answering (VQA) or image-captioning applications [94]. People who are blind can benefit from image captioning or VQA to alleviate their inconvenience of understanding the indoor/outdoor environments [95]. VizWiz dataset [96, 97] has been prepared to develop image-captioning methods [94] and VQA frameworks [98] for people who are blind. The dataset consists of 39 181 images generated by people who are blind, each paired with five captions. Image captioning is the process of generating a textual description of an image [99]. Image-captioning models [94, 95, 99] uses both natural-language processing and computer vision to generate captions. Another application that can provide information about an image to people with visual impairments is VQA, which aims at learning a multi-modal model to answer the question according to a given image [98]. The accuracy of the image captioning and VQA models depend on extracted text features [100], and these models require considering the text-detection and recognition techniques for better performance.

As described above, many indoor and outdoor environments are equipped with text; extracting accurate text information from these environments could be informative for many innovative assistive tools for people with visual impairment. During the past decade, researchers have proposed many techniques for recognizing text in images captured in the wild [44, 48, 65, 70, 101]. These techniques first localize text regions by predicting bounding boxes and then recognizing every detected region's contents. Thus, the process of interpreting text from images can be divided into two successive tasks, namely *text detection* (Figure 10.2) and *text recognition*, as shown in

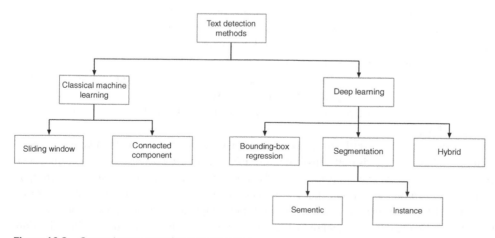

Figure 10.2 General taxonomy for text detection.

Figure 10.3 General schematic of scene text detection and recognition. Source: Veit et al. [102], from Cornell University.

Figure 10.3. The rest of this section will discuss the conventional and recent algorithms for text detection and recognition.

10.2.1.1 Text Detection

As illustrated in Figure 10.2, scene text detection methods can be categorized into those based on *classical machine learning* [30, 32–34, 70, 103–108] and those based on *deep learning* [45–52, 54, 55, 59, 62, 71]. We will review both categories.

Classical Machine Learning-Based Methods This section summarizes traditional methods used for scene text detection, which can be categorized into two main approaches of sliding windows and connected component.

In *sliding window* methods, such as [29–34], a given test image is used to construct an image pyramid to be scanned over all possible text locations and scales, using a sliding window of a certain size. From each window, certain image features are extracted, such as mean difference and standard deviation [31], histogram of oriented gradients (HOGs) [32, 109, 110], and edge regions [33]. These features are then classified by a classical approach, such as random ferns [32, 111], adaptive boosting (AdaBoost) [112], decision trees [33], or log-likelihoods [30, 31]. For example, in early work by Chen and Yuille [30], intensity histograms, intensity gradients, and gradient direction features were obtained at each sliding window location, after which several weak log-likelihood classifiers were used to construct a strong classifier using the AdaBoost framework for text detection. In [32], HOG features were extracted at every sliding window location and a random fern classifier [113] was used for multi-scale character detection, where the non-maximal suppression (NMS) in [114] was performed to detect each character separately. However, these methods [30, 32, 33] are only applicable to detect horizontal text and have a low detection performance on-scene images, in which text can appear at any orientation [115].

Connected-component methods aim to extract image regions of similar properties (such as color [35–39], texture [116], boundary [117–120], and corner points [121]) to create candidate components that can be categorized as text or non-text using traditional classifiers, such as support vector machine (SVM) [103], random forests [107], and nearest-neighbor [122]. These methods detect characters of a given image and then combine the extracted characters into a word [70, 103, 104] or a text-line (TL) [123]. In comparison to sliding-window based methods, connected-components are more efficient, more robust, and they typically offer a lower false positive rate, which is important in scene text detection [85].

Maximally stable extremal regions (MSERs) [69] and stroke width transform (SWT) [70] are the two main representatives connected component based methods that constitute the basis of many subsequent text detection works [42, 84, 103, 104, 107, 108, 120, 123–125]. However, these methods aim to detect individual characters or components that may easily cause regions with ambiguous characters to be discarded or may cause a large number of false detections to be generated [126]. Furthermore, their use of multiple complicated sequential steps leads to the propagation of errors. In addition, these methods may fail in difficult situations, such as non-uniform illumination and text with multiple connected characters [127].

Deep Learning-Based Methods The emergence of deep learning [128] has changed the way researchers approach text detection and has greatly enlarged the scope of research in this field. Deep learning techniques have been widely used [50, 51, 129] because of their advantages over classical machine learning, such as faster and simpler pipelines [130], detecting text at different aspect ratios [71], and the ability to be better trained on synthetic data [44]. Table 10.1 summarizes a comparison among some of the current state-of-the-art techniques in this field.

Earlier deep learning-based text detection methods [42–45] usually consist of multiple stages. For instance, Jaderberg et al. [45] extended the architecture of a CNN to train a supervised learning model of text saliency,

Table 10.1 Deep learning text detection methods, where W, word; T, text-line; C, character; D, detection; R, recognition; BB, bounding box regression-based; SB, segmentation-based; ST, synthetic text; IC13, ICDAR13; IC15, ICDAR15; M500, MSRA-TD500; IC17, ICDAR17MLT; TOT, total-text; CTW, CTW-1500 and the rest of the abbreviations used in this table are presented in Table 10.2.

Method	Year	IF			Neural network		Detection target	Challenges		Task	Code	Model name
		BB	SB	Hy	Architecture	Backbone		Quad	Curved			
Jaderberg et al. [45]	2014	–	–		CNN	–	W	–	–	D,R	–	DSOL
Huang et al. [42]	2014	–	–		CNN	–	W	–	–	D	–	RSTD
Tian et al. [46]	2016	✓	–		Faster R-CNN	VGG-16	T, W	–	–	D	✓	CTPN
Zhang et al. [51]	2016	–	✓		FCN	VGG-16	W	✓	–	D	✓	MOTD
Yao et al. [52]	2016	–	✓		FCN	VGG-16	W	✓	–	D	✓	STDH
Shi et al. [71]	2017	✓	–		SSD	VGG-16	C, W	✓	–	D	✓	SegLink
He et al. [126]	2017	–	✓		SSD	VGG-16	W	✓	–	D	✓	SSTD
Hu et al. [131]	2017	–	✓		FCN	VGG-16	C	✓	–	D	–	Wordsup
Zhou et al. [47]	2017	✓	–		FCN	VGG-16	W, T	✓	–	D	✓	EAST
He et al. [129]	2017	✓	–		DenseBox	–	W,T	✓	–	D	–	DDR
Ma et al. [50]	2018	✓	–		Faster R-CNN	VGG-16	W	✓	–	D	✓	RRPN
Jiang et al. [132]	2018	✓	–		Faster R-CNN	VGG-16	W	✓	–	D	✓	R2CNN
Long et al. [54]	2018	–	✓		U-Net	VGG-16	W	✓	✓	D	✓	TextSnake
Liao et al. [49]	2018	✓	–		SSD	VGG-16	W	✓	–	D,R	✓	Text-Boxes++
He et al. [62]	2018	–	✓		FCN	PVA	C,W	✓	–	D,R	✓	E2ET
Lyu et al. [60]	2018	–	✓		Mask-R-CNN	ResNet-50	W	✓	–	D,R	✓	MTSpotter
Liao et al. [133]	2018	✓	–		SSD	VGG-16	W	✓	–	D	✓	RRD
Lyu et al. [56]	2018	–	✓		FCN	VGG-16	W	✓	–	D	✓	MOSTD
Deng et al.[a] [55]	2018	✓	–		FCN	VGG-16	W	✓	–	D	✓	PixelLink[a]
Liu et al. [61]	2018	✓	–		CNN	ResNet-50	W	✓	–	D,R	✓	FOTS
Baek et al.[a] [58]	2019	–	✓		U-Net	VGG-16	C, W, T	✓	✓	D	✓	CRAFT[a]
Wang et al.[a] [134]	2019	–	✓		FPEM+FFM	ResNet-18	W	✓	✓	D	✓	PAN[a]
Liu et al.[a] [59]	2019	–	–	✓	Mask-R-CNN	ResNet-50	W	✓	✓	D	✓	PMTD[a]
Xu et al. [135]	2019	–	✓		FCN	VGG-16	W	✓	✓	D	✓	Textfield
Liu et al.[a] [136]	2019	–	✓		Mask-R-CNN	ResNet-101	W	✓	✓	D	✓	MB[a]
Wang et al.[a] [137]	2019	–	✓		FPN	ResNet	W	✓	✓	D	✓	PSENet[a]

a) The method has been considered for evaluation in this chapter, where all the selected methods have been trained on ICDAR15 (IC15) dataset to compare their results in a unified framework.

Table 10.2 Supplementary table of abbreviations.

Attribution	Description
FCN [138]	Fully convolutional neural network
FPN [139]	Feature pyramid networks
PVA-Net [140]	Deep but lightweight neural networks for real-time object detection
RPN [141]	Region proposal network
SSD [142]	Single shot detector
U-Net [143]	Convolutional networks developed for biomedical image segmentation
FPEM [134]	Feature pyramid enhancement module
FFM [134]	Feature fusion module

then combining bounding boxes at multiple scales by undergoing filtering and NMS. Huang et al. [42] utilized both conventional connected components and deep learning to improve the precision of the final text detector. A classical MSER [69] high contrast region detector was employed to seek character candidates, followed by a CNN classifier to filter out non-text candidates by generating a confidence map, also later used to obtain detection results. In [44] the aggregate channel feature (ACF) detector [144] was used to generate text candidates, followed by a CNN for bounding box regression to reduce false-positive candidates. However, these earlier deep-learning methods [42, 43, 45] aim mainly to detect characters; thus, their performance may decline for text characters on a complicated background or characters affected by geometric variations [51].

Recent deep learning-based text detection methods [46–50, 62, 71] inspired by object detection pipelines [138, 141, 142, 145, 146] can be categorized as *bounding-box regression*, *segmentation*, and *hybrid* approaches, as illustrated in Figure 10.4.

Bounding-box regression-based methods [45–50] regard the text as an object and aim to predict the candidate bounding boxes directly. For example, TextBoxes [48] modified the single-shot descriptor (SSD) [142] kernels by applying long default anchors and filters to handle the significant variation of aspect ratios within text instances. Shi et al. [71] utilized an architecture inherited from SSD to decompose a text into smaller segments and then link them into text instances, so-called SegLink, by using spatial relationships or linking predictions between neighboring text segments, which enabled SegLink to detect long lines of Latin and non-Latin text. The connectionist

(a)   (b)

Figure 10.4 Illustrative example for semantic versus instance segmentation. Compare the ground-truth annotations for (a) semantic segmentation, where very close characters are linked, and (b) instance segmentation. Source: Karatzas et al. [147], from IEEE.

text proposal network (CTPN) [46], a modified version of Faster-recurrent convolutional neural network (R-CNN) [141], used an anchor mechanism to predict the location and score of each fixed-width proposal simultaneously, and then connected the sequential proposals by a recurrent neural network (RNN). Gupta et al. [148] proposed a fully-convolutional regression network inspired by the YOLO network [145], together with a random-forest classifier to reduce false positives. However, these methods [46, 48, 148], which were inspired from general object detection, may fail to handle multi-orientated text and require further steps to group text components into text lines, since word or text regions require bounding boxes of larger aspect ratios [71, 129].

Since scene text can appear in a variety of shapes, several works have focused on detecting multi-orientated text [47, 49, 50, 71, 129]. He et al. [129] proposed multi-oriented text (OT) detection based on direct regression to generate arbitrary quadrilaterals, particularly beneficial for scene text, in which the constituent characters are hard to identify due to significant variations in scale and perspective. In EAST [47], FCN is applied to detect text regions directly without using candidate aggregation and word partition, and then NMS is used to detect word or line text. This method predicts rotated boxes or quadrangles of words or text lines at each point in the text. Ma et al. [50] introduced rotation region proposal networks (RRPNs), based on Faster-R-CNN [141], to detect arbitrarily-oriented text in scene images. Later, Liao et al. [49] extended TextBoxes to TextBoxes++ by improving the network structure and training process and generalizing from rectangular bounding boxes to quadrilaterals. Although bounding-box-based methods [46, 47, 49, 50, 71, 129] have simple architectures, they require complex anchor design, are hard to tune during training, and may fail to deal with curved text.

Segmentation-based methods in [51–57, 59] cast text detection as a semantic segmentation problem, which aims to classify text regions in images at the pixel level, as shown in Figure 10.4a. These methods first extract text blocks from the segmentation map generated by a FCN [138] and then obtain bounding boxes by post-processing. For example, Zhang et al. [51] adopted FCN to predict the salient map of text regions, as well as for predicting the center of each character. Yao et al. [52] modified FCN to produce three kinds of score maps: text/non-text, character classes, and character linking orientations of the input images. Then a word partition post-processing is applied to obtain word bounding boxes. Although these segmentation-based methods [51, 52] perform well on rotated and irregular text, they may fail to accurately separate adjacent-word instances that tend to connect.

To address the problem of linked neighbor characters, PixelLinks [55] leveraged eight-directional information for each pixel to highlight the text margin, and Lyu et al. [56] proposed corner detection to produce a position-sensitive score map. In [54], TextSnake was proposed to detect text instances by predicting the text regions and the center-line together with geometry attributes. This method does not require character-level annotation and is capable of reconstructing the precise shape and regional strike of text instances. Inspired by [131], character affinity maps were used in [58] to connect detected characters into a single word and a weakly-supervised framework was used to train a character-level detector. In [137] a progressive scale expansion network (PSENet) was introduced to find kernels with multiple scales and accurately separate proximate text instances. However, the methods in [58, 137] require large numbers of images for training, increasing run time, and presenting difficulties on resource-limited platforms.

Recently, [59, 60, 149, 150] have treated scene text detection as an *instance segmentation problem*, an example of which is shown in Figure 10.4b, and many of them have applied the Mask R-CNN [146] framework to improve the performance of scene- text detection. For example, SPCNET [150] uses a text context module to detect text of arbitrary shapes and a re-score mechanism to suppress false positives. However, the methods in [60, 149, 150] do have drawbacks, in that they suffer from the errors of bounding box handling in a complicated background, where the predicted bounding box fails to cover the whole text image, and the related problem of mislabeled pixels at text borders [59].

Hybrid methods [56, 126, 133, 151] use segmentation-based approaches to predict score maps of text and aim to obtain text bounding-boxes through regression. For example, the single-shot text detector (SSTD) [126] used an attention mechanism to enhance text regions in image and reduce background interference. Liao et al. [133] proposed rotation-sensitive regression for oriented scene text detection, which makes full use of rotation-invariant features by actively rotating the convolutional filters. However, this method is incapable of capturing all possible text shapes [58]. Lyu et al. [56] presented a method that detects and groups corner points of text regions to generate text boxes. Besides detecting long-oriented text and handling considerable variations in aspect ratio, this method also requires simple post-processing. Liu et al. [59] proposed a new Mask R-CNN-based framework, pyramid mask text detector (PMTD) that assigns a soft pyramid label for each pixel in text, and then reinterprets the obtained two-dimensional (2D) soft mask into the 3D space. A novel plane clustering algorithm is employed on the soft pyramid to infer the optimal text box that helped this method achieve the state-of-the-art performance on several recent datasets [23, 152, 153]. Although the PMTD framework is designed explicitly for handling multi-oriented text, it is still underperforming on curved-text datasets [24, 154].

10.2.1.2   Text Recognition

Scene text recognition aims to convert detected text regions into characters or words. Case-sensitive character classes often consist of digits, lowercase and uppercase letters, and 32 ASCII punctuation marks. However, proposed text recognition models have used different choices of character classes, as can be seen in Table 10.3.

Since the properties of scene text will generally be different from those of scanned documents, it is difficult to develop an effective text-recognition method based on a classical OCR or handwriting recognition method, such as [158–163]. As mentioned in Section 10.1, this difficulty stems from images captured in the wild including various challenging conditions such as images of low resolution (LR) [101, 164], lightning extremes [101, 164], environmental conditions [23, 147], fonts [23, 147, 165], orientation [24, 165], languages [153], and lexicons [101, 164]. Researchers have proposed different techniques to address these challenges, which can be categorized into *classical machine learning* [32, 40, 41, 101, 119, 160] and *deep learning* [44, 64–67, 72, 83] methods, which are discussed in the rest of this section.

Classical Machine Learning-Based Methods   In the past two decades, traditional scene text recognition methods [40, 41, 160] have used standard image features, such as HOG [166] and SIFT [167], with a classical machine learning classifier, such as SVM [168] or k-nearest neighbors [169], followed by a statistical language model or visual structure prediction to prune-out misclassified characters [10, 115].

Most classical machine learning-based methods follow a *bottom-up* approach, such that classified characters are linked up into words. For example, in [32, 101] HOG features are first extracted from each sliding window, and then a pre-trained nearest neighbor or SVM classifier is applied to classify the characters of the input word. Neumann and Matas [119] proposed a set of handcrafted features with an SVM classifier for text recognition. However, these methods [32, 34, 101, 119] cannot achieve either an effective recognition accuracy, due to the low representation capability of handcrafted features, or building models that are able to handle text recognition in the wild.

Other works adopted a *top-down* approach, where a word is directly recognized from the entire input images, rather than detecting and recognizing individual characters [170]. For example, Almazán et al. [170] treated word recognition as a content-based image retrieval problem, where word images and word labels are embedded into an Euclidean space and the embedding vectors are used to match images and labels. One of the main problems of using these methods [170–172] is that they fail to recognize input word images outside of the word-dictionary dataset.

Deep Learning-Based Methods   With the recent advances in deep neural networks [73, 138, 173, 174], many researchers proposed deep-learning methods [28, 34, 64–67, 72–83, 155–157] to tackle the challenges of recognizing text in the wild, of which the state of the art are compared in Table 10.3. For example, Wang et al. [105] proposed a CNN-based feature extraction for character recognition, then applied the NMS technique [175] to obtain the final word predictions. Bissacco et al. [34] employed a fully connected network (FCN) for character feature representation, and then an n-gram approach to recognize characters. Similarly, [72] designed a deep CNN framework with multiple softmax classifiers, trained on a new synthetic text dataset. These early deep CNN-based

Table 10.3 Comparison among state-of-the-art of deep learning-based text recognition methods. TL, text-line; C, character; Seq, sequence recognition; PD, private dataset; HAM, hierarchical attention mechanism; ACE, aggregation cross-entropy. Other abbreviations are introduced in Table 10.4.

Method	Model	Year	Feature extraction	Sequence modeling	Prediction	Training dataset[a]	Irregular recognition	Task	# classes	Code
Wang et al. [105]	E2ER	2012	CNN	–	SVM	PD	–	C	62	–
Bissacco et al. [34]	PhotoOCR	2013	HOG,CNN	–	–	PD	–	C	99	–
Jaderberg et al. [72]	SYNTR	2014	CNN	–	–	MJ	–	C	36	✓
Jaderberg et al. [72]	SYNTR	2014	CNN	–	–	MJ	–	W	90k	✓
He et al. [155]	DTRN	2015	DCNN	LSTM	CTC	MJ	–	Seq	37	–
Shi et al.[b][65]	RARE	2016	STN+VGG16	BLSTM	Attn	MJ	✓	Seq	37	✓
Lee et al. [73]	R2AM	2016	Recursive CNN	LTSM	Attn	MJ	–	C	37	–
Liu et al.[b][66]	STARNet	2016	STN+RSB	BLSTM	CTC	MJ+PD	✓	Seq	37	✓
Shi et al[b][64]	CRNN	2017	VGG16	BLSTM	CTC	MJ	–	Seq	37	✓
Wang et al. [74]	GRCNN	2017	GRCNN	BLSTM	CTC	MJ	–	Seq	62	–
Yang et al. [75]	L2RI	2017	VGG16	RNN	Attn	PD+CL	✓	Seq	–	–
Cheng et al. [76]	FAN	2017	ResNet	BLSTM	Attn	MJ+ST+CL	–	Seq	37	–
Liu et al. [77]	Char-Net	2018	CNN	LTSM	Att	MJ	✓	C	37	–
Cheng et al. [78]	AON	2018	AON+VGG16	BLSTM	Attn	MJ+ST	✓	Seq	37	–
Bai et al. [79]	EP	2018	ResNet	–	Attn	MJ+ST	–	Seq	37	–
Liao et al. [156]	CAFCN	2018	VGG	–	–	ST	✓	C	37	–
Borisyuk et al.[b][67]	ROSETTA	2018	ResNet		CTC	PD	–	Seq	–	–
Shi et al.[b][28]	ASTER	2018	STN+ResNet	BLSTM	Attn	MJ+ST	✓	Seq	94	✓
Liu et al. [80]	SSEF	2018	VGG16	BLSTM	CTC	MJ	✓	Seq	37	–
Baek et al.[b][68]	CLOVA	2018	STN+ResNet	BLSTM	Attn	MJ+ST	✓	Seq	36	✓
Xie et al. [81]	ACE	2019	ResNet	–	ACE	ST+MJ	✓	Seq	37	✓
Zhan et al. [82]	ESIR	2019	IRN+ResNet, VGG	BLSTM	Attn	ST+MJ	✓	Seq	68	–
Wang et al. [83]	SSCAN	2019	ResNet, VGG	–	Attn	ST	✓	Seq	94	–
Wang et al. [157]	2D-CTC	2019	PSPNet	–	2D-CTC	ST+MJ	✓	Seq	36	–

a) Trained dataset/s used in the original paper. We used a pre-trained model of MJ+ST datasets for evaluation to compare the results in a unified framework.

b) This method has been considered for evaluation.

character recognition methods [34, 72, 105] require localizing each character, which may be challenging due to the complex background (CB), irrelevant symbols, and the short distance between adjacent characters in scene text images.

For word recognition, Jaderberg et al. [44] conducted a 90k English word classification task with CNN architecture. Although this method showed better word recognition performance compared to just individual character recognition [34, 72, 105], it has drawbacks that the method cannot recognize out-of-vocabulary words, and the deformation of long-word images may affect their recognition.

Considering that scene text generally appears in the form of a *sequence* of characters, many recent works [64–66, 76, 78–83, 155] have mapped every input sequence into an output sequence of variable length. Inspired by speech recognition, several sequence-based text recognition methods [64, 66, 67, 73, 74, 80, 155] have used *Connectionist*

Table 10.4 Abbreviation descriptions.

Attribution	Description
Attn	Attention-based sequence prediction
BLSTM	Bidirectional LTSM
CTC	Connectionist temporal classification
CL	Character-labeled
MJ	MJSynth
ST	SynthText
PD	Private data
STN	Spatial transformation network [176]
TPS	Thin-plate spline
PSPNet	Pyramid scene parsing network [177]

Temporal Classification (CTC) [178] for the prediction of character sequences. Figure 10.5 illustrates three main CTC-based text recognition frameworks that have been used in the literature.

Figure 10.5a [67, 179] shows the use of CNN models (such as, VGG [173], R-CNN [73], and ResNet [174]) with CTC. For instance, in [179], a sliding window is first applied to the text-line image in order to effectively capture contextual information, and then a CTC prediction is used to predict the output words. ROSETTA [67] used only the extracted features from a CNN by applying ResNet as a backbone to predict the feature sequences. Despite

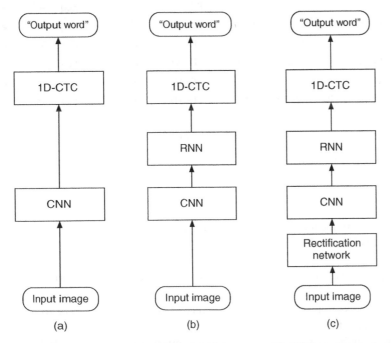

Figure 10.5 Comparison among some of the recent 1D CTC-based scene text recognition frameworks: (a) baseline frame of CNN with 1D-CTC. Source: Borisyuk et al. [67]. (b) Adding RNN to the baseline of (a). Source: Shi et al. [64]. (c) Using a rectification network before (b). Source: Liu et al. [66].

reducing the computational complexity, these methods [67, 179] suffered from a lack of contextual information and poor recognition accuracy.

Figure 10.5b shows the inclusion of RNN [75], combined with CTC [64, 74, 155] to identify the conditional probability between the predicted and the target sequences to better extract contextual information. For example, in [64] a VGG model [180] is employed as a backbone to extract features of input images, followed by a bidirectional long-short-term-memory (BLSTM) [181] for contextual information, and then a CTC loss is applied to identify character sequences. Later, Wang and Hu [74] proposed a new architecture based on R-CNN, namely gated recurrent convolutional neural network (GRCNN). However, as illustrated in Figure 10.6a these techniques [64, 74, 155] are insufficient to recognize irregular text [81] as characters are arranged on a 2D image plane and the CTC-based methods are only designed for one-dimensional (1D) sequential alignment, requiring a conversion of 2D image features to 1D, leading to a loss of relevant information [157].

Figure 10.5c introduces a spatial transform network (STN) [176] to handle text irregularities and distortions. Liu et al. [66] proposed a spatial-attention residue network (STAR-Net) that leveraged STN for tackling text distortions. Recently, Wan et al. [157] introduced a 2D-CTC technique (Figure 10.6b), generalizing 1D-CTC, which can be directly applied to 2D probability distributions to produce more precise recognition, considering all possible paths over height in order to better align the search space and focus on relevant features.

The *attention mechanism* first used for machine translation in [182] has been also adopted for scene text recognition [28, 65, 66, 73, 75, 77, 78, 82], where an implicit attention is learned automatically to enhance deep features in the decoding process. Figure 10.7 illustrates five main attention-based text recognition frameworks that have been used in the literature.

Figure 10.7a presents a basic 1D-attention-based encoder and decoder framework [73, 185, 186]. For example, Lee and Osindero [73] proposed a recursive RNN with attention modeling (R2AM), however directly training R2AM on irregular text is difficult due to the on-horizontal character placement [75].

Figure 10.7b shows attention-based methods [28, 68, 77, 82, 187] using image rectification modules to control distorted text images, similar to CTC-based recognition methods in handling irregular text. For instance, Shi et al. [28, 65] proposed a text recognition system that combined attention-based sequence and a STN module to rectify irregular text, followed by RNN recognition. However, training a STN-based method without considering human-designed geometric ground truth is difficult, especially, in complicated arbitrarily-oriented or strongly-curved text images. Instead of rectifying entire word images, Liu et al. [77] presented a Character-Aware

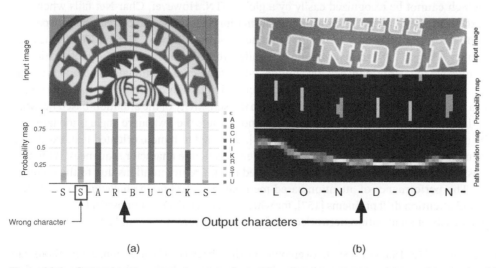

(a)   (b)

Figure 10.6 Comparing the processing steps for tackling the character recognition problem using (a) 1D-CTC. Source: (a) Ken Wolter/Shutterstock.com (b) [157]. Wan et al. (2006), from Cornell University.

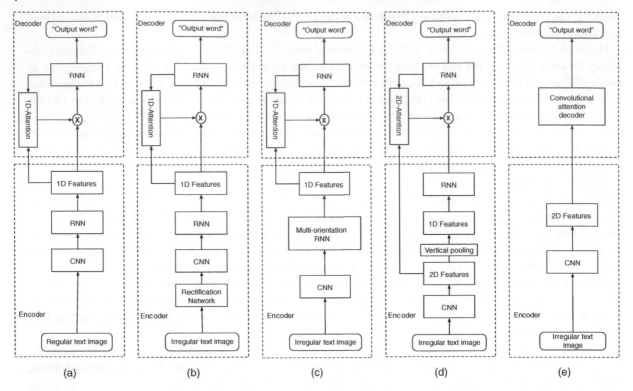

Figure 10.7 Comparison of recent attention-based scene text recognition frameworks: (a) 1D basic model. Source: Lee and Osindero [73]. (b) 1D rectification network of ASTER. Source: Shi et al. [28]. (c) Multi-orientation encoding of AON. Source: Cheng et al. [78]. (d) 2D-attention-based decoding. Source: Li et al. [183]. (e) Convolutional attention-based decoding in SRCAN (Source: Wang et al. [83]) and FACLSTM (Source: Wang et al. [184]).

Neural Network (Char-Net) for recognizing distorted scene characters. Unlike STN, Char-Net can detect and rectify individual characters using a simple local spatial transforms, leading to the detection of more complex forms of distorted text, which cannot be recognized easily by a global STN. However, Char-Net fails where the images contain blurry text, so in [82] a robust line-fitting transformation is proposed to correct the prospective and curvature distortion of scene text images in an iterative manner. For this purpose, an iterative rectification network using the thin plate spline (TPS) transformation is applied in order to increase the rectification of curved images, and thus improve the performance of recognition, but at high-computational cost due to multiple rectification steps. Luo et al. [187] proposed a new multi-object rectified attention network (MORAN) to rectify irregular text images and a fractional pickup mechanism to enhance the sensitivity of the attention-based network in the decoder, however, the method fails in complicated backgrounds, where the text curve angle is too large.

Figure 10.7c shows the arbitrary orientation network (AON) [78], designed to handle oriented text, extracting deep features in four orientations, and a 1D attention-based decoder to generate character sequences. Although AON can be trained by using word-level annotations, it leads to redundant representations due to the complex four-directional network. The performance of attention-based methods may decline in challenging conditions, leading to misalignment and attention drift problems [157], for which Cheng et al. [76] proposed a focusing attention network (FAN) that consists of an attention network (AN) for character recognition and a focusing network (FN) for adjusting attention.

Figure 10.7d used 2D attention [75, 156, 183, 188] to overcome the drawbacks of 1D attention. 2D methods can learn to focus on individual character features in the 2D space during decoding, which can be trained using either character-level [75] or word-level [183] annotations. For example, Yang et al. [75] introduced an auxiliary dense character detection task using a fully convolutional network (FCN) for encouraging the learning of visual representations to improve the recognition of irregular scene text. Later, Liao et al. [156] proposed character attention FCN (CA-FCN), modeling the irregular scene text recognition problem in a 2D space instead of 1D, in which a character attention module [189] is used to predict multi-orientation characters in an arbitrary shape of an image. In contrast, Li et al. [183] proposed a model that used word-level annotations which enables this model to utilize both real and synthetic data for training without using character-level annotations, however, two-layer RNNs are adopted which preclude computation parallelization and suffer from heavy computational burden.

Figure 10.7e addresses these computational costs, eliminating the RNN stage of 2D-attention techniques [83, 184], and a convolution-attention network [190] was used instead, enabling irregular text recognition and fully parallel computation. For example, Wang et al. [83] proposed a simple and robust convolutional-attention network (SRACN), where a convolutional attention network decoder is directly applied to 2D CNN features. Wang et al. [184] considered the scene text recognition as a spatio-temporal prediction problem and proposed the focus attention convolution long-short-term-memory (FACLSTM) for scene text recognition. It is shown in [184] that FACLSTM is more effective for text recognition, specifically for curved scene text datasets, such as CUT80 [165] and Street View Text-Perspective (SVT-P) [191].

10.3   Experimental Results

In this section, we present an extensive evaluation for some selected state-of-the-art scene-text detection [47, 55, 58, 59, 134, 136, 137] and recognition [28, 64–68] techniques on recent public datasets [23, 102, 147, 162, 164, 165, 191] that include wide variety of challenges. One of the important characteristics of a scene text detection or recognition scheme is to be generalizable, which shows how a trained model on one dataset is capable of detecting or recognizing challenging text instances on other datasets. This evaluation strategy is an attempt to close the gap in evaluating text detection and recognition methods that are used to be mainly trained and evaluated on a specific dataset. Therefore, to evaluate the generalization ability for the methods under consideration, we propose to compare both detection and recognition models on unseen datasets.

Specifically, we selected the following methods for evaluation of the recent advances in the deep learning-based schemes for scene text detection: PMTD[1] [59], CRAFT[2] [58], EAST[3] [47], PAN[4] [134], MB[5] [136], PSENET[6] [137], and PixelLink[7] [55]. For each method except MB [136], we used the corresponding pre-trained model directly from the authors' GitHub page that was trained on ICDAR15 [23] dataset. While for MB [136], we trained the algorithm on ICDAR15 according to the code that was provided by the authors. For testing the detectors in consideration, the ICDAR13 [147], ICDAR15 [23], and COCO-Text [192] datasets have been used. This evaluation strategy avoids an unbiased evaluation and allows assessment for the generalizability of these techniques. Table 10.5 illustrates the number of test images for each of these datasets.

For conducing evaluation among the scene text recognition schemes the following deep-learning-based techniques have been selected: CLOVA [68], ASTER [28], CRNN [64], ROSETTA [67], STAR-Net [66], and RARE [65]. Since recently the SynthText (ST) [148] and MJSynth (MJ) [72] synthetic datasets have been used extensively for building recognition models, we aim to compare the state-of-the-arts methods when using these synthetic datasets. All recognition models have been trained on combination of SynthText [148] and MJSynth [72] datasets, while for evaluation we have used ICDAR13 [147], ICDAR15 [23], and COCO-Text [192] datasets, in addition to four mostly

---

1 https://github.com/jjprincess/PMTD
2 https://github.com/clovaai/CRAFT-pytorch
3 https://github.com/ZJULearning/pixel_link
4 https://github.com/WenmuZhou/PAN.pytorch
5 https://github.com/Yuliang-Liu/Box_Discretization_Network
6 https://github.com/WenmuZhou/PSENet.pytorch
7 https://github.com/argman/EAST

Table 10.5 Comparison among some of the recent text detection and recognition datasets.

Dataset	Year	# Detection images			# Recognition words		Orientation			Properties		Task	
		Train	Test	Total	Train	Test	H	MO	Cu	Language	Annotation	D	R
IC03a)[162]	2003	258	251	509	1 156	1 110	cm	—	—	EN	W, C	cm	cm
SVTa)[101]	2010	100	250	350	—	647	cm	—	—	EN	W	cm	cm
IC11 [193]	2011	100	250	350	211	514	cm	—	—	EN	W, C	cm	cm
IIIT 5K-wordsa)[164]	2012	—	—	—	2 000	3 000	cm	—	—	EN	W	—	cm
MSRA-TD500 [107]	2012	300	200	500	—	—	cm	cm	—	EN, CN	TL	cm	—
SVT-Pa)[191]	2013	—	238	238	—	639	cm	cm	—	EN	W	cm	cm
ICDAR13a)[147]	2013	229	233	462	848	1 095	cm	—	—	EN	W	cm	cm
CUT80a)[165]	2014	—	80	80	—	280	cm	cm	cm	EN	W	cm	cm
COCO-Texta)[192]	2014	43 686	20 000	63 686	118 309	27 550	cm	cm	cm	EN	W	cm	cm
ICDAR15a)[23]	2015	1 000	500	1 500	4 468	2 077	cm	cm	—	EN	W	cm	cm
ICDAR17 [153]	2017	7 200	9 000	18 000	68 613	—	cm	cm	cm	ML	W	cm	cm
TotalText [24]	2017	1 255	300	1 555	—	11 459	cm	cm	cm	EN	W	cm	cm
CTW-1500 [154]	2017	1 000	500	1 500	—	—	cm	cm	cm	CN	W	cm	cm
SynthText [148]	2016	800k	—	800k	8M	—	cm	cm	cm	EN	W	cm	cm
MJSynth [72]	2014	—	—	—	8.9M	—	cm	cm	cm	EN	W	—	cm

a) This dataset has been considered for evaluation. H, horizontal; MO, multi-oriented; Cu, curved; EN, English; CN, Chinese; ML, multi-language; W, word; C, character; TL, textline; D, detection; R, recognition.

used datasets, namely, III5k [164], CUT80 [165], Street View Text (SVT) [101], and SVT-P [191] datasets. As shown in Table 10.5, the selected datasets cover datasets that mainly contain regular or horizontal text images and other datasets that include curved, rotated, and distorted, or so-called irregular, text images. Throughout this evaluation also, we used 36 classes of alphanumeric characters, 10 digits (0–9) + 26 capital English characters (A–Z) = 36.

In the remaining part of this section, we will start by summarizing the challenges within each of the utilized datasets (Section 10.3.1) and then presenting the evaluation metrics (Section 10.3.2). Next, we present the quantitative and qualitative analysis, as well as discussion on scene text detection methods (Section 10.3.3), and on scene text recognition methods (Section 10.3.4). Finally, we discuss open investigations for scene text detection and recognition (Section 10.3.5).

10.3.1 Datasets

There exist several datasets that have been introduced for scene text detection and recognition [23, 24, 72, 101, 107, 147, 148, 153, 154, 162, 164, 165, 191, 192]. These datasets can be categorized into synthetic datasets that are used mainly for training purposes, such as [72, 194], and real-word datasets that have been utilized extensively for evaluating the performance of detection and evaluation schemes, such as [23, 101, 107, 147, 154, 192]. Table 10.5 compares some of the recent text detection and recognition datasets, and the rest of this section presents a summary of each of these datasets.

10.3.1.1 MJSynth

The *MJSynth* [72] dataset is a synthetic dataset that is specifically designed for scene text recognition. Figure 10.8a shows some examples of this dataset. This dataset includes about 8.9 million word-box gray synthesized images,

Figure 10.8 Sample images of synthetic datasets used for training in scene text detection and recognition. (a) [148]. Gupta et al. (2016), from IEEE" and (b) SynthText. Source: [72]. Jaderberg et al. (2014), from Cornell University".

which have been generated from the Google fonts and the images of ICDAR03 [162] and SVT [101] datasets. All the images in this dataset have been annotated in word-level groundtruth and 90k common English words have been used for generating these text images.

10.3.1.2 SynthText

The *SynthText* in the wild dataset [194] contains 858 750 synthetic scene images with 7 266 866 word-instances, and 28 971 487 characters. Most of the text instances in this dataset are multi-oriented and annotated with word and character-level rotated bounding boxes, as well as text sequences (see Figure 10.8b). They are created by blending natural images with text rendered with different fonts, sizes, orientations, and colors. This dataset has been originally designed for evaluating scene text detection [194] and leveraged in training several detection pipelines [58]. However, many recent text recognition methods [28, 78, 81, 82, 157] have also combined the cropped word images of the mentioned dataset with the MJSynth dataset [72] for improving their recognition performance.

10.3.1.3 ICDAR03

The *ICDAR03* dataset [162] contains horizontal camera-captured scene text images. This dataset has been mainly used by recent text recognition methods, which consists of 1156 and 110 text instances for training and testing, respectively. In this chapter, we have used the same test images of [68] for evaluating the state-of-the-art text recognition methods.

10.3.1.4 ICDAR13

The *ICDAR13* dataset [147] includes images of horizontal text (the ith groundtruth annotation is represented by the indices of the top left corner associated with the width and height of a given bounding box as $G_i = \left[x_1^i, y_1^i, x_2^i, y_2^i\right]^\top$ that have been used in ICDAR 2013 competition, and it is one of the benchmark datasets used in many detection and recognition methods [47, 55, 58, 59, 61, 64–66, 68, 187]. The detection part of this dataset consists of 229 images for training and 233 images for testing; the recognition part consists of 848 word-image for training and 1095 word images for testing. All text images of this dataset have good quality and text regions are typically centered in the images.

10.3.1.5 ICDAR15

The *ICDAR15* dataset [23] can be used for the assessment of text detection or recognition schemes. The detection part has 1500 images in total consists of 1000 training and 500 testing images for detection, and the recognition part consists of 4468 images for training and 2077 images for testing. This dataset includes text at the word level of various orientation and captured under different illumination and complex backgrounds conditions than that included in ICDAR13 dataset [147]. However, most of the images in this dataset are captured for an indoor environment. In scene text detection, rectangular groundtruth used in the ICDAR13 [147] is not adequate for the representation of multi-oriented text because: (i) they cause unnecessary overlap, (ii) they cannot precisely localize marginal text, and (3) they provide unnecessary noise of background [195]. Therefore to tackle the mentioned issues, the annotations of this dataset are represented using quadrilateral boxes (the ith groundtruth annotation can be expressed as $G_i = \left[x_1^i, y_1^i, x_2^i, y_2^i, x_3^i, y_3^i, x_4^i, y_4^i \right]^\top$ for four corner vertices of the text).

10.3.1.6 COCO-Text

This dataset firstly was introduced in [192], and so far, it is the largest and the most challenging text detection and recognition dataset. As shown in Table 10.5, the dataset includes 63 686 annotated images, where the dataset is partitioned into 43 686 training images, and 20 000 images for validation and testing. In this chapter, we use the second version of this dataset, COCO-Text, as it contains 239 506 annotated text instances instead of 173 589 for the same set of images. As in ICDAR13, text regions in this dataset are annotated at a word level using rectangle bounding boxes. The text instances of this dataset also are captured from different scenes, such as outdoor scenes, sports fields, and grocery stores. Unlike other datasets, COCO-Text dataset also contains images with low resolution, special characters (SCs), and partial occlusion (PO).

10.3.1.7 SVT

The SVT dataset [101] consists of a collection of outdoor images with scene text of high variability of blurriness and/or resolutions, which were harvested using Google Street View. As shown in Table 10.5 [47, 55, 58, 59, 61, 64–66, 68, 187], this dataset includes 250 and 647 testing images for evaluation of detection and recognition tasks, respectively. We utilize this dataset for assessing the state-of-the-art recognition schemes.

10.3.1.8 SVT-P

The *SVT-P* dataset [191] is specifically designed to evaluate recognition of perspective distorted scene text. It consists of 238 images with 645 cropped text instances collected from non-frontal angle snapshot in Google Street View, which many of the images are perspective distorted.

10.3.1.9 IIIT 5K-Words

The *IIIT 5K-words* dataset contains 5000 word-cropped scene images [164], that is used only for word-recognition tasks, and it is partitioned into 2000 and 3000-word images for training and testing tasks, respectively. In this chapter, we use only the testing set for assessment.

10.3.1.10 CUT80

The *Curved Text* (CUT80) dataset is the first dataset that focuses on curved text images [165]. This dataset contains 80 full and 280 cropped word images for evaluation of text detection and text recognition algorithms, respectively. Although CUT80 dataset was originally designed for curved text detection, it has been widely used for scene text recognition [165].

10.3.2 Evaluation Metrics

The ICDAR standard evaluation metrics [23, 147, 162, 196] are the most commonly used protocols for performing quantitative comparison among the text detection techniques [10, 11].

10.3.2.1 Detection

In order to quantify the performance of a given text detector, as in [47, 55, 58, 59], we utilize the Precision (P) and Recall (R) metrics that have been used in information retrieval field. In addition, we use the H-mean or $F1$-score (Eq. (10.1)) that can be obtained as follows.

$$H - \text{mean} = 2 \times \frac{P \times R}{P + R} \tag{10.1}$$

where calculating the precision and recall are based on using the ICDAR15 intersection over union (IoU) metric [23], which is obtained for the jth groundtruth and ith detection bounding box (Eq. (10.2)) as follows:

$$\text{IoU} = \frac{\text{Area}(G_j \cap D_i)}{\text{Area}(G_j \cup D_i)} \qquad (10.2)$$

and a threshold of IoU ≥ 0.5 is used for counting a correct detection.

10.3.2.2 Recognition

Word recognition accuracy (WRA) is a commonly used evaluation metric, due to its application in our daily life instead of character recognition accuracy, for assessing the text recognition Schemes [28, 64–66, 68]. Given a set of cropped word images, WRA is defined as follows in the equation (Eq. (10.3)):

$$\text{WRA (\%)} = \frac{\text{No. of correctly recognized words}}{\text{Total number of words}} \times 100 \qquad (10.3)$$

10.3.3 Evaluation of Text Detection Techniques

10.3.3.1 Quantitative Results

To evaluate the generalization ability of detection methods, we compare the detection performance on ICDAR13 [147], ICDAR15 [23], and COCO-Text [102] datasets. Table 10.6 illustrates the detection performance of the selected state of the art text detection methods, namely, PMTD [59], CRAFT [58], PSENet [137], MB [136], PAN [134], PixelLink [55], and EAST [47]. From this table, although the ICDAR13 dataset includes less challenging conditions than that included in the ICDAR15 dataset, the detection performances of all the methods in consideration have been decreased in this dataset. Comparing the same method performance on ICDAR15 and ICDAR13, PMTD offered a minimum performance decline of ~0.60% in H-mean, while PixelLink which ranked the second-best on ICDAR15 had the worst H-mean value on ICDAR13 with decline of ~20.00%. Further, all methods experienced a significant decrease in detection performance when tested on COCO-Text dataset, which indicates that these models do not yet provide a generalization capability on different challenging datasets.

10.3.3.2 Qualitative Results

Figure 10.9 illustrates sample detection results for the considered methods [47, 55, 58, 59, 134, 136, 137] on some challenging scenarios from ICDAR13, ICDAR15, and COCO-Text datasets. Even though, the best text detectors, PMTD and CRAFT detectors, offer better robustness in detecting text under various orientation and partial

Table 10.6 Quantitative comparison among some of the recent text detection methods on ICDAR13 [147], ICDAR15 [23], and COCO-Text [102] datasets using precision (P), recall (R), and H-mean.

Method	ICDAR13			ICDAR15			COCO-Text		
	P (%)	R (%)	H-mean (%)	P (%)	R (%)	H-mean (%)	P (%)	R (%)	H-mean (%)
EAST [47]	*84.86*	*74.24*	*79.20*	84.64	77.22	80.76	55.48	32.89	41.30
PixelLink [55]	62.21	62.55	62.38	82.89	81.65	*82.27*	*61.08*	33.45	43.22
PAN [134]	83.83	69.13	75.77	*85.95*	73.66	79.33	59.07	43.64	50.21
MB [136]	72.64	60.36	65.93	85.75	76.50	80.86	55.98	48.45	51.94
PSENet [137]	81.04	62.46	70.55	84.69	77.51	80.94	60.58	49.39	54.42
CRAFT [58]	72.77	*77.62*	75.12	82.20	*77.85*	79.97	56.73	*55.99*	*56.36*
PMTD [59]	**92.49**	**83.29**	**87.65**	**92.37**	**84.59**	**88.31**	**61.37**	**59.46**	**60.40**

Note: The best and second-best methods are highlighted in bold and italics, respectively.
Source: Karatzas et al. [147]. Karatzas et al. [23]. Veit et al. [102].

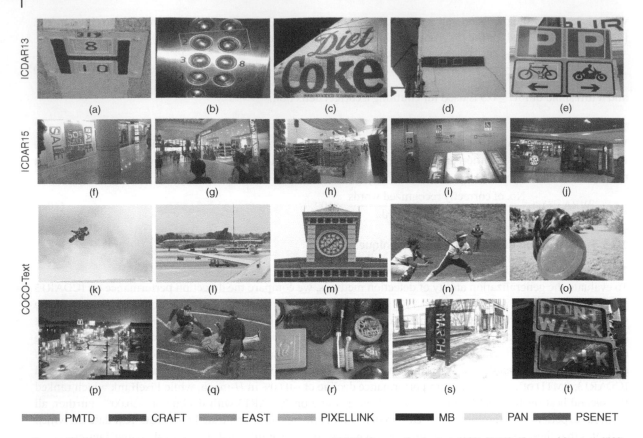

Figure 10.9 Qualitative detection results comparison among CRAFT (Source: Baek et al. [58]), PMTD (Source: Liu et al. [59]), MB (Source: Liu et al. [136]), PixelLink (Source: Deng et al. [55]), PAN (Source: Wang et al. [134]), EAST (Source: Zhou et al. [47]), and PSENET (Source: Wang et al. [137]) on some challenging examples, where PO, partial occlusion; DF, difficult fonts; LC, low contrast; IV, illumination variation; IB, image blurriness; LR, low resolution; PD, perspective distortion; IPR, in-plane rotation; OT, oriented text; and CT, curved text. Note: since we used pre-trained models on ICDAR15 dataset for all the methods in comparison, the results of some methods may differ from those reported in the original papers. (a) IPR; (b) IV; (c) DF; (d) LC; (e) PO; (f) OT; (g) OT, IV; (h) LR, DF; (i) IB, LR, IV; (j) CT, DF, LC; (k) DF, OT; (l) IL, PD; (m) CT; (n) OT, LC; (o) OT, IV; (p) IV, IPR; (q) PO; (r) DF, CT, IV; (s) OT, IV; and (t) LC, IV, IB.

occlusion levels, these detection results illustrate that the performances of these methods are still far from perfect. Especially, when text instances are affected by challenging cases like the text of difficult fonts (DFs), colors, backgrounds, and illumination variation (IV) and in-plane rotation (IPR), or a combination of challenges. Now we categorize the common difficulties in scene text detection as follows:

Diverse Resolutions and Orientations Unlike the detection tasks, such as the detection of pedestrians [197] or cars [198, 199], text in the wild usually appears on a wider variety of resolutions and orientations, which can easily lead to inadequate detection performances [11, 12]. For instance on ICDAR13 dataset, as can be seen from the results in Figure 10.9a, all the methods failed to detect the low and high resolution text using the default parameters of these detectors. The same conclusion can be drawn from the results in Figures 10.9h,q on ICDAR15 and COCO-Text datasets, respectively. As well as, this conclusion can also be confirmed from the distribution of word height in pixels on the considered datasets as shown in Figure 10.10. Although the considered detection models have focused on handling multi-oriented text, they still lack the robustness in tackling this challenge as well as facing difficulty in detecting text subjected to in-plan rotation or high curvature. For example, the low detection performance noted can be seen in Figure 10.9a,j,p on ICDAR13, ICDAR15, and COCO-Text, respectively.

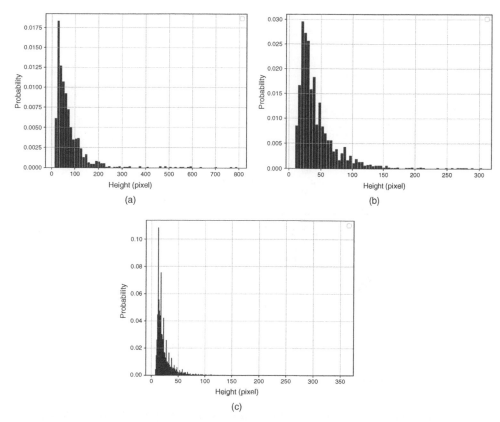

Figure 10.10 Distribution of word height in pixels computed on the test set of (a) ICDAR13, (b) ICDAR15, and (c) COCO-Text detection datasets.

Occlusions Similar to other detection tasks, text can be occluded by itself or other objects, such as text or object superimposed on a text instance as shown in Figure 10.9. Thus, it is expected for text-detection algorithms to at least detect partially occluded text. However, as we can see from the sample results in Figure 10.9e,f, the studied methods failed in the detection of text mainly due to the partial occluded effect.

Degraded Image Quality text images captured in the wild are usually affected by various illumination conditions (as in Figure 10.9b,d), motion blurriness (as in Figure 10.9g,h), and low contrast (LC) text (as in Figure 10.9o,t). As we can see from Figure 10.9, the studied methods perform weakly on these types of images. This is due to existing text detection techniques have not tackled explicitly these challenges.

10.3.3.3 Discussion
In this section, we present an evaluation of the mentioned detection methods with respect to their robustness and speed.

Detection Robustness As we can see from Figure 10.10, most of the target words that existed in the three target scene text detection datasets are of low resolutions, which makes the text detection task more challenging. To compare the robustness of the detectors under the various IoU values, Figure 10.11 illustrates the H-mean computed at IoU $\in [0, 1]$ for each of the studied methods. From this figure, it can be noted that increasing the IoU > 0.5 causes rapidly reducing the H-mean values achieved by the detectors on all three datasets, which indicates that the considered schemes are not offering adequate overlap ratios, i.e. IoU, at higher threshold values.

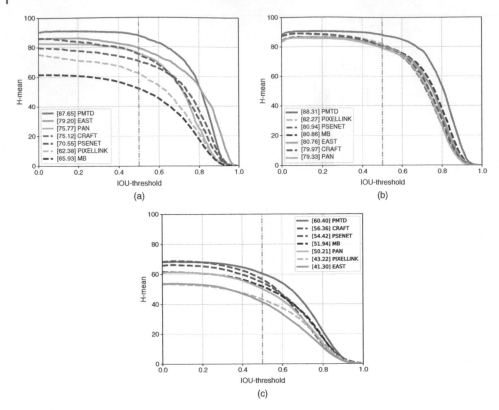

Figure 10.11 Evaluation of the text detection performance for CRAFT (Source: Baek et al. [58]), PMTD (Source: Liu et al. [59]), MB (Source: Liu et al. [136]), PixelLink (Source: Deng et al. [55]), PAN (Source: Wang et al. [134]), EAST (Source: Zhou et al. [47]), and PSENET (Source: Wang et al. [137]) using *H*-mean versus IoU $\in [0, 1]$ computed on (a) ICDAR13 (Source: Karatzas et al. [147]), (b) ICDAR15 (Source: Karatzas et al. [23]), and (c) COCO-Text (Source: Veit et al. [102]) datasets.

More specifically, on ICDAR13 dataset (Figure 10.11a) EAST [47] detector outperforms the PMTD [59] for IoU > 0.8; this can be attributed to that EAST detector uses a multi-channel FCN network that allows detecting more accurately text instances at different scales that are abundant in ICDAR13 dataset. Further, PixelLink [55] which ranked second on ICDAR15 (Figure 10.11b) has the worst detection performance on ICDAR13. This poor performance is also can be seen in challenging cases of the qualitative results in Figure 10.9. For COCO-Text [102] dataset, all methods offer poor *H*-mean performance on this dataset (Figure 10.11c). In addition, generally, the *H*-means of the detectors declined to half, from ~60% to below ~30%, for IoU $\geq$ 0.7.

In summary, PMTD and CRAFT show better *H*-mean values than that of EAST and PixelLink for IoU < 0.7. Since CRAFT is character-based method, it performed better in localizing difficult-font words with individual characters. Moreover, it can better handle text with different size due to its property of localizing individual characters, not the whole word detection has been used in the majority of other methods. Since COCO-Text and ICDAR15 datasets contain multi-oriented and curved text, we can see from Figure 10.11 that PMTD shows robustness to imprecise detection compared to other methods for precise detection for IoU > 0.7, which means this method can predict better arbitrary shape of text. This is also obvious in challenging cases like curved text, difficult fonts with various orientations, and in-plane rotated text of Figure 10.9.

Detection Speed To evaluate the speed of detection methods, Figure 10.12 plots *H*-mean versus the speed in frame per second (FPS) for each detection algorithm for IoU $\geq$ 0.5. The slowest and fastest detectors are PSENet [137] and PAN [134], respectively. PMTD [59] achieved the second-fastest detector with the best *H*-mean. PAN

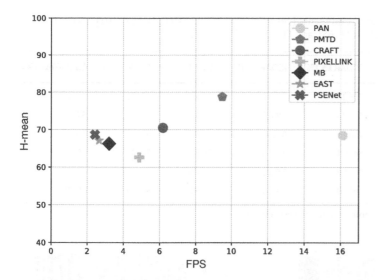

Figure 10.12 Average *H*-mean versus frames per second (FPS) computed on ICDAR13 (Source: Karatzas et al. [147]), ICDAR15 (Source: Karatzas et al. [23]), and COCO-Text (Source: Veit et al. [102]) detection datasets using IoU ≥ 0.5.

utilizes a segmentation-head with low computational cost using a lightweight model, ResNet [174] with 18 layers (ResNet18), as a backbone and a few stages for post-processing that results in an efficient detector. On the other hand, PSENet uses multiple scales for predicting text instances using a deeper (ResNet with 50 layer) model as a backbone, which causes it to be slow during the test time.

10.3.4 Evaluation of Text Recognition Techniques

10.3.4.1 Quantitative Results
In this section, we compare the selected scene text recognition methods [28, 64–67] in terms of the WRA defined in (10.3) on datasets with regular [147, 162, 164] and irregular [23, 102, 165, 191] text, and Table 10.7 summarizes these quantitative results. It can be seen from this table that all methods have generally achieved higher WRA values on datasets with regular text [147, 162, 164] than that achieved on datasets with irregular text [23, 102, 165, 191]. Furthermore, methods that contain a rectification module in their feature extraction stage for spatially transforming text images, namely, ASTER [28], CLOVA [68], and STAR-Net [66], have been able to perform better on datasets with irregular text. In addition, attention-based methods, ASTER [28], and CLOVA [68] outperformed the CTC-based methods, CRNN [64], STAR-Net [66], and ROSETTA [67] because attention methods better handle the alignment problem in irregular text compared to CTC-based methods.

It is worth noting that despite the studied text recognition methods having used only synthetic images for training, as can be seen from Table 10.3, they have been able to handle recognizing text in the wild images. However, for COCO-Text dataset, each of the methods has achieved a much lower WRA values than that obtained on the other datasets, this can be attributed to the more complex situations that exist in this dataset that the studied models are not able to fully encounter. In Section 10.3.4.2, we will highlight the challenges that most of the state-of-the-art scene text recognition schemes are currently facing.

10.3.4.2 Qualitative Results
In this section, we present a qualitative comparison among the considered text recognition schemes, as well as conduct an investigation on challenging scenarios that still cause partial or complete failures to the existing techniques. Figure 10.13 highlights a sample of qualitative performances for the considered text recognition methods

Table 10.7 Comparing some of the recent text recognition techniques using WRA on IIIT5k [164], SVT [101], ICDAR03 [162], ICDAR13 [147], ICDAR15 [23], SVT-P [191], CUT80 [165], and COCO-Text [102] datasets.

Method	IIIT5k (%)	SVT (%)	ICDAR03 (%)	ICDAR13 (%)	ICDAR15 (%)	SVT-P (%)	CUT80 (%)	COCO-Text (%)
CRNN	82.73	82.38	93.08	89.26	65.87	70.85	62.72	48.92
RARE	83.83	82.84	92.38	88.28	68.63	71.16	66.89	54.01
ROSETTA	83.96	83.62	92.04	89.16	67.64	74.26	67.25	49.61
STAR-Net	86.20	86.09	94.35	90.64	72.48	76.59	71.78	55.39
CLOVA	*87.40*	*87.01*	**94.69**	**92.02**	**75.23**	*80.00*	*74.21*	*57.32*
ASTER	**93.20**	**89.20**	*92.20*	*90.90*	*74.40*	**80.90**	**81.90**	**60.70**

Note: Best and second-best methods are highlighted by bold and italics, respectively.

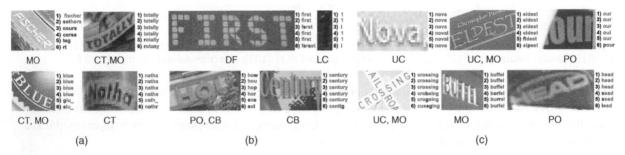

Figure 10.13 Qualitative results for challenging examples of scene text recognition: (a) multi-oriented and curved text, (b) complex backgrounds and fonts, and (c) text with partial occlusions, sizes, colors, or fonts. Note: The original target word images and their corresponding recognition output are on the left and right-hand sides of every sample results, respectively, where the numbers denote: (1) ASTER (Source: Shi et al. [28]), (2) CLOVA (Source: Baek et al. [68]), (3) RARE (Source: Shi et al. [65]), (4) STAR-Net (Source: Liu et al. [66]), (5) ROSETTA (Source: Borisyuk et al. [67]), and (6) CRNN (Source: Shi et al. [64]), and the resulted characters highlighted by light gray and dark gray denote correctly and wrongly predicted characters, respectively, where MO, multi-oriented; VT, vertical text; CT, curved text; DF, difficult font; LC, low contrast; CB, complex background; PO, partial occlusion; and UC: unrelated character. It should be noted that the results of some methods may differ from that reported in the original papers. because we used pre-trained models on MJSynth (Source: Jaderberg et al. [72]) and SynthText (Source: Gupta et al. [148]) datasets for each specific method.

on ICDAR13, ICDAR15, and COCO-Text datasets. As shown in Figure 10.13a, methods in [28, 65, 66, 68] performed well on multi-oriented and curved text because these methods adopt TPS as a rectification module in their pipeline that allows rectifying irregular text into a standard format, and thus, the subsequent CNN can better extract features from this normalized images. Figure 10.13b illustrates text subject to complex backgrounds or unseen fonts. In these cases, the methods that utilized ResNet, which has a deep CNN architecture, as the backbone for feature extraction, as in ASTER, CLOVA, STAR-Net, and ROSETTA, outperformed the methods that used VGG as in RARE and CRNN.

Although the considered state-of-the-art methods have shown the ability to recognize text under some challenging examples, as illustrated in Figure 10.13c, there are still various challenging cases that these methods do not explicitly handle, such as recognizing text of calligraphic fonts and text subject to heavy occlusion, low resolution, and illumination variation. Figure 10.14 shows some challenging cases from the considered benchmark datasets that all of the studied recognition methods failed to handle. In the rest of this section, we will analyze these failure cases and suggest future works to tackle these challenges.

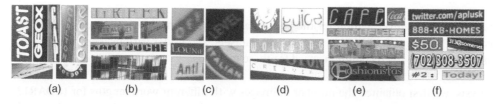

Figure 10.14 Illustration for challenging cases on scene text recognition that still cause recognition failure, where (a) vertical text (VT), multi-oriented (MO) text, and curved text (CT), (b) occluded text (OT), (c) low resolution (LR), (d) illumination variation (IV), (e) complex font (CF), and (f) special characters (SC).

Oriented Text Existing state-of-the-art scene text recognition methods have focused more on recognizing horizontal [64, 73], multi-oriented [76, 78] and curved text [28, 65, 68, 82, 157, 187], which leverage a spatial rectification module [82, 176, 187] and typically use sequence-to-sequence models designed for reading text. Despite these attempts to solve recognizing text of arbitrary orientation, there are still types of orientated text in the wild images that these methods could not tackle, such as highly curved text, IPR text, vertical text (VT), and text stacked from bottom-to-top and top-to-down demonstrated in Figure 10.14a.

In addition, since horizontal and vertical text have different characteristics, researchers have recently attempted [200, 201] to design techniques for recognizing both types of text in a unified framework. Therefore, further research would be required to construct models that are able to simultaneously recognize different orientations.

Occluded Text Although existing attention-based methods [28, 65, 68] have shown the capability of recognizing text subject to partial occlusion, their performance declined in recognizing text with heavy occlusion, as shown in Figure 10.14b. This is because the current methods do not extensively exploit contextual information to overcome occlusion. Thus, future researchers may consider superior language models [202] to utilize context maximally for predicting the invisible characters due to occluded text.

Degraded Image Quality It can be noted also that the state-of-the-art text recognition methods, as in [28, 64–68], did not specifically overcome the effect of degraded image quality, such as low resolution and illumination variation, on the recognition accuracy. Thus, inadequate recognition performance can be observed from the sample qualitative results in Figure 10.14c,d. As a suggested future work, it is important to study how image enhancement techniques, such as image super-resolution [203], image denoising [204, 205], and learning through obstructions [206], can allow text recognition schemes to address these issues.

Complex Fonts There are several challenging text of graphical fonts (e.g. Spencerian Script, Christmas, and Subway Ticker) in the wild images that the current methods do not explicitly handle (see Figure 10.14e). Recognizing text of complex fonts in the wild images emphasizes designing schemes that are able to recognize different fonts by improving the feature extraction step of these schemes or using style transfer techniques [207, 208] for learning the mapping from one font to another.

Special Characters In addition to alphanumeric characters, special characters (e.g. the $, /, -, !, :, @, and # characters in Figure 10.14f) are also abundant in the wild images, however, existing text recognition methods [64, 65, 68, 156, 177] have excluded them during training and testing. Therefore, these pre-trained models suffer from the inability to recognize special characters. Recently, CLOVA in [58] has shown that training the models on special characters improves the recognition accuracy, which suggests further study in how to incorporate special characters in both training and evaluation of text recognition models.

10.3.4.3 Discussion

In this section, we conduct empirical investigation for the performance of the considered recognition methods [64, 65, 68, 156, 177] using ICDAR13, ICDAR15, and COCO-Text datasets, and under various word lengths and aspect ratios. In addition, we compare the recognition speed for these methods.

Word-Length In this analysis, we first obtained the number of images with different word lengths for ICDAR13, ICDAR15, and COCO-Text datasets as shown in Figure 10.15. As can be seen from Figure 10.15, most of the words have a word length between two to seven characters, so we will focus this analysis on short and intermediate words. Figure 10.16 illustrates the accuracy of the text recognition methods at different word lengths for ICDAR13, ICDAR15, and COCO-Text datasets.

On ICDAR13 dataset, shown in Figure 10.16a, all the methods offered consistent accuracy values for words with length larger than two characters. This is because all the text instances in this dataset are horizontal and of high resolution. However, for words with two characters, RARE offered the worst accuracy (~58%), while CLOVA offered the best accuracy (~83%). On ICDAR15 dataset, the recognition accuracies of the methods follow a consistent trend similar to ICDAR13 [147]. However, the recognition performance is generally lower than that obtained on ICDAR13, because this dataset has more blurry, low resolution, and rotated images than ICDAR13. On COCO-Text dataset, ASTER and CLOVA achieved the best, and the second-best accuracies, and overall, except for some fluctuations at word length more than 12 characters, all the methods followed a similar trend.

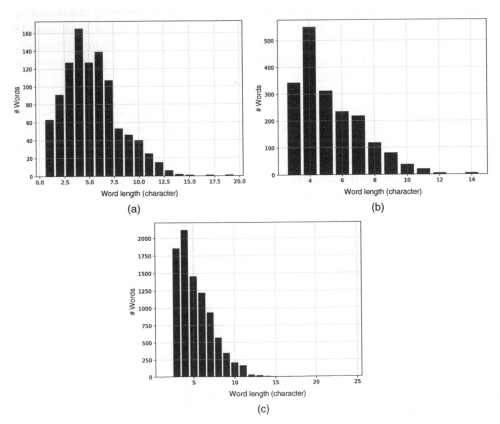

Figure 10.15 Statistics of word length in characters computed on (a) ICDAR13, (b) ICDAR15, and (c) COCO-Text recognition datasets.

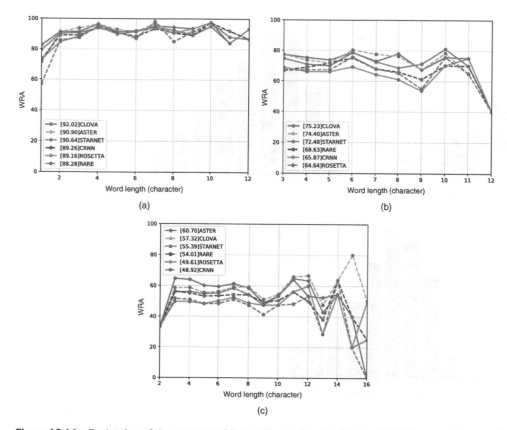

Figure 10.16 Evaluation of the average WRA at different word length for ASTER (Source: Shi et al. [28]), CLOVA (Source: Baek et al. [68]), STARNET (Source: Liu et al. [66]), RARE (Source: Shi et al. [65]), CRNN (Source: Shi et al. [64]), and ROSETTA (Source: Borisyuk et al. [67]) computed on (a) ICDAR13, (b) ICDAR15, and (c) COCO-Text recognition datasets.

Aspect-Ratio In this experiment, we study the accuracy achieved by the studied methods on words with different aspect ratio (height/width). As can be seen from Figure 10.17 most of the word images in the considered datasets are of aspect ratios between 0.3 and 0.6. Figure 10.18 shows the WRA values of the studied methods [28, 64–68] versus the word aspect ratio computed on ICDAR13, ICDAR15, and COCO-Text datasets. From this figure, for images with aspect-ratio < 0.3 the studied methods offer low WRA values on the three considered datasets. The main reason for this is that this range mostly includes the text of long words that face an assessment challenge of correctly predicting every character within a given word. For images within $0.3 \leq$ aspect-ratio ≤ 0.5, which include images of medium word length (four to nine characters per word), it can be seen from Figure 10.18 that the highest WRA values are offered by the studied methods. It can be observed from Figure 10.18 also that when evaluating the target state-of-the-art methods on images with aspect ratio ≥ 0.6, all the methods have experienced a decline in the WRA value. This is due to those images being mostly of words of short length and of low resolution.

Recognition Time We also conducted an investigation to compare the recognition time versus the WRA for the considered state-of-the-art scene text recognition models [28, 64–68]. Figure 10.19 shows the inference time per word image in milliseconds, when the test batch size is one, where the inference time could be reduced by using larger batch size. The fastest and the slowest methods are CRNN [64] and ASTER [28]. They achieve time/word of ~2.17 and ~23.26 ms, respectively and illustrate the big gap in the computational requirements of these models. Although attention-based methods, ASTER [28] and CLOVA [68], provide higher WRA than that of CTC-based

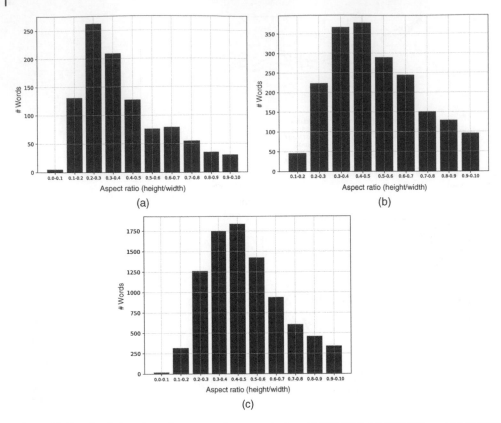

Figure 10.17 Statistics of word aspect-ratios computed on (a) ICDAR13, (b) ICDAR15, and (c) COCO-Text recognition datasets.

methods, CRNN [64], ROSETTA [67], and STAR-Net [66], however, they are much slower compared to CTC-based methods. This slower speed of attention-based methods comes back to the deeper feature extractor and rectification modules utilized in their architectures.

10.3.5 Open Investigations for Scene Text Detection and Recognition

Following the recent development in object detection and recognition problems, deep learning scene text detection and recognition frameworks have progressed rapidly such that the reported H-mean performance and recognition accuracy are about 80% to 95% for several benchmark datasets. However, as we discussed in Sections 10.3.3 and 10.3.4, there are still many open issues for future works.

10.3.5.1 Training Datasets

Although the role of synthetic datasets cannot be ignored in the training of recognition algorithms, detection methods still require more real-world datasets to fine-tune. Therefore, using generative adversarial network [209] based methods or 3D proposal based [210] models that produce more realistic text images can be a better way of generating synthetic datasets for training text detectors.

10.3.5.2 Richer Annotations

For both detection and recognition, due to the annotation shortcomings for quantifying challenges in the wild images, existing methods have not explicitly evaluated tackling such challenges. Therefore, future annotations of

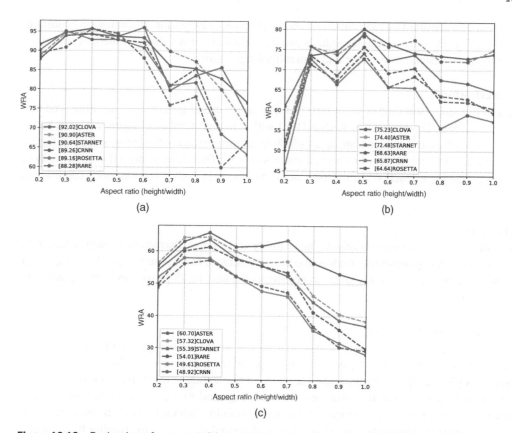

Figure 10.18 Evaluation of average WRA at various word aspect-ratios for ASTER (Source: Shi et al. [28]), CLOVA (Source: Baek et al. [68]), STARNET (Source: Liu et al. [66]), RARE (Source: Shi et al. [65]), CRNN (Source: Shi et al. [64]), and ROSETTA (Source: Borisyuk et al. [67]) using (a) ICDAR13, (b) ICDAR15, and (c) COCO-Text recognition datasets.

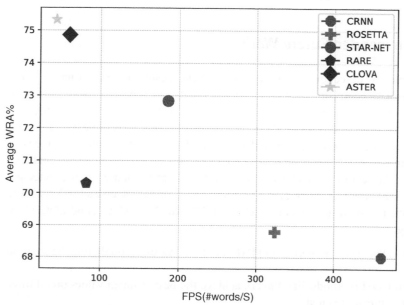

Figure 10.19 Average WRA versus average recognition time per word in milliseconds computed on (Source: Karatzas et al. [147]), ICDAR15 (Source: Karatzas et al. [23]), and COCO-Text (Source: Veit et al. [102]) datasets.

benchmark datasets should be supported by additional meta descriptors (e.g. orientation, illumination condition, aspect ratio, word length, and font type) such that methods can be evaluated against those challenges, and thus, it will help future researchers to design more robust and generalized algorithms.

10.3.5.3 Novel Feature Extractors

It is essential to have a better understanding of what type of features are useful for constructing improved text detection and recognition models. For example, a ResNet [174] with a higher number of layers will give better results [58, 211], while it is not clear yet what can be an efficient feature extractor that allows differentiating text from other objects and recognizing the various text characters as well. Therefore, a more thorough study of the dependents on different feature extraction architecture as the backbone in both detection and recognition is required.

10.3.5.4 Occlusion Handling

So far, existing methods in scene text recognition rely on the visibility of the target characters in images; however, text affected by heavy occlusion may significantly undermine the performance of these methods. Designing a text recognition scheme based on a strong natural language processing model like, BERT [202], can help in predicting occluded characters in a given text.

10.3.5.5 Complex Fonts and Special Characters

Images in the wild can include text with a wide variety of complex fonts, such as calligraphic fonts, and/or colors. Overcoming those variability's can be possible by generating images with more real-world like text using style transfer-learning techniques [207, 208] or improving the backbone of the feature extraction methods [212, 213]. As we mentioned in Section 10.3.4, special characters (e.g. $, /, -, !, :, @, and #) are also abundant in the wild images, but the research community has been ignoring them during training, which leads to incorrect recognition for those characters [214–216]. Therefore, including images of special characters in training future scene text detection/recognition methods as well will help in evaluating these models for detecting/recognizing these characters [217].

10.4 Conclusions and Recommended Future Work

It has been noticed that in recent scene text detection and recognition surveys, despite the performance of the analyzed deep learning-based methods have been compared on multiple datasets, the reported results have been used for evaluation, which makes the direct comparison among these methods difficult. This is due to the lack of common experimental settings, groundtruth and/or evaluation methodology. In this survey, we have first presented a detailed review on the recent advancement in scene text detection and recognition fields with a focus on deep learning-based techniques and architectures. Next, we have conducted extensive experiments on challenging benchmark datasets for comparing the performance of a selected number of pre-trained scene text detection and recognition methods, which represent the recent state-of-the-art approaches, under adverse situations. More specifically, when evaluating the selected scene text detection schemes on ICDAR13, ICDAR15, and COCO-Text datasets we have noticed the following:

- Segmentation-based methods, such as PixelLink, PSENET, and PAN, are more robust in predicting the location of irregular text.
- Hybrid regression and segmentation-based methods, like PMTD, achieve the best *H*-mean values on all three datasets, as they can handle better multi-oriented text.
- Methods that detect text at the character level, as in CRAFT, can perform better in detecting irregular shape text.

- In images with text affected by more than one challenge, all the studied methods performed weakly.

With respect to evaluating scene text recognition methods on challenging benchmark datasets, we have noticed the following:

- Scene text recognition methods that only use synthetic scene images for training have been able to recognize text in real-world images without fine-tuning their models.
- In general, attention-based methods, as in ASTER and CLOVA, that benefit from a deep backbone for feature extraction and transformation network for rectification have performed better than that CTC-based methods, as in CRNN, STARNET, and ROSETTA.

It has been shown that there are several unsolved challenges for detecting or recognizing text in the wild images, such as in-plane-rotation, multi-oriented and multi-resolution text, perspective distortion (PD), shadow and illumination reflection, image blurriness, partial occlusion, complex fonts, and special characters, which we have discussed throughout this survey, and which open more potential future research directions. This study also highlights the importance of having more descriptive annotations for text instances to allow future detectors to be trained and evaluated against more challenging conditions.

While the research community has come a long way, solutions to these challenging scenarios are still needed to further enhance the capabilities of text detection and recognition in the wild, and subsequently, ease the daily struggles of people with visual impairments.

Acknowledgments

The authors would like to thank the Ontario Centres of Excellence (OCE) – Voucher for Innovation and Productivity II (VIP II) – Canada program, and ATS Automation Tooling Systems Inc., Cambridge, ON, Canada, for supporting this research work.

References

1 World Health Organization (WHO) (2021). https://www.who.int/news-room/fact-sheets/detail/blindness-and-visual-impairment (accessed 15 June 2021).

2 Chen, X., Jin, L., Zhu, Y. et al. (2021). Text recognition in the wild: a survey. *ACM Computing Surveys (CSUR)* 54 (2): 1–35.

3 Hiratsuka, Y. (2021). Visual impairment as a public health problem in 2021. *Juntendo Medical Journal* 63 (3): 201, JMJ21-LN02–203.

4 Geetha, M.N., Sheethal, H., Sindhu, S. et al. (2019). Survey on smart reader for blind and visually impaired (BVI). *Indian Journal of Science and Technology* 12: 48.

5 Islam, M.M., Sadi, M.S., Zamli, K.Z., and Ahmed, M.M. (2019). Developing walking assistants for visually impaired people: a review. *IEEE Sensors Journal* 19 (8): 2814–2828.

6 Tapu, R., Mocanu, B., and Zaharia, T. (2020). Wearable assistive devices for visually impaired: a state of the art survey. *Pattern Recognition Letters* 137: 37–52.

7 Virmani, D., Gupta, C., Bamdev, P., and Jain, P. (2020). iSeePlus: a cost effective smart assistance archetype based on deep learning model for visually impaired. *Journal of Information and Optimization Sciences* 41 (7): 1741–1756.

8 El-Taher, F.E.Z., Taha, A., Courtney, J., and Mckeever, S. (2021). A systematic review of urban navigation systems for visually impaired people. *Sensors* 21 (9): 3103.

9 Yi, C. and Tian, Y. (2015). Assistive text reading from natural scene for blind persons. In: *Mobile Cloud Visual Media Computing* (ed. G. Hua and X.-S. Hua), 219–241. Springer.

10 Ye, Q. and Doermann, D. (2015). Text detection and recognition in imagery: a survey. *IEEE Transactions on Pattern Analysis and Machine Intelligence* 37 (7): 1480–1500.

11 Long, S., He, X., and Yao, C. (2018). Scene text detection and recognition: the deep learning era. *International Journal of Computer Vision* 129 (1): 161–184.

12 Lin, H., Yang, P., and Zhang, F. (2019). Review of scene text detection and recognition. *Archives of Computational Methods in Engineering* 27 (2): 433, 1–454, 22.

13 Case, C., Suresh, B., Coates, A., and Ng, A.Y. (2011). Autonomous sign reading for semantic mapping. In: *Proceedings of the IEEE International Conference on Robotics and Automation, 9–13 May 2011*, 3297–3303. Shanghai, China: IEEE.

14 Kostavelis, I. and Gasteratos, A. (2015). Semantic mapping for mobile robotics tasks: a survey. *Robotics and Autonomous Systems* 66: 86–103.

15 Ham, Y.K., Kang, M.S., Chung, H.K. et al. (1995). Recognition of raised characters for automatic classification of rubber tires. *Optical Engineering* 34 (1): 102–110.

16 Chandrasekhar, V.R., Chen, D.M., Tsai, S.S. et al. (2011). The Stanford mobile visual search data set. In: *Proceedings of the ACM Conference on Multimedia Systems*, 117–122. SPIE.

17 Tsai, S.S., Chen, H., Chen, D. et al. (2011). Mobile visual search on printed documents using text and low bit-rate features. In: *Proceeding of IEEE International Conference on Image Processing*, 2601–2604. Brussels, Belgium: IEEE.

18 Ma, D., Lin, Q., and Zhang, T. (2000). *Mobile Camera Based Text Detection and Translation*. Stanford University.

19 Cheung, E. and Purdy, K.H. (2008). System and method for text translations and annotation in an instant messaging session. US Patent 7,451,188, issued 11 November 2008.

20 Wu, W., Chen, X., and Yang, J. (2005). Detection of text on road signs from video. *IEEE Transactions on Intelligent Transportation Systems* 6 (4): 378–390.

21 Messelodi, S. and Modena, C.M. (2013). Scene text recognition and tracking to identify athletes in sport videos. *Multimedia Tools and Applications* 63 (2): 521–545.

22 Huang, Z., Chen, K., He, J. et al. (2019). ICDAR2019 competition on scanned receipt OCR and information extraction. In: *Proceedings of the International Conference on Document Analysis and Recognition (ICDAR)*, 1516–1520. Sydney, NSW, Australia: IEEE.

23 Karatzas, D., Gomez-Bigorda, L., Nicolaou, A. et al. (2015). ICDAR 2015 competition on robust reading. In: *Proceedings of the International Conference on Document Analysis and Recognition (ICDAR)*, 1156–1160.

24 Ch'ng, C.K. and Chan, C.S. (2017). Total-text: A comprehensive dataset for scene text detection and recognition. In: *Proceedings of the IAPR International Conference on Document Analysis and Recognition (ICDAR)*, vol. 1, 935–942. Kyoto, Japan: IEEE.

25 Somerville, P.J. (1991). Method and apparatus for barcode recognition in a digital image. US Patent 4,992,650, 12 February 1991.

26 Chen, D., Odobez, J.M., and Bourlard, H. (2003). Text detection and recognition in images and video frames. *Pattern Recognition* 37 (3): 595–608. ISSN 0031-3203.

27 Chaudhuri, A., Mandaviya, K., Badelia, P., and Ghosh, S.K. (2017). *Optical Character Recognition Systems for Different Languages with Soft Computing*, Studies in Fuzziness and Soft Computing, vol. 352. Springer.

28 Shi, B., Yang, M., Wang, X. et al. (2018). Aster: an attentional scene text recognizer with flexible rectification. *IEEE Transactions on Pattern Analysis and Machine Intelligence* 41 (9): 2035–2048.

29 Kim, K.I., Jung, K., and Kim, J.H. (2003). Texture-based approach for text detection in images using support vector machines and continuously adaptive mean shift algorithm. *IEEE Transactions on Pattern Analysis and Machine Intelligence* 25 (12): 1631–1639.

30 Chen, X. and Yuille, A.L. (2004). Detecting and reading text in natural scenes. In: *Proceedings of the IEEE Conference on Computer Vision and Pattern Recognition (CVPR)*, vol. 2, II–II. Washington, DC, USA: IEEE.

31 Hanif, S.M. and Prevost, L. (2009). Text detection and localization in complex scene images using constrained adaboost algorithm. In: *Proceedings of the International Conference on Document Analysis and Recognition*, 1–5. Barcelona, Spain: IEEE.

32 Wang, K., Babenko, B., and Belongie, S. (2011). End-to-end scene text recognition. In: *Proceedings of the International Conference on Computer Vision*, 1457–1464. Barcelona: IEEE.

33 Lee, J.J., Lee, P.H., Lee, S.W. et al. (2011). Adaboost for text detection in natural scene. In: *Proceedings of the International Conference on Document Analysis and Recognition*, 429–434. Beijing, China: IEEE.

34 Bissacco, A., Cummins, M., Netzer, Y., and Neven, H. (2013). PhotoOCR: reading text in uncontrolled conditions. In: *Proceedings of the IEEE International Conference on Computer Vision*, 785–792. Sydney, NSW, Australia: IEEE.

35 Wang, K. and Kangas, J.A. (2003). Character location in scene images from digital camera. *Pattern Recognition* 36 (10): 2287–2299.

36 Mancas Thillou, C. and Gosselin, B. (2006). Spatial and color spaces combination for natural scene text extraction. In: *Proceeding of IEEE International Conference on Image Processing*, 985–988. IEEE.

37 Mancas-Thillou, C. and Gosselin, B. (2007). Color text extraction with selective metric-based clustering. *Computer Vision and Image Understanding* 107 (1–2): 97–107.

38 Song, Y., Liu, A., Pang, L. et al. (2008). A novel image text extraction method based on k-means clustering. In: *Proceeding of the IEEE/ACIS International Conference on Computer and Information Science*, 185–190. Portland, OR, USA: IEEE.

39 Kim, W. and Kim, C. (2008). A new approach for overlay text detection and extraction from complex video scene. *IEEE Transactions on Image Processing* 18 (2): 401–411.

40 De Campos, T.E., Babu, B.R., Varma, M. et al. (2009). Character recognition in natural images. In: *Proceedings of the International Conference on Computer Vision Theory and Applications (VISAPP)*, vol. 7. Lisbon, Portugal: VISAPP.

41 Pan, Y.F., Hou, X., and Liu, C.L. (2009). Text localization in natural scene images based on conditional random field. In: *Proceedings of the International Conference on Document Analysis and Recognition*, 6–10. Barcelona, Spain: IEEE.

42 Huang, W., Qiao, Y., and Tang, X. (2014). Robust scene text detection with convolution neural network induced MSER trees. In: *Proceedings of the European Conference on Computer Vision*, 497–511. Springer.

43 Zhang, Z., Shen, W., Yao, C., and Bai, X. (2015). Symmetry-based text line detection in natural scenes. In: *Proceedings of the IEEE Conference on Computer Vision and Pattern Recognition*, 2558–2567. Boston, MA: IEEE.

44 Jaderberg, M., Simonyan, K., Vedaldi, A., and Zisserman, A. (2016). Reading text in the wild with convolutional neural networks. *International Journal of Computer Vision* 116 (1): 1–20.

45 Jaderberg, M., Simonyan, K., Vedaldi, A., & Zisserman, A. (2019). Deep structured output learning for unconstrained text recognition. International Conference on Learning Representations, 1–10.

46 Tian, Z., Huang, W., He, T. et al. (2016). Detecting text in natural image with connectionist text proposal network. In: *Proceedings of the European Conference on Computer Vision*, 56–72. Springer.

47 Zhou, X., Yao, C., Wen, H. et al. (2017). East: an efficient and accurate scene text detector. In: *Proceedings of the IEEE Conference on Computer Vision and Pattern Recognition*, 5551–5560. Honolulu, HI, USA: IEEE.

48 Liao, M., Shi, B., Bai, X. et al. (2017). Textboxes: a fast text detector with a single deep neural network. *Proceedings of AAAI Conference on Artificial Intelligence*, San Francisco, California USA, 4–9 February 2017.

49 Liao, M., Shi, B., and Bai, X. (2018). Textboxes++: a single-shot oriented scene text detector. *IEEE Transactions on Image Processing* 27 (8): 3676–3690.

50 Ma, J., Shao, W., Ye, H. et al. (2018). Arbitrary-oriented scene text detection via rotation proposals. *IEEE Transactions on Multimedia* 20 (11): 3111–3122.

51 Zhang, Z., Zhang, C., Shen, W. et al. (2016). Multi-oriented text detection with fully convolutional networks. In: *Proceedings of the IEEE Conference on Computer Vision and Pattern Recognition*, 4159–4167. Las Vegas, Nevada: IEEE.

52 Yao, C., Bai, X., Sang, N. et al. (2016). Scene text detection via holistic, multi-channel prediction. arXiv preprint arXiv:1606.09002.

53 Wu, Y. and Natarajan, P. (2017). Self-organized text detection with minimal post-processing via border learning. In: *Proceedings of the IEEE International Conference on Computer Vision*, 5000–5009. Venice, Italy: IEEE.

54 Long, S., Ruan, J., Zhang, W. et al. (2018). TextSnake: a flexible representation for detecting text of arbitrary shapes. In: *Proceedings of the European Conference on Computer Vision (ECCV)*, 20–36. Munich, Germany: IEEE.

55 Deng, D., Liu, H., Li, X. et al. (2018). PixelLink: detecting scene text via instance segmentation. *Proceedings of the AAAI Conference on Artificial Intelligence*, New Orleans, Lousiana, USA, 2–7 February 2018.

56 Lyu, P., Yao, C., Wu, W. et al. (2018). Multi-oriented scene text detection via corner localization and region segmentation. In: *Proceedings of the IEEE Conference on Computer Vision and Pattern Recognition*, 7553–7563. Salt Lake City, UT, USA: IEEE.

57 Qin, H., Zhang, H., Wang, H. et al. (2019). An algorithm for scene text detection using multibox and semantic segmentation. *Applied Sciences* 9 (6): 1054.

58 Baek, Y., Lee, B., Han, D. et al. (2019). Character region awareness for text detection. *Proceedings of the IEEE Conference on Computer Vision and Pattern Recognition*, Long Beach, CA, USA, 15–20 June 2019.

59 Liu, J., Liu, X., Sheng, J. et al. (2019, arXiv). Pyramid mask text detector. *CoRR* abs/1903.11800: 1903.

60 Lyu, P., Liao, M., Yao, C. et al. (2018). Mask textspotter: an end-to-end trainable neural network for spotting text with arbitrary shapes. In: *Proceedings of the European Conference on Computer Vision (ECCV)*, 67–83. Munich, Germany: IEEE Computer Society.

61 Liu, X., Liang, D., Yan, S. et al. (2018). FOTS: fast oriented text spotting with a unified network. In: *Proceedings of the IEEE Conference on Computer Vision and Pattern Recognition*, 5676–5685. Salt Lake City, UT, USA: IEEE.

62 He, T., Tian, Z., Huang, W. et al. (2018). An end-to-end textspotter with explicit alignment and attention. In: *Proceedings of the IEEE Conference on Computer Vision and Pattern Recognition*, 5020–5029. Salt Lake City, UT, USA: IEEE.

63 Busta, M., Neumann, L., and Matas, J. (2017). Deep textspotter: an end-to-end trainable scene text localization and recognition framework. In: *Proceedings of the IEEE International Conference on Computer Vision*, 2204–2212. Venice, Italy: IEEE.

64 Shi, B., Bai, X., and Yao, C. (2016). An end-to-end trainable neural network for image-based sequence recognition and its application to scene text recognition. *IEEE Transactions on Pattern Analysis and Machine Intelligence* 39 (11): 2298–2304.

65 Shi, B., Wang, X., Lyu, P. et al. (2016). Robust scene text recognition with automatic rectification. In: *Proceedings of the IEEE Conference on Computer Vision and Pattern Recognition*, 4168–4176.

66 Liu, W., Chen, C., Wong, K.-Y.K. et al. (2016). STAR-Net: A spatial attention residue network for scene text recognition. In: *Proceedings of the British Machine Vision Conference (BMVC)*, 43.1–43.13. BMVA Press.

67 Borisyuk, F., Gordo, A., and Sivakumar, V. (2018). Rosetta: large scale system for text detection and recognition in images. In: *Proceedings of the ACM SIGKDD International Conference on Knowledge Discovery and Data Mining*, 71–79. New York, NY, USA: Association for Computing Machinery.

68 Baek, J., Kim, G., Lee, J. et al. (2019). What is wrong with scene text recognition model comparisons? Dataset and model analysis. *Proceedings of the International Conference on Computer Vision (ICCV)*, October 27 to November 2, 2019, in Seoul, Korea.

69 Matas, J., Chum, O., Urban, M., and Pajdla, T. (2004). Robust wide-baseline stereo from maximally stable extremal regions. *Image and Vision Computing* 22 (10): 761–767.

70 Epshtein, B., Ofek, E., and Wexler, Y. (2010). Detecting text in natural scenes with stroke width transform. In: *Proceedings of the IEEE Conference on Computer Vision and Pattern Recognition*, 2963–2970. San Francisco, CA, USA: IEEE.

71 Shi, B., Bai, X., and Belongie, S. (2017). Detecting oriented text in natural images by linking segments. In: *Proceedings of the IEEE Conference on Computer Vision and Pattern Recognition*, 2550–2558. Honolulu, HI, USA: IEEE.

72 Jaderberg, M., Simonyan, K., Vedaldi, A. et al. (2014). Synthetic data and artificial neural networks for natural scene text recognition. arXiv preprint arXiv:1406.2227 [Online]. Available: https://www.robots.ox.ac.uk/~vgg/data/text.

73 Lee, C.Y. and Osindero, S. (2016). Recursive recurrent nets with attention modeling for OCR in the wild. In: *Proceedings of the IEEE Conference on Computer Vision and Pattern Recognition*, 2231–2239. Las Vegas, NV, USA: IEEE.

74 Wang, J. and Hu, X. (2017). Gated recurrent convolution neural network for OCR. In: *Proceedings of the Advances in Neural Information Processing Systems*, 335–344.

75 Yang, X., He, D., Zhou, Z. et al. (2017). Learning to read irregular text with attention mechanisms. In: *Proceedings of the International Joint Conference on Artificial Intelligence*, 3. Melbourne, Australia, vol. 1, no. 2: IJCAI.

76 Cheng, Z., Bai, F., Xu, Y. et al. (2017). Focusing attention: towards accurate text recognition in natural images. In: *Proceedings of the IEEE International Conference on Computer Vision*, 5076–5084.

77 Liu, W., Chen, C., and Wong, K.Y.K. (2018). Char-net: a character-aware neural network for distorted scene text recognition. *Proceedings of the AAAI Conference on Artificial Intelligence*.

78 Cheng, Z., Xu, Y., Bai, F. et al. (2018). AON: towards arbitrarily-oriented text recognition. In: *Proceedings of the IEEE Conference on Computer Vision and Pattern Recognition*, 5571–5579. Salt Lake City, UT, USA: IEEE.

79 Bai, F., Cheng, Z., Niu, Y. et al. (2018). Edit probability for scene text recognition. In: *Proceedings of the IEEE Conference on Computer Vision and Pattern Recognition*, 1508–1516. Salt Lake City, UT, USA: IEEE.

80 Liu, Y., Wang, Z., Jin, H., and Wassell, I. (2018). Synthetically supervised feature learning for scene text recognition. In: *Proceedings of the European Conference on Computer Vision (ECCV)*, 435–451. Munich, Germany: Springer.

81 Xie, Z., Huang, Y., Zhu, Y. et al. (2019). Aggregation cross-entropy for sequence recognition. In: *Proceedings of the IEEE Conference on Computer Vision and Pattern Recognition*, 6538–6547. Long Beach, CA, USA: IEEE.

82 Zhan, F. and Lu, S. (2019). ESIR: end-to-end scene text recognition via iterative image rectification. In: *Proceedings of the IEEE Conference on Computer Vision and Pattern Recognition*, 2059–2068. Long Beach, CA, USA: IEEE.

83 Wang, P., Yang, L., Li, H. (2019). A simple and robust convolutional-attention network for irregular text recognition. ArXiv, vol. abs/1904.01375.

84 Yin, X., Zuo, Z., Tian, S., and Liu, C. (2016). Text detection, tracking and recognition in video: a comprehensive survey. *IEEE Transactions on Image Processing* 25 (6): 2752–2773.

85 Liu, X., Meng, G., and Pan, C. (2019). Scene text detection and recognition with advances in deep learning: a survey. *International Journal on Document Analysis and Recognition (IJDAR)* 22 (2): 1431–1162.

86 Kang, M.C., Chae, S.H., Sun, J.Y. et al. (2017). An enhanced obstacle avoidance method for the visually impaired using deformable grid. *IEEE Transactions on Consumer Electronics* 63 (2): 169–177.

87 Afif, M., Said, Y., Pissaloux, E. et al. (2020). Recognizing signs and doors for indoor wayfinding for blind and visually impaired persons. In: *Proceeding of the International Conference on Advanced Technologies for Signal and Image Processing (ATSIP)*, 1–4. Sousse, Tunisia: IEEE.

88 González, Y., Millán, A., Sánchez, Y. et al. (2020). Sistema de reconocimiento de señalamientos en entornos abiertos para la orientación de personas con discapacidad visual. *Memoria Investigaciones en Ingeniería* (19): 43–62.

89 Afif, M., Ayachi, R., Said, Y., and Atri, M. (2021). Indoor sign detection system for indoor assistance navigation. In: *International Multi-Conference on Systems, Signals & Devices (SSD)*, 1383–1387. Monastir, Tunisia: IEEE.

90 Marzullo, G.D., Jo, K.H., and Cáceres, D. (2021). Vision-based assistive navigation algorithm for blind and visually impaired people using monocular camera. In: *IEEE/SICE International Symposium on System Integration (SII)*, 640–645. Iwaki, Fukushima, Japan: IEEE.

91 Hersh, M. and Johnson, M.A. (2010). *Assistive Technology for Visually Impaired and Blind People*. Springer Science & Business Media.

92 Zhang, H. and Ye, C. (2017). An indoor wayfinding system based on geometric features aided graph slam for the visually impaired. *IEEE Transactions on Neural Systems and Rehabilitation Engineering* 25 (9): 1592–1604.

93 Jafri, R., Campos, R.L., Ali, S.A., and Arabnia, H.R. (2017). Visual and infrared sensor data-based obstacle detection for the visually impaired using the Google project tango tablet development kit and the unity engine. *IEEE Access* 6: 443–454.

94 Dognin, P., Melnyk, I., Mroueh, Y. et al. (2020). Image captioning as an assistive technology: lessons learned from VizWiz 2020 challenge. arXiv preprint arXiv:2012.11696.

95 Kim, D.W., Hwang, J.W., Lim, S.H. et al. (2021). An improved feature extraction approach to image captioning for visually impaired people. *Proceedings of the IEEE/CVF Conference on Computer Vision and Pattern Recognition*, Virtual, 19–25 June .

96 Gurari, D., Li, Q., Stangl, A.J. et al. (2018). VizWiz grand challenge: Answering visual questions from blind people. In: *Proceedings of the IEEE Conference on Computer Vision and Pattern Recognition*, 3608–3617. Salt Lake City, UT, USA: IEEE.

97 Gurari, D., Zhao, Y., Zhang, M., and Bhattacharya, N. (2020). Captioning images taken by people who are blind. In: *European Conference on Computer Vision*, 417–434. Springer.

98 Cho, J.W., Kim, D.J., Choi, J. et al. (2021). Dealing with missing modalities in the visual question answer-difference prediction task through knowledge distillation. In: *Proceedings of the IEEE/CVF Conference on Computer Vision and Pattern Recognition*, 1592–1601.

99 Bujimalla, S., Subedar, M., and Tickoo, O. (2021). Data augmentation to improve robustness of image captioning solutions. arXiv preprint arXiv:2106.05437.

100 Mu, W. and Liu, Y. (2021). Deep co-attention model for challenging visual question answering on VizWiz. *Proceedings of IEEE Conference on Computer Vision and Pattern Recognition*, Virtual, 19–25 June.

101 Wang, K. and Belongie, S. (2010). Word spotting in the wild. In: *Proceedings of the European Conference on Computer Vision*, 591–604. Springer.

102 Veit, A., Matera, T., Neumann, L. et al. (2016). Coco-text: dataset and benchmark for text detection and recognition in natural images. arXiv preprint arXiv: 1601.07140, 2016. [Online]. Available: https://bgshih.github.io/cocotext/#h2-explorer.

103 Neumann, L. and Matas, J. (2010). A method for text localization and recognition in real-world images. In: *Proceedings of the Asian Conference on Computer Vision*, 770–783. Queenstown, New Zealand: Springer.

104 Yi, C. and Tian, Y. (2011). Text string detection from natural scenes by structure-based partition and grouping. *IEEE Transactions on Image Processing* 20 (9): 2594–2605.

105 Wang, T., Wu, D.J., Coates, A., and Ng, A.Y. (2012). End-to-end text recognition with convolutional neural networks. In: *Proceedings of the International Conference on Pattern Recognition(ICPR)*, 3304–3308. Tsukuba, Japan: IEEE.

106 Mishra, A., Alahari, K., and Jawahar, C. (2012). Top-down and bottom-up cues for scene text recognition. In: *Proceedings of the IEEE Conference on Computer Vision and Pattern Recognition*, 2687–2694. Providence, RI, USA: IEEE.

107 Yao, C., Bai, X., Liu, W. et al. (2012). Detecting texts of arbitrary orientations in natural images. In: *Proceedings of the IEEE Conference on Computer Vision and Pattern Recognition*, 1083–1090. Providence, RI, USA: IEEE.

108 Yin, X.C., Yin, X., Huang, K., and Hao, H.W. (2014). Robust text detection in natural scene images. *IEEE Transactions on Pattern Analysis and Machine Intelligence* 36 (5): 970–983.

109 Pan, Y.F., Hou, X., and Liu, C.L. (2010). A hybrid approach to detect and localize texts in natural scene images. *IEEE Transactions on Image Processing* 20 (3): 800–813.

110 Tian, S., Pan, Y., Huang, C. et al. (2015). Text flow: a unified text detection system in natural scene images. In: *Proceedings of the IEEE International Conference on Computer Vision*, 4651–4659. Santiago, Chile: IEEE.

111 Bosch, A., Zisserman, A., and Munoz, X. (2007). Image classification using random forests and ferns. In: *Proceedings of the IEEE International Conference on Computer Vision*, 1–8. Rio de Janeiro, Brazil: IEEE.

112 Schapire, R.E. and Singer, Y. (1999). Improved boosting algorithms using confidence-rated predictions. *Machine Learning* 37 (3): 297–336.

113 Ozuysal, M., Fua, P., and Lepetit, V. (2007). Fast keypoint recognition in ten lines of code. In: *Proceedings of the IEEE Conference on Computer Vision and Pattern Recognition*, 1–8. Minneapolis, MN, USA: IEEE.

114 Felzenszwalb, P.F., Girshick, R.B., McAllester, D., and Ramanan, D. (2009). Object detection with discriminatively trained part-based models. *IEEE Transactions on Pattern Analysis and Machine Intelligence* 32 (9): 1627–1645.

115 Zhu, Y., Yao, C., and Bai, X. (2016). Scene text detection and recognition: recent advances and future trends. *Frontiers of Computer Science* 10 (1): 19–36.

116 Ye, Q., Huang, Q., Gao, W., and Zhao, D. (2005). Fast and robust text detection in images and video frames. *Image and Vision Computing* 23 (6): 565–576.

117 Li, M. and Wang, C. (2008). An adaptive text detection approach in images and video frames. In: *International Joint Conference on Neural Networks*, 72–77. Hong Kong, China: IEEE.

118 Shivakumara, P., Phan, T.Q., and Tan, C.L. (2009). A gradient difference based technique for video text detection. In: *Proceedings of the International Conference on Document Analysis and Recognition*, 156–160. IEEE.

119 Neumann, L. and Matas, J. (2012). Real-time scene text localization and recognition. In: *Proceedings of the IEEE Conference on Computer Vision and Pattern Recognition*, 3538–3545. Providence, RI, USA: IEEE.

120 Cho, H., Sung, M., and Jun, B. (2016). Canny text detector: fast and robust scene text localization algorithm. In: *Proceedings of the IEEE Conference on Computer Vision and Pattern Recognition*, 3566–3573. Las Vegas, NV, USA: IEEE.

121 Zhao, X., Lin, K.-H., Fu, Y. et al. (2010). Text from corners: a novel approach to detect text and caption in videos. *IEEE Transactions on Image Processing* 20 (3): 790–799.

122 Neumann, L. and Matas, J. (2013). Scene text localization and recognition with oriented stroke detection. In: *Proceedings of the IEEE International Conference on Computer Vision*, 97–104.

123 Chen, H., Tsai, S.S., Schroth, G. et al. (2011). Robust text detection in natural images with edge-enhanced maximally stable extremal regions. In: *Proceeding of IEEE International Conference on Image Processing*, 2609–2612. Brussels, Belgium: IEEE.

124 Huang, W., Lin, Z., Yang, J., and Wang, J. (2013). Text localization in natural images using stroke feature transform and text covariance descriptors. In: *Proceedings of the IEEE International Conference on Computer Vision*, 1241–1248. Sydney, NSW, Australia: IEEE.

125 Busta, M., Neumann, L., and Matas, J. (2015). FASText: efficient unconstrained scene text detector. In: *Proceedings of the IEEE Conference on Computer Vision and Pattern Recognition*, 1206–1214. Santiago, Chile: IEEE Computer Society.

126 He, P., Huang, W., He, T. et al. (2017). Single shot text detector with regional attention. In: *Proceedings of the IEEE International Conference on Computer Vision*, 3047–3055. Venice, Italy: IEEE.

127 Zhang, S., Lin, M., Chen, T. et al. (2016). Character proposal network for robust text extraction. In: *Proceedings of the IEEE International Conference on Acoustics, Speech and Signal Processing (ICASSP)*, 2633–2637. Shanghai, China: IEEE.

128 Krizhevsky, A., Sutskever, I., and Hinton, G.E. (2012). ImageNet classification with deep convolutional neural networks. In: *Advances in Neural Information Processing Systems*, 1097–1105. Harrahs and Harveys, Lake Tahoe: Curran Associates, Inc.

129 He, W., Zhang, X.-Y., Yin, F., and Liu, C.-L. (2017). Deep direct regression for multi-oriented scene text detection. In: *Proceedings of the IEEE International Conference on Computer Vision*, 745–753. Venice, Italy: IEEE.

130 Zhou, Y., Ye, Q., Qiu, Q., and Jiao, J. (2017). Oriented response networks. In: *Proceedings of the IEEE Conference on Computer Vision and Pattern Recognition*, 519–528. Honolulu, HI, USA: IEEE.

131 Hu, H., Zhang, C., Luo, Y. et al. (2017). Wordsup: exploiting word annotations for character based text detection. In: *Proceedings of the IEEE International Conference on Computer Vision*, 4940–4949. Venice, Italy: IEEE.

132 Jiang, Y., Zhu, X., Wang, X. (2017). R2CNN: rotational region CNN for orientation robust scene text detection. arXiv preprint arXiv:1706.09579.

133 Liao, M., Zhu, Z., Shi, B. et al. (2018). Rotation-sensitive regression for oriented scene text detection. In: *Proceedings of the IEEE Conference on Computer Vision and Pattern Recognition*, 5909–5918. Salt Lake City, UT, USA: IEEE.

134 Wang, W., Xie, E., Song, X. et al. (2019). Efficient and accurate arbitrary-shaped text detection with pixel aggregation network. In: *Proceedings of the IEEE International Conference on Computer Vision*, 8440–8449. Seoul, Korea (South): IEEE.

135 Xu, Y., Wang, Y., Zhou, W. et al. (2019). Textfield: learning a deep direction field for irregular scene text detection. *IEEE Transactions on Image Processing* .

136 Liu, Y., Zhang, S., Jin, L. et al. (2019). Omnidirectional scene text detection with sequential-free box discretization. arXiv preprint arXiv:1906.02371.

137 Wang, W., Xie, E., Li, X. et al. (2019). Shape robust text detection with progressive scale expansion network. arXiv preprint arXiv:1903.12473.

138 Long, J., Shelhamer, E., and Darrell, T. (2015). Fully convolutional networks for semantic segmentation. In: *Proceedings of the IEEE Conference on Computer Vision and Pattern Recognition*, 3431–3440. Boston, Massachusetts, USA: IEEE.

139 Lin, T.Y., Dollar, P., Girshick, R. (2017). Feature pyramid networks for object detection. *Proceedings of the IEEE Conference on Computer Vision and Pattern Recognition (CVPR)*, Honolulu, Hawaii (July 2017).

140 Kim, K.H., Hong, S., Roh, B. et al. (2016). PVANET: deep but lightweight neural networks for real-time object detection. arXiv preprint arXiv:1608.08021.

141 Ren, S., He, K., Girshick, R., and Sun, J. (2015). Faster R-CNN: Towards real-time object detection with region proposal networks. In: *Proceedings of the Advances in Neural Information Processing Systems*, 91–99. Montréal, Canada: Curran Associates, Inc.

142 Liu, W., Anguelov, D., Erhan, D. et al. (2016). SSD: single shot multibox detector. In: *European Conference on Computer Vision*, 21–37. Springer.

143 Ronneberger, O., Fischer, P., and Brox, T. (2015). U-Net: convolutional networks for biomedical image segmentation. In: *Proceedings of the International Conference on Medical Image Computing and Computer-Assisted Intervention*, 234–241. Springer.

144 Dollár, P., Appel, R., Belongie, S., and Perona, P. (2014). Fast feature pyramids for object detection. *IEEE Transactions on Pattern Analysis and Machine Intelligence* 36 (8): 1532–1545.

145 Redmon, J., Divvala, S., Girshick, R., and Farhadi, A. (2016). You only look once: unified, real-time object detection. In: *Proceedings of the IEEE Conference on Computer Vision and Pattern Recognition*, 779–788. Las Vegas, NV, USA: IEEE.

146 He, K., Gkioxari, G., Dollár, P., and Girshick, R. (2017). Mask R-CNN. In: *Proceedings of the IEEE International Conference on Computer Vision*, 2961–2969. Venice, Italy: IEEE.

147 Karatzas, D., Shafait, F., Uchida, S. et al. (2013). ICDAR 2013 robust reading competition. In: *Proceedings of the International Conference on Document Analysis and Recognition*, 1484–1493. Washington, DC, USA: IEEE.

148 Gupta, A., Vedaldi, A., and Zisserman, A. (2016). Synthetic data for text localisation in natural images. In: *Proceedings of the IEEE Conference on Computer Vision and Pattern Recognition*, 2315–2324. Las Vegas, NV, USA: IEEE.

149 Huang, Z., Zhong, Z., Sun, L., and Huo, Q. (2019). Mask R-CNN with pyramid attention network for scene text detection. In: *Proceedings of the IEEE Winter Conference on Applications of Computer Vision (WACV)*, 764–772. Waikoloa, HI, USA: IEEE.

150 Xie, E., Zang, Y., Shao, S. et al. (2019). Scene text detection with supervised pyramid context network. In: *Proceedings of the AAAI Conference on Artificial Intelligence*, vol. 33, 9038–9045. Hawaii, USA: AAAI Press.

151 Zhang, C., Liang, B., Huang, Z. et al. (2019). Look more than once: an accurate detector for text of arbitrary shapes. In: *Proceedings of the IEEE Conference on Computer Vision and Pattern Recognition*, 10552–10561. Long Beach, CA, USA: IEEE.

152 Sun, Y., Ni, Z., Chng, C.-K. et al. (2019). ICDAR 2019 competition on large-scale street view text with partial labeling – RRC-LSVT. arXiv preprint arXiv:1909.07741.

153 Iwamura, M., Morimoto, N., Tainaka, K. et al. (2017). ICDAR2017 robust reading challenge on omnidirectional video. In: *Proceedings of the IAPR International Conference on Document Analysis and Recognition (ICDAR)*, vol. 1, 1448–1453. Kyoto, Japan: IEEE.

154 Yuliang, L., Lianwen, J., Shuaitao, Z. et al. (2017). Detecting curve text in the wild: new dataset and new solution. arXiv preprint arXiv:1712.02170.

155 He, P., Huang, W., Qiao, Y. et al. (2016). Reading scene text in deep convolutional sequences. *Proceedings of the AAAI Conference on Artificial Intelligence*, Phoenix, Arizona, USA, 12–17 February 2016.

156 Liao, M., Zhang, J., Wan, Z. (2018). Scene text recognition from two-dimensional perspective. ArXiv, vol. abs/1809.06508.

157 Wan, Z., Xie, F., Liu, Y. (2019). 2D-CTC for scene text recognition..

158 Bunke, H. and Wang, P.S.-P. (1997). *Handbook of Character Recognition and Document Image Analysis*. World Scientific.

159 Zhou, J. and Lopresti, D. (1997). Extracting text from WWW images. In: *Proceedings of the International Conference on Document Analysis and Recognition*, vol. 1, 248–252. Ulm, Germany: IEEE.

160 Sawaki, M., Murase, H., and Hagita, N. (2000). Automatic acquisition of context-based images templates for degraded character recognition in scene images. In: *Proceedings of the International Conference on Pattern Recognition (ICPR)*, vol. 4, 15–18. Barcelona, Spain: IEEE.

161 Arica, N. and Yarman-Vural, F.T. (2001). An overview of character recognition focused on off-line handwriting. *IEEE Transactions on Systems, Man, and Cybernetics Part C: Applications and Reviews* 31 (2): 216–233.

162 Lucas, S.M., Panaretos, A., Sosa, L. et al. (2003). ICDAR 2003 robust reading competitions. In: *Proceedings of the International Conference on Document Analysis and Recognition*, 682–687. Edinburgh, UK: IEEE.

163 Lucas, S.M. (2005). ICDAR 2005 text locating competition results. In: *Proceedings of the International Conference on Document Analysis and Recognition (ICDAR)*, vol. 1, 80–84. Seoul, Korea (South): IEEE.

164 Mishra, A., Alahari, K., and Jawahar, C.V. (2012). Scene text recognition using higher order language priors. In: *BMVC – British Machine Vision Conference, Sep 2012, Surrey*. United Kingdom: BMVA.

165 Risnumawan, A., Shivakumara, P., Chan, C.S., and Tan, C.L. (2014). A robust arbitrary text detection system for natural scene images. *Expert Systems with Applications* 41 (18): 8027–8048.

166 Dalal, N. and Triggs, B. (2005). Histograms of oriented gradients for human detection. In: *International Conference on Computer Vision and Pattern Recognition (CVPR)*, vol. 1, 886–893. San Diego, CA, USA: IEEE.

167 Lowe, D.G. (2004). Distinctive image features from scale-invariant keypoints. *International Journal of Computer Vision* 60 (2): 91–110.

168 Suykens, J.A. and Vandewalle, J. (1999). Least squares support vector machine classifiers. *Neural Processing Letters* 9 (3): 293–300.

169 Altman, N.S. (1992). An introduction to kernel and nearest-neighbor nonparametric regression. *The American Statistician* 46 (3): 175–185.

170 Almazán, J., Gordo, A., Fornés, A., and Valveny, E. (2014). Word spotting and recognition with embedded attributes. *IEEE Transactions on Pattern Analysis and Machine Intelligence* 36 (12): 2552–2566.

171 Gordo, A. (2015). Supervised mid-level features for word image representation. In: *Proceedings of the IEEE Conference on Computer Vision and Pattern Recognition*, 2956–2964. Boston, MA, USA: IEEE.

172 Rodriguez-Serrano, J.A., Perronnin, F., and Meylan, F. (2013). Label embedding for text recognition. In: *British Machine Vision Conference*, 5–1. England: BMVA Press, University in Bristol.

173 K. Simonyan and A. Zisserman (2014), "Very deep convolutional networks for large-scale image recognition," 3rd International Conference on Learning Representations (ICLR 2015), [pp. 1–14.].

174 He, K., Zhang, X., Ren, S., and Sun, J. (2015). Deep residual learning for image recognition. In: *Proceedings of the IEEE Conference on Computer Vision and Pattern Recognition (CVPR)*, 770–778. Las Vegas, NV, USA: IEEE.

175 Neubeck, A. and Van Gool, L. (2006). Efficient non-maximum suppression. In: *Proceedings of the International Conference on Pattern Recognition (ICPR)*, vol. 3, 850–855. Hong Kong, China: IEEE.

176 Jaderberg, M., Simonyan, K., Zisserman, A., and Kavukcuoglu, K. (2015). Spatial transformer networks. In: *Proceedings of the International Conference on Neural Information Processing Systems*, vol. 2, 2017–2025. MIT Press.

177 Zhao, H., Shi, J., Qi, X. et al. (2017). Pyramid scene parsing network. In: *Proceedings of the IEEE Conference on Computer Vision and Pattern Recognition*, 2881–2890. Hong Kong, China: IEEE.

178 Graves, A., Fernández, S., Gomez, F., and Schmidhuber, J. (2006). Connectionist temporal classification: labelling unsegmented sequence data with recurrent neural networks. In: *Proceedings of the 23rd International Conference on Machine Learning*, 369–376. Pennsylvania, USA: Association for Computing Machinery.

179 Yin, F., Wu, Y.-C., Zhang, X.-Y. et al. (2017). Scene text recognition with sliding convolutional character models. arXiv preprint arXiv:1709.01727.

180 Su, B. and Lu, S. (2014). Accurate scene text recognition based on recurrent neural network. In: *Asian Conference on Computer Vision*, 35–48. Springer.

181 Hochreiter, S. and Schmidhuber, J. (1997). Long short-term memory. *Neural Computation* 9 (8): 1735–1780.

182 Bahdanau, D., Cho, K., and Bengio, Y. (2014). Neural machine translation by jointly learning to align and translate. arXiv preprint arXiv:1409.0473.

183 Li, H., Wang, P., Shen, C., and Zhang, G. (2019). Show, attend and read: a simple and strong baseline for irregular text recognition. In: *Proceedings of the AAAI Conference on Artificial Intelligence*, vol. 33, 8610–8617. Honolulu, Hawaii, USA: AAAI Press.

184 Wang, Q., Jia, W., He, X. (2019). FACLSTM: ConvLSTM with focused attention for scene text recognition. arXiv preprint arXiv:1904.09405.

185 Wojna, Z., Gorban, A.N., Lee, D.-S. et al. (2017). Attention-based extraction of structured information from street view imagery. In: *International Conference on Document Analysis and Recognition (ICDAR), Kyoto, Japan*, vol. 1, 844–850.

186 Deng, Y., Kanervisto, A., Ling, J., and Rush, A.M. (2017). Image-to-markup generation with coarse-to-fine attention. In: *International Conference on Machine Learning*, 980–989. PMLR.

187 Luo, C., Jin, L., and Sun, Z. (2019). Moran: a multi-object rectified attention network for scene text recognition. *Pattern Recognition* 90: 109–118.

188 Xu, K., Ba, J., Kiros, R. et al. (2015). Show, attend and tell: neural image caption generation with visual attention. In: *Proceedings of the International Conference on Machine Learning*, 2048–2057. lille grand palais, France: PMLR.

189 Wang, F., Jiang, M., Qian, C. et al. (2017). Residual attention network for image classification. In: *Proceedings of the IEEE Conference on Computer Vision and Pattern Recognition (CVPR)*, 6450–6458. Honolulu, Hawaii, USA: IEEE.

190 Xingjian, S., Chen, Z., Wang, H. et al. (2015). Convolutional LSTM network: a machine learning approach for precipitation nowcasting. In: *Proceedings of the Advances in Neural Information Processing Systems*, 802–810. Montréal Canada: Curran Associates, Inc.

191 Quy Phan, T., Shivakumara, P., Tian, S., and Lim Tan, C. (2013). Recognizing text with perspective distortion in natural scenes. In: *Proceedings of the International Conference on Document Analysis and Recognition*, 569–576.

192 Lin, T.Y., Maire, M., Belongie, S. et al. (2014). Microsoft COCO: Common objects in context. In: *Proceedings of the European Conference on Computer Vision*, 740–755. Springer.

193 Shahab, A., Shafait, F., and Dengel, A. (2011). ICDAR 2011 robust reading competition challenge 2: reading text in scene images. In: *Proceedings of the International Conference on Document Analysis and Recognition*, 1491–1496. Beijing, China: IEEE.

194 Gupta, A., Vedaldi, A., and Zisserman, A. (2016). Synthetic data for text localisation in natural images. *Proceedings of the IEEE Conference on Computer Vision and Pattern Recognition*, Las Vegas, NV, USA, 27–30 June 2016, last retrieved 11 March 2020 [Online]. Available: https://www.robots.ox.ac.uk/~vgg/data/scenetext.

195 Liu, Y. and Jin, L. (2017). Deep matching prior network: Toward tighter multi-oriented text detection. In: *Proceedings of the IEEE Conference on Computer Vision and Pattern Recognition*, 1962–1969. Honolulu, HI, USA: IEEE.

196 Wolf, C. and Jolion, J.M. (2006). Object count/area graphs for the evaluation of object detection and segmentation algorithms. *International Journal on Document Analysis and Recognition (IJDAR)* 8 (4): 280–296.

197 Dollar, P., Wojek, C., Schiele, B., and Perona, P. (2011). Pedestrian detection: an evaluation of the state of the art. *IEEE Transactions on Pattern Analysis and Machine Intelligence* 34 (4): 743–761.

198 Du, X., Ang, M.H., and Rus, D. (2017). Car detection for autonomous vehicle: LIDAR and vision fusion approach through deep learning framework. In: *Proceeding of the IEEE/RSJ International Conference on Intelligent Robots and Systems (IROS)*, 749–754. Vancouver, BC, Canada: IEEE.

199 Ammour, N., Alhichri, H., Bazi, Y. et al. (2017). Deep learning approach for car detection in UAV imagery. *Remote Sensing* 9 (4): 312.

200 Choi, C., Yoon, Y., Lee, J., and Kim, J. (2018). Simultaneous recognition of horizontal and vertical text in natural images. In: *Proceedings of the Asian Conference on Computer Vision*, 202–212. Springer, Perth, Australia.

201 Ling, O.Y., Theng, L.B., Chai, A., and McCarthy, C. (2018). A model for automatic recognition of vertical texts in natural scene images. In: *Proceedings of the IEEE International Conference on Control System, Computing and Engineering (ICCSCE)*, 170–175. Penang, Malaysia: IEEE.

202 Devlin, J., Chang, M.W., Lee, K. et al. (2018). Bert: pre-training of deep bidirectional transformers for language understanding." arXiv preprint arXiv:1810.04805.

203 Lyn, J. and Yan, S. (2020). Image super-resolution reconstruction based on attention mechanism and feature fusion. arXiv preprint arXiv:2004.03939.

204 Anwar, S. and Barnes, N. (2019). Real image denoising with feature attention. In: *Proceedings of the International Conference on Document Analysis and Recognition*, 3155–3164. Seoul, Korea: IEEE.

205 Hou, Y., Xu, J., Liu, M. et al. (2020). NLH: a blind pixel-level non-local method for real-world image denoising. *IEEE Transactions on Image Processing* 29: 5121–5135.

206 Liu, Y.L., Lai, W.S., Yang, M.H. (2020). Learning to see through obstructions. arXiv preprint arXiv:2004.01180.

207 Gomez, R., Biten, A.F., Gomez, L. et al. (2019). Selective style transfer for text. In: *International Conference on Document Analysis and Recognition (ICDAR)*, 805–812. Sydney, NSW, Australia: IEEE.

208 T. Karras, S. Laine, M. Aittala et al. (2019). Analyzing and improving the image quality of styleGAN. arXiv preprint arXiv:1912.04958.

209 Goodfellow, I., Pouget-Abadie, J., Mirza, M. et al. (2014). Generative adversarial nets. In: *Proceedings of the Advances in Neural Information Processing Systems*, 2672–2680. Palais des Congrès de Montréal, Montréal, Canada: Curran Associates, Inc.

210 Chen, X., Kundu, K., Zhu, Y. et al. (2015). 3D object proposals for accurate object class detection. In: *Advances in Neural Information Processing Systems*, 424–432. Palais des Congrès de Montréal, Montréal, Canada: Curran Associates, Inc.

211 Zhao, Z.-Q., Zheng, P., Xu, S.-t., and Wu, X. (2019). Object detection with deep learning: a review. *IEEE Transactions on Neural Networks and Learning Systems* 30 (11): 3212–3232.

212 Xie, S., Girshick, R., Dollár, P. et al. (2017). Aggregated residual transformations for deep neural networks. In: *Proceedings of the IEEE Conference on Computer Vision and Pattern Recognition*, 1492–1500. Honolulu, HI, USA: IEEE.

213 Sabour, S., Frosst, N., and Hinton, G.E. (2017). Dynamic routing between capsules. In: *Proceedings of the Advances in Neural Information Processing Systems*, 3856–3866. Long Beach, CA, USA: Curran Associates, Inc.

214 Ghai, D. and Jain, N. (2013). Text extraction from document images-a review. *International Journal of Computer Applications* 84: 40–48.

215 Ghai, D. and Jain, N. (2016). A new approach to extract text from images based on DWT and K-means clustering. *International Journal of Computational Intelligence Systems* 9: 900–916.

216 Ghai, D. and Jain, N. (2019). Comparative analysis of multi-scale wavelet decomposition and k-means clustering based text extraction. *Wireless Personal Communications* 109: 455–490.

217 Raisi, Z., Naiel, M.A., Zelek, J. et al. (2020). Text detection and recognition in the wild: a review https://www.arxiv-vanity.com/papers/2006.04305?> (3 August 2022).

11

Machine-Learning Techniques for Deaf People

Yogini D. Borole[1] and Roshani Raut[2]

[1]*Department of E & TC Engineering, G H Raisoni College of Engineering and Management, SPPU Pune University, Pune, India*
[2]*Department of Information Technology, Pimpri Chinchwad College of Engineering, Pune, India*

11.1 Introduction

Talk affirmation in establishment uproar is apparently hard for people with hearing handicaps. Numerous clients utilizing customary cochlear embed (CE) gadgets in calm acoustic conditions have accomplished close to-ordinary discourse considerate conditions [1]. Though, discourse getting capacity about CE clients is contrarily influenced through the existence of contending communicators then ecological sounds (e.g. foundation commotions). Discourse gathering edge (SRT) is applied to quantify the presentation corruption. At the end of the day, tear shot hindered listeners can be characterized as sign to-clamor proportion (signal-to-noise ratio [SNR]), where discourse coherence accomplished is 50%. When contrasted with the SRTs of normal audible range, audience members, the SRTs of CE clients can ordinarily change in ear shot hindered listeners (more awful) range from 10 to 25 dB [2, 3]. The research from [4, 5] is moderate plentifulness variances or transient holes of fixed commotion imitator profits appeared somewhat appreciated by the CE beneficiaries contrasted with the discourse clarity of ordinary listeners audience members. Regardless, discharge from concealing has characterized tear shot hindered listeners process. Most remarkable unearthly channels [6, 7] in little amount are shaped by lessening the otherworldly data proposed by a CE. Worldly data is to a great extent utilized by the CE clients, however, when contrasted with ordinary listeners and audience members, the regulated covering commotion is profoundly powerless through CE workers [4, 8] may be, CE client's discourse kind execution is diminished through consolidating expanded regulation obstruction and diminished otherworldly goal. Ear shot hindered listeners is apparent with CE recreations tried for ordinary listeners [4, 5, 9] clatter part about loud blend feasible constricted utilizing the planned discourse improvement calculations; henceforth, the apparent value also comprehensibility about discourse segment is expanded. Restricted ear shot hindered listeners apparent has been accomplished by embracing ordinary single channel strategies. The acknowledgment capacity of ordinary hearing and ear shot hindered listeners [10] is expanded by utilizing the "hear-able concealed limit commotion concealment" technique [10] under boisterous situations [11]. Equally, prattle commotion for ear shot ear shot hindered listeners [12] discourse formed superiority also discourse clarity in discourse molded clamor was further developed utilizing the research facility tried inadequate cryptogram reduction calculation. Work done lastly has detailed that utilized single-channel improvement calculations for ear shot hindered listeners audience members publicized horrible showing in word acknowledgment yet not for listeners fondness [13–16]. In the latest tear shot hindered listeners, artificial intelligence (AI) approaches [17] have demonstrated their incredible strength in discourse clarity improvement for cochlear implant (CI) users [18], ordinary listeners, audience members, ear

Machine Learning Algorithms for Signal and Image Processing, First Edition.
Edited by Deepika Ghai, Suman Lata Tripathi, Sobhit Saxena, Manash Chanda, and Mamoun Alazab.
© 2023 The Institute of Electrical and Electronics Engineers, Inc. Published 2023 by John Wiley & Sons, Inc.

shot hindered listeners [19–21], and clinical signals [6, 22–24]. Ordinarily, in light of the appraiser shot hindered listeners ng sign, acquire work for clamor and discourse insights is assessed. In any case, the ideal addition work was assessed utilizing the customary AI approaches through joining earlier information on discourse and commotion designs. Tear shot hindered listeners assessment is done to apply for the appraiser shot hindered listeners ng sign. Discourse coherence of both CI users [25] and ordinary listeners [26] is improved by applying Gaussian-blend models. Ordinary listeners and ear shot hindered listeners audience's discourse coherence scores are exceptionally further developed utilizing a profound neural-organization algorithm [20].

11.2 Literature Survey

Many of the exploration fields, an AI technique called profound learning is applied generally because of its further developed order execution. As such, profound learning is for the most part utilized for discourse detachment, SE, and discourse recognized [27–31] dependent on period–recurrence components, the paired or delicate characterization choice is made utilizing a few information-driven strategies to accomplish SE. Model for tear shot hindered listeners is binaural discourse variation accomplished through assessing leveled model proportion cover or great parallel cover [32, 33]. Discourse coherence can be adequately further developed utilizing the hard targets IBM. On the other hand, discourse value can be upgraded by giving a well forecast in contradiction to easy objectives ideal ratio marks. All recurrence units, a concealment expansion could be construed on ideal ratio marks in the scope of [1–10]. Rough and noisy components furthermore, assessed IRM are increased in a component astute way to acquire the further developed provisions as the last yield. Decreased discourse twisting is accomplished by stifling commotion to certain degrees by utilizing effective delicate covering calculations. Aside from tear shot hindered listeners ideal ratio mask straight expectation, joint advancement of deep neural networks (DNNs) taking an extra covering layer and veiling capacities is inspected and broken down in Huang et al. [34]. Furthermore, discourse phantom is planned straightforwardly utilizing profound learning draws near for overcoming the time–recurrence cover expectation. Authors [35] have utilized a huge assortment of mixed information to prepare profound and broad neural organization design just as to recommend a DNN depending on relapse structure. The justification for the presentation of DNN-based relapse method is that it has gained the capacity to keep up with both the ear shot hindered listeners non-fixed commotions and non-straight clamors and to keep away from the superfluous suspicions on factual features of signs [36, 37]. Assessed clean discourse signal sometimes has experienced mutilations due to the impressive commotion expulsion from the loud discourse utilizing the relapse DNN. Be that as it may, eliminating such contortions goes to be a monotonous and testing task. To tackle tear shot hindered listeners monotonous undertaking, different balance of elements was applied for additional post-handling of the relapse DNN. In ear shot hindered listeners, the contortions remembered for assessed clean elements are eliminated successfully. Inconspicuous clamor conditions are not really settled utilizing the speculation limit of profound learning draws near. Authors [38] have utilized dynamic clamor mindful preparing strategy to improve the speculation ability of profound learning draws near.

Prominently, the multi-target structure has been created through expanding the DNN design [39]. For discourse acknowledgment, Kim and Smaragdis [40] have utilized the variation techniques and further developed the noise canceling AI execution at the test stage by applying a tweaking plan. In small SNR condition, corruption of execution is viewed as an additional and most important testing task. Under basic commotion conditions, the discourse coherence was upgraded work by authors [41] through incorporating SE and voice movement location into a future combined DNNs structure. Considering study fact, speech enhancement utilizing the extra perplexing nervous organization is as yet deprived to be considered and investigated. Extended transient recollection is investigated by authors in their works [42, 43] have broken down the ConvNets. Deriving basis otherworldly and mark basis ghostly are getting the hang of utilizing the design with double yields by Tu et al. [44]. In tear shot hindered listeners work, we contemplated the discourse clarity improvement of CI clients in various foundation commotions through joining profound ConvNets with a SE calculation. The boisterous discourse gesture is disintegrated hooked on ear shot hindered listeners outlines utilizing a SE calculation. Then, at that point, a DCNN is taken care of with deteriorated uproarious discourse signal edges to create a recurrence channel assessment. In any case, an ear shot hindered listeners SNR data are contained in created recurrence channel assessment. Utilizing tear shot hindered listeners gauge, discourse overwhelmed circular implantation networks considered to create electrical incitement. Tear shot hindered listeners cycle is equivalent to traditional no FM CI programming procedures. To decide the discourse in-commotion execution of 12 CI clients, the flicker and harmony sound practical are careful as foundation commotions.

11.3 Objectives

Execution of the planned calculation is assessed by tear shot hindered listeners thinking about these foundation commotions. The short synopses of tear shot hindered listeners works are as per the following:

Our principle commitments are summed up as given below:

- The proposed design has a nine-layer DCNN to assess the language improvement for further developing discourse comprehensibility in a clamor for CI clients.
- Cluster investment is complete with 12 CI clients having circular navel CI embedded, altogether are local language, Tamilian talkers.
- Presentation of the future calculation is assessed through leading broad reproduction, looked at, and examined the results with the ordinary strategies.

Rest of tear shot hindered listeners article is coordinated as given below: profound knowledge-based discourse improvement-based speech enhancement calculation is clarified in Section 11.4. Planned calculation depiction with resources and strategies is elaborated in Section 11.5. Trial arrangement of the planned strategy is depicted in Section 11.6 with the name Assessment. Section 11.7 gives an idea about outcomes and conversations. Section 11.8 discusses about discourse coherence. Conclusion is concluded in Section 11.9.

11.4 Proposed Calculation Depiction

CI discourse preparing system execution of progressive mixture encoder is coordinated with the deep learning speech enhancement calculation. The proposed framework is shown in Figure 11.1.

11.4.1 Reference System

An execution of an examination ACE system goes about as the reference methodology. To a programmed acquire control (AGC), the boisterous discourse flags that are passing utilizing a pre-accentuation channel and down sampled to 16 kHz are communicated. Basically, the acoustic unique reach is packed utilizing AGC. Then, at that point, to a CI beneficiary having a more modest electrical powerful reach, tear shot hindered listeners can be passed on without any problem. Consequently, to the packed sign, a channel bank dependent on a quick fast Fourier was sent. For every one of the M recurrence stations (commonly, $M = 22$), a gauge of the packet was given by the yield of the composite fast Fourier transform size. Then, at that point, just for one anode, individual network was dispensed. To hold the subgroup of N networks having more noteworthy packet sizes (wear shot hindered listeners position the topics of circular implantation, an audiology set $N < M$), the packet for every network is planned promptly with tumult development work. Likewise, for electrical incitement, the subject's unique reach sear shot hindered listeners between the edge level and greatest agreeable clamor level. The boundaries edge level

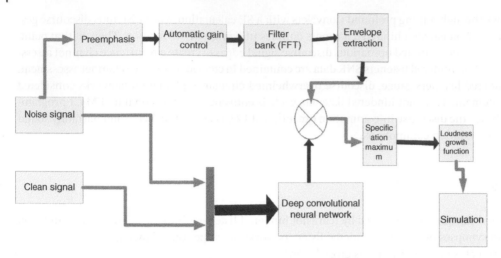

Figure 11.1 Flowchart of proposed design.

and clamor level are utilized from the theme's CI workstation. Finally, the pattern of incitement can be finished through consecutively invigorating the terminals identified with the chosen channels. Outstandingly, the station incitement amount is characterized as an amount of sequences each another, and the network incitement rate is N times the complete incitement amount.

Profound convolutional neural network (CNN) sorter for speech enhancement to an electrical yield, the recurrence network packet is changed straightforwardly with CI handling and furthermore, it doesn't request any of the recreation periods. Rather than undertaking preparing of the uproarious sign, we like to consolidate the deep learning-based speech enhancement (DLSE) straightforwardly into the CI sign way. In any case, the excess combination stage is kept away from and furthermore, deferral and intricacy of the framework are expanded with the presentation of extra clamor. Basically, pre-preparing and deep CNNs are the two primary parts of the deep learning speech enhancement calculation.

Component change during pre-handling, end of quiet edges after the sign, and down sampling of the discourse signs to 8 kHz is completed. Plentifulness scale subjects tended to conclude standardizing individually edge utilizing Z-value standardization. Therefore, preparing what's more, the trying phase in 1D profound education CNN ought to be directed solely afterward taking care of the discourse portions by eliminating the offset. Fundamentally, 10-ms hamming window remembered for 256-point brief period frame frequency range change is practical to process the unearthly paths. Notwithstanding, the window sear shot hindered listeners is fixed to 2 ms (64-point). For every recurrence receptacle, the recurrence goal was fixed to 5 kHz/128 = 39.06 Hz. Proportional half was taken out to 129-point by decreasing 256-point rule invariant component change size vectors. Regardless, eight successive boisterous filter size paths with term 100 ms and size 129 × 8 remembered for the info included are utilized by CNN for additional preparation. To acquire the unit, change and zero mean, normalize both the info ear shot hindered listeners highlights. CNN engineering. Among counterfeit neural organizations, CNN plays a typical role [45]. A multi-facet neural network is looked like a CNN. An actuation work remembered for each and every neuron of multi-layered perception can complete the planning of subjective contributions to the yields. In the organization, more than one covered up layers are coordinated to change over a solitary multi-layered perception into profound MLP. Likely, utilizing a unique design, the CNN is accepted to act equivalent to that of multi-layered perception. Next design model, pivot invariant and interpretation properties can be displayed by CNN with the help of extraordinary structure [46]. CNN engineering incorporates a corrected direct initiation work and extra three normal stages, for example, a convolutional layer, pooling layer, and completely associated layer [47]. The engineering of the planned CNN model is summed up in Table 11.1.

Table 11.1 Proportions of planned convolutional neural network model.

Layers	Nature	No. of nerves cells (yield stage)	Portion scope	Step
0 to 1	Complication	258×5	4	2
1 to 2	Maximum sharing	129×5	3	1
2 to 3	Complication	126×10	4	2
3 to 4	Maximum sharing	63×10	2	1
4 to 5	Complication	60×20	4	2
5 to 6	Maximum sharing	30×20	3	1
6 to 7	Completely associated	30	—	—
7 to 8	Completely associated	20	—	—
8 to 9	Completely associated	5	—	—

Essentially, CNN design is comprised of nine stages (e.g. three CNN stages [48], three maximum sharing stages, and three completely associated stages). Utilizing condition (1), the piece sizes (4, 3, and 3) are utilized by their separate layer for convolution. Tear shot hindered listeners is accomplished for each convolution layer (layers 1, 3, and 5). To the component maps, a maximum pooling activity is utilized close to each convolution layer. Quiet, the component map size is diminished by the maximum sharing activity. In tear shot hindered listeners article, the savage power method is applied to acquire the channel (bit) size boundary. Be that as it may, 1 and 2 are the step fixed for convolution and max-pooling activity. Layers 1, 3, 5, 7, and 8 have gotten actuation work utilizing the defective rectifier direct unit (LeakyRelu) [49]. Output neurons (e.g. 30, 20, and 5) are remembered for the completely associated layers. Besides, layer 9 (last layer) comprises of w yield neurons. Wiener acquires work is applied to figure the SoftMax work. Equation (11.1) is utilized to decide the assessed acquire:

$$G(b, \tau) = \sqrt{\frac{\text{SNR}_p(b, \tau)}{1 + \text{SNR}_p(b, \tau)}} \tag{11.1}$$

where τ is the casing, b is the Gamma tone recurrence channel. Here, prior SNR is demonstrated as SNRp wear shot hindered listeners characteristics are displayed in the underneath condition (Eq. (11.2)) as follows:

$$\text{SNR}_p(b, \tau) = \frac{\alpha |\overline{X}(b, \tau - 1)|^2}{\lambda_D(b, \tau - 1)} + (1 - \alpha) \times \max \left[\frac{|S(b, \tau)|^2}{\lambda_D(b, \tau)} \right] - 1,0 \tag{11.2}$$

where $\alpha = 0.98$ is a smoother shot hindered listeners ng consistent, λ_D is the gauge of the foundation clamor difference. Utilizing the underneath condition (Eq. (11.3)), the convolution activity is determined

$$\sum_{k=0}^{N-1} y_k f_{n-k} \tag{11.3}$$

where N, n, and y address the quantity of components in y, channel, and sign, individually. The term x means the yield vector. Considering the list of capabilities (input test size) in the back-engendering algorithm, 48 the preparation phase of CNN is directed. Parametric measure for the hyper-boundaries, specifically energy, learning rate, and regularization (λ) is fixed to 0.7, 3×10^{-3}, and 0.2, individually. Advantages of utilizing these hyper-boundaries in the preparation stage are the following: (a) controlling the learning speed, (b) helping the information combination, (c) and overseeing overfitting of information. Ideal execution can be accomplished by changing the boundaries relying upon the animal power strategy. Utilizing conditions (Eqs. (11.4) and (11.5)), the loads and inclinations are refreshed:

$$\Delta W_l(t+1) = -\frac{x_\lambda}{r} W_l - \frac{x}{n} \frac{\partial C}{\partial W_l} + m \Delta W_l(t) \tag{11.4}$$

$$\Delta B_l(t+1) = -\frac{x}{n} \frac{\partial C}{\partial B_l} + m \Delta B_l(t) \tag{11.5}$$

where C, t, m, n, x, λ, l, B, and W demonstrate the expense work, refresher shot hindered listeners ng advance, energy, the aggregate number of preparing tests, knowledge percentage, regularization boundary, layer number, inclination, and weight, individually.

11.5 Resources and Strategies

11.5.1 Equipment/Programming

By software simulation, deep learning speech enhancement calculation and exploration advanced combination encoder technique are carried out. The ACE technique carried out in a PC is applied to handle the upgrades. Then, at that point, the embed clients are straightforwardly given the handled boosts.

L34 exploratory mainframe is utilized to associate the circular NIC3 interface for conveying electrical incitement. In tear shot hindered listeners manner, a loop gets radio recurrence yield conveyed by the framework. At last, the subjects embed contain the boost information that is sent by a curl.

11.5.2 Topics

Gathering cooperation is made with 12 CI clients taking circular center CI embedded, and all are local Tamilian talkers. A sum of 12 CI clients (documents) is remembered for the Tamil discourse data set. Table 11.2 demonstrates the topic's segment information.

Nonetheless, 12 vowels sound (each solid reach ear shot hindered listeners going out to 3 seconds) and 18 consonant flickers are included in each record. The size of inspecting recurrence is $16 \times^3$ Hz and every one holds 16 pieces for each example. During the underlying examination, 60 years was fixed as the mean age of the gathering; accordingly, it might run from 20 to 80 years. For testing, single ear of each subject is thought of.

Table 11.2 Segment information.

Boundary	Qualities
Energy boundaries	0.8
Learning rate	4×10^{-3}
Regularization (λ)	0.3
p esteem	0.0056
Vowels	12
Consonants	8
Identification of gender	5 male, 9 female
Speakers	13
Age bunch	20–80
First language	Tamilian, Marathi

Nonetheless, turn off the CI or listening device fixed to the contra lateral side of each subject. At the underlying stage of the examination, 10 years was fixed as the inserts mean span going from 1.3 to 3.5 years. In advanced computation encoder technique, all topics have behaved like clients.

11.5.3 Handling Conditions and Improvements

Target discourse material is accepted to the vowels and consonants of Tamilian talkers. Uproarious climate applied in tear shot hindered listeners examination is isolated into two structures specifically, fan and music foundation. Then, at that point, vowels and consonants (e.g. two discourse supplies) were utilized to examine all CI clients under these uproarious conditions. Handling conditions are of two structures:

- **Naturalstate:** Advanced computation encoder, natural condition.
- **DLSE:** Deep learning-based speech enhancement, handled condition.

11.5.4 Utilized Convention

A versatile method is utilized to gauge the sort I and type II blunder rates for two ecological conditions (e.g. two handling conditions × two maskers [flicker and sound foundation]). Tear shot hindered listeners calculation is acted in a music room with the assistance of an audiovisual. Throughout examination, the preparing circumstances, both the audiologist and topic, remained kept in an inconspicuous state (e.g. dazzle).

11.6 Assessment

For every handling state, the target examination is completed earlier scientific verification. At various SNRs, the electrograms were assessed. Then, at that point, a reference electrogram is applied to tear shot hindered listeners about the registered electrograms utilizing type I and type II mistake rates. Standardized qualities are contained in the improvements of an electrogram. These qualities demonstrate the electrical insight fluctuating from solace level and edge remembered for the recurrence channel and edges, consistently.

In calm climate, with the presence of advanced computation encoder, the handling discourse was made utilizing reference electrogram. In any case, excluding DLSE in tear shot hindered listeners climate can deliver great clamor decrease result through applying preparing discourse for creating reference electrogram. Type I and type II mistake rates were acquired by partitioning the complete number of potential blunders to the added type I and type II mistakes across all edges and recurrence channels. With ACE maxima choice (11 picked channels), the normal blunder rates processed more than 11 dB SNR and 21 characters at −5, 0, 5 are considered as the blunder paces of handling condition. Along these lines, both the blunder types incorporate the equivalent number of potential blunders and furthermore control the presentation of a predisposition toward 2.

UN giving the sort I and type II blunder rates for music foundation varies from 37% to 67% and 96% to 16% wear shot hindered listeners channel implies SNR = −6 and 11 dB, individually. Comparable mistake rates are furnished by the DLSE conditions with the cost of to some degree ear shot hindered listeners sort II mistake rates (≤15% and ≤21% at −5 and 11 dB SNR, individually) also, with exceptionally decreased sort I mistake rates (≤7% and ≤18% at −5 and 11 dB SNR, separately). UN giving sort I blunder rates and type II mistake rates changing from 21% to 43% and 5% to 11% (e.g. SNR = −5 and 11 dB, individually) is shown for fan foundation commotion. With the cost of ear shot hindered listeners sort II mistake rates, type I blunder rates are decreased utilizing both the DLSE conditions. At last, not just commotion is exceptionally decreased with the deep learning speech enhancement calculation, yet additionally, some discourse evacuation twists are presented. Contrasted with UN, the all-out blunder is likewise diminished for all SNRs furthermore, clamor. These outcomes support CI clients in the verification of clinical discourse execution and furthermore accomplished an improvement in discourse insight.

11.7 Outcomes and Conversations

The tests performed for consonant letters and Tamilian vowel for 11 maxima holding Pro system. With the presence of two sorts of foundation commotions (fan and music), the discourse was disturbed in the subsequent board. DLSE handling strategy circumstances are shown in tear shot hindered listeners board. The proposed deep learning speech enhancement strategy is so gainful for CI beneficiaries by method for giving viable and prevalent comprehensibility execution. Proposed DLSE-based methodology productivity is breaking down by contrasting two neural organization-based approaches: (a) correlation ear shot hindered listeners highlight set (N-COM), (b) hear-able list of capabilities (NN-AIM), and Wiener separating (WF). Correlation ear shot hindered listeners highlight set: Using similar arrangement of provisions (contingent upon a reciprocal arrangement of features) [50] N-COM (correlation include set) is produced. In addition, from every 20 ms long time period concerning the loud discourse mixture [50], the mel-recurrence cepstral coefficients relative-otherworldly change and perceptual direct expectation coefficients [51], and the sufficiency adjustment range [52] were extricated to deliver the ear shot hindered listeners highlight set. A dimensionality of 445 for each time span (AMS (25×15) + RASTA-PLP (3×13) + MFCC) contained the linked elements. The delta–delta ear shot hindered listeners highlights for RASTA-PLP just (as portrayed by Healy et al. [50]) and delta (back to back outline ear shot hindered listeners highlights contrast) are utilized for linking the current time period wear shot hindered listeners character was extricated from the N-COM. Hear-able list of capabilities: The hear-able picture model50 was utilized to the N-AIM hear-able list of capabilities. For better contrast the two handling conditions, the factual examination was done separately for each clamor type. Table 11.1 portrays the segment information for tear shot hindered listeners work. In view of the test, distinctive likelihood circulations are utilized to process the p esteems. In the event that they chose importance level (by and large 0.05) is more noteworthy than the p esteem then, at that point a decent outcome is created.

11.8 Discourse Coherence

For four speech enhancement calculations, the S/Ns of 0 and +5 dB was fixed to decide the discourse coherence of both sound and flicker foundation clamor circumstances. Along these lines, the correlation is made with unchanged commotion circumstances. Experimental consequences of 12 circular implantation clients introduced for the listening test are announced in tear shot hindered listeners learning. Results got for 12 circular implantation clients are tried on four speech enhancement calculations by setting the SNR levels from 0 to +5 dB, and the mean scores of commotion discourse are shown in Figures 11.2–11.4. Arrived at the midpoint of character correct rate action utilized for announcing the presentation and it is displayed in Figure 11.3. Correct rate demonstrates the proportion of an all-out number not really settled on each testing condition to the quantity not set in stone person CC. Figure 11.4 shows that ear shot hindered listeners coherence grades are acquired with the planned DLSE-method contrasted with additional current methodologies. Language excellence for both SNR of 0 and +5 dB and two clamor conditions, the discourse quality evaluations were recognized as far as every calculation. Regardless, in a large portion of the circumstances, the typical dispersion of information was unrealistic. Accordingly, nonparametric statistic insights are utilized and hair and container schemes are shown. At the ear shot hindered listeners SNR, the discourse quality is better-quality for accomplisher shot hindered listeners ng the discourse comprehensibility. Likewise, when contrasted with the music foundation clamor, the proposed calculation has exposed better upgrades in fan foundation clamor. Worth of N-AIM ($p = 0.018$) fixed at +5 dB SNR has exposed critical excellence evaluation enhancement in the sound foundation commotion conditions (e.g. superiority grade acquired was 0.82). When fixed, N-AIM ($p = 0.0074$), N-COM ($p = 0.018$), and scanty coding ($p = 0.0024$) at 0 dB S/N have exposed enormous ordinary listener advancements in the flicker foundation conditions (e.g. improvement accomplished is 0.45, 0.53, and 0.70, individually). Additionally, worth of N-AIM ($p = 0.0014$) and N-COM

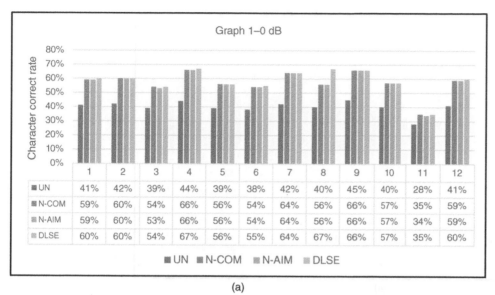

(a)

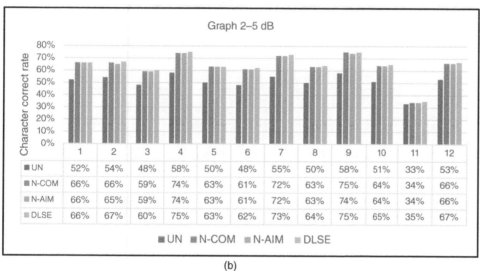

(b)

Figure 11.2 Acquired gathering mean level of effectively perceived characters as far as every calculation for: (a) higher level (flicker sound) and (b) lesser level (melody sound).

($p < 0.0013$) fixed at 4 dB S/N has shown huge ordinary listeners advancements (e.g. 0.99 and 0.87, respectively). Properties of perceptibility for every member and condition, their discourse perceptibility were estimated through registering the discourse clarity list. Table 11.3 portrays the SII esteems. The tracker separating, commotion after the sending of improvement acquires capacity, and discourse spectra were embraced for ascertaining SII to repay with the upgraded conditions. Notwithstanding, an increase work is not utilized on inadequate code preparing yet, contingent upon the distinction between the first level and 33% octave groups and upgraded signals in 10 ms outlines, an increase work was figured for ascertaining SII.

Besides, benefits on changing the discourse range level are controlled by registering the SII an incentive for unique clamor range and upgraded discourse range. Notwithstanding, because of ordinary listeners conditions there experienced just little abatement in the SII esteems; yet the SII esteems somewhat increment for Wiener

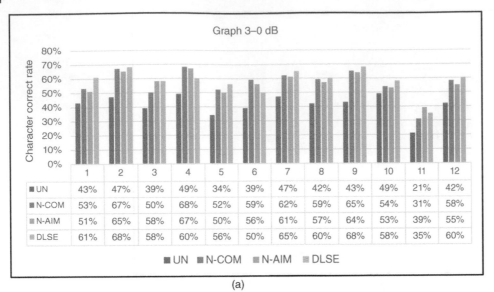

(a)

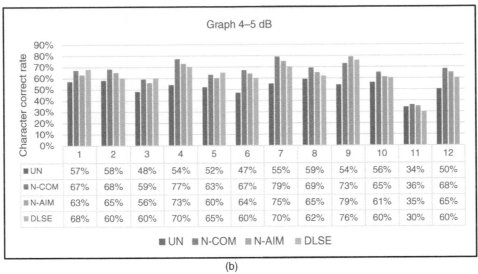

(b)

Figure 11.3 Mean clarity outcome of two examination circumstances: (a) 0 dB SNR and (b) +5 dB SNR. The blunder bars demonstrate the standard blunders of mean qualities.

sifting concerning most preparing and clamor conditions. Here, jibber jabber clamor likewise measured by current two commotion conditions. Likewise, for N-AIM and N-COM, both discourse molded clamor circumstances are shown (mean qualities increment in discourse clarity record of 0.018, 0.014, and 0.013, separately, at 0 dB S/N and 0.00810, 0.00429, and 0.0038 at +5 dB S/N). Tables 11.3–11.5 show a presentation correlation of various strategies for three commotion conditions, such as aerate environment, melody environment, and hum noise with various S/Ns on the exam set, respectively. Thoughtful assessment of discourse superiority portion, brief period frame target coherence amount, and minor preparing delay-based assessment in the previously mentioned, not really set in stone the HA framework segments, independently. To distinguish the parametric qualities, tear shot hindered listeners interaction upholds a great deal and under pragmatic circumstances of these segments, an adequate ear

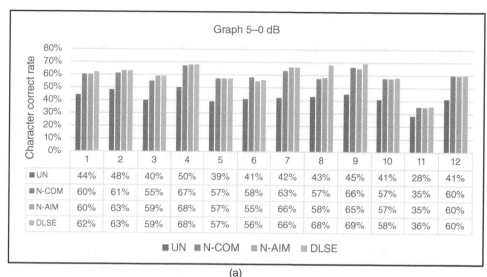

(a)

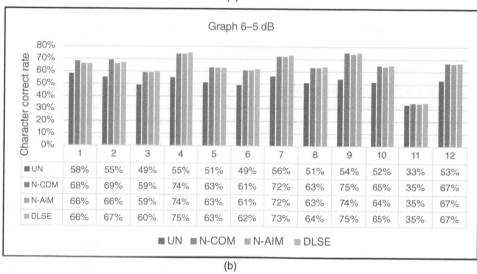

(b)

Figure 11.4 Case and bristle plans of discourse superiority evaluations for various calculations in: (a) the flicker foundation condition and (b) melody foundation clamor conditions.

shot hindered listeners conditions is accomplished with the distinguished boundary esteems. In tear shot hindered listeners segment, the presentation of the proposed strategy is assessed as far as the perceptual assessment of discourse quality measure, brief time frame target coherence, and the low preparing interruption. In view of sound perception standards, after transient arrangements with their separate sign, a test discourse signal is investigated utilizing a target quantity termed "perceptual evaluation speech quality" [53]. In the examination, the above discussed about 63 seconds since a long time ago information discourse signs were utilized. To create 5 dB SNR, the fan commotion was considered to be the discourse signs. The conventional WF technique, N-COM, furthermore, N-AIM was utilized for breaking down the extremely shot hindered listeners between the planned DLSE technique. Each and every technique. The emotional assessment was directed on the music value yield in the consistent stage.

Table 11.3 Each subject and conditions discourse comprehensibility record esteems on aerate environment.

Sr. No.	UN	N-COM	N-AIM	DLSE	UN	N-COM	N-AIM	DLSE
Aerate environment								
1	0.41	0.59	0.59	0.60	0.52	0.66	0.66	0.66
2	0.42	0.60	0.60	0.60	0.54	0.66	0.65	0.67
3	0.39	0.54	0.53	0.54	0.48	0.59	0.59	0.60
4	0.44	0.66	0.66	0.67	0.58	0.74	0.74	0.75
5	0.39	0.56	0.56	0.56	0.50	0.63	0.63	0.63
6	0.38	0.54	0.54	0.55	0.48	0.61	0.61	0.62
7	0.42	0.64	0.64	0.64	0.55	0.72	0.72	0.73
8	0.40	0.56	0.56	0.67	0.50	0.63	0.63	0.64
9	0.45	0.66	0.66	0.66	0.58	0.75	0.74	0.75
10	0.40	0.57	0.57	0.57	0.51	0.64	0.64	0.65
11	0.28	0.35	0.34	0.35	0.33	0.34	0.34	0.35
12	0.41	0.59	0.59	0.60	0.53	0.66	0.66	0.67

Table 11.4 Each subject and conditions discourse comprehensibility record esteems on melody environment.

Sr. No.	UN	N-COM	N-AIM	DLSE	UN	N-COM	N-AIM	DLSE
Melody environment								
1	0.43	0.53	0.51	0.61	0.57	0.67	0.63	0.68
2	0.47	0.67	0.65	0.68	0.58	0.68	0.65	0.60
3	0.39	0.50	0.58	0.58	0.48	0.59	0.56	0.60
4	0.49	0.68	0.67	0.60	0.54	0.77	0.73	0.70
5	0.34	0.52	0.50	0.56	0.52	0.63	0.60	0.65
6	0.39	0.59	0.56	0.50	0.47	0.67	0.64	0.60
7	0.47	0.62	0.61	0.65	0.55	0.79	0.75	0.70
8	0.42	0.59	0.57	0.60	0.59	0.69	0.65	0.62
9	0.43	0.65	0.64	0.68	0.54	0.73	0.79	0.76
10	0.49	0.54	0.53	0.58	0.56	0.65	0.61	0.60
11	0.21	0.31	0.39	0.35	0.34	0.36	0.35	0.30
12	0.42	0.58	0.55	0.60	0.50	0.68	0.65	0.60

Table 11.6 portrays the handling deferrals, brief time frame target understandability and perceptual assessment of discourse quality upsides, every ear shot hindered listeners ng being equal, and the client rating on 95% certainty stretches in mean qualities. Client evaluations of whole hearing techniques and profiles are accounted for utilizing the PESQ and STOI esteems. Tear shot hindered listeners aides HA calculations for the frame determination. After contrasted with the NN-AIM, N-COM, and WF technique, the planned circular implant strategy has acquired

Table 11.5 Each subject and conditions discourse comprehensibility record esteems on hum noise.

Sr. No.	UN	N-COM	N-AIM	DLSE	UN	N-COM	N-AIM	DLSE
Hum noise								
1	0.44	0.60	0.60	0.62	0.58	0.68	0.66	0.66
2	0.48	0.61	0.63	0.63	0.55	0.69	0.66	0.67
3	0.40	0.55	0.59	0.59	0.49	0.59	0.59	0.60
4	0.50	0.67	0.68	0.68	0.55	0.74	0.74	0.75
5	0.39	0.57	0.57	0.57	0.51	0.63	0.63	0.63
6	0.41	0.58	0.55	0.56	0.49	0.61	0.61	0.62
7	0.42	0.63	0.66	0.66	0.56	0.72	0.72	0.73
8	0.43	0.57	0.58	0.68	0.51	0.63	0.63	0.64
9	0.45	0.66	0.65	0.69	0.54	0.75	0.74	0.75
10	0.409	0.574	0.573	0.578	0.516	0.645	0.641	0.650
11	0.281	0.351	0.349	0.355	0.334	0.346	0.345	0.350
12	0.412	0.598	0.595	0.600	0.530	0.668	0.665	0.670

Table 11.6 All strategies wide-ranging value and short preparing postpone execution.

	Technique			
Events	DLSE	N-COM	N-AIM	WF
Maximum suspension (ms)	1.78	2.8	3.89 ± 0.33	411 114 444 444 444 433 335
PESQ	3.95	3.6	3.9	3.5
STOI user ratings	0.97 4.01 ± 0.4	0.86 3.7 ± 0.3	0.75 3.35 ± 0.3	0.66 3.7 ± 0.3

healthier quality. In addition, proposed deep learning speech enhancement strategy preparing delay is contrasted and different techniques. It is apparent from Table 11.6 that the proposed technique has created a greater number of relics than different techniques, like N-AIM, N-COM, and WF strategy. At last, the future deep learning speech enhancement technique has accomplished low handling deferrals and ear shot hindered listeners unearthly goals than great as different techniques.

In the presence test, six topics available hooked on support for surveying the wide-ranging value. A calm climate is kept up with and a couple of earphones are utilized for setting them equally in both the ears and afterward the handled noises stayed played. Differing the rule from 1 to 5, every framework's topic is appraised; fresh discourse is paid attention to straightforwardly with the rule 5, and low-quality sign compared to the rule 1 is measured as unsatisfactory. All the more significantly, persistent uproarious ancient rarities are demonstrated utilizing grade 1, nonstop short level curios as grade 2, discontinuous short level recognizable antiques as evaluation 3, and awesome to phenomenal discourse quality is shown utilizing the grade [4–5]. To acquire a conventional correlation of "awesome" superiority, the prepared yield then the info could be exchanged by the topics. Preceding the beginning of the test, the subjects were set in the perfect sound climate. Throughout evaluation of the handled noises, fresh sound might be gotten to consistently during the topics.

11.9 Conclusion

SE issue in CI clients was evaluated utilizing DCNNs on tear shot hindered listeners thinking about the two significant components to be specific, discourse quality appraisals and SE. Utilizing tear shot hindered listeners calculation, edges or elements were shaped through breaking down the boisterous discourse signal. Consequently, a DCNN is taken care of with deteriorated boisterous discourse signal edges to deliver recurrence channel assessment. In any case, ear shot hindered listeners SNR data were contained in created recurrence channel assessment. Utilizing tear shot hindered listeners gauge, discourse overwhelmed CI networks stay to deliver electric incitement. The ear shot hindered listeners interaction was similar to that of the ordinary no FM circular implantation programming systems. To decide the discourse in-commotion execution of 12 circular implantation clients, the flicker and harmony noises applied are measured as foundation commotions. Execution of the proposed calculation is assessed through considering these foundation commotions. Small handling interruption with solid engineering are the top qualities of DLSE calculation; thus, tear shot hindered listeners can be reasonably provided to all uses of audible range gadgets. N-COM and N-AIM are the two semantic network organization constructed techniques utilized for examination. Test results show that the proposed DLSE strategy has beaten the current techniques as far as emotional listening tests also, subjective assessments. These discoveries have demonstrated that SE execution of CI clients can be essentially further developed utilizing tear shot hindered listeners promising DLSE technique.

References

1 Lai, Y.H. and Zheng, W.Z. (2019). Multi-objective learning based speech enhancement method to increase speech quality and intelligibility for hearing aid device users. *Biomedical Signal Processing and Control* 48: 35–45.

2 Nossier, S.A., Rizk, M.R.M., Moussa, N.D. et al. (2019). Enhanced smart hearing aid using deep neural networks. *Alexandria Engineering Journal* 58 (2): 539–550.

3 Martinez, A.M.C., Gerlach, L., Payá-Vayá, G. et al. (2019). DNN-based performance measures for predicting error rates in automatic speech recognition and optimizing hearing aid parameters. *Speech Communication* 106: 44–56.

4 Spille, C., Ewert, S.D., Kollmeier, B. et al. (2018). Predicting speech intelligibility with deep neural networks. *Computer Speech & Language* 48: 51–66.

5 Chiea, R.A., Costa, M.H., and Barrault, G. (2019). New insights on the optimality of parameterized Wiener filters for speech enhancement applications. *Speech Communication* 109: 46–54.

6 Kolbæk, M., Tan, Z.H., and Jensen, J. (2018). On the relationship between short-time objective intelligibility and short-time spectral-amplitude mean-square error for speech enhancement. *IEEE/ACM Transactions on Audio, Speech and Language Processing* 27 (2): 283–295.

7 Friesen, L.M., Shannon, R.V., Baskent, D. et al. (2001). Speech recognition in noise as a function of the number of spectral channels: comparison of acoustic hearing and cochlear implants. *The Journal of the Acoustical Society of America* 110 (2): 1150–1163.

8 Fu, Q.J., Shannon, R.V., and Wang, X. (1998). Effects of noise and spectral resolution on vowel and consonant recognition: acoustic and electric hearing. *The Journal of the Acoustical Society of America* 104 (6): 3586–3596.

9 Jin, S.H., Nie, Y., and Nelson, P. (2013). Masking release and modulation interference in cochlear implant and simulation listeners. *American Journal of Audiology* 22: 135–146.

10 Tsoukalas, D.E., Mourjopoulos, J.N., and Kokkinakis, G. (1997). Speech enhancement based on audible noise suppression. *IEEE/ACM Transactions on Audio, Speech and Language Processing* 5 (6): 497–514.

11 Xia, B. and Bao, C. (2013). Speech enhancement with weighted denoising auto-encoder. In: *Proceedings of the 14th Annual Conference of the International Speech Communication Association, INTERSPEECH*, 3444–3448. Lyon, (25–29 August).

12 Sang, J., Hu, H., Zheng, C. et al. (2015). Speech quality evaluation of a sparse coding shrinkage noise reduction algorithm with normal hearing- and hearing-impaired listeners. *Hearing Research* 327: 175–185.

13 Bentler, R., Wu, Y.H., Kettel, J. et al. (2008). Digital noise reduction: outcomes from laboratory and field studies. *International Journal of Audiology* 47 (8): 447–460.

14 Zakis, J.A., Hau, J., and Blamey, P.J. (2009). Environmental noise reduction configuration: effects on preferences, satisfaction, and speech understanding. *International Journal of Audiology* 48: 853–867.

15 Luts, H., Eneman, K., Wouters, J. et al. (2010). Multicenter evaluation of signal enhancement algorithms for hearing aids. *The Journal of the Acoustical Society of America* 127 (3): 1491–1505.

16 Loizou, P.C. and Kim, G. (2010). Reasons why current speech-enhancement algorithms do not improve speech intelligibility and suggested solutions. *IEEE/ACM Transactions on Audio, Speech and Language Processing* 19 (1): 47–56.

17 Khaleelur Rahiman, P.F., Jayanthi, V.S., and Jayanthi, A.N. (2019). Deep convolutional neural network-based speech enhancement to improve speech intelligibility and quality for hearing-impaired listeners. *Medical and Biological Engineering and Computing* 57: 757.

18 Goehring, T., Bolner, F., Monaghan, J.J.M. et al. (2017). Speech enhancement based on neural networks improves speech intelligibility in noise for cochlear implant users. *Hearing Research* 344: 183–194.

19 Healy, E.W., Yoho, S.E., Chen, J. et al. (2015). An algorithm to increase speech intelligibility for hearing-impaired listeners in novel segments of the same noise type. *The Journal of the Acoustical Society of America* 138 (3): 1660–1669.

20 Healy, E.W., Yoho, S.E., Wang, Y. et al. (2013). An algorithm to improve speech recognition in noise for hearing impaired listeners. *The Journal of the Acoustical Society of America* 134 (4): 3029–3038.

21 Bolner, F., Goehring, T., Monaghan, J. et al. (2016). Speech enhancement based on neural networks applied to cochlear implant coding strategies. In: *Proceedings of the IEEE International Conference on Acoustics, Speech and Signal Processing (ICASSP)* (20–25 March), 6520–6524. New York: IEEE.

22 Acharya, U.R., Oh, S.L., Hagiwara, Y. et al. (2017). A deep convolutional neural network model to classify heartbeats. *Computers in Biology and Medicine* 89: 389–396.

23 Gelderblom, F.B., Tronstad, T.V., and Viggen, E.M. (2018). Subjective evaluation of a noise-reduced training target for deep neural network-based speech enhancement. *IEEE/ACM Transactions on Audio, Speech and Language Processing* 27 (3): 583–594.

24 Kavalekalam, M.S., Nielsen, J.K., Boldt, J.B. et al. (2019). Model-based speech enhancement for intelligibility improvement in binaural hearing aids. *IEEE/ACM Transactions on Audio, Speech and Language Processing* 27 (1): 99–113.

25 Hu, Y. and Loizou, P.C. (2010). Environment-specific noise suppression for improved speech intelligibility by cochlear implant users. *The Journal of the Acoustical Society of America* 127 (6): 3689–3695.

26 Kim, G., Lu, Y., Hu, Y. et al. (2009). An algorithm that improves speech intelligibility in noise for normal-hearing listeners. *The Journal of the Acoustical Society of America* 126 (3): 1486–1494.

27 Dahl, G.E., Yu, D., Deng, L. et al. (2011). Context-dependent pre-trained deep neural networks for large-vocabulary speech recognition. *IEEE/ACM Transactions on Audio, Speech and Language Processing* 20 (1): 30–42.

28 Geoffrey, H., Li, D., Dong, Y. et al. (2012). Deep neural networks for acoustic modeling in speech recognition: the shared views of four research groups. *IEEE Signal Processing Magazine* 29 (6): 82–97.

29 Yang, D. and Mak, C.M. (2018). An investigation of speech intelligibility for second language students in classrooms. *Applied Acoustics* 134: 54–59.

30 Diliberto, G.M., Lalor, E.C., and Millman, R.E. (2018). Causal cortical dynamics of a predictive enhancement of speech intelligibility. *NeuroImage* 166: 247–258.

31 Rennies, J. and Kidd, G. Jr., (2018). Benefit of binaural listening as revealed by speech intelligibility and listening effort. *The Journal of the Acoustical Society of America* 144: 2147–2159.

32 Wang, Y. and Wang, D.L. (2013). Towards scaling up classification-based speech separation. *IEEE/ACM Transactions on Audio, Speech and Language Processing* 21 (7): 1381–1390.

33 Wang, Y., Narayanan, A., and Wang, D.L. (2014). On training targets for supervised speech separation. *IEEE/ACM Transactions on Audio, Speech and Language Processing* 22 (12): 1849–1858.

34 Huang, P.S., Kim, M., Hasegawa-Johnson, M. et al. (2015). Joint optimization of masks and deep recurrent neural networks for monaural source separation. *IEEE/ACM Transactions on Audio, Speech and Language Processing* 23 (12): 2136–2147.

35 Xu, Y., Du, J., Dai, L.R. et al. (2015). A regression approach to speech enhancement based on deep neural networks. *IEEE/ACM Transactions on Audio, Speech and Language Processing* 23 (1): 7–19.

36 Sundararaj, V. (2019). Optimized denoising scheme via opposition-based self-adaptive learning PSO algorithm for wavelet-based ECG signal noise reduction. *International Journal of Biomedical Engineering and Technology* 31 (4): 325–345.

37 Vinu, S. (2016). An efficient threshold prediction scheme for wavelet-based ECG signal noise reduction using variable step size firefly algorithm. *International Journal of Intelligent Systems* 9 (3): 117–126.

38 Xu, Y., Du, J., Dai, L.R. et al. (2014). Dynamic noise aware training for speech enhancement based on deep neural networks. *Proceedings of the 15th Annual Conference of the International Speech Communication Association*. https://pdfs.semanticscholar.org/62e7/ec2f188764b3be45f0a64ea55ab963d7185f.pdf.

39 Xu, Y., Du, J., Huang, Z., et al. (2017). Multi-objective learning and mask-based post-processing for deep neural network based speech enhancement. arXiv: 1703.07172.

40 Kim, M. and Smaragdis, P. (2015). Adaptive denoising autoencoders: a fine-tuning scheme to learn from test mixtures. In: *Proceedings of the International Conference on Latent Variable Analysis and Signal Separation*, Liberec (25–28 August), 100–107. Cham: Springer.

41 Gao, T., Du, J., Xu, Y. et al. (2015). Improving deep neural network-based speech enhancement in low SNR environments. In: *Proceedings of the International Conference on Latent Variable Analysis and Signal Separation*, Liberec (25–28 August), 75–82. Cham: Springer.

42 Weninger, F., Erdogan, H., Watanabe, S. et al. (2015). Speech enhancement with LSTM recurrent neural networks and its application to noise-robust ASR. In: *Proceedings of the International Conference on Latent Variable Analysis and Signal Separation*, Liberec (25–28 August), 91–99. Cham: Springer.

43 Fu, S.W., Tsao, Y., and Lu, X. (2016). SNR-aware convolutional neural network modeling for speech enhancement. In: *Proceedings of the Interspeech*, San Francisco, CA, 8–12 September, 3768–3772.

44 Tu, Y., Du, J., Xu, Y. et al. (2014). Speech separation based on improved deep neural networks with dual outputs of speech features for both target and interfering speakers. In: *Proceedings of the 9th International Symposium on Chinese Spoken Language Processing*, Singapore (12–14 September), 250–254. New York: IEEE.

45 Schmid, H.J. (2015). Deep learning in neural networks: an overview. *Neural Networks* 61: 85–117.

46 Zhao, B., Lu, H., Chen, S. et al. (2017). Convolutional neural networks for time series classification. *Journal of Systems Engineering and Electronics* 28 (1): 162–169.

47 LeCun, Y., Bengio, Y., and Hinton, G. (2015). Deep learning. *Nature* 521: 436–444.

48 Bouvrie, J. (2006). Notes on convolutional neural networks. http://cogprints.org/5869.

49 He, K., Zhang, X., Ren, S. et al. (2015). Delving deep into rectifiers: surpassing human-level performance on ImageNet classification. In: *Proceedings of the IEEE International Conference on Computer Vision*, Santiago (7–13 December), 1026–1034.

50 Healy, E.W., Yoho, S.E., Wang, Y. et al. (2014). Speech-cue transmission by an algorithm to increase consonant recognition in noise for hearing-impaired listeners. *The Journal of the Acoustical Society of America* 136 (6): 3325–3336.

51 Bleeck, S., Ives, T., and Patterson, R.D. (2004). Aim-mat: the auditory image model in MATLAB. *Acta Acustica united with Acustica* 90 (4): 781–787.

52 Jürgen, T. and Kollmeier, B. (2003). SNR estimation based on amplitude modulation analysis with applications to noise suppression. *IEEE/ACM Transactions on Audio, Speech and Language Processing* 11 (3): 184–192.

53 Rix, A.W., Beerends, J.G., Hollier, M.P. et al. (2001). Perceptual evaluation of speech quality (PESQ)-a new method for speech quality assessment of telephone networks and codecs. In: *Proceedings of the 2001 IEEE International Conference on Acoustics, Speech, and Signal Processing*, Salt Lake City, UT (7–11 May) , 749–752. New York: IEEE.2

12

Design and Development of Chatbot Based on Reinforcement Learning

Hemlata M. Jadhav[1], Altaf Mulani[2], and Makarand M. Jadhav[3]

[1]*Electronics and Telecommunication Department, Marathwada Mitra Mandal's College of Engineering, Pune, India*
[2]*Electronics and Telecommunication Department, SKNSCOE, Pandharpur, India*
[3]*Electronics and Telecommunication Department, NBN Sinhgad School of Engineering, Pune, India*

12.1 Introduction

Natural language processing (NLP) facilitates PCs to manage human languages smartly. It is a multidomain technology traversing soft computing as well as cognitive learning and linguistics to interact with human and machine languages. The following are examples of NLP applications:

(i) **Speech Recognition:** NLP has a sub-field called speech recognition. Furthermore, a lack of ability to learn from data and to overcome reasoning uncertainty continues to drive research in speech recognition, soft computing, and NLP.

(ii) **Dialog Systems:** Substantial advances in speech, as well as language areas in understanding to build human–machine dialog systems, are still demanding in the market. Dialog systems are also known as chatbots. It finds products varying from technical assistance and language learning tools to entertainment. In general, new success in deep-neural networks has provoked research in developing data-driven dialog models.

(iii) **Information Retrieval:** Structured information regarding the real world is opening more research in NLP as well as information retrieval. These are web search, question answering, and speech recognition. It has applications for academia and industry sectors. Presently, the natural language process and information-retrieval technique with machine learning (ML) is exploited in social computing. However, solving the challenges raised by social media remains a thrust area.

(iv) **Question Answering:** Image captioning is a sub-area in the intelligence field. Currently, visual question answering, story-telling, and grounded dialogs as well as image synthesis from text descriptions are becoming a new set of problems. Thus, multimodal intelligence progress that involves natural language is significant to construct artificial intelligence (AI) capabilities in the time to come. Furthermore, categorization refers to the contemporary deep-learning algorithms used for question answering and machine comprehension. It provides a foundation for ethical ways to answer the queries, text, and common-sense interpretation.

(v) **Machine Translation:** An artificial intelligence branch that translates any natural language to a different natural language. Nowadays, commercial machine-translation systems are often used in skilled interpretation departments. In general, fully automatic translators are not doable, whereas human post-editing gives acceptable results. All the applications have changed the way people identify, collect, and use data.

(vi) **Rationalism:** In the first phase, rationalistic approaches promoted the design of handmade rules to compile knowledge into NLP methods based on the idea that knowledge of the language is fixed in advance in the human mind through genetic inheritance.

Machine Learning Algorithms for Signal and Image Processing, First Edition.
Edited by Deepika Ghai, Suman Lata Tripathi, Sobhit Saxena, Manash Chanda, and Mamoun Alazab.
© 2023 The Institute of Electrical and Electronics Engineers, Inc. Published 2023 by John Wiley & Sons, Inc.

(vii) **Empiricism:** In the second phase, empirical approaches assume that rich sensory input and superficially observable language data are necessary and sufficient to allow the mind to learn the detailed structure of natural language. Consequently, probabilistic models were designed to find the laws of languages from large masses.

(viii) **Deep Learning:** In the third phase, deep learning uses hierarchical non-linear processing models, inspired by biological neural methods, to learn intrinsic representations of language data to simulate human beings' cognitive abilities. The intersection of deep learning and NLP has resulted in remarkable success in practical tasks. Speech recognition is the first industrial application of NLP that was strongly influenced by deep learning. With the availability of extensive training data, deep-neural networks achieved dramatically fewer detection errors than traditional empirical approaches. Another well-known and successful application of deep learning in NLP is machine translation. End-to-end neural machine translation, which models mapping between human languages using neural networks, is a significant improvement in translation quality. As a result, neural machine translation has quickly become the de facto new technology in large online business translation services offered by major tech companies – Google, Microsoft, Facebook, Baidu, and more. Many other areas of NLP, including language understanding and dialogue, lexical analysis and analysis, knowledge graph vii, information retrieval, answering questions from text, social computing, voice generation, and text sentiment analysis have also made great strides through deep learning and also riding the third wave of NLP. Today, deep learning is a dominant method that applies practically to all NLP tasks.

Chatbots are known as online human–computer dialogue systems that use natural language. The first conception of the chatbot is attributed to machines that can think. Nowadays, chatbot technology has advanced with NLP as well as ML techniques. Presently, chatbot adoption has increased with the launch of chatbot platforms, such as Facebook and Telegram [1]. A deep reinforcement chatbot system consists of a set of models of natural generation and recovery [2]. Traditional chatbots adopt the retrieved model. It requires a huge amount of data for classification. A method is presented to construct a chatbot with a sentence generation model. It generates sequence sentences using a generative adversarial network [3]. A chatbot trust hand-crafted rules. A goal-oriented approach is presented for end-to-end training of the chatbot [4]. Recent neural models used for dialogue generation systems are utterances one at a time. Whereas they ignore their influence on upcoming outcomes. To model, dialogue is a challenge to generating coherent as well as thought-provoking dialogs to use reinforcement learning (RL) [5]. Presently, designing and implementing chatbots with artificial intelligence (AI) is considered to process human language with NLP and neural-network techniques [6]. Reinforcement-learning techniques with social means allow for achieving human-level performance. Here, the social impact of agents that are interacting with a human is a crucial factor, whereas most of the available system has not been investigated properly. A learning mechanism considering a contention of social impact is presented [7]. Chatbots are considered a substantial technological increase in the field of colloquial services. It allows the device to communicate with a user after receiving user requests in natural language. It uses artificial intelligence and ML to respond to the user with automatic responses [8]. An intelligent chatbot system handles real-time chat with human life by applying natural language techniques. A variety of chatbots communicating in Indian languages is in demand applying deep-reinforcement learning methods [9]. A situation vagueness conscious chatbot along with a reinforcement- learning model is used to train the chatbot. The user simulator was developed to mimic the uncertainty of the statements of users based on real data [10]. The rapid growth to consider chatbots in fields such as the healthcare, educational, and entertainment sectors is increasing exponentially. In general, the educational chatbot involves a specific architecture design and model development to deal with communication and provide adequate answers to student questions.

The main goal of this chapter is to provide a comprehensive analysis of the recent advances in deep learning applied to NLP. It presents the latest techniques of deep learning associated with NLP. A chatbot is designed to access college information easier. It helps college administration as well as students to reduce website handling excessively.

12.2 Student Guide Using Chatbot

In the twenty-first century, colleges are adopting the digital route to explore education facilities at one glance society. Because of this, college query chatbots are making waves in the globe of higher education. It can transform the approach students interact with their colleges [11–15]. It works as a support system for students. It provides immediate responses for the campus, learning, and IT services as well as details of enrollment, and assistance regarding exams and scholarships. The chatbot designed centered on ML techniques can learn real-time queries from conversations with students. It constantly addresses the needs of students for 24 hours and seven days. Automatic assistance is used for campus facilities and locations. The chatbots analyze users' queries. It also understands the user's message. The response rule matches a user's input judgment [16–24]. In the end, it is observed that users can ask a query about college-concerned events without physical presence. A graphical user interface is provided so that the user can log into the system to obtain credentials.

The application is aimed at college students and parents to save their valuable time. It provides an easy way to communicate. It is revealed that chatbot communication is much easier than talking to humans regarding significant issues. The reason beyond is that chatbots provide automatic responses. This chapter aims to develop college inquiry chatbots designed using NLP and RL algorithms to analyze and solve users' queries as well as understand users' messages. The primary intention of this work is to reduce the time it takes to resolve college-related queries and also gives a response to the user-centered on college-related queries to simplify communication between user and machine. The system provides an accessible path for students to connect with college administration using a chatbot. Here, students' participation is online to avail feedback, share ideas, as well as receive information using minimum dialogs.

There are two existing systems available for discussion with their application. The first is known as the Eliza chatbot. The other is an artificial linguistic internet computer entity (ALICE) chatbot. Mr. Joseph Weizenbaum, a German scientist in computers, developed the Eliza program in 1966. It is the first chatbot in the history of computing. It act as a therapist by rephrasing users' statements. It is then posed back as queries. He simulated conversation using a substitution methodology. It gave users an inkling of understanding **of** the program. On the other hand, there is no integrated framework to contextualize events. The designed scripts dictate rules as well as directions to process user inputs. ALICE is an open-source chatbot with artificial intelligence in natural language. This chatbot is run by Dr. Wallace created. It is based on natural language and pattern matching. Generate responses to user queries by applying identical rules to patterns. However, it cannot pass the turing test, as even the casual user often reveals its weaknesses in short conversations.

Advantages

- Provides accurate results on less data volume.
- Trial and error model that learns to do well by taking environment feedback.

Disadvantages

- The Q-learning approach engenders random results.
- The trial and error model generates few abusive replies.
- Production of long as well as complex sentences.

12.3 Implementation of Chatbot System

The chatbot system architecture for academics is shown in Figure 12.1. Here, students send out queries to avail institute information and details of faculty as well as administration, placements, and hostel. The query input is

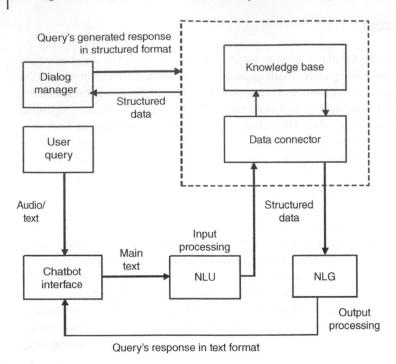

Figure 12.1 Chatbot system architecture for academics.

either text or speech form. First, the user enters the query in the type of plain text. The input processing is done with stop word and string similarity NLU algorithms. It converts plain text into a structured format. NLG fetches correlated answers to the query from the database management system. Q-learning algorithm gives reward points for each answer. The best answer is presented to the user in text form.

12.3.1 Data-Flow Diagram

It is a graphic illustration of the data flow through an information system for modeling process aspects. It helps to create a system outline to elaborate on later without going into details. It is used to visualize data processing such as the type of information to input to avail system output, data progress, and storage in the scheme. The data-flow diagram for handling the queries in academics is shown in Figure 12.2.

12.3.2 Use-Case Diagram

The use-case diagram represents users' interaction with the system. It shows various types of system users and their interactions. It is also used to gather the design requirements of a system. Following are the use-cases description at the admin end as well as at the user end.

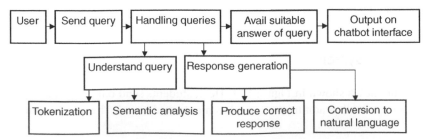

Figure 12.2 Data-flow diagram for handling the queries in academics.

12.3.2.1 At the Admin End

- **Add Circulars:** Allows adding the circulars for the course semester-wise in the form of .jpg or .pdf.
- **Add Admission Detail:** Add the admission details for the semester-wise course in the form of .csv file format.
- **Add Placement Detail:** Add the placement details in the form of .csv file format.
- **Add Hostel Detail:** Add the hostel details year-wise in the form of a .csv file.
- **Add Fee Structure Detail:** Add the fee structure details yearly in the form of .csv file format.

12.3.2.2 At User End for Student/Parent

- **Login:** The user has to log in to access the functionality of the chatbot.
- **Query Type:** The user is permitted to type a query.
- **View Circulars:** The user is permitted to view the circulars for the course semester-wise.
- **View Admission Detail:** The user is permitted to view the admission details for the course semester-wise.
- **View Placement Detail:** The user is permitted to view the placement details month-wise.
- **View Hostel Detail:** The user is permitted to view the hostel details year-wise.
- **View Fee Structure Detail:** The user is permitted to view the fee structure details year-wise.

12.3.3 Class Diagram

It is a diagram representing the static view of the application. This demonstrates the attributes and operations of a class. It also illustrates the limitations that are placed on the system. It constructs the executable code for the forward process and the reverse process. The class- diagram design and development of chatbot based on reinforcement learning is shown in Figure 12.3.

The class diagram with eight classes is constructed for the chatbot system such as user, admin, chatbot, database, and admission as well as placement details to get general information of the department as shown in Figure 12.3. It includes various attributes and operations. Multiplicity and aggregation relations are also included. A multiplicity

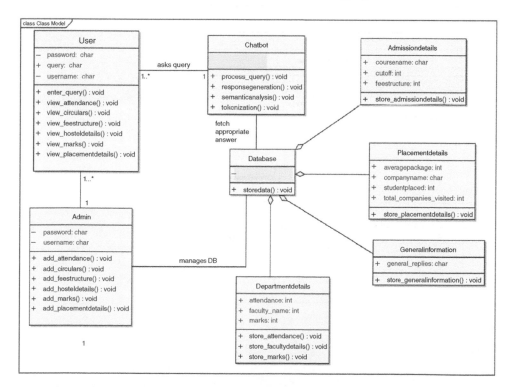

Figure 12.3 Class-diagram design and development of chatbot based on reinforcement learning.

relationship between user and chatbot is many to one. Further, aggregation presents its relationship. In this system, the database class owns classes such as admission, department, as well as placement details.

12.3.4 Sequence Diagram

It is a user-interface diagram. It articulates interaction among objects and navigation through user interfaces. It represents the objects and classes engaged on the stage. It also describes the sequence of messages that are swapped between the objects necessary to perform the functions of the scenario. In this scheme, the user has to log in. Thereafter, login credentials are authenticated. If validation is successful, then the user can ask the queries. Further, an appropriate answer is fetched from the database. It is displayed on the chatbot interface to the user. On the other hand, unsuccessful login demands correct login credentials as shown in Figure 12.4.

12.3.5 Activity Diagram

An important diagram in unified modeling language (UML) explains the dynamic aspects of the chatbot system. It is represented with a flowchart to show the control flow from one to another activity control. The activity control flow of the chatbot system is shown in Figure 12.5.

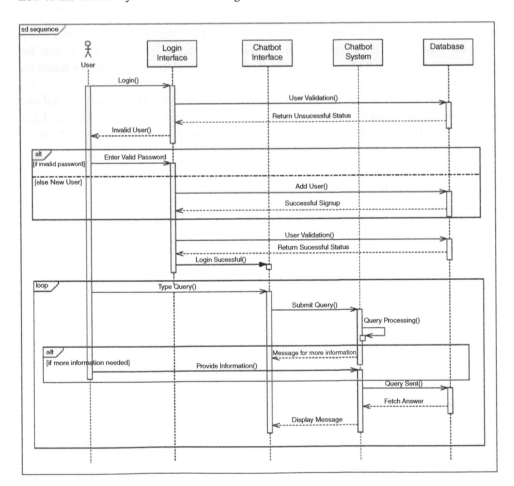

Figure 12.4 Sequence diagram.

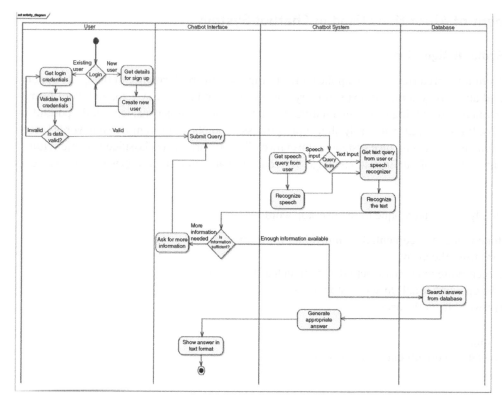

Figure 12.5 Activity diagram of chatbot system.

The user first logins to the system with available credentials. Further, login credentials are authenticated to ask the queries in text or speech form. The suitable answer is fetched from the database and displayed on the chatbot interface. On the other hand, unsuccessful validation requests for sign-up with accurate login credentials.

12.3.6 State Diagram

It is a transition diagram that defines the states of an object. Events in which an object changes state are called transitions. Also, the conditions that must be met before the transition takes place are called watchdogs. In the end, the activities that take place throughout the life of an object are called actions. In this work, four states are named as idle, query processing, display message, and search database. The transition from idle to query processing state occurs when the user query reaches the system. Further, transition to the database occurs after fetching suitable from the database. Thereafter, displaying the correct answer to the user on the chatbot interface will have a transition to display the message state as shown in Figure 12.6.

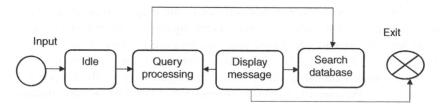

Figure 12.6 State transition diagram of chat.

12.4 Development of Algorithms Used in Chatbot System

12.4.1 Stop Word Removal Algorithm

It is a pre-processing technique preferred in NLP applications. It improves the performance of the query system, the word processing and analysis system, the text summary, the question and answer system, and word formation. Stop words are frequently occurring words in a natural language, whereas it is considered unimportant in certain NLP applications. It is observed that nearly all text pre-processing applications remove stop words before processing documents and queries to enhance system performance. They are primarily classified in conjunctions, prepositions, adverbs, and articles of the English language. To reduce impacts on results, stop words are excluded from the original text.

Algorithm 12.1 To apply search technique to remove stop word

Step 1: Generate the token for the document. Save individual words in an array.
Step 2: Read a stop word from the array.
Step 3: Use the search technique to compare actual and predicted stop words.
Step 4: If the predicted stop word equals to actual stop word
    then remove the word from the array
    else
    continue searching
Step 5: Statistics of the scanned document are displayed
Step 6: Stop

12.4.2 String Similarity Algorithm

In text similarity approach, words are similar lexically. It is also found to be semantic. Words are said lexically similarly only if they have a similar string of characters. While words are only said to be semantically similar if they have the same in front of each other and are used in the same way and context. Lexical similarity is presented through string-centered algorithms. On the other hand, semantic similarity through with corpus as well as knowledge-founded algorithms. The string-centered algorithm runs on string sequences as well as character composition. Further, text similarity is segregated into three methods such as string similarity measures, corpus similarity measures, and knowledge similarity measures. In this context, the knowledge-centered similarity is also preferred for the semantic similarity. It identifies the similarity degree among the words in the document. It works on hypotheses obtained from the semantic networks. WordNet is the most preferred semantic network for measuring knowledge-centric word similarity. It is an extensive lexical database. This database is used for the English language. In addition, nouns, verbs, adjectives, and adverbs are sets of cognitive synonyms. It makes the idea distinctive.

Knowledge-based similarity measures techniques are divided into two groups. One group measures semantic similarity and relatedness. Semantically similar models are related to their likeness. Semantic relatedness, on the other hand, is a general impression of relatedness. It is not explicitly tied to the shape or concept. In other words, semantic similarity is a sort of relatedness between two words. It encompasses a clearer range of relationships between concepts that includes extra similarity relations. Few examples are is-a-kind-of, is-a-specific example-of, is-a-part-of, and is-the-opposite-of. Knowledge-based techniques for measuring similarity fall into almost two groups. One group measures semantic similarity and relationship. Semantically similar models are related to their similarity. Semantic kinship, on the contrary, is a general impression of kinship. It is not explicitly tied to structure or concept. Particularly, semantic similarity is a kind of relationship among words. It covers a clearer range of relations among concepts, including additional similarity relationships. One of the examples can be is-a-art-of.

12.4.3 Q-Learning Algorithm

The dialogue manager takes the attention of the dialogue. Depending on the framework, prior user, and system activities, the manager produces later system activity. In general, the dialogue agent helps to adopt an optimum policy to develop a useful and lucrative dialogue with the user. It is employed by utilizing a Q-learning reward-penalty strategy method. Here, a reward is given for any action occurring from one state to reach another state. The expected reward or penalty is generated based on the input, state as well as agent's action. This sequence of actions eventually generates a total reward known as the Q-value given by the equation (Eq. (12.1)).

$$Q(B, A) = r(B, A) + G_{max} [Q(B', A)] \tag{12.1}$$

The equation (Eq. (12.1)) affirms that returned Q-value being at B state and performs A action. It is the summation of the direct reward $r(B, A)$ added to the maximum Q-value likely from the subsequent state B'. G_{max} is the discount factor. It decides the assignment of future rewards. In general, it can be stated that $Q(B', A)$ depends on $Q(B'', A)$ that has a coefficient of G_{max} squared. Thus, Q-value alters on future states Q-values given by the equation (Eq. (12.2)).

$$Q(B, A) \rightarrow G_{max} [Q(B', A) + G_{max}^2 [Q(B'', A) + \cdots + G_{max}^m [Q(B''^{\cdots n}, A)]] \tag{12.2}$$

It can be observed from equation (Eq. (12.2)) that adjusting the G_{max} value increases the payment of imminent rewards. Such a recursive approach forces to make arbitrary rules for each Q-value known as covers optimal policy given by the equation (Eq. (12.3)).

$$Q^{new}(B_t, A_t) \leftarrow Q(B_t, A_t) + G[R_{t+1} + G_{max} [Q(B_{t+1}, A_t) - Q(B_t, A_t)] \tag{12.3}$$

Here, $Q^{new}(B_t, A_t)$ represents a new Q-value for that state as well as action, whereas, $Q(B_t, A_t)$ engenders a recent Q-value and G represents the rate of learning. It helps to determine to what level newly learned knowledge supersedes old evidence. Further, $G_{max} Q(B_{t+1}, A_t)$ represents the highest probable reward assigned to the new state and all actions at that new state. In the end, R_{t+1} is a reward for getting action at state G referred to as discounting factor.

12.5 Conclusion

Design and development of the proposed chatbot system in this chapter help students as well as parents to access college information efficiently. It is found to be beneficial to students and the administration section of the institute. It reduces website handling in a superfluous manner. It also reduces the time gap between students and administration by adopting a chatbot for the college administration and marketing section. Thus, it can be concluded that the proposed system makes information access easier for students. The input processing brought out within the system competently processes queries related to avail information in the education sector. Further, a system consigned with ML techniques can train the chatbot to think, learn, and remember in providing a suitable answer to a query raised by the user in real-time. Thus, the development of a chatbot system with a Q-learning algorithm can access and manage information sharing competently between the students/parents and academic institutes.

References

1 Cahn, J. (2017). CHATBOT: architecture, design, & development. Computer Science thesis. University of Pennsylvania School of Engineering and Applied Science Department of Computer and Information Science.

2 Serban, I.V., Sankar, C., Germain, M. et al. (2018). A deep reinforcement learning chatbot (short version). arXiv preprint arXiv:1801.06700.

3 Chou, T.L. and Hsueh, Y.L. (2019). A task-oriented chatbot based on LSTM and reinforcement learning. In: *Proceedings of the 2019 3rd International Conference on Natural Language Processing and Information Retrieval*, 87–91. https://doi.org/10.1145/3342827.3342844. ACM.

4 Liu, J., Pan, F., and Luo, L. (2020). Gochat: goal-oriented chatbots with hierarchical reinforcement learning. In: *Proceedings of the 43rd International ACM SIGIR Conference on Research and Development in Information Retrieval*, 1793–1796.

5 Li, J., Monroe, W., Ritter, A. et al. (2016). Deep reinforcement learning for dialogue generation. arXiv preprint arXiv:1606.01541.

6 Kumar, R. and Ali, M.M. (2020). A review on chatbot design and implementation techniques. *International Research Journal of Engineering and Technology* 7: 2791–2800.

7 Barros, P., Tanevska, A., Yalcin, O. et al. (2020). Incorporating rivalry in reinforcement learning for a competitive game. arXiv preprint arXiv:2011.01337.

8 Suhaili, S.M., Salim, N., and Jambli, M.N. (2021). Service chatbots: a systematic review. *Expert Systems with Applications* 184: 115461.

9 George, A.S., Muralikrishnan, G., Ninan, L.R. et al. (2021). Survey on the design and development of Indian language chatbots. In: *IEEE International Conference on Communication, Control and Information Sciences*, vol. 1, 1–6.

10 Yin, C., Zhang, R., Qi, J. et al. (2018). Context-uncertainty-aware chatbot action selection via parameterized auxiliary reinforcement learning. In: *Pacific-Asia Conference on Knowledge Discovery and Data Mining*, 500–512. Cham: Springer.

11 Clarizia, F., Colace, F., Lombardi, M. et al. (2018). Chatbot: An education support system for student. In: *International Symposium on Cyberspace Safety and Security*, 291–302. Cham: Springer.

12 Jadhav, M.M., Durgude, Y., and Umaje, V.N. (2019). Design and development for generation of real object virtual 3D model using laser scanning technology. *International Journal of Intelligent Machines and Robotics* 1: 273–291.

13 Jadhav, M.M. and Markande, S.D. (2020). Performance optimization of polar code based OFDM volte system using Taguchi method. *International Journal of Advanced Science and Technology* 29: 792–803.

14 Jadhav, M.M., Dongre, G.G., and Sapkal, A.M. (2019). Seamless optimized LTE based mobile polar decoder configuration for efficient system integration, higher capacity, and extended signal coverage. *International Journal of Applied Metaheuristic Computing* 10: 68–90.

15 Mulani, A.O. and Mane, P.B. (2017). Watermarking and cryptography based image authentication on reconfigurable platform. *Bulletin of Electrical Engineering and Informatics* 6: 181–187. https://doi.org/10.11591/eei.v6i2.651.

16 Kulkarni, P.R., Mulani, A.O., and Mane, P.B. (2017, 2017). Robust invisible watermarking for image authentication. In: *Emerging Trends in Electrical, Communications and Information Technologies*, Lecture Notes in Electrical Engineering, vol. 394, 193–200. Singapore: Springer https://doi.org/10.1007/978-981-10-1540-3_20.

17 Ghai, D., Gianey, H.K., Jain, A., and Uppal, R.S. (2020). Quantum and dual-tree complex wavelet transform-based image watermarking. *International Journal of Modern Physics B* 33: 1–10.

18 Swami, S.S. and Mulani, A.O. (2017). An efficient FPGA implementation of discrete wavelet transform for image compression. In: *2017 International Conference on Energy, Communication, Data Analytics and Soft Computing*, 3385–3389.

19 Mulani, A.O. and Mane, P.B. (2016). Area efficient high speed FPGA based invisible watermarking for image authentication, Indian. *Journal of Science and Technology* 9: 1–6. https://doi.org/10.17485/ijst/2016/v9i39/101888.

20 Mulani, A.O. and Mane, P.B. (2017). An efficient implementation of DWT for image compression on reconfigurable platform. International. *Journal of Control Theory and Applications* 10: 1006–1011.

21 Marne, H., Mukherji, P., Jadhav, M., and Paranjape, S. (2021). Bio-inspired hybrid algorithm to optimize pilot tone positions in polar-code-based orthogonal frequency-division multiplexing–interleave division multiple access system. *International Journal of Communication Systems* 34 (3): e4676.

22 Alazab, M. and Tang, M. (ed.) (2019). *Deep learning applications for cyber security*. Springer.

23 Asghar, M.Z., Subhan, F., Ahmad, H. et al. (2021). Senti-eSystem: a sentiment-based eSystem-using hybridized fuzzy and deep neural network for measuring customer satisfaction. *Software: Practice and Experience* 51 (3): 571–594.

24 Bhattacharya, S., Somayaji, S.R.K., Gadekallu, T.R. et al. (2022). A review on deep learning for future smart cities. *Internet Technology Letters* 5 (1): e187.

13

DNN Based Speech Quality Enhancement and Multi-speaker Separation for Automatic Speech Recognition System

Ramya[1] and Siva Sakthi[2]

[1]*Department of Electronics and Communication Engineering, Sri Ramakrishna Engineering College, Anna University (Autonomous), Coimbatore, India*
[2]*Department of Biomedical Engineering, Sri Ramakrishna Engineering College, Anna University (Autonomous), Coimbatore, India*

13.1 Introduction

Communication is the one that makes all living beings alive in terms of physical and social aspects of life. It improves the good relationship between different persons. In humans' communication is mainly established along with speech, therefore it is important to converse with the people through clear and efficient speech. Speech is widely mixed with many kinds of other noises and makes the receiver in dilemma about the conversation. One such example is a party cocktail problem [1]. In a complex acoustic environment, listening and understanding of speech signal is challenging since multiple speakers and the surrounding environment noise overlap with one another and thus degrading the signal of interest. To overcome those problems, speech-processing techniques like speech enhancement and separation are important. It is necessary to increase the quality of speech in such a complex environment by using some signal-processing algorithms. Increasing speech intelligibility by decreasing the noisy signal is a goal for almost all researchers. Their idea comes true with the application of artificial intelligence, machine learning, and deep learning in various speech-processing algorithms. The aim of this chapter is to apply deep learning concepts in speech enhancement and separation techniques to enhance the speech quality even in complex environments [2].

This chapter is organized as follows: Section 13.2 deals with overview of deep-learning algorithm in speech recognition system. Section 13.3 discusses various speech enhancement and separation problems. Section 13.4 deals with the traditional speech enhancement algorithms, such as spectral subtraction and statistical methods, and their drawbacks in effectively enhancing the speech signals. Section 13.5 deals with various traditional speech separation algorithms like harmonic models, auditory scene analysis (ASA), and so on and their drawbacks. Section 13.6 deals with deep learning-based speech enhancement. Section 13.7 deals with application of deep learning in speech separation, and various algorithms used to effectively separate speech signals from multi-speaker speech separation problems. Section 13.8 deals with results and discussion of speech enhancement and separation techniques with the quality parameters. Section 13.9 concludes the book chapter by evaluating the speech signal using perceptual evaluation of speech quality and short-time objective intelligibility.

13.2 Deep Learning

The subset of machine learning is deep learning, which has three to more layers of neural network. It is developed by inspiring the activity of human brain. It has been dated back more than half a century but its application

Machine Learning Algorithms for Signal and Image Processing, First Edition.
Edited by Deepika Ghai, Suman Lata Tripathi, Sobhit Saxena, Manash Chanda, and Mamoun Alazab.
© 2023 The Institute of Electrical and Electronics Engineers, Inc. Published 2023 by John Wiley & Sons, Inc.

is more prone in recent days. Artificial neural network (ANN) is the first prototypical learning model of deep learning; this is developed by inspiring the neuroscience and the neurons connectivity inside the brain with perceptron-learning model. This perceptron model is limited only to simple linear input–output mapping that is it could solve only a simple task. To apply the same concept for solving a complex non-linear mapping, the layers of perceptron are stacked to develop a multi-layer perceptron (MLP). This deeper and advanced model of ANN is capable of solving complex problems in a matter of time by applying a suitable computational process. Back-propagation algorithm is a more efficient algorithm for training MLP. The theoretical result obtained from deep learning using back-propagation algorithm is called as universal approximation theorem [3]. It uses any one of the following learning algorithms: (i) supervised learning algorithm, (ii) unsupervised learning algorithm, and (iii) semi-supervised learning algorithm. In Sections 13.2.1 and 13.2.2 recurrent neural network (RNN) and long short-term memory (LSTM) were discussed respectively.

13.2.1 Recurrent Neural Network

RNN is the most powerful neural network, as it processes sequential data, and it is the only neural network that has memory in it. This makes it unique in processing time-series data such as speech, audio, and video more efficiently. It can able to predict the results since it has knowledge about current and past inputs. The information in RNN cycles through a loop. It has one or more hidden layer. Every layer in the network receives weight from the previous network layer. It uses sigmoid bipolar as an activation function for hidden layer. The delay between input layer and the first hidden layer is represented by $t-1$ and the current time is represented by t [4]. Since it processes information through loops, the noise in the previous input is accommodated into the next input by adjusting the weighting function. The architecture of RNN is shown in Figure 13.1.

Let $x(t)$ be the input and $y(t)$ be the output of the time-series data. The connection matrices of input, hidden, and output layer can be W_{IH}, W_{HH}, and W_{OH}, respectively. The f_{H} and f_{O} are the hidden and output activation functions.

Where $h(t)$ is the state of dynamical system. It provides information about the previous behavior of the network. The behavior of the RNN is described by non-linear matrix equations (Eqs. (13.1) and (13.2)).

$$h(t+1) = f_{\text{H}}(W_{\text{IH}}x(t) + W_{\text{HH}}h(t)) \tag{13.1}$$

$$y(t) = f_{\text{O}}(W_{\text{OH}}h(t+1)) \tag{13.2}$$

The time index and weight sharing property of RNN makes it unique in processing subsequent data with strong temporal structure such as speech. RNN capacity can be increased by stacking RNN into deeper RNNs. Any dynamical system can be modeled using RNN with expected accuracy, this is known as the universal approximation theorem for RNNs. This is trained using back-propagation algorithm.

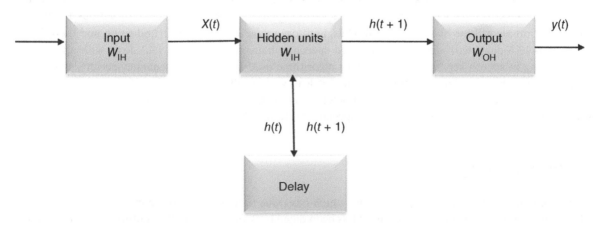

Figure 13.1 RNN architecture.

13.2.2 Long Short-Term Memory (LSTM) Networks

The extension of RNN network in terms of its memory is LSTM network. The LSTM network is framed by adding the units of LSTM to the layer of RNN network. It has a special function called memory blocks, and it enables to remember the input for a longer period of time. The temporal state of the network is stored using a memory block that contain a memory cell with self-connections. The information flow is controlled using a special multiplicative cell called gates. Using gated cell, LSTM can be able to read, write, and delete information from its memory. Input, forget, and output gate are the three gates of LSTM. The input gate let the new input in, whereas the forget gate deletes the information based on its importance and output gate lets the output from the network. To learn the precise timing of an output, the modern LSTM architecture has a peephole connection from its internal cells to the gate [5].

Let x be the input signal, h is recurrent signal, i is input gate signal, f is forget gate signal, and o is output gate signal.

An LSTM network computes a mapping from input sequence to output sequence by using the series of equations (Eqs. (13.3)–(13.8)).

$$i_t = \sigma(W_i x(t) + U_i h_{t-1} + b_i) \tag{13.3}$$

$$z_t = \tanh(W_z x_t + U_z h_{t-1} + b_z) \tag{13.4}$$

$$f_t = \sigma(W_f x_t + U_f h_{t-1} + b_f) \tag{13.5}$$

$$C_t = i_t \times z_t + f_t \times C_{t-1} \tag{13.6}$$

$$O_t = \sigma(W_o x_t + U_o h_{t-1} + V_o C_t + b_o) \tag{13.7}$$

$$h_t = O_t \times \tanh(C_t) \tag{13.8}$$

where W_i, W_z, W_f, W_o, U_i, U_z, U_f, and U_o are model parameters estimated during model training σ and tanh are activation functions and b are a bias.

13.2.3 Convolutional Neural Network

Convolutional neural network (CNN) is a special type of neural network which works well on images, and it also has applications in speech recognition, image segmentation, and text processing. It has much advancement, i.e. weight sharing, convolutional filters, and pooling. CNN are widely used in automatic speech recognition (ASR) system. It has three main layers: convolutional layer, pooling layer, and fully-connected layer [6].

The building block of CNN is convolution layer, where majority of computation occurs. It needs a few blocks such as input data, filter, and feature map. The pooling layer performs down sampling, that is, it reduces the number of parameters in the input and reducing the dimensionality. It is similar to convolutional layer in terms of filtering, but it doesn't have any weights. Instead, the output array is populated using the kernel aggregation within the receptive field. Most of the information is reduced in polling layer, but it reduces the complexity, improves efficiency, and limits the risk of overfitting [7]. The classification in fully connected layer is based on the features extracted from the previous layers and different filters. This layer uses softmax as an activation function, whereas the convolutional layer and pooling layer uses ReLu as an activation function. The architecture of CNN is shown in Figure 13.2.

Figure 13.2 CNN architecture.

13.3 Speech Enhancement and Separation

Speech enhancement is a pre-processing technique in many applications especially in speech recognition systems since it reduces the noise in the raw speech signal, thus increasing the intelligibility and quality of the signal. The speech signal can be recorded using single-microphone or multi-microphones. The single-microphone algorithms process the signal acquired using single microphone. The applications of single microphone algorithms are used in hearing aids, where arrays of microphone cannot be used. It does not depend on spatial location of the signal of interest and interference signals. Multi-microphones use microphone arrays to record the speech signal, and it uses multi-microphone algorithms. These microphone arrays are used when the speech signal has to cover large surface areas and doesn't have any hardware restrictions. It is used as a post-processing step for beamforming, as this technique is effective when target and interference signals are separated spatially. Hence algorithms capable of enhancing and separating noise from single-channel speech enhancement are highly desirable, and it is also a highly challenging task [8].

Depends on the number of target signals, the speech processing is separated as simply speech enhancement and multi-speaker speech separation algorithms. In speech enhancement task, the recording may contain many signals but the target signal or signal of interest is a single speech signal and all other signals including other speech signals were considered as noise. Multi-speaker speech separation is a special case of speech enhancement, in which the recording has both speech and non-speech signal, where the signal of interest may be two or more speech signal.

The research on speech signal enhancement and separation is increasing since it finds application in mobile communication systems, hearing aids, speech recognition system, and so on. Wearing mask in this pandemic situation also degraded the quality and intelligibility of speech signals widely. This has also increased the research on speech signal enhancement and separation to the greater levels.

13.4 Speech Enhancement Algorithms

Speech enhancement is a technique that deals noisy speech input signal and producing an enhanced output signal which is devoid of noise. The main objective of speech enhancement is to improve the intelligibility of sound signal from both single microphone and multi-microphone systems. Some of the algorithms used for speech enhancement technique are shown in Figure 13.3.

13.4.1 Basic Principles of Spectral Subtraction

In spectral subtraction algorithm, additive noise is assumed to be mixed with clean speech signal. Here the clean signal spectrum estimate is obtained from subtracting the noise spectrum which is obtained during the non-speech intervals. The clean spectrum is subjected to inverse Fourier transform.

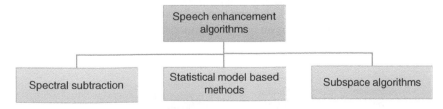

Figure 13.3 Speech enhancement algorithms.

Let the noisy speech signal be $y(n)$ which contains the clean speech signal $x(n)$ and the additive noise $d(n)$. It is given as (Eq. (13.9)).

$$y(n) = x(n) + d(n) \tag{13.9}$$

Taking discrete Fourier transform for Eq. (13.9), gives Eq. (13.10)

$$Y(\omega) = X(\omega) + D(\omega) \tag{13.10}$$

$Y(\omega)$ can be represented in polar form as per Eq. (13.11),

$$Y(\omega) = |Y(\omega)|e^{j\varphi_y(\omega)} \tag{13.11}$$

Here $|Y(\omega)|$ gives the magnitude spectrum and $\varphi_y(\omega)$ gives the phase spectrum of the noisy signal. The estimated clean spectrum after spectral subtraction is given by Eq. (13.12) as,

$$\widehat{X}(\omega) = [|Y(\omega)| - |\widehat{D}(\omega)|]e^{j\varphi_y(\omega)} \tag{13.12}$$

where $|\widehat{D}(\omega)|$ is the magnitude of noise spectrum obtained during non-speech activity. Now inverse Fourier transform of $\widehat{X}(\omega)$ is taken to get the enhanced speech signal.

13.4.1.1 Spectral Subtraction Using Over-Subtraction

During spectral subtraction musical noise is created due to non-linear processing of negative values which arise as isolated peaks in each frame at random frequencies. Over-subtraction given in Eq. (13.13) deals with setting the negative spectral components to zero.

$$|\widehat{X}_i(\omega)| = \begin{cases} |Y_i(\omega)| - |\widehat{D}(\omega)| & \text{if } |Y_i(\omega)| - |\widehat{D}(\omega)| > \max |\widehat{D}(w)| \\ \min\limits_{j=i-1,i,i+1} |\widehat{X}_j(w)| & \text{else} \end{cases} \tag{13.13}$$

Another method of spectral subtraction [9] which does not requires information of ω is given by Eq. (13.14).

$$|\widehat{X}_i(\omega)| = \begin{cases} |Y(\omega)|^2 - \alpha|\widehat{D}(\omega)|^2 & \text{if } |Y_i(\omega)|^2 > (\alpha + \beta)|\widehat{D}(\omega)|^2 \\ \beta|\widehat{D}(\omega)|^2 & \text{else} \end{cases} \tag{13.14}$$

where α is the over-subtraction factor and β is the spectral floor parameter which has the range of $0 < \beta \ll 1$. In case of broadband peaks available in estimated spectrum, we can reduce its amplitude using the over-subtraction factor. Still the peaks are surrounded by deep valleys and this can be removed by the spectral floor parameter.

13.4.1.2 Nonlinear Spectral Subtraction

Spectral subtraction assumes that the additive noise affects the spectral components equally. Also, the over-subtraction factor α equally subtracts the noise estimation with the whole spectrum. The real-world noise like train noise, restaurant noise, etc. affects the low-frequency region more than the high-frequency region. Therefore, a non-linear subtraction is necessary for higher values at low SNR values frequency and vice versa. The non-linear subtraction is given as per Eq. (13.15),

$$|\widehat{X}(\omega)| = \begin{cases} |\overline{Y}(\omega)| - a(\omega)N(\omega) & \text{if } |\overline{Y}(\omega)| > a(\omega)N(\omega) + \beta \cdot |\overline{D}(\omega)| \\ \beta|\overline{Y}(\omega)| & \text{else} \end{cases} \tag{13.15}$$

Here, $a(\omega)$ is a frequency-dependent subtraction factor, and $N(\omega)$ is a nonlinear function of the noise spectrum. $a(\omega)$ in Eq. (13.16) remains same for all frequencies within the frame but depending on the *posteriori* SNR it varies from frame to frame.

$$\alpha(\omega) = \frac{1}{1 + \gamma\rho(\omega)} \tag{13.16}$$

$$\rho(\omega) = \frac{\overline{Y}(\omega)}{\overline{D}(\omega)} \tag{13.17}$$

Here γ is the scaling factor and $\rho(\omega)$ in Eq. (13.17) is the square root of the a posteriori SNR estimate.

13.4.2 Statistical Model Based Methods

13.4.2.1 Maximum-Likelihood Estimators

Let us consider the noisy speech magnitude spectrum y with N-data points $y = \{y(0), y(1), \ldots, y(N-1)\}$ and θ be clean speech magnitude spectrum which is the parameter of interest. The probability density function of y needs to be maximized with respect to θ to give the maximum-likelihood estimate of θ given as $\hat{\theta}_{ML}$ in Eq. (13.18). Here θ is assumed to be deterministic.

$$\hat{\theta}_{ML} = \arg\max_{\theta} p(y; \theta) \tag{13.18}$$

Polar representation of noisy speech signal, $y(n) = x(n) + d(n)$ is given by Eq. (13.19)

$$Y_k e^{j\theta_y(k)} = X_k e^{j\theta_x(k)} + D_k e^{j\theta_d(k)} \tag{13.19}$$

where Y_k, X_k, and D_k denote the magnitude of noisy speech, clean speech, and noise, respectively. $\theta_y(k)$, $\theta_x(k)$, and $\theta_d(k)$ denote phases of noisy speech, clean speech, and noise, respectively. The pdf of $Y(\omega_k)$ which is DFT of $y(n)$ is Gaussian with variance $\lambda_D(k)$ and mean $X_k e^{j\theta_x(k)}$.

$$p(Y(\omega_k); X_k; \theta_k(k)) = \frac{1}{\pi \lambda_d(k)} \exp\left[-\frac{\left|Y(\omega_k) - X_k e^{j\theta_x(k)}\right|^2}{\lambda_d(k)}\right] \tag{13.20}$$

The maximum of pdf of $Y(\omega_k)$ in Eq. (13.20) needs to be computed with respect to X_k, to obtain the maximum-likelihood estimate of X_k. The maximum-likelihood estimate of the magnitude spectrum [10] is given in Eq. (13.21).

$$\hat{X}_k = \frac{1}{2}\left[Y_k + \sqrt{Y_k^2 - \lambda_d(k)}\right] \tag{13.21}$$

13.4.2.2 Minimum Mean Square Error (MMSE) Estimator

Short-time spectral amplitude plays a major role in intelligibility and quality of speech. This needs optimal estimators to get a good estimator of noise power spectral density. An optimal estimator always minimizes the mean square error in Eq. (13.22) between the true and predicted magnitude values.

$$e = E\left\{(\hat{X}_k - X_k)^2\right\} \tag{13.22}$$

where X_k is the true magnitude and $\hat{X}_k$ is the estimated magnitude. Bayesian MSE based minimization depicted in Eqs. (13.23) and (13.24) gives an optimal minimum mean square error (MMSE) estimator [11].

$$\hat{X}_k = \int X_k p(X_k|Y) dX_k \tag{13.23}$$

$$E[X_k|Y(\omega_0)Y(\omega_1)\ldots Y(\omega_{N-1})] \tag{13.24}$$

Equation (13.16) gives the mean of posteriori probability density function of X_k. $P(X_k)$ is the a priori probability density function of X_k.

MMSE Magnitude Estimator The posteriori probability density function $p(X_k \mid Y(\omega_k))$ can be given using Bayes' rule [11], in Eq. (13.25).

$$p(X_k|Y(\omega_k)) = \frac{p(Y(\omega_k)|X_k)p(X_k)}{p(Y(\omega_k))} \tag{13.25}$$

The MMSE estimator is given by Eq. (13.26) as,

$$\hat{X}_k = \frac{\sqrt{\pi}}{2} \frac{\sqrt{v_k}}{\gamma_k} \exp\left(-\frac{v_k}{2}\right) \left[(1 + v_k)I_0\left(\frac{v_k}{2}\right) + v_k I_1\left(\frac{v_k}{2}\right)\right] Y_k \tag{13.26}$$

Here $I_0(.)$ and $I_1(.)$ represent the zeroth and first-order modified Bessel function. v_k, γ_k, ξ_k are defined as

$$v_k = \frac{\xi_k}{1 + \xi_k} \gamma_k \tag{13.27}$$

$$\gamma = \frac{Y_k^2}{\lambda_d(k)} \tag{13.28}$$

$$\xi_k = \frac{\lambda_x(k)}{\lambda_d(k)} \tag{13.29}$$

$\xi_k, \gamma_k,$ and $\lambda_d(k)$ are the a priori SNR, a posteriori SNR, and noise variance, respectively which are given by Eqs. (13.27)–(13.29).

Estimating a Priori SNR by Decision-Directed Approach In MMSE estimator, a priori SNR and noise variance are assumed to be known provided the noise is stationary. For non-stationary noise situations decision-directed approach helps in computing a priori SNR. A priori SNR is related to posteriori SNR as given by [12],

$$\xi_k(m) = \frac{E\left\{X_k^2(m)\right\}}{\lambda_d(k \cdot m)} \tag{13.30}$$

$$\xi_k(m) = E\{\gamma_k(m) - 1\} \tag{13.31}$$

Combining Eqs. (13.30) and (13.31), we get the final estimator of $\hat{\xi}_k(m)$ as per Eq. (13.32) using recursion procedure.

$$\hat{\xi}_k(m) = a\frac{\hat{X}_k^2(m-1)}{\lambda_d(k, m-1)} + (1 - a)\max\left[\gamma_k(m) - 1, 0\right] \tag{13.32}$$

where a is the weighting factor having the range of $0 < a < 1$. Optimal value of a is given as 0.98 as per [12]. To ensure the $\hat{\xi}_k(m)$ is non-negative, max(.) operator is used.

13.4.3 Subspace Algorithms

Subspace algorithms are evolved on the fact that clean signal might be in the subspace of the Euclidean space of the noise. These algorithms work in the principle of decomposing the whole signal vector space as noise and clean signal vector subspace. After nullifying the noisy components in the initial estimated subspace, the final clean speech is recovered. Decomposition of matrix can be done using singular value decomposition (SVD) or eigenvector factorization. Karhunen–Loève transform (KLT) is obtained by taken eigenvector matrix for the covariance matrix of the signal [13].

13.4.3.1 Definition of SVD

As per [14], for any matrix X with dimension $a \times b$ in Eq. (13.33) has rank denoted as rank(X) = r, there is a diagonal matrix $A_{r \times r} = \text{diag}(\sigma_1, \sigma_2, ..., \sigma_r)$ orthogonal matrices $S_{a \times a}$ and $R_{b \times b}$ such that

$$X = S\begin{pmatrix} A & 0 \\ 0 & 0 \end{pmatrix}_{a \times b} R^T \tag{13.33}$$

with $\sigma_1 \geq \sigma_2 \geq ... \geq \sigma_r \geq 0$ where σ_i are the non-zero singular values of X. SVD of matrix X gives column matrix S and R which are the left- and right-hand singular vectors of X.

13.4.3.2 Subspace Decomposition Method

Consider that $x(n)$, is a noise-free signal where n has the range of $0 \leq n \leq N-1$. The matrix representation of X is given in Eq. (13.34) as [13],

$$
X = \begin{bmatrix}
x(L-1) & x(L-2) & \cdots & x(0) \\
x(L) & x(L-1) & \cdots & x(1) \\
\vdots & \vdots & \vdots & \vdots \\
x(N-1) & x(N-2) & \cdots & x(N-L)
\end{bmatrix}
\tag{13.34}
$$

where $L < N$ and X is non-symmetric. The Hankel structured of matrix X is given as per Eq. (13.35),

$$
X_h = \begin{bmatrix}
x(0) & x(1) & \cdots & x(L-1) \\
x(1) & x(2) & \cdots & x(L) \\
\vdots & \vdots & \vdots & \vdots \\
x(N-L) & x(N-L+1) & \cdots & x(N-1)
\end{bmatrix}
\tag{13.35}
$$

Each frame has N samples in the Hankel matrix with M column and L rows. The total frame samples are given as $[y(1)\ldots y(N)]$. The decomposition of H_y is given by Eq. (13.36) as in [14],

$$
H_y = S \sum R^T = [S_{1,k} S_{k+1,M}] \begin{bmatrix} \Sigma_{1,k}^{\text{signal}} & 0 \\ 0 & \Sigma_{K+1,M}^{\text{noise}} \end{bmatrix} \begin{bmatrix} R_{1,k}^T \\ R_{k+1,M}^T \end{bmatrix}
\tag{13.36}
$$

where $\Sigma_{1,K}^{\text{signal}}$ represents the signal subspace and $\Sigma_{K+1,M}^{\text{noise}}$ gives the noise subspace.

13.4.3.3 Eigen Value Decomposition

A symmetric matrix X can be converted as a diagonal matrix using its eigens value and eigenvectors as given by Eq. (13.37), [15, 16],

$$
X = [q_1, q_2, \ldots, q_n] \begin{bmatrix} \lambda_1 & \cdots & 0 \\ \vdots & \ddots & \vdots \\ 0 & \cdots & \lambda_n \end{bmatrix} \begin{bmatrix} q_1^T \\ q_2^T \\ \vdots \\ q_n^T \end{bmatrix} = Q \Lambda Q^T
\tag{13.37}
$$

where Λ is the diagonal matrix and Q is the ortho-normal matrix with eigenvectors as its column.

Let y be the noise speech vector with clean signal vector added with noise vector containing K samples. The covariance matrix in Eq. (13.38) of y is given as [12]:

$$
R_y = R_x + R_d = R_x + \sigma^2 I
\tag{13.38}
$$

In this σ^2 is the noise variance. Rx has the M positive eigen values $\lambda_x(k)$.

$$
\lambda_y(k) = \begin{cases} \lambda_x(k) + \sigma^2 & \text{if} \ldots k = 1, 2, \ldots, M \\ \sigma^2 & \text{if} \ldots k = M+1, \ldots, K \end{cases}
\tag{13.39}
$$

Based on Eq. (13.39) eigenvector matrix can be partitioned assignal+noise subspace and noise subspace.

13.5 Speech Separation Algorithms

13.5.1 Classical Speech Separation Algorithms

The multi-speaker speech separation can be done by some traditional speech separation algorithms. Let $x_s(n)$ is the time domain speech signal sample from the user and $y(n)$ is the mixture of signal from the speakers, then the signal can be represented by Eq. (13.40).

$$
y(n) = \sum_{s=1}^{S} x_s(n)
\tag{13.40}
$$

where s is the speakers count and s takes the value $1, 2, 3, \ldots, S$.

Figure 13.4 Speech separation algorithms.

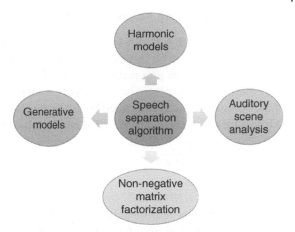

The speech enhancement technique in Section 13.4 cannot be applied here since it deals with multi-speaker problem that is two or more signals are considered to be signal of interest. Therefore, the enhancement techniques discussed above does not produce the accurate result for this problem. Hence a few traditional algorithms are provided for mono channel multi-speaker speech separation. Such algorithms are shown in Figure 13.4.

13.5.2 Harmonic Models

The early techniques available for speech processing were only speech enhancement technique rather than speech separation techniques. The main aim of that technique is to suppress the interfering speaker rather than separating it. When compared to speech enhancement technique, the speech separation technique was more used. This technique is less successful and complicated, whereas a good performance can be achieved by prior knowledge about the signals. The two speech signals mixture can be modeled into summation of two sinusoids, where the speakers have varied frequencies and harmonics. If the frequency and harmonics of the interfering speaker is a known parameter, then those frequencies can be suppressed so is the harmonics. This approach needs a multi-pitch tracking as individual pitch can be estimated using the noise signals and this approach works only when the frequency of the individual speakers is known and separated efficiently. But unvoiced speech signal doesn't have any particular harmonics, this algorithm is partially successful and cannot be applied for single-microphone multi-speaker speech separation [17].

13.5.3 Computational Auditory Scene Analysis

ASA can be used as a different approach to separate signal of interest from the single microphone multi-speaker separation task. The auditory system decomposes the acoustic signal into auditory streams based on ASA. These auditory streams can be used to isolate a signal of interest. Acoustic signal either voiced or unvoiced signal that impinges on the eardrum produces an auditory stream that helps the subconscious mind to focus on the individual signal of interest apart from the prevailing surrounding noise. This auditory stream is processed in two steps: (i) decomposing the time domain acoustic signals into frequency domain by segmentation and each frequency component corresponds to the single source of speech, and (ii) these frequency components are grouped into auditory streams using either sequential or simultaneous grouping. The frequency component which is similar across time series is grouped under similar sources and is called sequential grouping. The similarity across frequency in time–frequency components such as harmonics is grouped under simultaneous grouping. The algorithm which is designed based on ASA to separate speech signals is called computational auditory scene analysis (CASA).

CASA-based single microphone multi-speaker speech separation is successful but it has its own drawbacks. It is designed only for voiced speech signals and cannot be applied for unvoiced signals. The algorithm is designed by manual grouping and segmentation techniques which are not optimal and this algorithm also lacks capability to learn from real-time data [18].

13.5.4 Non-negative Matrix Factorization (NMF)

A different approach to CASA-based technique is non-negative matrix factorization (NMF). The non-negative data matrix $Q \in R^{K \times M}$ can be factorized as $Q \approx DH$, where $D \in R^{K \times L}$ is the dictionary and $H \in R^{L \times M}$ is the activation matrix and $L \ll M$ are non-negative matrix. The set of basis vectors for the data matrix Q is represented by the D and columns of H represent how much of each basis vector is needed to represent each column of Q. The accuracy of the approximation is controlled by the tuning parameter L. If $L = M$, the solution $Q = DH$ can always be found. Q is factorized into D and H, if $L \ll M$ can be achieved by Eqs. (13.41) and (13.42) respectively.

$$H = H \cdot \frac{D^T Q}{D^T DH'} \tag{13.41}$$

$$D = D \cdot \frac{QH^T}{DHH^{T\prime}} \tag{13.42}$$

where element-wise multiplication and division take place.

The dictionary D_s is used for speech separation problem for different speaker, where $s = 1, 2, \ldots, S$ is used for each speaker and any noise source. STFT is employed to split the time and frequency frames in the signal [17].

Although NMF is the most aggressive technique for mono microphone multi-speaker speech separation problem, it has their own drawbacks. It is a linear model, which limits the capacity of the model. It requires talker-dependent dictionaries, to make NMF less suitable for speaker, or noise-type independent applications. Finally due to computational complexity NMF doesn't suit well for large datasets.

13.5.5 Generative Models

Voice signals are highly packed with temporal dynamics on various levels. The speech separation algorithms discussed so far will consider the temporal dynamics partially. The drawbacks of these algorithms motivated to find a new algorithm based on hidden Markov models (HMMs). It is a generative model; it learns the temporal dynamics of all sequential data [19]. Moreover, the finite state machine HMM converts the discrete number of states in a contemporaneous manner. The variable changes from one state to another relying on transition probabilities for each unit increment in time. Transition and emission probabilities of HMM are found to maximize the likelihood function with respect to the provided dataset. For ASR system, the phoneme based HMM is designed to model distribution over speech signals. To identify the unknown phoneme from voice signal, the conditional probability of HMM is evaluated.

In case of multi-speaker speech separation problem, where more than one speaker is present, the standard HMM is extended to factorial HMM. More than one latent variable can be used in factorial HMM, which helps in modeling speech mixtures containing multiple simultaneous sources. Factorial HMM has achieved a reasonable milestone in speech separation systems to separate a two-talker speech [18].

Although factorial HMM is successful, it also has some drawbacks. To compute conditional probabilities, the number of training and testing dataset is more, and these probabilities have to be estimated, which is complex and scales poor with an increase in speakers. Factorial HMM for multi-speaker speech separation require speaker dependent HMMs for each speaker in the mixture. These are the limitations that make HMMs not suitable for real-world applications. Therefore, different techniques are required for multi-speaker speech separation algorithms for general applications, which ultimately led to the application of deep neural network (DNN) in speech enhancement and separation techniques.

13.6 Deep Learning Based Speech Enhancement

Deep learning-based single-microphone speech enhancement techniques can be divided into two types: mask approximation-based techniques and signal approximation-based techniques.

13.6.1 Mask Approximation

Consider $x(n)$ and $y(n)$ be the clean speech signal and noise speech signal in the time domain with time frame m. Along with this, the additive noise $v(m)$ is mixed. Also, $a(m)$ and $r(m)$ be the STFT spectral magnitude vectors of $x(n)$ and $y(n)$.

The STFT representation of $y(n)$ is given in Eqs. (13.43) and (13.44) as [20],

$$Y_{k,l} = X_{k,l} + V_{k,l} \tag{13.43}$$

$$\sum_{n=0}^{N-1} y(lL + m)w(m)e^{-j\Omega_k m} \tag{13.44}$$

Here l, k, Ω_k is the segment index, frequency index, and angular frequency to the corresponding center frequencies of the STFT bands, respectively. Using STFT representation in Eq. (13.45) we obtain the amplitudes $R_{k,l}, A_{k,l},$ $D_{k,l}$ and phases $\varphi_{k,l}^Y, \varphi_{k,l}^X, \varphi_{k,l}^V$ of the complex spectral coefficients $y, x,$ and v respectively [20].

$$Y_{k,l} = R_{k,l}e^{j\,\phi_{k,l}^Y}; S_{k,l} = A_{k,l}e^{j\,\phi_{k,l}^S}; V_{k,l} = D_{k,l}e^{j\,\phi_{k,l}^V} \tag{13.45}$$

The feature transformation of $y(m)$ in DNN is $h[y(m)]$,

$$\hat{g}_m = f_{\text{DNN}}(h(y_m), \theta) \tag{13.46}$$

where $\hat{g}(m)$ in Eq. (13.46) denotes gain vector estimated by DNN with parameter θ. Mask approximation-based techniques help to find the DNN parameter θ as given in Eq. (13.47).

$$\theta^* = \arg\min_{\theta} \sum_{D_{\text{train}}} \Im(\hat{g}_m, g_m), (y_m, g_m) \in D_{\text{train}} \tag{13.47}$$

$D_{\text{train}}, J(.), \hat{g}(m)$ denotes the training dataset, cost function, and target gain vector. Mask approximation techniques try to minimize the difference that is measured by the cost function.

13.6.1.1 Complex Ideal Ratio Mask

Complex ratio mask helps in retrieving the STFT of clean speech when it is applied to the noisy speech. The complex spectrum of clean speech $S_{t,f}$ and noisy speech $M_{t,f}$ is given in Eqs. (13.48)–(13.51) as [21],

$$S_{t,f} = M_{t,f} \times Y_{t,f} \tag{13.48}$$

$$Y = Y_r + iY_i \tag{13.49}$$

$$S = S_r + iS_i \tag{13.50}$$

$$M = M_r + iM_i \tag{13.51}$$

where r and i indicate the real and imaginary components, respectively. The real and imaginary components of M are defined in Eqs. (13.52) and (13.53) respectively as:

$$M_r = \frac{Y_r S_r + Y_i S_i}{Y_r^2 + Y_i^2} \tag{13.52}$$

$$M_i = \frac{Y_r S_i - Y_i S_r}{Y_r^2 + Y_i^2} \tag{13.53}$$

Finally Eq. (13.54) defines the complex ideal ratio mask as:

$$M = \frac{Y_r S_r + Y_i S_i}{Y_r^2 + Y_i^2} + i\frac{Y_r S_i - Y_i S_r}{Y_r^2 + Y_i^2} \tag{13.54}$$

The real and imaginary components of cIRM can have large values between $(-\infty, \infty)$ and this can be compressed using hyperbolic tangent given in Eq. (13.55).

$$\text{cIRM}_x = K \frac{1 - e^{-C \cdot M_x}}{1 + e^{-C \cdot M_x}} \tag{13.55}$$

To train the DNN, the empirical values of $K = 10$ and $C = 0.1$ suits well as said in [21].

DNN Based cIRM Estimation In DNN, it is necessary to update the weights. Back-propagation algorithm based on mean square error of the data is used to accomplish this. Equation (13.56) gives the cost function as [21],

$$\frac{1}{2N} \sum_t \sum_f \left[(O_r(t,f) - M_r(t,f))^2 + (O_i(t,f) - M_i(t,f))^2 \right] \tag{13.56}$$

where N is the number of time frames, $O_r(t,f)$ and $O_i(t,f)$ are the real and imaginary outputs of DNN.

13.6.1.2 Ideal Binary Mask

CASA-based speech separation system undergoes peripheral analysis to give the time–frequency representation. This uses energy or auto-correlation for T–F segmentation. A binary matrix is constructed based on the energy known as ideal binary mask (IBM) [22]. Speech energy is compared with noise energy in a particular T–F unit. If speech energy is dominant compared to noise energy it is labeled as "1" else as "0" is denoted in Eq. (13.57).

$$\text{IBM}(t,f) = \begin{cases} 1, \text{if } s(t,f) - n(t,f) > 0 \\ \quad 0, \quad \text{Otherwise} \end{cases} \tag{13.57}$$

The major problem with IBM is noise dominant T–F unit is labeled as "0" even though there is sufficient speech energy. This makes to lose the necessary speech.

13.6.2 Signal Approximation

Mask approximation-based technique focus on minimizing the gap between target gain and estimated gain. The cost function defined in Eq. (13.5) is used to minimize the target and the estimated gain vector. The DNN is trained to achieve nearer to target gain with the cost function, whereas the target vector is not defined explicitly. In signal approximation techniques, it works on reducing the magnitude of clean speech from estimated speech obtained using STFT [20, 23]. They are used in situations like noise type or SNR is known as priori.

13.7 Deep Learning Based Speech Separation

Single microphone multi-speaker speech separation using conventional speech separation algorithm has some measurable drawbacks, which makes single microphone multi-speaker speech separation based on DNN as popular. The mixed signal can be separated by using multi-layer feed-forward neural network which is trained by unsupervised learning method to calculate log power spectrum of the single talker from the log power spectrum of a mixed signal consisting of two speakers. This is same as DNN-based enhancement technique except for the noise signal. In DNN-based enhancement, the environment signal is considered as interference whereas in DNN-based separation, it is speech signal from different speaker. Speaker dependency is one of the major problems in this method and this can be overcome by some more assumptions. It is assumed that the average energy level of one speech signal is always higher than the other speech signal, i.e. a mixture SNR different from 0 dB [24]. With this assumption, the speech signal was extracted to get high average energy and low average energy using multi-layer FNN. With this modification, the two different speech signals can be extracted somewhat successfully. Gender based speech separation can be done without some prior knowledge of the speech signal, where a separate output stream is assumed for both male and female. But still, this gender-based separation has some limitation as it is suitable only for two speakers and it cannot be scaled for more than two speakers.

13.7.1 Label Permutation Problem (LPP)

DNN-based speaker individualistic multiple speaker speech separation has gained only limited success rate because most techniques considered were applicable only for two known speakers and this is due to label permutation problem (LPP). Especially for dual speaker separation tasks, let P_1 and P_2 represent the target vector for speaker one and two, respectively. Let $P = \left[P_1^T P_2^T \right]^T$ denote a concatenated super vector. Finally, $\hat{P}$ is the result of DNN, representing the estimate of P. The target vector can take any one of the two configurations during training: $P = \left[P_1^T P_2^T \right]^T$ or as $P = \left[P_2^T P_1^T \right]^T$. It is observed that P_1 and P_2 are similar speakers if trained with predefined permutation. The LPP training fails when the training set comprises multiple samples with large utterances of both genders [25].

13.7.2 Deep Clustering

The first successful technique which solved the LPP was deep clustering. In this method, speech separation is identified as a cumulative problem. The time–frequency unit in the mixture signal is mapped to multi-dimensional units using LSTM-RNN algorithm. The embeddings of time–frequency unit of the same speaker are closely located and can be clustered into a single group, which is then used to segregate the speech signals [26].

Let $Y \in R^N$ denote a feature vector of a mixed signal, the feature representation is given by STFT, such that $N = K \times M$ indicates the sum of time–frequency units, where K and M are the amount of frequency bins and time frames. It uses K-means clustering technique to identify the speaker by grouping the closely spaced time–frequency units. It can be used as a binary mask to separate mixture signal. By using clustering technique, the LPP is avoided. Furthermore, it is modeled using LSTM-RNN.

Although, this technique has its own disadvantage. It is important to know the number of speakers in prior to use K-means clustering algorithm. It uses binary gain therefore; it is not optimal for multiple speakers or environmental noise in the mixture. Each time–frequency unit is embedded in D dimensional embedding vector; therefore, the output of the clustering model should be D times greater than the input, which is complex for long signals [27–30].

13.8 Results and Discussions

The phonetically balanced speech database, such as TIMIT, NOIZEUS, and Danish speech corpus Akustiske Databaser for Dansk are generally selected to work on speech enhancement-based research activities. Speech signals are sampled at a sampling frequency of 16 kHz and are normalized using root mean square of unity value. The generalization capability of DNN-based ideal ratio mask is given in [3]. Speech enhancement is measured in relation with noise type, speaker type, and SNR. DNN is trained to predict IRM in a supervised manner. Gamma tone filter bank is used for T–F representation of IRM, which gives 20 ms speech frames with overlap of 10 ms. A feed-forward DNN architecture with three hidden layers and rectified linear units is used as activation function for hidden layer. Output layer is activated with sigmoid activation function. DNN-based speech enhancement shows good improvement in short-time objective intelligibility (STOI) for babble at 5 bB and perceptual evaluation of speech quality (PSEQ) at 10 and 15 dB. When concerned with speaker type, DNN works good for same gender and not for the opposite gender. DNN trained with female speaker and tested with male speaker shows reduced improvement.

The key problem in speech separation for an ASR is to identify the keywords like numbers and letters of the target speaker from multiple speakers. Speech separation based on instantaneous energy and average pitch is investigated in [4]. The DNN has two training sets, named high energy training set (6, 3, 0 dB, and clean) and low energy training set (0, −3, −6, −9 dB). DNN exhibits a refinement in word error rate (WER) percentage when using multi-style training set rather than single training set.

13.9 Conclusion

The objective measure used for estimating speech quality is PSEQ. The grade of speech quality is given on a linear scale of 1 to 5, where 5 denotes excellent audio quality and 1 denotes poor audio quality. The second objective measure is the STOI, which estimates the speech intelligibility. STOI is considered to be binary, as it measures whether a word is understood or not. The value ranges from 0 to 1. PSEQ and STOI can be applied for single source of speech signal. In case of multi-speaker environment blind source separation (BSS), source-to-distortion ratio (SDR), source-to-interference ratio (SIR), source-to-artifact ratio (SAR), and SNR are used as measurement parameters. IBM/IRM-based speech enhancement DNN training shows good STOI score for babble and cafeteria noise for −5 to 5 dB SNR values. cIRM shows good STOI score for cafeteria and factory noise for SNR ranging −6 to 6 dB compared to IRM. LPP in DNN training is addressed. Deep clustering uses K-means clustering algorithm to accumulate similar frequency components by changing to T–F domain. The cluster formed is considered to be one single source. Speech separation method called CASA helps to separate signal of interest from multi-speakers, whereas harmonic models help to model individual signals in multiple signal environment. Spectral subtraction shows less speech quality improvement compared to statistical and subspace algorithm. Statistical methods deal with non-linear estimators of signal magnitude. Subspace algorithm helps in decomposing a signal into its subspace and identifies the necessary signal.

DNN/IRM and cIRM-based speech enhancement systems can be trained with large number of speakers, noise type, and different SNR. It shows good improvement in STOI and PSEQ when compared with NMF based approach. DNN-based speech separation holds good in WER and SAR compared to HMM and Gaussian mixture model-based ASR systems.

References

1 Bronkhorst, W. (2000). The cocktail party phenomenon: a review of research on speech intelligibility in multiple-talker conditions. *Acta Acustica united with Acustica* 86 (1): 117–128.

2 McDermott, J.H. (2009). The cocktail party problem. *Current Biology* 19 (22): R1024–R1027.

3 Kolbæk, M., Tan, Z.H., and Jensen, J. (2017). Speech intelligibility potential of general and specialized deep neural network based speech enhancement systems. *IEEE/ACM Transactions on Audio, Speech and Language Processing* 25 (1): 153–167.

4 Weng, C., Yu, D., Seltzer, M.L., and Droppo, J. (2015). Deep neural networks for single-channel multi-speaker speech recognition. *IEEE/ACM Transactions on Audio, Speech and Language Processing* 23 (10): 1670–1679.

5 Hochreiter, S. and Schmidhuber, J. (1997). Long short-term memory. *Neural Computation* 9 (8): 1735–1780.

6 Erdogan, H., Hershey, J.R., Watanabe, S., and Roux, J.L. (2017). Deep recurrent networks for separation and recognition of single channel speech in non-stationary background audio. In: *New Era for Robust Speech Recognition: Exploiting Deep Learning* (ed. S. Watanabe, M. Delcroix, F. Metze and J. Hershey), 165–186. Springer.

7 Yoshioka, T., Erdogan, H., Chen, Z. et al. (2018). Recognizing overlapped speech in meetings: A multichannel separation approach using neural networks. arXiv e-prints.

8 G. Saon, G. Kurata, T. Sercu, K. Audhkhasi, S. Thomas, D. Dimitriadis, X. Cui, B. Ramabhadran, M. Picheny, L. Lim, B. Roomi, P. Hall (2017) "English conversational telephone speech recognition by humans and machines," INTERSPEECH.

9 Bharathi, S., Kavitha, S., and Priya, K.M. (2016). Speaker verification in a noisy environment by enhancing the speech signal using various approaches of spectral subtraction. In: *2016 10th International Conference on Intelligent Systems and Control (ISCO),Coimbatore, India*, 1–5.

10 Vanambathina, S., Sivaprasad, N., and Kumar, T.K. (2014). A survey on statistical based single channel speech enhancement techniques. *International Journal of Intelligent Systems and Applications* 6: 69–85.

11 Zhang, Q., Nicolson, A., Wang, M. et al. (2020). DeepMMSE: a deep learning approach to MMSE-based noise power spectral density estimation. *IEEE/ACM Transactions on Audio, Speech and Language Processing* 28: 1404–1415.

12 Loizou, P.C. (2013). *Speech Enhancement: Theory and Practice.* CRC Press.

13 Riadh, A., Salim, S., Massaoud, H. et al. (2018). Subspace Approach for Enhancing Speech based on SVD. In: *International Conference on Communications and Electrical Engineering (ICCEE), El Oued, Algeria*, 1–8.

14 Patil, N.M. and Nemade, M.U. (2017). Audio signal deblurring using singular value decomposition (SVD). In: *IEEE International Conference on Power, Control, Signals and Instrumentation Engineering (ICPCSI), Chennai, India*, 1272–1276.

15 Ephraim, Y. and Trees, H.L.V. (1995). A signal subspace approach for speech enhancement. *IEEE Transactions on Speech and Audio Processing* 3 (4): 251–266.

16 Zhou, L. and Li, C. (2016). Outsourcing eigen-decomposition and singular value decomposition of large matrix to a public cloud. *IEEE Access* 4: 869–879.

17 Souden, M., Benesty, J., and Affes, S. (2010). An optimal frequency domain multichannel linear filtering for noise reduction. *IEEE Transactions on Audio, Speech and Language Processing* 18 (2): 260–276.

18 Yu, J., Zhang, S.X., Wu, B. et al. (2021). Audio-visual multi-channel integration and recognition of overlapped speech. *IEEE/ACM Transactions on Audio, Speech and Language Processing* 29: 2067–2082.

19 Barker, J., Ricard, M., Vincent, V. et al. (2015). The third 'CHiME' speech separation and recognition challenge: dataset, task and baselines. *2015 IEEE Workshop on Automatic Speech Recognition and Understanding (ASRU), Scottsdale, AZ, USA*, pp. 504–511.

20 Krawczyk, M. and Gerkmann, T. (2014). STFT phase reconstruction in voiced speech for an improved single-channel speech enhancement. *IEEE/ACM Transactions on Audio, Speech and Language Processing* 22 (12): 1931–1940.

21 Williamson, D.S., Wang, Y., and Wang, D. (2016). Complex ratio masking for monaural speech separation. *IEEE/ACM Transactions on Audio, Speech and Language Processing* 24 (3): 483–492.

22 Minipriya, T.M. and Rajavel, R. (2018). Review of ideal binary and ratio mask estimation techniques for monaural speech separation. In: *4th International Conference on Advances in Electrical, Electronics, Information, Communication and Bio-Informatics*, 1–5.

23 Xu, Y., Du, J., Dai, L.R., and Lee, C.H. (2015). A regression approach to speech enhancement based on deep neural networks. *IEEE/ACM Transactions on Audio, Speech and Language Processing* 23 (1): 7–19.

24 Kolbæk, M., Tan, Z.H., and Jensen, J. (2016). Speech enhancement using long short-term memory based recurrent neural networks for noise robust speaker verification. In: *IEEE Spoken Language Technology Workshop*, 305–311.

25 Wang, Y., Narayanan, A., and Wang, D. (2014). On training targets for supervised speech separation. *IEEE/ACM Transactions on Audio, Speech and Language Processing* 22 (12): 1849–1858.

26 Hershey, J.R., Chen, Z., Roux, J.L., and Watanabe, S. (2016). Deep clustering: Discriminative embeddings for segmentation and separation. In: *2016 IEEE International Conference on Acoustics, Speech and Signal Processing (ICASSP), Shanghai, China*, 31–35.

27 Huang, P.S., Kim, M., Hasegawa-Johnson, M., and Smaragdis, P. (2015). Joint optimization of masks and deep recurrent neural networks for monaural source separation. *IEEE/ACM Transactions on Audio, Speech and Language Processing* 23 (12): 2136–2147.

28 Rix, W., Beerends, J.G., Hollier, M.P., and Hekstra, A.P. (2001). Perceptual evaluation of speech quality (PESQ) – a new method for speech quality assessment of telephone networks and codecs. In: *2001 IEEE*

International Conference on Acoustics, Speech, and Signal Processing (ICASSP), Salt Lake City, UT, USA, vol. 2, 749–752.

29 Taal, C.H., Hendriks, R.C., Heusdens, R., and Jensen, J. (2010). A short-time objective intelligibility measure for time–frequency weighted noisy speech. In: *2010 IEEE International Conference on Acoustics, Speech and Signal Processing (ICASSP), Dallas, TX, USA*, 4214–4217.

30 Divya, D., Kumar, S., and Ghai, D. (2019). A review on speech enhancement techniques. *Think India Journal* 22: 25–34.

14

Design and Development of Real-Time Music Transcription Using Digital Signal Processing

Thummala Reddychakradhar Goud[1], Koneti Chandra Sekhar[1], Gannamani Sriram[1], Gadamsetti Narasimha Deva[1], Vuyyuru Prashanth[1], Deepika Ghai[1], and Sandeep Kumar[2]

[1]*School of Electronics and Electrical Engineering, Lovely Professional University, Phagwara, Punjab, India*
[2]*Department of Electronics and Communications, Sreyas Institute of Engineering and Technology, Hyderabad, Telangana, India*

14.1 Introduction

Automatic music transcription (AMT) has been one of the most emerging and challenging research fields in music information retrieval (MIR). All these factors originate from audio content analysis (ACA) [1]. There are many proposed methods and techniques for developing a music transcribers [2]. Though there are various music signal analysis methods, they are different from monophonic music to polyphonic music. Music transcription becomes much more complicated when it comes to polyphonic music, as it possesses much more parameters to be determined [3]. Several researchers had worked on both polyphonic and monophonic music transcribers by proposing various algorithms and techniques. The current chapter focuses on transcribing monophonic music to a list of keys (i.e. notes), combining pitch class and octave as shown in Figure 14.1. For this, the input signal is processed using digital signal processing (DSP), image processing (IP), and musical instrument digital interface (MIDI) conversion.

Figure 14.1 Representation of visualized sample input (in the form of symbols) and sample output, i.e. keys in chromatic form (pitch class and octave).

The rest of the chapter is presented as follows. Section 14.2 describes related work. Section 14.3 discusses the motivation of the proposed work for real-time music transcription using DSP. Section 14.4 explains the mathematical expressions of signal processing. Section 14.5 explains the proposed methodology for music transcription. Results and discussion are included in Section 14.6 and conclusions are concluded in Section 14.7.

14.2 Related Work

Many algorithms are used for the MIR such as predicting polyphonic music prediction using long-short term memory (LSTM) network and construction of piano-roll based on music language model (MLM) proposed by Ycart and Benetos in [4], various techniques in signal processing like Fourier transform (FT), short-time Fourier transform

Machine Learning Algorithms for Signal and Image Processing, First Edition.
Edited by Deepika Ghai, Suman Lata Tripathi, Sobhit Saxena, Manash Chanda, and Mamoun Alazab.

(STFT), wavelet transform (WT), Q-transform (QT) for determining music parameters such as melody, harmony, rhythm, and timbre as proposed by M. Müller et al. in [5], techniques for estimation of pitch, timing, duration of each note from written music with symbols and notations proposed by A. Klapuri et al. in [6], several approaches proposed by Alm and Walker for time-frequency analysis of musical instruments using Fourier transform, spectrogram and scalogram analysis [7], piano frequency detection and determining respective key note using Goertzel algorithm proposed by J. Jiang et al. in [8], C. Dittmar and Abeßer proposed construction of toolbox based on graphical user interface (GUI) with various algorithms for transcribing drum and bass music by finding pitch, onsets, offsets using multi-resolution fast-Fourier transform (MRFFT) in [9], estimation and extraction of pitch for polyphonic music using modified-Morlet wavelet (MMW) proposed by N. Kumar et al. in [10].

There are other ways as proposed by N. Wats and Patra that is spectral factorization technique for which non-negative matrix factorization (NMF) has been used and a variant of NMF, which is based on accelerated multiplicative update with some predefined templates [11], evaluation of multi-pitch detection, audio to MIDI conversion, note value detection as proposed by A. McLeod and Steedman in [12], A. Sadekar and Mahajan proposed several methods like convolutional neural networks (CNNs), recurrent neural networks (RNNs), LSTM and constant-Q transform (CQT) for multi-pitch detection, note estimation in polyphonic music [13], formation of piano rolls for multi-instrument music to compute pitchgram using deep neural network (DNN) and separate each pitchgram for specific instrument using deep spherical clustering technique proposed by K. Tanaka et al. in [14], music detection system that include pitch detection and transcription using support vector machine (SVM) proposed by X. Pan et al. in [15].

Other methods like audio to score represent monophonic music to estimate pitch, time, and symbolic representation using DNN proposed by M. Román et al. in [16]. Monophonic transcription using probabilistic and deterministic methods, and their comparison by N. Silva et al. in [17], more methods like conventional K-nearest neighbor (KNN) algorithm for monophonic music pitch estimation proposed by F. Pishdadian and Nelson in [18], construction of online transcription using auto-regressive (AR) LSTM and multi-note-state by D. Jeong [19], online applications like [20] contribute to ACA and processing.

14.3 Motivation of the Proposed Work

The most challenging part of music transcription is determining onsets and offsets following the pitch. Few ways to determine these parameters are by converting the audio file into MIDI as in [21], which involves forming probabilities for musically relevant regions in the audio file. The previous works like [22] propose algorithms for finding relevant regions that are to be transcribed. Even image processing can be used in music transcription as in [23] for transcribing music sheets and other musical symbols, there are few ways to be worked on like applying image processing on spectrogram, scalogram, and chroma to extract audio features like pitch class, time sequence, and other parameters.

In this chapter, we work with python that includes libraries librosa, OpenCV, and pretty midi. Input audio file undergoes signal processing and results in spectrogram, scalograms, and chroma. Later the same input file is converted to a MIDI file by analyzing onsets and offsets, this gives out MIDI pitch which can be later converted to octave-class. Previously created chroma undergoes image processing for extracting pitch class. The combination of octave and pitch classes gives keys (notes) as a result. There is a way to extract pitch class from MIDI files instead of image processing but use MIDI to get pitch class has a high computational cost because if a piano of 88 keys is considered, it is needed to verify among all 88 MIDI pitch numbers. On the other hand, chroma image processing gives out only those pitch classes involved in the input audio file. The proposed technique is evaluated on a publically available dataset documented in [24]. The proposed technique in this chapter focuses on the following objectives:

(i) To construct spectrogram, scalogram, and chroma using STFT, continuous wavelet transform (CWT) for signal visualization and further processing.

(ii) To find probabilities for musically relevant regions, build a transition matrix with silence, onsets, and sustain notes to form piano-roll, which can help the construction MIDI file.

(iii) To extract audio features like the beat, MIDI pitch, and frequencies. Later, the MIDI pitch can be classified into octaves.

(iv) To determine contours on previously produced chroma, which is modified using a threshold for generating pitch class.

(v) To determine keys in a timely sequence as it is in the input audio.

14.4 Mathematical Expressions of Signal Processing

Signal processing of the input audio signal starts with STFT as we want to know at what time a particular frequency event occurs in the signal [25–31]. STFT in Eq. (14.1) has a window that slides along the time series and performs Fourier transform on every time-dependent segment. Hanning window is used as it touches zero at both ends eliminating all discontinuity.

$$X(\tau, \omega) = \int_{-\infty}^{\infty} x(t)\omega(t - \tau)e^{-i\omega t}dt \qquad (14.1)$$

In Eq. (14.1), $X(\tau, \omega)$ is STFT of the input audio signal $x(t)$, along time axis τ and frequency axis ω.

$\omega(\tau)$ is the Hanning window function given in Eq. (14.2) and we use it as $\omega(t - \tau)$ as it slides through the input signal along with the time(t)-axis.

$$\omega(\tau) = 0.5[1 - \cos(2\pi * (\tau/T))], \text{ where } 0 \leq \tau \leq T \qquad (14.2)$$

For STFT, the analyzing function is a window. On the contrary, CWT uses wavelet as the analyzing function. CWT compares the signal to shifted and scale (compressed or stretched) versions of a wavelet. STFT performs Fourier transform on patches of the signal. Unlike STFT, CWT in Eq. (14.3) needs only an orthogonal filter bank like a wavelet to work on the overall signal.

$$C(a, b; x(t), \Psi(t)) = \int_{-\infty}^{\infty} x(t)\frac{1}{a}\Psi^*\left(\frac{t - b}{a}\right) dt \qquad (14.3)$$

$C(a, b)$ is CWT of the input signal $x(t)$ is the input signal, a is the scale parameter, b is the position or shifting parameter, and Ψ^* is the complex conjugate of the wavelet function $\Psi(t)$ given in Eq. (14.4). Here the wavelet used is the Gaussian function.

$$\Psi(t) = C_p e^{-t^2} \qquad (14.4)$$

C_p is such that by taking pth derivative of the $\Psi(t)$. Gaussian wavelet would be good at signal visualization while capturing a short burst of changing notes in a music signal.

For image processing, OpenCV is used for finding moments on the chroma. In general, OpenCV uses Green's theorem for finding contours. Green's theorem relates line integral around a curve to double integral over a plane bounded by the curve. Consider a curve C (counter clock-wise) enclosing the region D in the x–y plane. F be the 2D vector field in the plane, $f(x, y)$ be the partial derivative of vector field F in the x–y plane. According to Green's theorem in Eq. (14.5), we get

$$\int_C F \cdot ds = \iint_D f(x, y)dA \qquad (14.5)$$

As $f(x, y)$ given in Eq. (14.6) is a partial derivative of F vector filed concerning x and y coordinates within the region of D in the plane, we get

$$f(x, y) = \frac{dF_2}{dx} - \frac{dF_1}{dy} \qquad (14.6)$$

By substituting Eq. (14.6) in Eq. (14.5), we get Eq. (14.7)

$$\int_C F \cdot ds = \iint_D \left(\frac{dF_2}{dx} - \frac{dF_1}{dy} \right) dA \tag{14.7}$$

14.5 Proposed Methodology

The flowchart of the proposed methodology for music transcription is shown in Figure 14.2. The steps involved in the proposed methodology for music transcription are enlisted as follows: (i) reading music file and visualizing amplitude and time parameters; (ii) processing the signal for extracting chroma and other features like spectrogram and scalogram; (iii) MIDI conversion; (iv) feature extraction from MIDI file; (v) image processing of chroma for finding contours, their centroids, and pitch class; and (vi) key extraction.

14.5.1 Reading and Visualization

Music file containing monophonic music is taken as input, then it is passed through Kaiser filter for fast resampling, extracting time duration and for data, visualization plotting the signal between amplitude and samples as shown in Figure 14.3. For processing at a fast pace, we use kaiser_best for resampling and sampling frequency of 11 025 Hz as the original sampling frequency is 44 100 Hz for typical music in a wave file.

14.5.2 Signal Processing

Signal processing is the base stage of the proposed methodology. In the signal processing, the first step after visualizing the input music file is to construct a scalogram (scale versus time) of the signal using CWT, then building a power spectrogram (frequency versus time) of the signal and chroma (pitch class versus time) using STFT, and finally forwarding the chroma of the signal to image processing for chroma feature extraction. Besides signal processing, the same input signal undergoes MIDI conversion.

Analyzing the signal CWT is one of the best ways, [10] as it uses wavelets for processing almost all samples throughout the signal. As the current chapter works on monophonic music, it goes well with Gaussian wavelet. So by using Eqs. (14.3) and (14.4), we get a scale versus time scalogram as shown in Figure 14.4. Scalogram is good at visualizing the data, but it possesses many bins and is difficult to extract features.

STFT is used to analyze the power spectrum of the music data [2, 13, 22] in terms of the frequency concerning time. The resulted valuable spectrogram is used for the analysis of power or amplitude on frequency versus time plot. By using Eq. (14.1) and Hanning window from Eq. (14.2), we get a frequency versus time power spectrogram as shown in Figure 14.5.

The same STFT used above can help create chroma of the music data using librosa. Chroma is a plot of musical notes between pitch class and time in Figure 14.6. It helps to extract pitch class during image processing in further processing. Pitch class include 12 pitches C, C#, D, D#, E, F, F#, G, G#, A, A#, B. The chroma defines the notes accordingly to their classes concerning the time from 0 to n seconds of the music clip. The chroma is created with actual dimensions 288 along the rows and 432 and the columns, useful in image processing.

14.5.3 MIDI Conversion

Converting audio files into MIDI is one of the challenging tasks in MIR. The basic requirements are to find the onsets, sustains, and offsets [21]. We need to find probabilities for musically relevant regions [22] for finding these parameters. In a music sequence, few parameters like silence, onsets, and sustains can be called states. The first step is to form a transition matrix (T) as Eq. (14.8) with states along the musical sequence using librosa and pretty

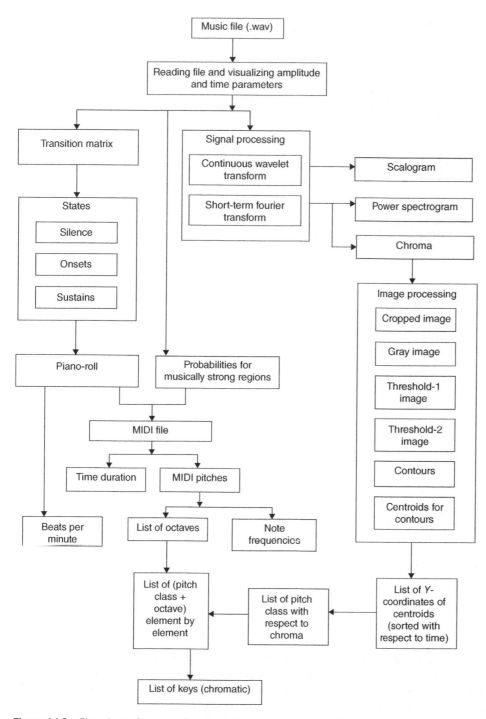

Figure 14.2 Flowchart of proposed methodology for music transcription system.

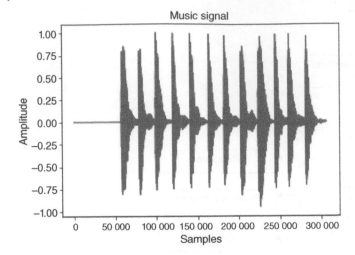

Figure 14.3 Input signal.

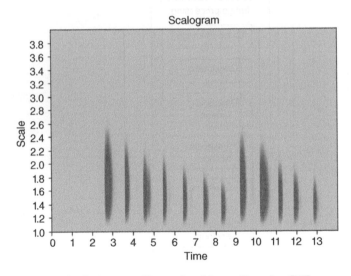

Figure 14.4 Scalogram of input signal formed by using CWT.

midi libraries. The transition matrix is a 2×2 NumPy array that contains states. This also needs probabilities for a note to existing at a specific time.

$$T[i, j], \text{ where } i, j \text{ are consecutive states} \tag{14.8}$$

$T[i, j]$ is the probability function for the transition matrix in finding states at a particular time in the musical sequence. Using Eq. (14.8) and librosa we get probabilities for silence, onsets, and sustains from starting note to ending note in the music file. Silence, onsets, and sustains form a transition matrix, which can be helpful in further processing. The second step is to find the probability (P) of note existing in a particular state (silence, onset, sustain) at some particular time in Eq. (14.9), let P be a 2D NumPy array then

$$P[j, t], \text{ where } j \text{ is the present state at time } t \tag{14.9}$$

$P[j, t]$ is the probability function for the note to exist at that particular time. Using Eq. (14.9) with librosa we get the probability for note existence which can be used further. Thirdly, we get a piano-roll by using previously

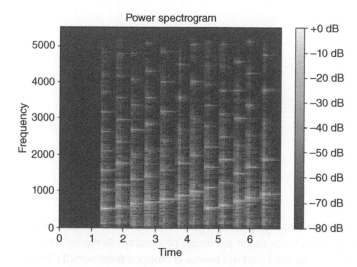

Figure 14.5 Power spectrogram of input formed by using STFT.

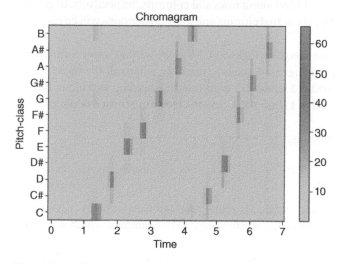

Figure 14.6 Chroma of the input signal using STFT.

built states from the transition matrix T. Piano-roll is an intermediate stage where we get a list of lists with each note parameter like onsets and offsets and the pitch. Final step is to convert this piano roll into a MIDI file using commands from librosa.

14.5.4 Feature Extraction

In this process, we extract beat, onsets, offsets, time duration, frequencies, and MIDI pitch from the previously created MIDI file using the pretty_midi library. From the MIDI pitch, we get the octave class as shown in Table 14.1. The tabular form has a MIDI pitch range resulting in octaves accordingly.

The obtained octaves are sorted in a list concerning time. This list is further valuable for crucial extraction and frequency extraction. Using commands from pretty midi, we get frequencies from the MIDI number that can be matched with notes. For the current music sample, we get

Octave list = ["5," "5," "5," "5," "5," "5," "5," "5," "5," "5," "5," "5"]

Table 14.1 Representation of octaves according to their MIDI pitch ranges.

S. No.	MIDI pitch (P)	Octave
1	$60 <= P <= 71$	4
2	$72 <= P <= 83$	5
3	$84 <= P <= 95$	6
4	$96 <= P <= 107$	7

14.5.5 Image Processing

In the image processing process (Figure 14.7), we use chroma for pitch-class extraction. As chroma is formed with actual dimensions during the signal processing, it becomes easier for image processing, as we only need a specific part of chroma for feature extraction. Firstly the required part of chroma is cropped from actual chroma using OpenCV, as shown in Figure 14.7a. The crop dimensions are 35 : 253 along rows and 54 : 323 along columns resulting dimensions of the required crop image will be 218 269 along rows and columns, respectively, to get the required region of chroma. Previously we used 288 432 for the actual chroma and these dimensions will be constant irrespective of the lengths of various music files.

Later the cropped image is converted into a grayscale image in Figure 14.7b, as we cannot apply a threshold on a colored image (colored chroma). The formed gray image undergoes a threshold where all the objects with threshold value 147 are upgraded to a maximum threshold value 255 with the thresholding technique binary inverse using Eqs. (14.10) and (14.11). Results are shown in Figure 14.7c, where objects are turned white and the rest of the image to black.

As we use the binary inversion technique for thresholding,

$$\text{dst}(x, y) = 0, \text{if src}(x, y) > \text{threshold} \tag{14.10}$$

$$\text{dst}(x, y) = \text{maximum threshold, if any other case} \tag{14.11}$$

where $\text{dst}(x, y)$ is updated pixel intensity at point (x, y) and $\text{src}(x, y)$ is actual pixel intensity at point (x, y). Here, the threshold value is 147 proven experimentally after successful attempts on 200 test samples, which works for octaves 4, 5, 6, and 7 correctly.

There is a problem using the threshold-1 image, the outline is also white and coincides with the objects along the border. As a result, we get the wrong contours. So we have to use the threshold-2 image in Figure 14.7d for further processing. In the threshold-1 image, we draw a rectangle of black color from starting point (0, 0) to (268, 217), i.e. (column, row) as we know the dimensions of the cropped image to create the threshold-2 image, where objects do not coincide with the border.

In OpenCV, by using Green's theorem in Eq. (14.7), we get contours along with their centroids plotted on the gray image as shown in Figure 14.7e. These centroid coordinates are stored in a list and sorted concerning x coordinates (as we know x-axis possesses time samples in actual chroma). So the updated list contains sort coordinates from lower value of x-coordinate to higher value of x-coordinate. After sorting the centroid list, we only need y-coordinates of the centroids (as we know the y-axis contains pitch classes in actual chroma), so we separate y-coordinates and form a new list with only y-coordinates.

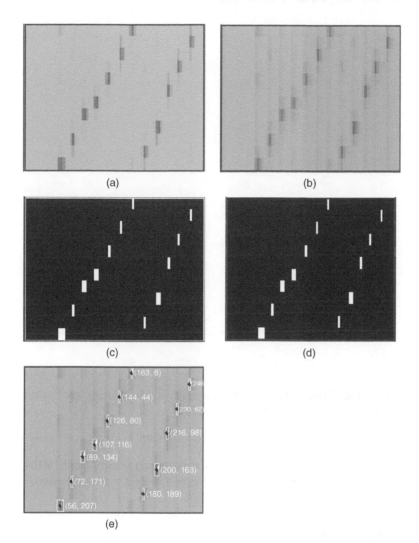

Figure 14.7 Steps involved in image-processing process: (a) cropped image; (b) gray image; (c) threshold-1; (d) threshold-2; and (e) centroids.

Sorted list with respect to *x*-axis = [(55, 207), (72, 171), (89, 134), (107, 116), (126, 80), (144, 44), (162, 8), (180, 189), (200, 153), (215, 98), (230, 62), (248, 25)].

The new list with *y*-coordinates resembles actual points on the *y*-axis where the same pitch classes exist. So the new list of pitch-class coordinates.

Pitch − class coordinates = [207, 171, 134, 116, 80, 44, 8, 189, 153, 98, 62, 25]

By using Table 14.2, we can classify pitch class irrespective of the octave.

After processing the *y*-coordinates with pitch class, we get a list of the pitch-class list with strings and crucial extraction.

Pitch−class list = ["C, " "D, " "E, " "F, " "G, " "A, " "B, " "C#, " "D#, " "F#, " "G#, " "A#"]

Table 14.2 Representation pitch classes according to y-coordinates.

S. No.	Y-coordinates	Pitch class
1	207	C
2	189	C#
3	171	D
4	153	D#
5	134	E
6	116	F
7	98	F#
8	80	G
9	62	G#
10	44	A
11	25	A#
12	8	B

14.5.6 Key Extraction

From the MIDI file created in MIDI conversion using Table 14.1, we get an octave list, and from image processing and Table 14.2, we get a note class list. By concatenating the zip of both lists, we get

Keys = ["C5," "D5," "E5," "F5," "G5," "A5," "B5," "C#5," "D#5," "F#5," "G#5," "A#5"]

Along with keys, we also get beat, onsets, offsets, time duration, MIDI pitch, and frequency using pretty MIDI in python. For the current sample, beats per minute (BPM) are 129.199. In Table 14.3, we get a detailed analysis of each note in the current sample music.

Table 14.3 Representation of obtained key and their features.

Keys	Onsets (s)	Offsets (s)	Time duration (s)	MIDI pitch	Frequency (Hz)
C5	1.311	1.799	0.488	72	523.251
D5	1.799	2.251	0.452	74	587.33
E5	2.252	2.693	0.441	76	659.255
F5	2.693	3.203	0.51	77	698.456
G5	3.203	3.703	0.499	79	783.991
A5	3.703	4.133	0.43	81	880
B5	4.133	4.597	0.464	83	987.767
C#5	4.632	5.061	0.429	73	554.365
D#5	5.061	5.537	0.476	75	622.254
F#5	5.537	5.908	0.372	78	739.989
G#5	5.909	6.397	0.488	80	830.69
A#5	6.397	6.965	0.568	82	932.328

From the beats, we can find the music's tempo, so by using Eq. (14.12), we get the tempo of the sample file is 2.15, which means that the beat is 2.15 times faster.

$$\text{Tempo} = \text{BPM}/60 \tag{14.12}$$

For extracting frequency from MIDI pitch, we use MIDI pitch of A4 is 69, and its frequency is 440 Hz. So considering m as MIDI pitch and F_m as the required frequency of MIDI pitch (m) by using Eq. (14.13), we get

$$F_m = (440) \times [2^{(m-69)/12}] \tag{14.13}$$

14.6 Experimental Results and Discussions

Experiments are conducted on a publicly available database for testing and evaluation. For testing, around 180 samples are used. The experiments were executed on Python 3.8.5 in Spyder IDE v4.1.5, Anaconda3, using Intel(R) Core(TM) i5-7200U CPU laptop with Windows 10, 64-bit operating system at 2.5 GHz with 4 GB RAM.

14.6.1 Benchmark Database

The proposed system is validated on a publicly available database called NSynth. This database consists of 305 979 music files split into 289 205 samples for training, 12 678 samples for validation, and 4096 for testing purposes. The dataset is classified into three types of sound production instruments acoustic, electric, and synthetic, then they are further classified into different musical instruments. Each note played in the sample possesses a definite pitch, onset, and offset. The current chapter focuses on piano samples and evaluated 200 test samples. The dataset was first documented in [24].

14.6.2 Evaluation Parameters

To extract pitch class, the chroma of the input signal undergoes image processing. In image processing, chroma is cropped up to the required dimensions and converted to a grayscale image for thresholding. After the first thresholding, a black rectangle is drawn on the threshold image's borders to form the second threshold image because, in the first threshold image, objects coincide with borders. From the second threshold, image contours are found and centroids are determined. The y-coordinates of the centroids correspond with pitch class on chroma. Soby using Table 14.2, we get a list of pitch class concerning the time after sorting. For octave extraction, the input file is converted to a MIDI file to extract the octave list using Table 14.1.

 (i) **Centroids:** To find centroids, the first step is to find contours using Green's theorem from Eq. (14.7). Later, centroid coordinates of contours are sorted concerning time and y-coordinates are separated. These y-coordinates are matched with pitch class to extract the list of pitch class.
Centroids sorted with respect to time (x-coordinates) = [(55, 207), (72, 171), (89, 134), (107, 116), (126, 80), (144, 44), (162, 8), (180, 189), (200, 153), (215, 98), (230, 62), (248, 25)]
 (ii) **Onsets:** Onset is the time at which a note started in an audio sample. Onsets are one of the basic features of music transcription. By using librosa and MIDI file onsets are extracted. The obtained onsets are classified in Table 14.3. Onsets are measured in seconds.
Onsets = [1.311, 1.799, 2.252, 2.693, 3.203, 3.703, 4.133, 4.632, 5.061, 5.537, 5.909, 6.397]
 (iii) **Offsets:** The offset is the time at which a note ends in an audio sample. After onsets, offsets are essential components in music transcription. The obtained offsets are classified in Table 14.3. Offsets are measured in seconds.
Offsets = [1.799, 2.251, 2.693, 3.203, 3.703, 4.133, 4.597, 5.061, 5.537, 5.908, 6.397, 6.965]

(iv) **Time Duration:** The time duration of a note is extracted using the onset and offset of the note. The arithmetic difference of offsets between onsets gives the time duration of the note. By using Eq. (14.14), we get time durations of each note and they are classified in Table 14.3. Time duration is measured in seconds.

$$\text{Time duration} = \text{offset} - \text{onset} \tag{14.14}$$

Time durations = [0.488, 0.452, 0.441, 0.51, 0.499, 0.43, 0.464, 0.429, 0.476, 0.372, 0.488, 0.568]

(v) **MIDI Pitch:** The note number for each note in a musical instrument can be used to extract the note's frequency component. They are classified in Table 14.3.

MIDI pitches = [72, 74, 76, 77, 79, 81, 83, 73, 75, 78, 80, 82]

(vi) **BPM:** BPM is used to measure the tempo of the music. Tempo is the speed of the beat in music. BPM for the experimented sample are 129.199.

14.6.3 Performance Evaluation

This system is evaluated on three types of piano available in the dataset (types: acoustic, electric, and synthetic). The accuracy of finding pitch class varies in each type for the proposed system.

Accuracy is measured based on y-coordinates. During image processing, the chroma has divided into 12 classes from C to B each class contains 18 y-coordinates. According to this proposed method y-coordinates given in Table 14.2 are the means of these 18 y-coordinates in each class and they represent actual coordinates. These actual coordinates are analyzed with obtained coordinates to get accuracy for pitch class as given in Eq. (14.15).

Actual coordinates sorted with respect pitch class = [207, 171, 134, 116, 80, 44, 8, 189, 153, 98, 62, 25]

Obtained coordinates sorted with respect pitch class = [207, 171, 134, 116, 80, 44, 8, 189, 153, 98, 62, 25]

$$\text{Accuracy} = \left\{ \frac{[(\text{Obtained coordinate of pitch } k - \text{Actual coordinate of pitch } k) + 18]}{18} \right\} \times 100 \tag{14.15}$$

In Eq. (14.15) pitch k can be C, C#, D, D#, E, F, F#, G, G#, A, A#, and B.

From Figure 14.8 and by using Eq. (14.15) we can observe using the proposed system, the accuracy of finding pitch on an acoustic piano is 100% for all pitch classes, electric piano is 94.44% accurate on G# and A classes, a synthetic piano is 94.44% accurate on G, G#, A, and B classes.

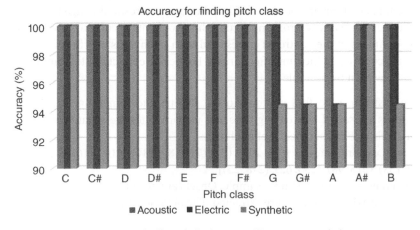

Figure 14.8 Accuracy for finding pitch class on different types of piano.

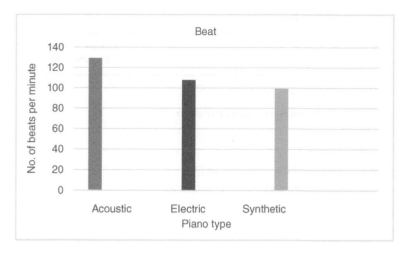

Figure 14.9   Beat of music for different types of piano.

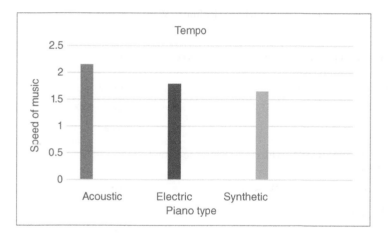

Figure 14.10   Tempo of music for different types of piano.

Beat also differs in all three types. In acoustic, BPM are 129.199, in electric BPM are 107.666 and in synthetic BPM are 99.384. Beats can further give out the music's tempo using Eq. (14.12), we get the tempo of acoustic is 2.15, for electric it is 1.79 and for synthetic it is 1.65. Both beat and tempo are represented in Figures 14.9 and 14.10, respectively.

The proposed method has 100% accuracy on acoustic piano, 94.44% accuracy on electric and synthetic pianos. Image processing to find the pitch class is the newest method proposed in this chapter. Table 14.4 shows the performance comparison of the proposed method with the existing methods. Though some of the existing techniques have 100% accuracy, the proposed technique has a less computational cost when compared to those techniques as this technique evaluates those MIDI pitches that exist in the music sample.

Table 14.4 Comparison of the proposed technique with existing techniques.

References	Techniques	Accuracy (%) for finding the pitch
[1]	Short-time Fourier transform (STFT), discrete cosine transform (DCT)	—
[2]	Short-time Fourier transform (STFT), multiresolution Fourier transform (MFT), convolutional neural network (CNN)	—
[3]	Non-negative matrix factorization (NMF)	—
[4]	Music language models (MLMs) and long short-time memory (LSTM)	96.7
[5]	Fast Fourier transform (FFT), STFT, wavelet transform (WT), Q-transform (QT)	—
[6]	FFT, continuous wavelet transform (CWT)	—
[7]	FFT, DFT	—
[8]	Goertzel algorithm	—
[9]	Multi-resolution fast Fourier transform (MRFFT)	—
[10]	Short-time Fourier transform (STFT), MRFFT, and Morlet wavelet from CWT	99
[11]	NMF	68
[12]	MV2H (multi-pitch detection, voice separation, metrical alignment, note value detection, and harmonic analysis)	100
[13]	Recurrent neural network (RNN) and bi-LSTM	85.37
[14]	Deep neural network (DNN)	57
[15]	Support vector machine (SVM)	72.5
[16]	Convolutional recurrent neural network (CRNN) with connectionist temporal classification (CTC) loss function	—
[17]	Probabilistic and deterministic methods using neural networks	100
[18]	*K*-nearest neighbor (KNN) algorithm	92.37
[19]	Auto-regressive (AR) RNN	—
[20]	FFT	90
[21]	Bi-LSTM	95
[22]	Short-time Fourier transform (STFT)	82
[23]	CNN	96
Proposed System	Short-time Fourier transform (STFT), CWT, and image processing	100

14.7 Conclusion

In this chapter, we propose a new approach to convert the monophonic notes in piano music into keys, i.e. the chromatic form of piano notes using STFT and image processing. This proposed system was evaluated on the NSynth database and achieved 100% accuracy on acoustic piano and 94.44% accuracy on both electric and synthetic pianos. This system is found to be robust for notes from C4 to A7 covering four octaves. In the future, this system can be improved to perform on the rest of the low-frequency octaves.

References

1 Burred, J.J., Haller, M., Jin, S. et al. (2008). Audio content analysis. In: *Semantic Multimedia and Ontologies*, 123–162. London: Springer https://doi.org/10.1007/978-1-84800-076-6_5.

2 Bello, J.P., Monti, G., and Sandler, M. (2000). Techniques for automatic music transcription. In: *International Symposium on Music*, 1–8. http://www.music.mcgill.ca/~ich/classes/mumt611_07/transcription/bello_paper.pdf.

3 Benetos, E., Dixon, S., Duan, Z., and Ewert, S. (2019). Automatic music transcription: an overview. *IEEE Signal Processing Magazine* 36 (1): 20–30. https://doi.org/10.1109/MSP.2018.2869928.

4 Ycart, A. and Benetos, E. (2020). Learning and evaluation methodologies for polyphonic music sequence prediction with LSTMs. *IEEE/ACM Transactions on Audio, Speech and Language Processing* 28: 1328–1341. https://doi.org/10.1109/TASLP.2020.2987130.

5 Müller, M., Ellis, D.P.W., Klapuri, A., and Richard, G. (2011). Signal processing for music analysis. *IEEE Journal of Selected Topics in Signal Processing* 5 (6): 1088–1110. https://doi.org/10.1109/JSTSP.2011.2112333.

6 Klapuri, A. and Virtanen, T. (2008). Automatic music transcription. In: *Handbook of Signal Processing in Acoustics*, 277–303. New York: Springer https://doi.org/10.1007/978-0-387-30441-0_20.

7 Alm, J.F. and Walker, J.S. (2002). Time-frequency analysis of musical instruments. *SIAM Review* 44 (3): 457–476. https://doi.org/10.1137/S00361445003822.

8 Jiang, J., Brewer, R., Jakubowski, R., and Tan, L. (2018). Development of a piano frequency detecting system using the Goertzel algorithm. In: *IEEE International Conference on Electro/Information Technology (EIT)*, vol. 2018, 346–349. https://doi.org/10.1109/EIT.2018.8500220.

9 Dittmar, C. and Abeßer, J. (2008). Automatic music transcription with user interaction. *Digital Media* 2: 567–568.

10 Kumar, N., Kumar, R., Murmu, G., and Sethy, P.K. (2021). Extraction of melody from polyphonic music using modified Morlet wavelet. *Microprocessors and Microsystems* 80: 103612. https://doi.org/10.1016/j.micpro.2020.103612.

11 Wats, N. and Patra, S. (2018). Automatic music transcription using accelerated multiplicative update for non-negative spectrogram factorization. In: *2017 International Conference on Intelligent Computing and Control (I2C2)*, vol. 2018, 1–5. https://doi.org/10.1109/I2C2.2017.8321812.

12 McLeod, A. and Steedman, M. (2018). Evaluating automatic polyphonic music transcription. In: *Proceedings of 19th International Society for Music Information Retrieval Conference ISMIR 2018*, 42–49.

13 Sadekar, A. and Mahajan, S.P. (2019). Polyphonic piano music transcription using Long Short-Term Memory. In: *2019 10th International Conference on Computing, Communication and Networking Technologies ICCCNT 2019*, 6–12. https://doi.org/10.1109/ICCCNT45670.2019.8944400.

14 Tanaka, K., Nakatsuka, T., Nishikimi, R. et al. (2020). Multi-instrument music transcription based on deep spherical clustering of spectrograms and pitchgrams. In: *ISMIR International Society for Music Information Retrieval Conference*, 327–334.

15 Pan, X., Song, B., Liu, C., and Zhang, H. (2019). Music detecting and recording system based on support vector machine. In: *Proceedings of 2019 International Conference on Communications, Information System and Computer Engineering (CISCE)*, 244–248. https://doi.org/10.1109/CISCE.2019.00063.

16 Román, M.A., Pertusa, A., and Calvo-Zaragoza, J. (2020). Data representations for audio-to-score monophonic music transcription. *Expert Systems with Applications* 162: 113769. https://doi.org/10.1016/j.eswa.2020.113769.

17 Silva, N., Fischione, C., and Turchet, L. (2020). Towards real-time detection of symbolic musical patterns: probabilistic vs. deterministic methods. *2020 27th Conference of Open Innovations Association (FRUCT)* 2020: 238–246. https://doi.org/10.23919/FRUCT49677.2020.9211010.

18 Pishdadian, F. and Nelson, J.K. (2013). On the transcription of monophonic melodies in an instance-based pitch classification scenario. IEEE Digital Signal Processing and Signal Processing Education Meeting, pp. 222–227.

19 Jeong, D. (2020). Real-time automatic piano music transcription system. Proceedings of Late Breaking / Demo of 21st International Society of Music Information Retrieval Conference, pp. 4–6.

20 Chis, L.G., Marcu, M., and Dragan, F. (2018). Software tool for audio signal analysis and automatic music transcription. In: *2018 IEEE 12th International Symposium on Applied Computational Intelligence and Informatics (SACI)*, 1–5. https://doi.org/10.1109/SACI.2018.8440966.

21 Grossman, A. and Grossman, J. (2020). Automatic music transcription: generating MIDI from audio. Reports of CS230 Deep Learning, Lecture at Stanford University. pp. 1–6.

22 Figueiredo, N. (2020). Detection of musically relevant regions in multiresolution time-frequency representations evaluated on piano recordings. Proceedings of the 17th Sound and Music Computing Conference, pp. 94–101. https://doi.org/10.5281/zenodo.3898701.

23 Sober-Mira, J., Calvo-Zaragoza, J., Rizo, D., and Inesta, J.M. (2018). Pen-based music document transcription. In: *2017 14th IAPR International Conference on Document Analysis and Recognition (ICDAR)*, vol. 2, 21–22. https://doi.org/10.1109/ICDAR.2017.258.

24 Engel, J., Resnick, C., Roberts, A. et al. (2009). Neural audio synthesis of musical notes with WaveNet autoencoders.

25 Tripathi, S.L., Prakash, K., Balas, V. et al. (2021). *Electronic Devices, Circuits, and Systems for Biomedical Applications: Challenges and Intelligent Approach*. Edited Book. Elsevier. ISBN: 9780323851725 https://doi.org/10.1016/C2020-0-02289-9.

26 Tripathi, S.L., Dhir, K., Ghai, D., and Patil, S. (2021). *Health Informatics and Technological Solutions for Coronavirus (COVID-19)*. Edited Book in Taylor and Francis Group,. CRC Press. ISBN: 9780367704179 https://doi.org/10.1201/9781003161066.

27 Singh, A.K., Pathak, J., and Tripathi, S.L. (2019). Frequency-based RO-PUF. In: *AI Techniques for Reliability Prediction for Electronic Components* (ed. C. Bhargava), 252–261. IGI global publishers. ISBN: 9781799814641 https://doi.org/10.4018/978-1-7998-1464-1.ch014.

28 Ghai, D., Gianey, H.K., Jain, A., and Uppal, R.S. (2020). Quantum and dual-tree complex wavelet transform-based image watermarking. *International Journal of Modern Physics B* 33: 1–10. https://doi.org/10.1142/S0217979220500095.

29 Ghai, D., Tiwari, S., and Das, N.N. (2021). Bottom-boosting differential evolution based digital image security analysis. *Journal of Information Security and Applications* 61: https://doi.org/10.1016/j.jisa.2021.102811.

30 Sameer, M. and Gupta, B. (2022). Time–frequency statistical features of delta band for detection of epileptic seizures. *Wireless Personal Communications* 122: 489–499. https://doi.org/10.1007/s11277-021-08909-y.

31 Beeraka, S.M., Kumar, A., Sameer, M. et al. (2022). Accuracy enhancement of epileptic seizure detection: a deep learning approach with hardware realization of STFT. *Circuits, Systems, and Signal Processing* 41: 461–484. https://doi.org/10.1007/s00034-021-01789-4.

Section III

Applications of Signal and Image Processing with Machine Learning and Deep Learning Techniques

15

Role of Machine Learning in Wrist Pulse Analysis

Sachin Kumar[1], Pooja[1], Sanjeev Kumar[2], and Karan Veer[1]

[1] *Department of Instrumentation and Control Engineering, Dr B R Ambedkar National Institute of Technology, Jalandhar, India*
[2] *Department of BioMedical Applications (BMA), Central Scientific Instruments Organisation (CSIO)-CSIR, Chandigarh, India*

15.1   Introduction

In ayurvedic and traditional Chinese medicinal (TCM) system, radial artery signal examination has been practiced for periods. It is a noninvasive, unique, suitable, and most effective disease identification methodology for analyzing the beat pattern of radial artery from the patient's wrist [1]. As per the western hones, the human body comprises a complex structure of organs that procured the particular imperative supplements and proteins through blood by the heart to arteries and then in veins. A wave (i.e. forward wave) is generated when the blood is pumped to arteries through the aortic circuit; a wave (i.e. forward wave) is generated. Arterial branches and peripheral organs reflect that reflected wave has been developed during the resting state [2]. The wrist pulse analysis observes these beats from the radial artery blood vessel (nadi) on the wrist to analyze the health status of each part of the body. The signal/pulse found on the radial arteries is a combination of reflected and forward waves (shown in Figure 15.1).

The ayurveda practitioners used this noninvasive system to measure these patterns of three doshas viz., vata, pitta, and kapha, called tri-dosha and similarly, in traditional Chines pulse analysis, the practitioners have been used as chun, guan, and chi that used to regulate the entire human body [3–6]. The location to obtain doshas through pulse waveform acquired from the wrist by putting the index, middle, and ring fingers. Pulse analysis needs considerable training, experience, and deep knowledge to recognize the different pulse patterns. As a result, combining ancient and modern practices is essential for better subjects' health status recognition [7]. Therefore, digital signal processing has been carried out for automated pulse signal evaluation.

Moreover, it reduces using high-appraised diagnostic evaluation and aids in the analysis of almost all disorders. Pulse preprocessing, characteristic extraction, and classifications are the three tactics that pulse analysis goes through [8]. Figure 15.1 shows different signal processing steps used for wrist pulse analysis.

In recent years, research into human pulse has gotten a lot of attention. Primarily, the pulse signal was processed in two ways, viz., time-domain features, frequency-domain features, and time–frequency domain analysis. Statistical metrics are taken from the temporal domain, such as kurtosis and skewness, percussion wave height (P), valley (V), dichotic wave (D), and tidal wave (T), which frequently characterizes the pulse signal [9–11]. However, noise can have a significant impact on time-domain features that cause a high random error. In addition, the time-domain analysis does not give all of the information confined in the pulse signal.

Recurrence space (i.e. frequency domain) investigation strategies, such as Fourier transformation viz., DFT, FFT, and envelop investigation, are essentially used to extract frequency spectrums of pulse signals [6, 12–16]. But it does not give the recurrence data at a few nearby times so a few valuable data may be lost when it is utilized to

Machine Learning Algorithms for Signal and Image Processing, First Edition.
Edited by Deepika Ghai, Suman Lata Tripathi, Sobhit Saxena, Manash Chanda, and Mamoun Alazab.

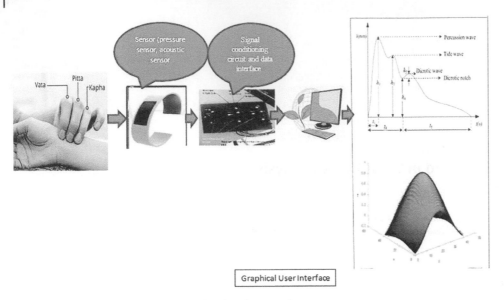

Figure 15.1 Block diagram of wrist pulse signal processing.

examine the beat–beat non-stationary signals. However, many authors used wavelet transformation for analyzing the non-stationary wrist pulse signal [17]. Some other pulse parameters, such as pulse wave velocity, rate variability, and augmented index retrieved from radial arteries, have been used to examine the physiological aspects of a wrist pulse signal. Further, machine learning (ML) and deep learning (DL) employed for data classification to determine what is expected and what is aberrant.

ML was becoming more prevalent and was being utilized in many researches in wrist pulse analysis for disease classification and detection. Classification models are critical for identifying and quantifying disorders with accuracy, precision, and sensitivity. Various feature extraction and selection algorithms are utilized in different ML applications for this goal. Yinghui Chen et al. [18] classified wrist pulse signal of healthy and unwell people with accuracy of 90% using the fuzzy C-means (F-CM) ML technique. support vector machine (SVM) has been the most commonly used ML technique for classifying disorders, including emotion, pain, organ inflammation, and hypertension [19–23]. Other ML and DL practices, such as convolutional neural network (CNN), *K*-nearest neighbor (KNN), Naïve-Bayes (NB), artificial neural network (ANN), linear discriminant analysis (LDA), random forest (RF), hidden Markov model (HMM), and so on, were also employed for wrist pulse signal based disease identification [18, 24–32]. ML delivers useful means for refining the accuracy, sensitivity, and other characteristics of medicinal system. ML has been mined several features, patterns, and forecasting from database to expect the appropriate findings [33, 34]. ML is also essential for diagnosing, managing, and limiting the disease's course. ML has been used for classification and forecast since the 1900s. The complete breakthrough of ML techniques is shown in Figure 15.2.

The main contributions of the study are threefold:

1. Present a detailed overview of ML and its performance analysis.
2. Present the detailed survey on ML-based wrist pulse signal classification.
3. Discuss which ML algorithms contribute the best to the categorization of radial pulse signals.

The chapter is organized into five sub-sections: Section 15.2 discusses ML and its many methodologies in the relevant literature. The performance analysis of ML algorithms is discussed in Section 15.3. Section 15.4 discusses ML in wrist pulse detection categorization. The conclusion is concluded in Section 15.5.

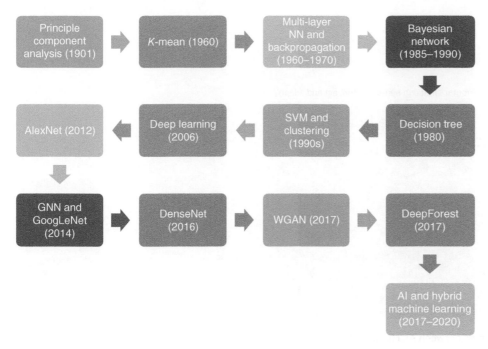

Figure 15.2 Timeline in the development of machine learning and deep learning.

15.2 Machine-Learning Techniques

ML is the subpart of artificial intelligence (AI) in which a model algorithm has been trained to recognize hidden characteristics/information of raw data. The relationships between AI, ML, and DL are depicted in Figure 15.3.

DL is a part of ML, and whole ML and DL are a sub-field of AI. ML is mainly divided into viz., unsupervised learning supervised and semi-supervised learning [35, 36].

Unsupervised learning does not require training datasets for respective output data and is categorized based on their inherent properties as the input data is not characterized. ML has been permitted to take care of many open-ended approaches to recognize hidden information or data that has been missed. In comparison, supervised learning required trained datasets for training the ML algorithm to classify new and similar datasets. Finally, semi-supervised learning hybrid supervised and unsupervised learning for classifying categorized datasets and uncategorized input datasets.

ML has been used for classification and forecast since the 1900s. The complete breakthrough of ML techniques is shown in Figure 15.2. DL and ML are a subpart of AI. The seven ML techniques are as follows:

- Regression
- Classification
- Clustering
- Dimensionality reduction
- Ensemble methods
- Neural networks and DL
- Reinforcement learning (RL)

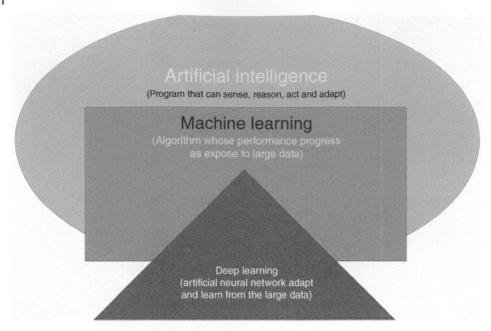

Figure 15.3 The relationships between AI, ML, and DL.

15.2.1 Regression

Regression is a supervised ML technique in which information has been detected based on previous datasets, for example, prediction of disease based on previous historical datasets for similar characteristics [37]. The most common and simple method of regression is linear regression, in which the equation of the line, i.e. $y = m \times x + b$ has been used to train the model datasets. Here, the model is trained to x and y data pairs by finding slope m and b for a line approximating the distance between line and points. The other regression techniques are random forest regression, decision tree, polynomial regression, and linear regression.

15.2.2 Classification

Classification is another class of the supervised ML method which forecasts class type. Logistic is the simplest classification method that estimates event occurrence probability based on one or many inputs [38]. For example, prediction of COVID-19 cases across the world. It helps to forecast whether patients will increase or decrease and the result can be positive or negative. The classification techniques can be divided into nonlinear classifiers such as SVM, random forest, and decision tree.

15.2.3 Clustering

It is an unsupervised ML that aims to group or cluster datasets with alike features without relying on the training output information. The ML algorithm, on the other hand, has defined the output. The most well-known clustering technique is K-mean, in which the value of K is user-defined (K can be chosen using many methods like as the elbow technique) and reflects the number of groups/clustered objects [39, 40].

The value of K is chosen at random from the datasets, and each data point is positioned near the randomly picked center. The value of the cluster center will change at each step, and once the numbers cease changing, the procedure will be terminated. On the other hand, if the center's value changes, the maximum number of repetitions in advance is established. Different clustering techniques include expectation–maximization clustering with

Gaussian mixture models (GMMs), density-based spatial cluster applications (DBSCAN), agglomerative hierarchical clustering, and mean shift clustering.

15.2.4 Dimensionality Reduction

Dimensional reduction is a process through which new features are created from the existing features by reducing the feature dimensions from the datasets. Many data with thousands of features are present in practice, so feature transformation is necessary for better classification. Feature transformation transforms the original space into lower dimension subspace by removing the less important information and make data manageable. Various methods are used for feature transformation, like independent component analysis (ICA), LDA, and principal component analysis (PCA) [41, 42]. PCA is the most popular data dimension technique that reduces the feature dimension space by calculating the new vector with maximizing linear deviation. PCA reduced dimension without losing data when the linear association is strong.

15.2.5 Ensemble Methods

The Ensemble technique is used to hybrid the several predictive methods to improve the quality of prediction that the individual could deliver [43]. For example, random forest, where the combination makes many decisions three random datasets, results in a higher prediction quality than a single DT. Another example is hybrid optimization ML techniques that combine the many heuristic techniques to improve detection accuracy. For example, hybrid particle swarm optimization and gravitational algorithm [44], evolutionary algorithms (EAs), and the gradient search technique [45]. The ensemble method can decrease the variance and biasness of a single ML.

15.2.6 Artificial Neural Networks and Deep Learning

ANN and DL are ML tools employed in many applications like image analysis, disease diagnosis, etc. It is unsupervised learning that learns from the unstructured and uncategorized dataset. ANN is inspired in the 1960s by the human brain and neuron connection in the human body connected in a network node, creating multiple layers [46]. Mathematically the AAB is expressed in term of nonlinear transformation equation i.e. $y = f(v)$ where $v =$ input [47].

DL was first introduced in the year 2006 as a deep belief network that used the greedy layer unsupervised learning [48]. However, DL has the same architecture as ANN with more hidden layers, increasing learning capacity, and feature extraction from complex input data. The basic architecture of DL includes four fundamental modules:

- Unsupervised retrained networks
- CNNs
- Recurrent neural networks (RNNs)
- Recursive neural networks

DL has improved the accuracy as compared to another algorithm for almost every type of dataset. The ANN is used for two primary purposes: regression and categorization. The classifier aims to input data organized into different classes using supervised/unsupervised learning. At the same time, regression is used to forecast the unknown parameter through trained datasets.

15.2.7 Reinforcement Learning

RL is a part of ML concerned with how agents learn in a specific environment by using a trial–error approach; RL helps to increase the same part with cumulative rewards. RL is a subpart of unsupervised ML techniques in which no historical information or data is needed, unlike other traditional ML methods. Still, it will learn from the data as applied [49]. However, the main drawback of RL is the long training period during complex problems. The most common RL technique is the Markov decision process [29] and the NB classifier [40].

15.3   Performance Analysis of ML Algorithms

Learning performance evaluation can be best determined with the highest classification accuracy (shown in Eq. (15.1)), precision (shown in Eq. (15.2)), recall (shown in Eq. (15.3)), *F*-score value (shown in Eq. (15.4)), area under the curve receiver operation characteristics (AROC) (Eq. (15.5)), etc.

$$\text{Accuracy} = \frac{\text{TP} + \text{TN}}{\text{TP} + \text{TN} + \text{FP} + \text{FN}} \tag{15.1}$$

$$\text{Precision} = \frac{\text{TP} + \text{TN}}{\text{TP} + \text{FP}} \tag{15.2}$$

$$\text{Recall} = \frac{\text{TP}}{\text{TP} + \text{FN}} \tag{15.3}$$

$$F - \text{score} = \frac{2 \times (\text{recall} + \text{precision})}{\text{recall} + \text{precision}} \tag{15.4}$$

$$\text{AROC} = \int_0^1 \left(\frac{\text{TP}}{\text{TP} + \text{FP}} \right) \times D \left(\frac{\text{TP}}{\text{Precision}} \right) \tag{15.5}$$

where TP = true positive, FP = false positive, TN = true negative, and FN = false negative.

The *F*-score defines the average weight precision and recall of datasets [50]. Finally, AROC is a curve that plots between true positive value and false-positive value [51]. For example, it will differentiate between healthy and unhealthy subjects. If the AROC curve is closer to the one, it shows better and more accurate classifier performance.

15.4   Role of the Machine and Deep Learning in Wrist Pulse Signal Analysis (WPA)

The major challenge in biomedical signal analysis is predicting illnesses noninvasively and with good features classification accuracy. Artificially intelligent computer procedures involved good learning behavior with the evolution of technology, enhance algorithm enactment with practice. ML and DL, often known as AI, have enabled autonomous disease diagnosis and detection early. ML approaches are usually categorized according to how they learn. There are two kinds of learning techniques: supervised learning and unsupervised learning.

15.4.1   Supervised Machine Learning in WPA

Previous datasets/features to predict future events were employed in supervised learning techniques. First, the algorithms are trained with known features/parameters, resulting in conditional functions that can be used to forecast future output. The oriented algorithm set targets for all inputs, compared the production to the actual and anticipated parameters, and calculated the error [52, 53].

SVM, KNN, and random forest (RF) are most common well-known supervised learning methods employed to categorize wrist pulse pattern. Dong Zhang et al. [19] have projected a Doppler ultrasonic-based radial pulse analysis expedient to diagnose gastric and cholecystitis subjects. The authors have used wavelet transformation for withdrawal of features and SVM as a forecasting tool for classification between diseased and healthy subjects.

Zhixing Jiang et al. [54] have used the SVM to classify diabetic, nephropathy subjects. The classifier is trained using KL-MGDCCA (Karhunen–Loève multiple generalized discriminative canonical correlation analysis) investigation attached feature vectors and attained the *F*-score of 79.02% and 75.25%.

Rangaprakash and Dutt [28] have investigated the effect of exercise and food habits on the radial artery signal. Signals were attained from the wrist artery of subjects, and further, the spatial feature has been calculated. SVM is further used for the classification of these groups. The classification accuracy of 99.71% and 99.94% has been achieved for exercise and lunch cases, respectively. Xiaodong Ding et al. [55] have developed the SVM-based ML technique to detect healthy and unhealthy subjects with a classification of 96% accuracy. Zhichao Zhang et al. [56] have proposed wrist pulse signal-based cancer recognition device in which 12 feature has been acquired from signals of three locations on the wrist.

Further, eight classifiers, viz., SVM, KNN, etc., were applied to categorize cancer patients from healthy one. Similarly, SVM-based writs pulse analysis has been used in many diseases diagnoses, such as anxiety [57], appendicitis ulcer, nephropathy, cardiac syndrome, hypertension and pancreatitis. [22, 30, 58, 59]. Other supervised tools like KNN, RF were also used to classify the patients based on wrist pulse examination. Zhang et al. [60] proposed IoT and KNN ML-based radial artery analysis. Signal was acquired from the patient's wrist, and then Amazon Web Services (AWS) based IoT was utilized to send data to remote places for further research. In other work, KNN has used edit distance with real penalty (ERP) based pulse waveform classification. Similarly, RF classifiers were used to investigate the pulse diagnosis data acquired from the pulse diagnostic instrument and attained a classification accuracy of 60%. In another study, RF was utilized to detect healthy and hypertension disease subjects and achieved a classification accuracy of 85% [61].

15.4.2 Unsupervised Machine Learning and Reinforcement Machine Learning in WPA

Unsupervised learning is employed when the model does not need to be trained or supervised using previous data. Still, the model can anticipate similar information and patterns that were not observed before [62, 63]. Clustering viz., fuzzy logic, density-based, and *K*-means are examples of unsupervised learning. In 2009, Chunming Xia et al. [64] proposed the *K*-mean cluster ML methodology to classify pulse features for the detection of cardiac disease. R-P and Q-R-A features have been extracted for further disease classification and diagnosis.

Similarly, in another research paper, wrist pulse signal was classified by the F-CM and Gaussian tools for disease diagnosis. This approach was tested on data of 100 healthy and 88 disease persons [18]. For reinforcement ML technique, Markov model, NB was used for wrist pulse classification. Ruqiang Yan et al. [29] demonstrated the HMM and wavelet packet transform (WPT)-based approach for identifying hypertension. The optimal feature viz., energy features and Fisher linear discriminant were used as input to the HMM classifier. Similarly, other reinforced ML, i.e. NB were utilized to identify hypertension subject in TCM-based pulse analysis [65].

15.4.3 Deep Learning in WPA

DL is also a subsection of ML, with more hidden layers that automatically extract more hidden characteristics from trained datasets [66–68]. Deep neural networks (DNNs), backpropagation neural networks (BNNs), RNNs, CNNs, and other DL approaches are popular. Cheng and coworkers [69] used the ANN to distinguish between different depths of the radial artery. Similarly, Xiaojuan Hu et al. [70] utilized Hilbert transform and deep convolution network classifier for the radial signal analysis. The data has been trained by the DCNN to increase the sampling size for excavating extracted features by adding noise. Moreover, CNN was used to detect the pregnancy pulse. The CNN method has acquired a very high classification accuracy of 97.08% [71]. According to the literature, the ML approaches and types mentioned above successfully forecast biological signal data acquired from various sources. ML predicts output values from learned features (datasets) and builds a model automatically or without human intervention.

Table 15.1 Machine-learning techniques used for analysis of radial pulse signal analysis.

ML tool	Motivation	Type of learning	References	ML tool	Motivation	Type of learning	References
SVM	• Reduces kernel function based overfitting problems. • Appropriate kernels were used to solve complex problem.	Supervised learning	[21, 22, 30, 54–59, 72–74]	ANN, CNN, RNN,	• Weight sharing feature. • Unable to solve overfitting issue. • High computation cost.	Deep learning	[58, 69–71, 75–77]
				NB	• Easiest way forecasting. • Low accuracy	Reinforcement machine learning	[31, 65, 78]
RF	• Reduce the overfitting problem. • Accurate findings with large time.		[30, 57, 58, 61, 79, 80]	HMM	• High flexibility fitting. • Required large memory and time.		[29, 81]
KNN	• The value "K" must be known. • Give good estimated finding with increase trained samples. • Large computation cost.		[55, 56, 58, 60, 80, 82]	K-means minimum distance classifier	• Low computational cost. • Ease to use. • Fast response time. • Less accurate during global clusters.	Unsupervised learning	[30, 64]
				Fuzzy C-means			[18]

ML approaches are now being utilized to categorize radial pulse signals using various feature extraction techniques to diagnose the fundamental cause of ill health. We've seen numerous publications confirming the effectiveness of this ML technology in diagnosing multiple health issues, including hypertension, diabetes, cancer, fatty liver, obesity, and other conditions. Other cardiovascular diseases depict the profound potential of this technique in the modern medicinal system. The ML techniques used for analyzing wrist pulse signals are shown in Table 15.1.

15.5 Discussion and Conclusion

This section summarizes the findings from wrist pulse investigations and provides an overview. The string search method was utilized, and 500 readings were obtained from the entire internet database. The articles were curtained using the tollgate tactic, which consists of five phases as follows:

(a) Use a string search in the chosen database to find relevant studies.
(b) Paper exclusion on the basis of title.
(c) After reading the abstract of the article, the third step is to eliminate the candidates.
(d) Introduction and conclusion are excluded from consideration.
(e) Exclusions after evaluation of the full paper and assessing the quality of the analysis are used for wrist pulse analysis.

An entire 400 fundamentals were mined after studying the analysis provided in the literature from 2004 to 2021 and specific readings detailing the wrist pulse history. The 360 items were found identical. About 180 papers were discarded because of their titles, whereas 125 were discarded on the basis of abstracts. The 50 papers were discarded after studying the outline and conclusion portion of the paper. About 20 studies were utilized after interpreting the full article and exploit the quality valuation requisite for the recommended analysis.

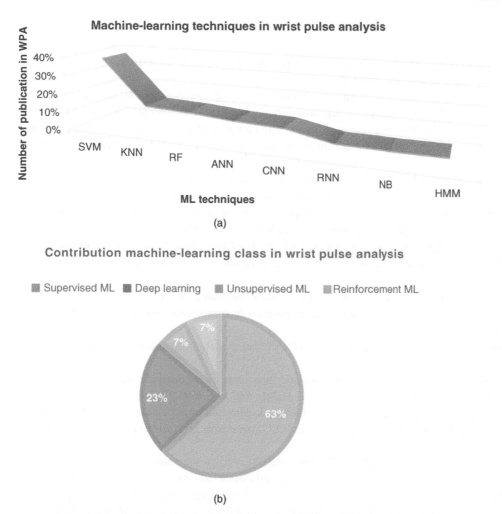

Figure 15.4 (a) Distribution of various machine-learning techniques used for WPA, (b) graphical presentation of percent type's machine-learning and deep-learning techniques in WPA.

According to the analysis of these studies, eight ML models have been implemented for recognizing healthy and ill people. Three ML techniques were utilized for wrist pulse analysis, with supervised ML being the most frequently used (see Figure 15.4b). As supervised ML models, KNN, SVM, and ANN ML models were used, with SVM being the most popular (see Figure 15.4a).

Wrist pulse diagnoses are turning into true AI diagnosis systems by applying ML to wrist pulse data, offering richer recommendations and visions for following conclusions and activities with the decisive goal of early disease detection. Therefore, ML models and DL will become even more common in the upcoming studies, allowing for low-cost, fast, and applicable solutions. All of the ML and DL techniques are now focused on individual approaches and results and are insufficiently linked to the decision-making process. This combination of automated data recording, feature extraction, and ML classification will provide the early biomarker of disease diagnosis using wrist pulse (radial artery) analysis.

References

1 Wang, D., Zhang, D., and Lu, G. (2016). A robust signal preprocessing framework for wrist pulse analysis. *Biomedical Signal Processing and Control* 23: 62–75.

2 Wang, P., Zuo, W., and Zhang, D. (2014). A compound pressure signal acquisition system for multichannel wrist pulse signal analysis. *IEEE Transactions on Instrumentation and Measurement* 63 (6): 1556–1565.

3 Fu, S.E. and Lai, S.P. (1989). A system for pulse measurement and analysis of Chinese medicine. In: *Images of the Twenty-First Century. Proceedings of the Annual International Engineering in Medicine and Biology Society, Seattle, WA, USA*, 1695–1696.

4 He, D., Wang, L., Fan, X. et al. (2017). A new mathematical model of wrist pulse waveforms characterizes patients with cardiovascular disease–a pilot study. *Medical Engineering & Physics* 48: 142–149.

5 Roopini, N., Shivaram, J.M., and Shridhar, D. (2015). Design & development of a system for Nadi Pariksha. *International Journal of Engineering Research Technology* 4 (2278–0181): 465–470.

6 Kalange, A.E., Mahale, B.P., Aghav, S.T., and Gangal, S.A. (2012). Nadi Parikshan Yantra and analysis of radial pulse. In: *2012 1st International Symposium on Physics and Technology of Sensors (ISPTS-1)*, 165–168. Pune, India: IEEE.

7 Dattatray, L.V. (2007). *Secrets of the Pulse*, 206. Motilal Banarsidass Publishing House.

8 Liu, Z., Yan, J.-Q., Tang, Q.-L., and Li, Q.-L. (2006). Recent progress in computerization of TCM. *Journal of Communication and Computer* 3 (7): 78–81.

9 Suguna, G.C. and Veerabhadrappa, S.T. (2019). A review of wrist pulse analysis. *Biomedical Research* 30 (4): 538–545.

10 Zhang, D.-Y., Zuo, W.-M., Zhang, D. et al. (2010). Wrist blood flow signal-based computerized pulse diagnosis using spatial and spectrum features. *Journal of Biomedical Science and Engineering* 3 (04): 361.

11 Xu, L., Meng, M.Q.-H., Liu, R., and Wang, K. (2008). Robust peak detection of pulse waveform using height ratio. In: *2008 30th Annual International Conference of the IEEE Engineering in Medicine and Biology Society*, 3856–3859. Vancouver, BC, Canada: IEEE.

12 Jianjun, Y., Yiqin, W., Fufeng, L. et al. (2008). Analysis and classification of wrist pulse using sample entropy. In: *2008 IEEE International Symposium on IT in Medicine and Education*, 609–612. Xiamen, China: IEEE.

13 Hu, C.-S., Chung, Y.-F., Yeh, C.-C., and Luo, C.-H. (2012). Temporal and spatial properties of arterial pulsation measurement using pressure sensor array. *Evidence-Based Complementary and Alternative Medicine* 2012.

14 Thakker, B. and Vyas, A.L. (2011). Suppressed dicrotic notch pulse classifier design. *International Journal of Machine Learning and Computing* 1 (2): 148.

15 Liu, L., Li, N., Zuo, W. et al. (2012). Multiscale sample entropy analysis of wrist pulse blood flow signal for disease diagnosis. In: *International Conference on Intelligent Science and Intelligent Data Engineering*, 475–482. Berlin, Heidelberg: Springer.

16 Chang, H., Chen, J., and Liu, Y. (2018). Micro-piezoelectric pulse diagnoser and frequency domain analysis of human pulse signals. *Journal of Traditional Chinese Medical Sciences* 5 (1): 35–42.

17 Xu, L., Zhang, D., Wang, K. et al. (2007). Baseline wander correction in pulse waveforms using wavelet-based cascaded adaptive filter. *Computers in Biology and Medicine* 37 (5): 716–731.

18 Chen, Y., Zhang, L., Zhang, D., and Zhang, D. (2009). Wrist pulse signal diagnosis using modified Gaussian models and fuzzy C-means classification. *Medical Engineering & Physics* 31 (10): 1283–1289.

19 Zhang, D., Zhang, L., Zhang, D., and Zheng, Y. (2008). Wavelet-based analysis of Doppler ultrasonic wrist-pulse signals. In: *2008 International Conference on BioMedical Engineering and Informatics*, vol. 2, 539–543. Sanya, China: IEEE.

20 Liu, L., Zuo, W., Zhang, D. et al. (2012). Combination of heterogeneous features for wrist pulse blood flow signal diagnosis via multiple kernel learning. *IEEE Transactions on Information Technology in Biomedicine* 16 (4): 598–606.

21 Wang, D., Zhang, D., and Lu, G. (2015). A novel multichannel wrist pulse system with different sensor arrays. *IEEE Transactions on Instrumentation and Measurement* 64 (7): 2020–2034.

22 Chui, K.T. and Lytras, M.D. (2019). A novel MOGA-SVM multinomial classification for organ inflammation detection. *Applied Sciences* 9 (11): 2284.

23 Garg, N., Kumar, A., and Ryan, H.S. (2021). Analysis of wrist pulse signal: emotions and physical pain. *IRBM*.

24 Aihua, Z. and Fengxia, Y. (2005). Study on recognition of sub-health from pulse signal. In: *2005 International Conference on Neural Networks and Brain*, vol. 3, 1516–1518. Beijing, China: IEEE.

25 Sun, Y., Shen, B., Chen, Y., and Xu, Y. (2010). Computerized wrist pulse signal diagnosis using kpca. In: *International Conference on Medical Biometrics*, 334–343. Berlin, Heidelberg: Springer.

26 Thakker, B., Vyas, A.L., Farooq, O. et al. (2011). Wrist pulse signal classification for health diagnosis. In: *2011 4th International Conference on Biomedical Engineering and Informatics (BMEI)*, vol. 4, 1799–1805. Shanghai, China: IEEE.

27 Huang, P.-Y., Lin, W.-C., Chiu, B.Y.-C. et al. (2013). Regression analysis of radial artery pulse palpation as a potential tool for traditional Chinese medicine training education. *Complementary Therapies in Medicine* 21 (6): 649–659.

28 Rangaprakash, D. and Dutt, D.N. (2015). Study of wrist pulse signals using time-domain spatial features. *Computers and Electrical Engineering* 45: 100–107.

29 Yan, R., Zhou, M., Sun, W., and Meng, J. (2017). Analyzing wrist pulse signals measured with polyvinylidene fluoride film for hypertension identification. *Sensors and Materials* 29 (9): 1339–1351.

30 Luo, Z., Cui, J., Hu, X.-j. et al. (2018). A study of machine-learning classifiers for hypertension based on radial pulse wave. *BioMed Research International* 2018.

31 Garg, N. and Kaur, G. (2021). Exploring wrist pulse signals using empirical mode decomposition: emotions. *IOP Conference Series: Materials Science and Engineering* 1033 (1): 12008.

32 Wang, B., Luo, J., Xiang, J., and Yang, Y. (2001). Power spectral analysis of human pulse and study of traditional Chinese medicine pulse diagnosis mechanism. *Journal of Northwest University(Natural Science Edition)* 31 (1): 22–25.

33 Morris, R., Martini, D.N., Madhyastha, T. et al. (2019). Overview of the cholinergic contribution to gait, balance and falls in Parkinson's disease. *Parkinsonism & Related Disorders* 63: 20–30.

34 Belić, M., Bobić, V., Badža, M. et al. (2019). Artificial intelligence for assisting diagnostics and assessment of Parkinson's disease—a review. *Clinical Neurology and Neurosurgery* 184: 105442.

35 Singh, A., Thakur, N., and Sharma, A. (2016). A review of supervised machine learning algorithms. In: *2016 3rd International Conference on Computing for Sustainable Global Development (INDIACom)*, 1310–1315. New Delhi, India: IEEE.

36 Zhu, X. and Goldberg, A.B. (2009). Introduction to semi-supervised learning. *Synthesis Lectures on Artificial Intelligence and Machine Learning* 3 (1): 1–130.

37 Aalen, O.O. (1989). A linear regression model for the analysis of lifetimes. *Statistics in Medicine* 8 (8): 907–925.

38 Wright, R.E. (1995). Logistic regression. In: *Reading and Understanding Multivariate Statistics* (ed. L.G. Grimm and P.R. Yarnold), 217–244. American Psychological Association.

39 Sridevi, S., Parthasarathy, S., and Rajaram, S. (2018). An effective prediction system for time series data using pattern matching algorithms. *International Journal of Industrial Engineering* 25 (2): 123–136.

40 Varuna, S. and Natesan, P. (2015). An integration of k-means clustering and naïve Bayes classifier for intrusion detection. In: *2015 3rd International Conference on Signal Processing, Communication and Networking (ICSCN)*, 1–5. Chennai, India: IEEE.

41 Karamizadeh, S., Abdullah, S.M., Manaf, A.A. et al. (2013). An overview of principal component analysis. *Journal of Signal and Information Processing* 4 (3B): 173.

42 Kuzilek, J., Kremen, V., Soucek, F., and Lhotska, L. (2014). Independent component analysis and decision trees for ECG Holter recording de-noising. *PLoS One* 9 (6).

43 Zhou, Z.-H. (2021). Ensemble learning. In: *Machine Learning*, 181–210. Singapore: Springer.

44 Mirjalili, S. and Hashim, S.Z.M. (2010). A new hybrid PSOGSA algorithm for function optimization. In: *2010 International Conference on Computer and Information Application*, 374–377. Tianjin, China: IEEE.

45 Woo, H.W., Kwon, H.H., and Tahk, M.-J. (2004). A hybrid method of evolutionary algorithms and gradient search. In: *2nd International Conference on Autonomous Robots and Agents, Palmerston North, New Zealand*, 115–123.

46 Hubel, D.H. and Wiesel, T.N. (2020). 8. Receptive fields of single neurones in the cat's striate cortex. In: *Brain Physiology and Psychology*, 129–150. University of California Press.

47 Hubel, D.H. and Wiesel, T.N. (1968). Receptive fields and functional architecture of monkey striate cortex. *The Journal of Physiology* 195 (1): 215–243.

48 Hinton, G.E., Osindero, S., and The, Y.-W. (2006). A fast learning algorithm for deep belief nets. *Neural Computation* 18 (7): 1527–1554.

49 Szepesvári, C. (2010). Algorithms for reinforcement learning. *Synthesis Lectures on Artificial Intelligence and Machine Learning* 4 (1): 1–103.

50 Hasan, M., Islam, M.M., Zarif, M.I.I., and Hashem, M.M.A. (2019). Attack and anomaly detection in IoT sensors in IoT sites using machine learning approaches. *Internet of Things* 7: 100059.

51 Othman, S.M., Ba-Alwi, F.M., Alsohybe, N.T., and Al-Hashida, A.Y. (2018). Intrusion detection model using machine learning algorithm on big data environment. *Journal of Big Data* 5 (1): 1–12.

52 Hendricks, D., Mazeika, M., Kadavath, S. et al. (2019). Using self-supervised learning can improve model robustness and uncertainty. In: *NeurIPS*. arXiv preprint arXiv1906.12340.

53 Zhou, Z.-H. (2018). A brief introduction to weakly supervised learning. *National Science Review* 5 (1): 44–53.

54 Jiang, Z., Guo, C., Zang, J. et al. (2020). Features fusion of multichannel wrist pulse signal based on KL-MGDCCA and decision level combination. *Biomedical Signal Processing and Control* 57: 101751.

55 Ding, X., Cheng, F., Morris, R. et al. (2020). Machine learning-based signal quality evaluation of single-period radial artery pulse waves: model development and validation. *JMIR Medical Informatics* 8 (6): e18134.

56 Zhang, Z., Umek, A., and Kos, A. (2017). Computerized radial artery pulse signal classification for lung cancer detection. *Facta Universitatis. Series: Mechanical Engineering* 15 (3): 535–543.

57 Jerath, H., Bisht, A., and Kour, H. (2020). Classification of boredom and anxiety in wrist pulse signals using statistical features. *Research Journal of Pharmacy and Technology* 13 (5): 2199–2206.

58 Jiang, Z., Lu, G., and Zhang, D. (2020). Sparse decomposition of pressure pulse wave signal based on time-frequency analysis. In: *2020 5th International Conference on Intelligent Informatics and Biomedical Sciences (ICIIBMS)*, 129–135. Okinawa, Japan: IEEE.

59 Chen, Y., Zhang, L., Zhang, D., and Zhang, D. (2011). Computerized wrist pulse signal diagnosis using modified auto-regressive models. *Journal of Medical Systems* 35 (3): 321–328.

60 Zhang, D., Zuo, W., Li, Y., and Li, N. (2010). Pulse waveform classification using ERP-based difference-weighted KNN classifier. In: *International Conference on Medical Biometrics*, 191–200. Berlin, Heidelberg: Springer.

61 Tago, K., Ogihara, A., Nishimura, S., and Jin, Q. (2018). Analysis of pulse diagnosis data from a TCM doctor and a device by random forest. In: *JSAI International Symposium on Artificial Intelligence*, 74–80. Cham: Springer.

62 Dalca, A.V., Guttag, J., and Sabuncu, M.R. (2018). Anatomical priors in convolutional networks for unsupervised biomedical segmentation. In: *Proceedings of the IEEE Conference on Computer Vision and Pattern Recognition*, 9290–9299.

63 Dou, Q., Ouyang, C., Chen, C. et al. (2018). Unsupervised cross-modality domain adaptation of convnets for biomedical image segmentation with adversarial loss. In: *International Joint Conference on Artificial Intelligence (IJCAI)*. arXiv preprint arXiv1804.10916.

64 Xia, C., Liu, R., Wang, Y. et al. (2009). Wrist pulse analysis based on RP and QRA. In: *2009 2nd International Conference on Biomedical Engineering and Informatics*, 1–5.

65 Lee, B.J., Jeon, Y.J., Ku, B. et al. (2015). Association of hypertension with physical factors of wrist pulse waves using a computational approach: a pilot study. *BMC Complementary and Alternative Medicine* 15 (1): 222.

66 Haque, I.R.I. and Neubert, J. (2020). Deep learning approaches to biomedical image segmentation. *Informatics in Medicine Unlocked* 18: 100297.

67 Baldi, P. (2018). Deep learning in biomedical data science. *Annual Review of Biomedical Data Science* 1: 181–205.

68 Tobore, I., Li, J., Yuhang, L. et al. (2019). Deep learning intervention for health care challenges: some biomedical domain considerations. *JMIR mHealth and uHealth* 7 (8): e11966.

69 Chung, C.-Y., Cheng, Y.-W., and Luo, C.-H. (2015). Neural network study for standardizing pulse-taking depth by the width of artery. *Computers in Biology and Medicine* 57: 26–31.

70 Hu, X., Zhu, H., Xu, J. et al. (2014). Wrist pulse signals analysis based on deep convolutional neural networks. In: *2014 IEEE Conference on Computational Intelligence in Bioinformatics and Computational Biology*, 1–7.

71 Li, N., Jiao, Y., Mao, X. et al. (2020). Analysis of pregnancy pulse discrimination based on wrist pulse by 1D CNN. In: *International Conference on Bio-Inspired Computing: Theories and Applications*, 336–346.

72 Zhang, Y., Wang, Y., Wang, W., and Yu, J. (2002). Wavelet feature extraction and classification of Doppler ultrasound blood flow signals. *Journal of Biomedical Engineering* 19 (2): 244–246.

73 Rangaprakash, D. and Dutt, D.N. (2014). Study of wrist pulse signals using a bi-modal Gaussian model. In: *2014 International Conference on Advances in Computing, Communications and Informatics (ICACCI)*, 2422–2425.

74 Zhang, D., Zuo, W., and Wang, P. (2018). Generalized feature extraction for wrist pulse analysis: from 1-D time series to 2-D matrix. In: *Computational Pulse Signal Analysis*, 169–189. Singapore: Springer.

75 Spulak, N., Foeldi, S., Koller, M. et al. (2016). Wrist pulse detection and analysis using three in-line sensors and linear actuators. In: *CNNA 2016; 15th International Workshop on Cellular Nanoscale Networks and Their Applications*, 1–2.

76 Chen, Z., Huang, A., and Qiang, X. (2020). Improved neural networks based on genetic algorithm for pulse recognition. *Computational Biology and Chemistry* 88: 107315.

77 Joshi, S. and Bajaj, P. (2021). Design & Development of portable vata, pitta & kapha [VPK] pulse detector to find Prakriti of an individual using artificial neural network. In: *2021 6th International Conference for Convergence in Technology (I2CT)*, 1–6. Maharashtra, India: IEEE.

78 Prochazka, A., Vyšata, O., Vališ, M. et al. (2015). Bayesian classification and analysis of gait disorders using image and depth sensors of Microsoft Kinect. *Digital Signal Processing* 47: 169–177. https://doi.org/10.1016/j.dsp.2015.05.011.

79 Pogadadanda, H., Shankar, U.S., and Jansi, K.R. (2021). Disease diagnosis using ayurvedic pulse and treatment recommendation engine. In: *2021 7th International Conference on Advanced Computing and Communication Systems (ICACCS)*, vol. 1, 1254–1258. Coimbatore, India: IEEE.

80 Zhang, Q., Zhou, J., and Zhang, B. (2020). Graph-based multichannel feature fusion for wrist pulse diagnosis. *IEEE Journal of Biomedical and Health Informatics* 25: 3732–3743.

81 Meng, J., Qian, Y., and Yan, R. (2013). Pulse signal analysis based on wavelet packet transform and hidden Markov model estimation. In: *2013 IEEE International Instrumentation and Measurement Technology Conference (I2MTC)*, 671–675. Minneapolis, MN, USA: IEEE.

82 Zhang, D., Zuo, W., Zhang, D. et al. (2010). Classification of pulse waveforms using edit distance with real penalty. *EURASIP Journal on Advances in Signal Processing* 2010: 1–8.

16

An Explainable Convolutional Neural Network-Based Method for Skin-Lesion Classification from Dermoscopic Images

Biswarup Ganguly[1], Debangshu Dey[2], and Sugata Munshi[2]

[1]*Department of Electrical Engineering, Meghnad Saha Institute of Technology, Maulana Abul Kalam Azad University of Technology, West Bengal, India*
[2]*Department of Electrical Engineering, Jadavpur University, Kolkata, West Bengal, India*

16.1 Introduction

16.1.1 Background, Motivation, and Literature

The most common class of cancer is skin cancer, which is increasing rapidly throughout the world mostly in the United States. According to the report in [1], five million cases are being diagnosed in the United States and more than 9000 deaths occurred in a year [2]. As skin cancer causes a global-health threat, early detection of skin cancer becomes a prime concern. Skin diseases, such as malignant, consist of melanocytic and benign consist of epidermal lesions. Melanocytic lesions consider melanoma (MEL) and nevus (NEV), whereas epidermal lesions comprise of basal cell carcinoma (BCC) and seborrheic keratoses (SEBK). Figure 16.1 displays the four types of dermoscopic images, including melanocytic and epidermal skin lesions. According to expert dermatologists, pigmentations with increased vascularity and irregular margin are found in melanocytic lesions. On the other hand, branching vessels and blue–gray areas are observed in epidermal lesions, especially in BCC, and comedo-type openings and/or brain-like appearance are found in the case of SEBK [7]. Some of the dermoscopic images with the annotations on the skin lesions have been provided by expert dermatologists as shown in Figure 16.2a–f.

Malignant melanoma is found to be the dangerous skin lesion abnormality responsible for skin cancer deaths of 75% and the premature diagnosis of melanoma increases the rate of survival up to 99% [8]. Though melanoma and NEV fall under the category of melanocytic lesions, very often either MEL or NEV or sometimes both of those have similar appearances that are displayed in Figure 16.2a–c. Hence, dermatologists employ dermoscopy, a gold standard approach, for the skin lesion segmentation, identification, and classification.

Dermoscopy, a non-invasive image acquiring method, is a tool to procure magnified and illuminated skin-lesion images directly from the skin by enhancing the visual effects of deeper skin levels with reduced surface reflectance and providing richer details of the skin lesions [9]. To enhance the skin-lesion diagnosis performance, dermoscopy has been widely used by dermatologists, as it produces accuracy higher than that evaluated by naked eyes leading toward error-prone and time-consuming results [10].

Effective research on skin-lesion diagnosis has been carried out over the years by employing hand-crafted features like color [11], texture, both color and texture [12], and shape. Researchers also employed various feature selection algorithms for proper melanoma classification [13]. Chatterjee et al. [7] proposed a cross-correlation-based feature extraction method in the space and frequency domain to classify skin lesions using ensemble classifiers. Xie et al. [14] extracted texture, color, and lesion border features and fed those to a network combining a fuzzy neural network and back propagation neural network for the classification of skin lesions.

Machine Learning Algorithms for Signal and Image Processing, First Edition.
Edited by Deepika Ghai, Suman Lata Tripathi, Sobhit Saxena, Manash Chanda, and Mamoun Alazab.
© 2023 The Institute of Electrical and Electronics Engineers, Inc. Published 2023 by John Wiley & Sons, Inc.

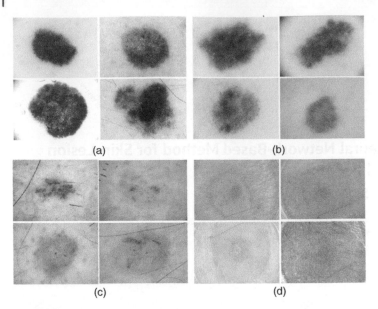

Figure 16.1 Four types of dermoscopic images taken from [3–6]: (a) MEL, (b) NEV, (c) BCC, and (d) SEBK.

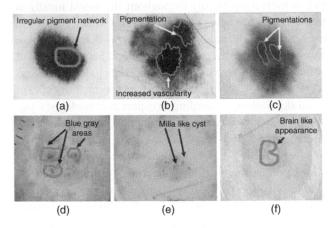

Figure 16.2 Expert annotations on dermoscopic images: (a), (b) MEL, (c) NEV, (d) BCC, and (e), (f) SEBK.

Recently, deep learning (DL) frameworks based on convolutional neural network (CNN) and recurrent neural network (RNN) have gained a lot of attention in medical signal and image analysis, such as detection, segmentation, and classification [15–19]. Yu et al. [20] have proposed a 16-layer deep residual CNN for melanoma recognition, where a convolutional residual network (ResNet) is used for melanoma segmentation and a deep ResNet has been used for classification. Zhang et al. [21] proposed a residual-CNN with an attention learning mechanism to classify three types of skin lesions. Sarkar et al. [22] introduced a deep ResNet employing a channel-wise separable convolution operation to classify melanoma from dermoscopic images. Gessert et al. [23] have addressed an attention model, where three pre-trained CNN models are modified to learn the global context among patches in an end-to-end manner to classify seven types of skin lesions. Xie et al. [24] have proposed a deep CNN, based on mutual bootstrapping, to segment and classify skin lesions.

Moreover, several approaches have been introduced to increase the classification as well as learning performance by combining handcrafted features with DL networks [25–28]. González-Díaz [29] incorporated the knowledge of dermatologists with CNN to diagnose skin lesions via eight dermoscopic structures, such as globules, dots, reticular patterns, and pigmented network. Yu et al. [30] have proposed an aggregated model where features have been extracted via residual-CNN followed by classification via a support vector machine. Here, ResNet-50 and ResNet-101 have been exploited and compared with recently published DL frameworks. Hagerty et al. [31] have presented a method using conventional image features, e.g. border, hair-ruler mask, and fed to ResNet-50 to diagnose melanoma images. Tang et al. [32] proposed a fusion of global and local scale CNN for extraction of global and local features from skin patches. Finally, all the features are fused by a weighted ensemble strategy to classify skin lesions.

16.1.2 Major Contributions

Summarizing the state-of-the-art dermoscopic image classification approaches, it is observed that DL frameworks employing CNN have achieved an impressive result with increased accuracy. But the trained CNN employed for classification does not explain the reason why the skin lesion has been correctly identified and classified into a certain category. To address this issue, an explainable deep learning (*x*-DL) framework has been implemented to classify and localize skin lesions from dermoscopic images and also to make an attempt to explain the decisions. This chapter aims to present an *x*-DL framework to generate layer-wise relevance maps (R-Maps) with positive and negative relevance scores (RSs) for automated skin-lesion classification as shown in Figure 16.3.

The major contributions of the chapter are highlighted as follows:

1. **Methodology:** An *x*-DL framework employing CNN has been proposed adopting pixel-wise decomposition (PwD) from classification layer to fully connected layer (FCL) and layer-wise relevance backpropagation (LRBP) from fully connected to the input layer (IL) through max-pooling and convolutional layers. The predictive output, obtained from the classification layer, is fully redistributed throughout the CNN layers to the input layer.
2. **Sensitivity Analysis:** Layer-wise relevance maps have been depicted, indicating both positive relevance (pos-Rel) and negative relevance (neg-Rel) scores, to interpret the classification decision obtained by CNN. The relevance maps (R-Maps) have been generated via PwD and LRBP approaches throughout the convolutional and the pooling layers to interpret significant regions in the activation maps. The final relevance map, obtained in the input layer, has been processed to regularize by a factor for possessing more relevance. Significant ablation analysis has been conducted for an improved visualization of the R-Maps considering four variants of CNN.
3. **Demonstration and Validation:** To validate the proposed *x*-DL framework, dermatological findings have been annotated by the medical expert, and it is found that the interpretation obtained from the relevance map successfully matches the expert annotations.

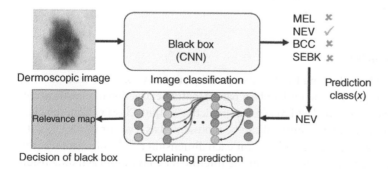

Figure 16.3 Explainable deep learning (*x*-DL) framework. The input dermoscopic image has been correctly classified as "nevus" (NEV). In order to visualize why the CNN (black box) has arrived at this particular classification decision, *x*-DL has been formulated and designed.

The chapter is organized as follows: Section 16.2 discusses the methods and materials used for the explainable CNN. The explainable CNN for dermoscopic image classification has been demonstrated in Section 16.3. Experimental results with input and output images have been illustrated in Section 16.4. At last, Section 16.5 concludes the chapter.

16.2 Methods and Materials

CNNs are applied effectively to identify dermoscopic images, containing ample information to classify skin lesions. The framework can possess either a simple or complex architecture having more number of layers, which may produce high classification accuracy by extracting richer information from successive layers. However, the information, used by the CNN to distinguish different skin lesions, have been obscured via PwD and layer-wise relevance backpropagation [33] stated in Sections 16.2.1 and 16.2.2.

16.2.1 Pixel-Wise Decomposition

The primary role of PwD is to yield the contribution of a pixel of a skin lesion to the classification score Class(x) obtained by CNN. For each lesion, the pixels are found out that contribute either a positive or negative result as given in Eq. (16.1).

$$\text{Class}(x) = \sum_{n=1}^{N} \text{RS}_n \tag{16.1}$$

where, RS_n is the relevance score of the nth neuron of a particular layer with N number of neurons.

Deep neural networks (DNNs) are organized in a layer-wise format with the help of an array of interconnected neurons. FCL represents mathematical functions, stated in Eqs. (16.2)–(16.4), which maps the fully connected neurons to the output neurons as follow:

$$y_{ij} = x_i w_{ij} \tag{16.2}$$

$$y_j = \sum_i y_{ij} + b_j \tag{16.3}$$

$$x_j = \emptyset(y_j) \tag{16.4}$$

where, w_{ij} denotes the weight between x_i and x_j, (i, j) are the indices of neurons in the former and later layer respectively, b_j is the bias, and $\emptyset(.)$ is the activation function.

16.2.2 Layer-Wise Relevance Back-Propagation

The LRBP is employed as a fundamental tool for achieving PwD as stated in Eq. (16.1). Also, LRBP assumes that a classifier Class(.) can be split into multiple layers [33] that might be a part of feature extraction or classification. The first layer consists of the skin-lesion pixels, and the last layer is the predictive output Class(x) from the classifier. Intermediate layers (InLs), i.e. max-pooling or convolutional can be represented as a vector $v \left(= v_{n=1}^{N}\right)$ with a dimension N. LRBP assumes to possess a relevance score $\text{RS}_n^{(\text{InL}+1)}$ for each dimension $v_n^{(\text{InL}+1)}$ of vector v at the (InL + 1)th layer. The target is to evaluate a relevance score $\text{RS}_n^{(\text{InL})}$ at the intermediate layer, for each dimension of the vector v nearer to the input layer, expressed in Eq. (16.5), such as:

$$\text{Class}(x) = \sum_{n \in \text{Out}} \text{RS}^{\text{Out}} = \sum_{n \in \text{FC}} \text{RS}^{\text{FC}} = \sum_{n \in \text{MP2}} \text{RS}^{\text{MP2}} = \sum_{n \in \text{C2}} \text{RS}^{\text{C2}} = \sum_{n \in \text{MP1}} \text{RS}^{\text{MP1}} = \sum_{n \in \text{C1}} \text{RS}^{\text{C1}} = \sum_{n \in \text{In}} \text{RS}^{\text{In}} \tag{16.5}$$

where, RS^{Out}, RS^{FC}, RS^{MP2}, RS^{C2}, RS^{MP1}, RS^{C1}, and RS^{In} denote the relevance scores (RS) of output layer (OL), FCL, max-pooling layer 2 (MPL-2), convolutional layer 2 (CL-2), max-pooling layer 1 (MPL-1), convolutional layer 1 (CL-1), and input layer (IL), respectively.

The intermediate layer relevance is split and forwarded to the previous layer neurons in terms of explanations or messages ($RS_{i \leftarrow j}$) to such an extent for holding the conservation property in Eq. (16.6) as follows:

$$\sum_i RS_{i \leftarrow j}^{InL,InL+1} = RS_j^{InL+1} \tag{16.6}$$

In the case of CNN, while applying ReLU activation function, satisfying ReLU(0) = 0, the pre-activation (y_{ij}) calculates the relevance score of neurons from x_i to x_j, where the relevance score (RS) decomposition can be expressed in Eq. (16.7) as:

$$RS_{i \leftarrow j}^{InL,InL+1} = \frac{y_{ij}}{y_j} RS_j^{InL+1} \tag{16.7}$$

where, y_{ij} denotes the local and y_j denotes the global pre-activations.

Here, the limitation of the backward propagation rule, i.e. Eq. (16.7) is that $RS_{i \leftarrow j}$ can hold unbounded values for low y_j. Hence, a regularization factor, i.e. $\gamma \geq 0$, is added to circumvent the limitation in Eq. (16.8) as follows:

$$RS_{i \leftarrow j}^{InL,InL+1} = \frac{y_{ij}}{y_j + \gamma} RS_j^{InL+1} y_j \geq 0$$
$$= \frac{y_{ij}}{y_j - \gamma} RS_j^{InL+1} y_j < 0 \tag{16.8}$$

It is observed that the relevance is completely absorbed when the regularization factor is high.

16.3 Explainable Deep-Learning (*x*-DL) Framework for Dermoscopic Image Classification

An *x*-DL framework employing CNN has been visualized to explain how the lesion patterns are learnt by the CNN to establish a proper skin-lesion classification. This section explains the dataset considered and its preprocessing, structure of the explainable CNN, and training implementation of the CNN. The overall framework of the *x*-DL is displayed in Figure 16.4.

16.3.1 Datasets and Image Preprocessing

For the analysis of *x*-DL, both melanocytic and epidermal lesion types have been chosen from various databases [3–6]. In this chapter a total of four classes of dermoscopic images, i.e. MEL (924), NEV (733), BCC (725), and SEBK (510) have been chosen for the proposed experimentation. The numbers in the parenthesis indicate the samples selected of the particular skin lesions for study.

As the resolution of images in the datasets differs widely, the resolution of size $224 \times 224 \times 3$ is selected as the input for explainable CNN. In this case, the resizing is performed along the shortest side keeping the aspect ratio unaltered.

16.3.2 Structure of Convolutional Neural Network (CNN)

The CNN is constructed by two convolution layers (CLs), two max-pooling layers (MPLs), one FCL, and an output layer (OL). The preprocessed skin lesions are fed to the CNN as inputs. As the inputs are three-dimensional, a 3D

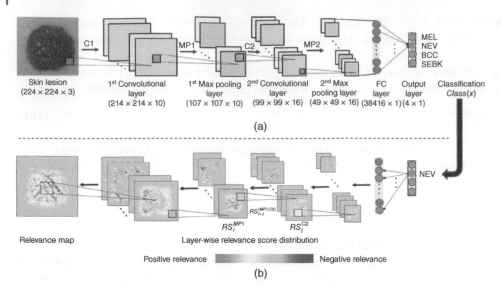

Figure 16.4 Explainable deep learning (*x*-DL) framework employing CNN with two convolution and two max-pooling layers.

convolution has been applied to the skin lesions as shown in Eq. (16.9) as follows:

$$\text{ConvOut} = \sum_q I * F_q \tag{16.9}$$

where, I, F, and q represent the skin image, image filter, and number of convolution filters, "*" is the convolution function, and ConvOut is the convolution output.

The convolutional outputs are passed through an activation function ReLU, stated in Eq. (16.10) such that:

$$\text{ReLU}(\theta) = \begin{cases} \theta & \theta \geq 0 \\ 0 & \theta < 0 \end{cases} \tag{16.10}$$

The outputs from CL are transferred through a MPL to diminish the feature dimension. The details of the hyper-parameters with output dimensions of the CNN topology are tabulated in Table 16.1.

The outputs from the MPL-2 are flattened and a FCL is formed. Xavier uniform initializes [34] the weights and biases that are connected from the FCL to classification layer (ClassL). The weighted resultant from the ClassL is transferred via an activation function, i.e. *softmax*, to possess the classification score Class(*x*) of dermoscopic images stated in Eq. (16.11) as:

$$\text{Class}(x) = \text{softmax}(\text{FC}^\circ \, W^\text{T} + B) \tag{16.11}$$

where, FC, W, and B represent the nodal elements of the FCL, weights connected from FCL to ClassL, and bias term, respectively. Element wise multiplication operator is denoted as "°," and the nonlinear function softmax is denoted in Eq. (16.12) as:

$$\text{softmax} = \frac{e^{\theta_k}}{\sum_{k=1}^{M} e^{\theta_k}} \tag{16.12}$$

where, θ, k, and M are the input to the softmax function, index, and number of output layer neurons, respectively.

Table 16.1 *x*-DL (CNN-2P) architecture details for dermoscopic image classification.

Layer	Operational Description	Output size	Parameters
Input	—	$224 \times 224 \times 3$	—
C1	$10\,F\,(11 \times 11 \times 3)$; St: 1, Pad: 1, ReLU	$214 \times 214 \times 10$	3640
MP1	Pool: 2×2, Pad: 1	$107 \times 107 \times 10$	—
C2	$16\,F\,(9 \times 9 \times 10)$; St: 1, Pad: 1, ReLU	$99 \times 99 \times 16$	12 976
MP2	Pool: 2×2, Pad: 1	$49 \times 49 \times 16$	—
FC	Flattened	$38\,416 \times 1$	—
Output	4 neurons; softmax	4×1	153 668

For both C1 and C2 layer, ReLU activation function has been applied after the convolution, and softmax is applied at the output layer.
F (kernel-size), number of kernels and kernel size; St, stride; Pad, padding; Pool, max pooling size.

16.3.3 Training Details and System Implementation

Out of 2892 dermoscopic images, a train-to-test ratio of 50 : 50 has been set for the *x*-DL implementation. Therefore, a total of 1446 images have been prepared for training of which 462 are MEL, 366 are NEV, 363 are BCC, and 255 are SEBK.

The explainable DL framework based on CNN has been performed on a PC with i5-4460T processor having NVIDIA GTX graphics card of 4 GB and size of RAM as 16 GB. "Adam" optimizer has been used for network training. The learning rate is set to 0.01 for the *x*-DL training. Batch of size 35 has been used to train all the CNN architectures with 80 epochs. The loss function chosen in the ClassL is "cross entropy". A total of 170 284 parameters have been trained during CNN training.

16.4 Experimental Results and Discussion

16.4.1 Analysis of Learnt Skin-Lesion Patterns from *x*-DL

Figure 16.4 demonstrates the proposed *x*-DL framework employing a CNN with two convolution and two max-pooling layers, i.e. (*x*-CNN-2P), achieving 100% training accuracy and 98.47% testing accuracy. This result demonstrates that the CNN-2P has learnt discriminatory patterns from the dermoscopic images effectively to discriminate four types of skin lesions in the testing data. PwD and LRBP have been used to investigate the lesion patterns learnt throughout the layers from the skin-lesion diagnosis and also to bring out the discriminatory information used for proper classification. The main observations obtained from the *x*-DL using CNN-2P are as follows:

(a) The relevance's are the feature-based information in the dermoscopic images used by the *x*-CNN-2P to discriminate skin lesions of various types.
(b) The way the input features for a skin lesion vary from other types of skin lesions of identical dermoscopic classes is revealed.
(c) The variational pattern for other explainable CNN modules trained and tested on the same dataset and testing images with different hyperparameters, such as convolution block, filters, and stride, are also exposed.

(d) The trained CNN-2P, shown in Figure 16.4, has been used to yield the relevance map of dermoscopic images at the input layer. In this chapter, the R-maps are represented as heat maps (H-maps). The pixel colors in the H-maps signify its corresponding RS. Here positive and negative relevances are highlighted as light gray and dark gray respectively as shown in Figure 16.4. A total of eight dermoscopic images (two images per class) with its relevance maps have been shown by varying the regularization factor. It is observed that the region of interest (RoI) of the dermoscopic images has been visualized in terms of positive and negative relevance, without applying any segmentation scheme.

16.4.2 Ablation Analysis Considering Regularization Factor

The performance of the relevance maps, obtained from x-DL has been compared by changing the regularization factor (γ). Figures 16.5–16.8 show MEL, NEV, BCC, and SEBK classes of dermoscopic images respectively, where γ is selected as 0.0, 0.001, 0.01, 10, and 100. It is noticed that the skin lesions have been localized prominently for higher values of γ in all the dermoscopic images of all classes. It is evident that the initial relevance map, obtained by Eq. (16.7), produces results with superior granularity with a greater number of negative relevance. In case of increasing γ, the relevance map possesses lesser granularity obtaining lesser negative relevance than those produced without applying γ. These results are displayed in Figures 16.5–16.8 (c) $\gamma = 0.001$, (d) $\gamma = 0.01$, (e) $\gamma = 10$, and (f) $\gamma = 100$. The granularity minimization and neg-Rel formation are because of the term $\frac{|y_{ij}|}{|y_j|+\gamma}$ in

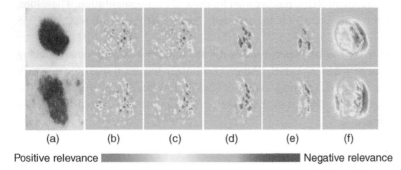

(a)   (b)   (c)   (d)   (e)   (f)

Positive relevance ▬▬▬▬▬▬▬▬ Negative relevance

Figure 16.5 Relevance maps for melanoma skin lesions (MEL) obtained via x-CNN-2P: (a) dermoscopic images; (b) initial relevance map ($\gamma = 0.0$) obtained by Eq. (16.7); (c)–(f) relevance maps obtained by varying the regularization factor (γ) as 0.001, 0.01, 10, and 100 respectively using Eq. (16.8).

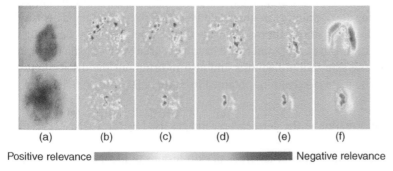

(a)   (b)   (c)   (d)   (e)   (f)

Positive relevance ▬▬▬▬▬▬▬▬ Negative relevance

Figure 16.6 Relevance maps for nevus skin lesions (NEV) obtained via x-CNN-2P: (a) dermoscopic images; (b) initial relevance map ($\gamma = 0.0$) obtained by Eq. (16.7); (c)–(f) relevance maps obtained by varying the regularization factor (γ) as 0.001, 0.01, 10, and 100 respectively using Eq. (16.8).

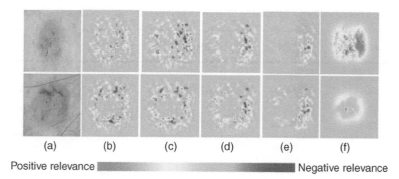

Positive relevance ▬▬▬▬▬▬▬▬▬▬ Negative relevance

Figure 16.7 Relevance maps for basal cell carcinoma skin lesions (BCC) obtained via *x*-CNN-2P: (a) dermoscopic images; (b) initial relevance map ($\gamma = 0.0$) obtained by Eq. (16.7); (c)–(f) relevance maps obtained by varying the regularization factor (γ) as 0.001, 0.01, 10, and 100 respectively using Eq. (16.8).

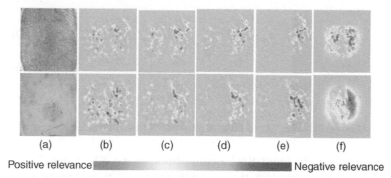

Positive relevance ▬▬▬▬▬▬▬▬▬▬ Negative relevance

Figure 16.8 Relevance maps for seborrheic keratoses (SEBK) skin lesions obtained via *x*-CNN-2P: (a) dermoscopic images; (b) initial relevance map ($\gamma = 0.0$) obtained by Eq. (16.7); (c)–(f) relevance maps obtained by varying the regularization factor (γ) as 0.001, 0.01, 10, and 100 respectively using Eq. (16.8).

Eq. (16.8), tending toward zero when $|y_j|$ approaches to zero. Hence, diminishing by this factor enhances whenever $|y_j|$ decreases. Moreover, the sign of the input neuron relevance (RS$_j$) will be reversed in RS$_{i \leftarrow j}^{\text{InL,InL+1}}$ for either positive inputs or negative inputs y_{ij}, based on the sign of $\frac{y_{ij}}{y_j}$.

Alternatively, it is expected from a CNN that the predicted score Class(x) > 0, i.e. majority of network inputs and their relevance scores (RS$_j$) are positive as their weighted sum in the classification layer is greater than zero. Hence, positive scores are hardly affected and eliminating out the negative scores incorporating a large γ in the InL that lead to a more positive score implemented in the final relevance map, which has a lesser granularity than the former.

16.4.3 Comparative Study with Other CNN Modules

The R-maps produced by employing *x*-CNN-2P have been compared with other explainable CNN modules as listed in Table 16.2. It lists the layer-wise architectures and performance metrics of the four CNN modules. The "C," "MP," "S," "FC," and "O" denote the convolution, max-pooling, sampling, fully connected, and output block, respectively. The parameters and hyper parameters have been chosen as per the ongoing explainable schematic. The stride and sample or pool size for "S" and "P" has been set to 2.

The performance of the CNN modules has been accessed in terms of five performance metrics, i.e. accuracy (ACC), sensitivity (SEN), specificity (SPE), positive predicted value (PPV), and negative predicted value (NPV),

Table 16.2 Architecture and performance metrics (%) of various explainable CNN modules used for ablation analysis.

Methods	Architectures	ACC	SEN	SPE	PPV	NPV
x-CNN-1S	C-S-FC-O	92.80	95.25	89.76	92.03	93.80
x-CNN-2S	C-S-C-S-FC-O	94.95	96.89	92.50	94.20	95.95
x-CNN-1P	C-C-MP-FC-O	97.02	98.16	95.56	96.62	97.56
x-CNN-2P	C-MP-C-MP-FC-O	98.47	99.02	97.75	98.31	98.70

described in Eqs. (16.13)–(16.17) as follows:

$$ACC = (TP + TN)/(TP + TN + FN + FP) \tag{16.13}$$

$$SPE = TN/(TN + FP) \tag{16.14}$$

$$SEN = TP/(TP + FN) \tag{16.15}$$

$$PPV = TP/(TP + FP) \tag{16.16}$$

$$NPV = TN/(TN + FN) \tag{16.17}$$

where, true positive (TP), true negative (TN), false positive (FP), and false negative (FN) are obtained from confusion matrices.

The relevance maps have been procured through all the four CNN-based modules, i.e. x-CNN-1S, x-CNN-2S, x-CNN-1P, and x-CNN-2P, with $\gamma = 100$, as shown in Figure 16.9. It is found that the x-CNN-2P (Figure 16.9f)

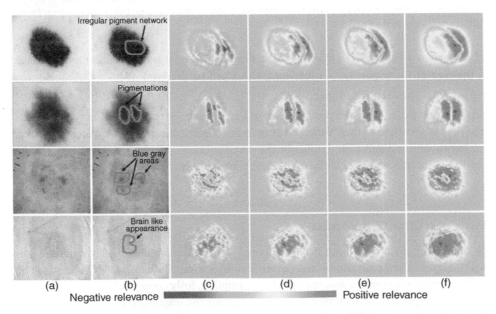

Figure 16.9 Relevance maps for skin lesions, i.e. MEL (first row), NEV (second row), BCC (third row), and SEBK (fourth row): (a) input dermoscopic images, (b) dermoscopic findings of the skin lesions referred by dermatologist, (c)–(f) final relevance map ($\gamma = 100$) obtained via x-CNN-1S, x-CNN-2S, x-CNN-1P, and x-CNN-2P, respectively.

produces an acceptable performance with the highest classification accuracy and outperforms the results obtained in Figure 16.9c–e according to the dermoscopic images shown in Figure 16.9a.

Comparing the *x*-DL results with that annotated by dermatologist (Figure 16.9b), it is observed that the regions with positive relevance have been visualized prominently and matched with the expert annotations. Also, the severity of the skin lesions has been analyzed by the saturation of the red color.

16.4.4 Discussion

The PwD along with the layer-wise relevance backpropagation distributes the class relevance iteratively from the classification layer to the input layer. For a correctly classified skin lesion, the relevance map has been produced that depicts pixels or regions that are solely dependent upon the membership of the predicted class. This is achieved without any segmented training images. Heatmapping has been considered as one of the most remarkable tools for the explanation of nonlinear classifiers. The primary aspect of heatmapping is that it hardly needs to change the learning algorithm or learn an extra model. The present scheme could be a great contender to be applied directly to any trained deep-learning network whenever applicable.

16.5 Conclusion

In this chapter, a thorough investigation has been implemented on the manner that how the skin lesions of the dermoscopic images have been learnt via an explainable DL framework using various schemes of CNNs. Four benchmark datasets have been considered for the present study. A pixel-wise decomposition with layer-wise relevance has been explored to generate the relevance maps of the dermoscopic images in terms of heat maps. Comprehensive ablation analysis has been conducted to judge the superiority of the proposed *x*-DL framework. The relevance map reveals that the proposed module can serve as an alternative decision next to the dermatologist in the computer-aided diagnosis (CAD) system.

Acknowledgments

The authors acknowledge with thanks the support and help extended by Dr. Tanmoy Kumar Mandal, MD (Internal Medicine), DM (Medical Oncology), and Consultant Medical & Hemato-Oncologist of Tata Memorial Hospital, Mumbai, India. Authors also acknowledge the grant received through the Visvesvaraya Young Faculty Research Fellowship, MeitY, Government of India, awarded to Jadavpur University.

References

1 American Cancer Society (2018). Cancer Facts & Figures 2018. [Online], https://www.cancer.org/research/cancer-facts-statistics/all-cancer-facts-figures/cancer-facts-figures-2018.html.

2 Kasmi, R. and Mokrani, K. (2016). Classification of malignant melanoma and benign skin lesions: implementation of automatic ABCD rule. *IET Image Processing* 10 (6): 448–455.

3 Mendonça, T., Ferreira, P.M., Marques, J.S. et al. (2013). PH^2 - a dermoscopic image database for research and benchmarking. In: *2013 35th Annual International Conference of the IEEE Engineering in Medicine and Biology Society (EMBC)*, 5437–5440.

4 Codella, N.C.F., Gutman, D., Celebi, M.E. et al. (2018). Skin lesion analysis toward melanoma detection: A challenge at the 2017 international symposium on biomedical imaging (ISBI), hosted by the international skin imaging collaboration (ISIC). In: *2018 IEEE 15th International Symposium on Biomedical Imaging (ISBI 2018)*, 168–172.

5 ISIC Archive (2019). The International Skin Imaging Collaboration: Melanoma Project. [Online]. Available: https://isic-archive.com.

6 Dermoscopy Atlas (2019). www.deroscopyatlas.com.

7 Chatterjee, S., Dey, D., Munshi, S., and Gorai, S. (2019). Extraction of features from cross correlation in space and frequency domains for classification of skin lesions. *Biomedical Signal Processing and Control* 53: 101581.

8 Silverberg, E., Boring, C.C., and Squires, T.S. (1990). Cancer Statistics, 1990. *CA: a Cancer Journal for Clinicians* 40 (1): 9–26.

9 Codella, N., Cai, J., Abedini, M. et al. (2015). Deep learning, sparse coding, and SVM for melanoma recognition in dermoscopy images. In: *Proceedings of Medical Image Computing and Computer Assisted Intervention*, 118–126.

10 Celebi, M.E., Codella, N., and Halpern, A. (2019). Dermoscopy image analysis: overview and future directions. *IEEE Journal of Biomedical and Health Informatics* 23 (2): 474–478.

11 Barata, C., Celebi, M.E., and Marques, J.S. (2015). Improving dermoscopy image classification using color constancy. *IEEE Journal of Biomedical and Health Informatics* 19 (3): 1146–1152.

12 Barata, C., Ruela, M., Francisco, M. et al. (2014). Two systems for the detection of melanomas in dermoscopy images using texture and color features. *IEEE Systems Journal* 8 (3): 965–979.

13 Barata, C., Celebi, M.E., and Marques, J.S. A survey of feature extraction in dermoscopy image analysis of skin cancer. *IEEE Journal of Biomedical and Health Informatics*, to be published.

14 Xie, F., Fan, H., Li, Y. et al. (2017). Melanoma classification on dermoscopy images using a neural network ensemble model. *IEEE Transactions on Medical Imaging* 36 (3): 849–858.

15 Ganguly, B., Biswas, S., Ghosh, S. et al. (2019). A deep learning framework for eye melanoma detection employing convolutional neural network. In: *2019 International Conference on Computer, Electrical & Communication Engineering (ICCECE)*, 1–4. Kolkata, West Bengal, India (18–19 January).

16 Zhang, J., Xie, Y., Wu, Q., and Xia, Y. (2019). Medical image classification using synergic deep learning. *Medical Image Analysis* 54: 10–19.

17 Goyal, M., Oakley, A., Bansal, P. et al. (2020). Skin lesion segmentation in dermoscopic images with ensemble deep learning methods. *IEEE Access* 8: 4171–4181.

18 Ganguly, B., Ghosal, A., Das, A. et al. (2020). Automated detection and classification of arrhythmia from ECG signals using feature-induced long short-term memory network. *IEEE Sensors Letters* 4 (8): 1–4.

19 Hu, H., Li, Q., Zhao, Y., and Zhang, Y. (2021). Parallel deep learning algorithms with hybrid attention mechanism for image segmentation of lung tumors. *IEEE Transactions on Industrial Informatics* 17 (4): 2880–2889.

20 Yu, L., Chen, H., Dou, Q. et al. (2017). Automated melanoma recognition in dermoscopy images via very deep residual networks. *IEEE Transactions on Medical Imaging* 36 (4): 994–1004.

21 Zhang, J., Xie, Y., Xia, Y., and Shen, C. (2019). Attention residual learning for skin lesion classification. *IEEE Transactions on Medical Imaging* 38 (9): 2092–2103.

22 Sarkar, R., Chatterjee, C.C., and Hazra, A. (2019). Diagnosis of melanoma from dermoscopic images using a deep depthwise separable residual convolutional network. *IET Image Processing* 13 (12): 2130–2142.

23 Gessert, N., Sentker, T., Madesta, F. et al. (2020). Skin lesion classification using CNNs with patch-based attention and diagnosis-guided loss weighting. *IEEE Transactions on Biomedical Engineering* 67 (2): 495–503.

24 Xie, Y., Zhang, J., Xia, Y., and Shen, C. (2020). A mutual bootstrapping model for automated skin lesion segmentation and classification. *IEEE Transactions on Medical Imaging* 39 (7): 2482–2493.

25 Majtner, T., Yildirim-Yayilgan, S., and Hardeberg, J.Y. (2016). Combining deep learning and hand-crafted features for skin lesion classification. In: *2016 Sixth International Conference on Image Processing Theory, Tools and Applications (IPTA)*, 1–6. Oulu, Finland.

26 Codella, N.C.F., Nguyen, Q.B., Pankanti, S. et al. (2017). Deep learning ensembles for melanoma recognition in dermoscopy images. *IBM Journal of Research and Development* 61 (4/5): 5:1–5:15.

27 Wei, L., Ding, K., and Hu, H. (2020). Automatic skin cancer detection in dermoscopy images based on ensemble lightweight deep learning network. *IEEE Access* 8: 99633–99647.

28 Matsunaga, K., Hamada, A., Minagawa, A. et al. (2017). Image classification of melanoma, nevus and seborrheic keratosis by deep neural network ensemble. [Online]. Available: https://arxiv.org/abs/1703.03108.

29 González-Díaz, I. (2019). DermaKNet: incorporating the knowledge of dermatologists to convolutional neural networks for skin lesion diagnosis. *IEEE Journal of Biomedical and Health Informatics* 23 (2): 547–559.

30 Yu, Z., Jiang, X., Zhou, F. et al. (2019). Melanoma recognition in dermoscopy images via aggregated deep convolutional features. *IEEE Transactions on Biomedical Engineering* 66 (4): 1006–1016.

31 Hagerty, J.R., Stanley, R.J., Almubarak, H.A. et al. (2019). Deep learning and handcrafted method fusion: higher diagnostic accuracy for melanoma dermoscopy images. *IEEE Journal of Biomedical and Health Informatics* 23 (4): 1385–1391.

32 Tang, P., Liang, Q., Yan, X. et al. (2020). GP-CNN-DTEL: global-part CNN model with data-transformed ensemble learning for skin lesion classification. *IEEE Journal of Biomedical and Health Informatics* 24 (10): 2870–2882.

33 Bach, S., Binder, A., Montavon, G. et al. (2015). On pixel-wise explanations for non-linear classifier decisions by layer-wise relevance propagation. *PLoS One* 10 (7): e0130140.

34 Glorot, X. and Bengio, Y. (2010). Understanding the difficulty of training deep feedforward neural networks. In: *Proceedings of the 13th International Conference on Artificial Intelligence and Statistics (AISTATS) 2010*, 249–256. Sardinia, Italy (13–15 May).

17

Future of Machine Learning (ML) and Deep Learning (DL) in Healthcare Monitoring System

Kanak Kumar[1], Kaustav Chaudhury[2], and Suman Lata Tripathi[3]

[1]Electronics Engineering Department, IEEE Member, Indian Institute of Technology (Banaras Hindu University), Varanasi, India
[2]Electronics and Communication Engineering, Heritage Institute of Technology, Anandapur, Kolkata, India
[3]School of Electronics & Electrical Engineering, Lovely Professional University, Phagwara, Punjab, India

17.1 Introduction

Smart devices use different ways to predict, discover, and understand their surroundings. In the pre-process, learn from data always. Moreover, if anyone looks into it, it contains several steps and processes to make such a device. Optimization of hardware and software and adding data security either on devices or cloud is essential. So, an approach to optimize each feature is made carefully so that each element is less dependent on its previous step. A promising approach for optimizing software for making decisions is to use machine learning (ML). ML algorithms have been classified into the following: (i) supervised learning, (ii) unsupervised learning, (iii) semi-supervised learning, and (iv) reinforcement learning. In the data-driven world, supervised learning is more common; they help us to classify classes or predict values. They derive the relationship between features and labels, so supervised-learning algorithms only work on labeled data [1–9]. Since they work entirely on data, data pre-processing is fundamental. Sometimes getting tagged information is problematic as it involves manually marking them by people.

The ratio between the amount of labeled data and the amount required to train is relatively minor. Unsupervised learning algorithms work on un-labeled data. They are used in denoising, clustering [10], and data augmentation. Still, they are not that much robust as supervised learning. A hybrid method known as semi-supervised learning is quite helpful. They learn the relationship between labeled data and use the learned associations to train on un-labeled data. All parameters get adjusted finally during training. The reinforcement-learning algorithm works on positive reinforcement theory. They work in particular environments; the reward is calculated when an agent makes a step. The primary purpose is to improve the compensation received [9].

The algorithms used to optimize parameters during training should run fast and not get stuck in local minima or maxima before even reaching their global minima/maxima. Several methods are utilized to ignore these pitfalls. One of the popular algorithms that are used for the optimization of parameters is stochastic optimization. Stochastic approximation, non-convex optimization, Laplace expansion of integral, Taylor expansion, and gradient descent are some examples of popular stochastic optimization techniques. Some people also use combinatorial algorithms to deal with operations research, computational complexity theory, and algorithm theory. New research methods are arising in the field of deep-neuro evolution. Genetic algorithms and particle swarm optimization (PSO) are some of the initial researches in this field. They help in faster optimization and can handle complex parameter functions very easily [9].

Apart from optimizing the algorithm, some optimization and research on data pre-processing are carried out every year. An internet-of-things (IoT) system streams everyday data in the cloud. Before moving toward data

Machine Learning Algorithms for Signal and Image Processing, First Edition.
Edited by Deepika Ghai, Suman Lata Tripathi, Sobhit Saxena, Manash Chanda, and Mamoun Alazab.
© 2023 The Institute of Electrical and Electronics Engineers, Inc. Published 2023 by John Wiley & Sons, Inc.

training, it is essential to look for unwanted and non-numeric data types. One hot encoding is a popular data pre-processing technique for handling categorical data [1–3, 27, 38]. It transforms information like yes/no, and true/false to 1 and 0, respectively. To handle missing data, all the values are used to replace missing values [12]. The K-Nearest Neighbor (KNN) algorithm is also used for pre-processing, as it is a light model that benefits a lot and gives a much closer generated value to the missing value. It can also be used to handle categorical data too [1]. Normalizing data is reasonably necessary as it helps to scale data within ranges [2, 3, 7, 15, 19, 25]. Two different types of climbing data include transforming data to ranges (0, 1) and (−1, 1). The label smoothing of data (−0.9, +0.9) is a reasonable adjustment for early convergence. It also helps to train networks faster. Self-organizing maps (SOMs) are used for clustering data based on a particular feature that data possesses. It creates a low-dimensional representation of data by securing the topologies of the components [1]. They are most likely used as unsupervised learning for clustering data. The architecture of SOM contains an input layer whose neurons are an equal number of features.

17.1.1 ML/DL Algorithms for Optimization

Optimized trainable variables connect the input and output or the Kohonen layer. SOM calculates the Euclidean distance between the input vector and the weight of nodes to find the best matching unit (BMU) and updates the networks for network evaluation. Now, reduce the data's dimensionality through principal component analysis (PCA) [1, 24, 25]. Principal components can be easily extracted using the PCA algorithm by sampling linearly uncorrelated orthogonal axes. Principal components (PCs) obtain from the eigen decomposition of the covariance matrix. SVD is a handy analysis tool for dealing with very high-data dimensions. Like PCA, it also reveals the minimum sizes necessary to represent the data [4, 24].

Fisher discriminant analysis (FDA) is a well-known linear technique for dimensionality reduction and pattern recognition. FDA maximizes the scatter between different classes while minimizing the scatter within a category to find Fisher optimal discriminant vectors. The transformation matrix in FDA minimizes the "within-class" scatter and maximizes the "between-class" scatter. The upgraded version of the FDA is local Fisher discriminant analysis (LFDA). As a result, they work much better in the presence of outliers, as they tend to preserve within-class scatter. Another way to pre-process data is using a fuzzy preprocessor. Fuzzy weight processing helps in generating new features from the old value. In the first stage, the mean value of the features is calculated across all samples, and triangles from the input and output membership functions are formed. Every input in the data is mapped into the input membership function. It gives two outputs, and the minimum value between the two results is fed to the output membership function, which offers another two output values. The mean of two values is calculated for the final output of the fuzzy weight system [21]. Image data is hard to collect and more challenging to label, so it is more favorable to develop a solution. To provide variability in data, different variations like brightness, sharpness, and data augmentation are required for proper training. Segmented cropping is a valuable augmentation feature [11, 25].

Other approaches to handling features are not reduced to a lower dimension but by selecting elements from the dataset, which creates the most significant prediction [20]. Feature selection is entirely dependent on data or the dataset. A dataset having *n* features can have *n−m* selected features, as *m* features are redundant. Several data gained from IoT devices can generate a dataset where all elements are necessary, as those devices are made with sensors that have the highest correlation with the environment. Some features that neither need to be encoded nor imputed are identity documents (IDs) or names, as they can create wrong predictions. Correlation-based feature selection uses heuristics to extract a smaller set of features from its parent set, rather than traditionally ranking them individually. The heuristic assigns high scores to the smaller feature sets that correlate highly with the class and are highly uncorrelated with each other [20]. Two important evolution strategy (ES) algorithms are genetic algorithms and PSO, and they can be used as feature selection and dynamic training [3]. Modified algorithms like ant colony optimization (ACO) are proposed by researchers who aim to find the best solution to some parameters. It can find the best features among all groups, so it is mainly used for searching operations through a dataset [4].

ACO uses the ant colony system, which offers high stability and directions for searching in three different manners. Initially, the transition state rules provide a better way to balance the exploration of new edges and the exploitation of prior ones. Pheromone updates through global pheromone updating rules are performed on the best ant tour edges, and local pheromone updates are carried out while ants prepare a solution. An updated version was introduced, dynamic ant colony system three levels (DACS3). During the local pheromone update, the intermediate pheromone update is carried out by extracting the best knowledge of the group of the best individual ants after completing a tour and then divides into the best group and worst group. A global pheromone update is carried out, after which the ant that has the best term being deemed and is divided into the best of the intermediate best and the worst of the global best. Another adaptation to PSO is multi-objective optimization (MOO).

MOO involves the simultaneous optimization of several objective functions. Instead of finding one optimal solution, find several optimal solutions. Determination of many solutions is needed to consider that no one answer can be regarded as better than others concerning all objective functions. PSO is a popular population-based optimization technique. Time variant multi-objective PSO is the upgraded version of PSO, which is adaptive and allows vital parameters like inertial weights and acceleration coefficient to adjust with iterations. This adaptiveness elevates the algorithm for better exploration of its search space. It incorporates the crowding distance computation into the PSO during its global best selection [23]. Other PSO includes artificial bee colony optimization [22].

17.1.2 Pre-processing Methods

To handle genetic data, a different approach to pre-processing is required. When we have data from multiple sources, we need numerous pre-processing techniques. Multi-array gene expression is dealt with using the MAS5 function in R [8]. The normalization, background adjustment, and summarization of affymetrix microarray probe-level data. Each gene expression is converted into quantile values of 1–5 (very low, low, typical, high, and very high) using several quantile and numeric conversions. They have collected OMIM database gene lists mapped into their microarray probe-id. This helps in the mapping of the targeted genes. Each patient is classified into adjuvant radiotherapy and no adjuvant chemotherapy for feature selection. The calculated chi-square test value identifies the highest or the top 10 survival correlated gene signatures across every dataset. It is estimated between variables or genes here. It also used the subset of gene signatures for assigning model variables for final prediction. This eventually helps to find the best fit of the model variables.

Apart from data pre-processing, the KNN algorithm is also used to predict labels [3, 14]. So, this has a huge benefit, but accuracy is an important factor; for this reason, it is kept mostly limited to data pre-processing. To find a good algorithm, it is necessary to look into different algorithms and understand how they work. One of the promising techniques is support vector machine (SVM) [1–4, 24, 25, 27, 30]. This is a supervised learning technique that is used for regression and classification. SVM works by constructing hyperplanes that help to separate classes. It uses the concept of support vectors, which are the data points closest to the hyperplane. I create a margin concerning the support vectors, and the hyperplane is iteratively calculated by minimizing the loss. For nonlinear data classification or regression, kernel-based SVM is used. The different kernels are linear kernel $(x^T{}_iX_j)$, polynomial kernel $(r+Yx^T{}_iX_j)^p$, Gaussian kernel $(\exp(-y\|x_i-x_j\|^2))$ and sigmoid kernel $(\tanh(r+yx^T{}_ix_j))$. Fuzzy SVM is an adaptation to SVM where the data samples are divided into positive and negative samples. Then, the SVM is carried out separately by changing the formulation function [1]. Other adaptations of SVM include C-SVM, nu-SVC, and linear SVM [3].

The decision tree, chi-squared automatic interaction detection, is a statistical model, which uses chi-squared intuitive interaction method. A recessive partitioning method can be achieved by providing predictors and predictive classes as input features. A chi-squared static test is performed between the predictive features and the target data [2]. Random forest (RF) is one of the robust and accurate models [2, 3, 19]. The intuition behind the random forest is an ensemble learning algorithm. Random forest's instinct is to select a subset of features from a set of features, and it calculates nodes using the best split point in the data. The node is then split into daughter nodes based on the best split. Thus, it creates more like a tree structure. It uses ensemble learning of many such trees for prediction. An adaptation to random forest (RF) is rotation forest, which also uses independent decision trees. However, rotation forest is trained with a different dataset with its feature space rotated. As new classifiers are built, it uses hyperplanes parallel to the feature axes, and a slight rotation of the axes leads to diversity in trees [20]. Another type of decision tree is the C5.0, and the model uses bagging techniques and boosting techniques to increase accuracy during prediction. The C5.0 model allows weighted different features and types of incorrect classifiers. Using C5.0 to test features is the ability to compute the gain ratio with information entropy reduced bias. Other prominent classifiers are the naive Bayesian classifier, and the probability given to it by the likelihood by Bayes theorem [3, 14, 30] is issued by Eq. (17.1):

$$P(C|V_i) = P(C) \prod P(C|V_i)/P(C) \tag{17.1}$$

Other algorithms may include reg log, linear discriminant analysis (LDA), quantitative descriptive analysis (QDA) [3], Bayes Net [19, 30], jRip [19], j48 [19, 30], and Hoeffding Tree [19]. Simple systems can also be a good way to predict disease. One of the optimizations in algorithms is dynamically collecting data to achieve prediction. A necessary amount of data is quite important for prediction during training. Moreover, to deal with epistemic uncertainty, a Fuzzy expert system can gather information from the user. The fact that makes the data mode certain can be solved through a question. The features are input to the fuzzy system through different questions asked by the user, and the risk level of the disease is output to the system in medical cases. When all the questions are finished, the fuzzy system processes the input data, and the output is given out. The system has four main components: fuzzily, inference, fuzzy rules, and denazify [17]. Simple algorithms like this can achieve necessary goals easily. Another system is the rule-based system (RBS). In RBS, the prediction is obtained dynamically based on user input; if the user-provided data to the system is sufficient and matches with the knowledge base, then the fuzzy system displays the actual prediction, or else if the information is not sufficient or missing in the knowledge base the system asks for more information. To find out the missing attributes in the dataset, the order of dataset theory has been introduced so that nearby predictions related to the actual prediction can be detected. The optimization algorithm is used along with the decision-making algorithm that provides predictions by RDS [22]. The simulated annealing technique is used to find the proper solution to the optimization problem by sampling from random variations of the current solution. This is the generalization of the Monte Carlo method examining the equations of state and frozen state of *n* body systems. The algorithm to simulate annealing is given as follows: (i) first initialization of the precession, the best solution, and the initial state done along with a constant value for the maximum score that needs to be achieved; (ii) some new neighbor is selected, and precision is calculated if the new precision generated it's better than old precision the state is changed; (iii) new precision is checked to be the new solution or not if it is the best solution it is taken into account [27]. Optimization for finding the optimal hyperparameters is carried out through evolution strategy algorithms [3].

PSO and genetic algorithms are two great evolutionary algorithms. The essential operation of evolutionary algorithms are (i) Initialization of the population of candidate solutions, and alphabet of cardinality is used to encode each solution; (ii) Fitness value is the critical way to evaluate solutions; (iii) The existing solutions are altered to create new solutions; (iv) Fitness values of new solutions are calculated and inserted into the population; (v) Better new solutions are used to replace the old ones; (vi) Continue this until reaching the maximum time allotted [9]. The use of genetic algorithms is for creating antibodies for a disease; say you have an infection, then the attributes of the organ can be denoted by *v* and *c* by the condition of the organ. It also generates untrained antigens, and when we apply a genetic algorithm, we get new antibodies. Like any other genetic algorithm that starts with mutation and then cloning followed by crossing operation, the fitness of the antibodies created is calculated if the created antibodies are not suitable when repeated. After the procedure is done, the antibodies are stored in the memory state, and with the given data of the particular diseases, the response is calculated [15]. With continuous repetition, the best antibody is extracted. Machine learning neural network (MLNN) or artificial neural network (ANN) is an alternative to conventional pattern recognition for disease diagnosis. A back-propagation algorithm is used to train the network [3, 5, 8, 14, 19, 29, 30]. Neural network (NN) works on the principle of many perceptrons connected internally. NN architecture may vary, but it has input, output, and hidden layers. The number of perceptrons in the information of the neural network depends on the number of feature inputs, and the number of perceptrons in the output of the neural network depends on the number of classes. The hidden can vary in perceptron count based on the user. Each perceptron has particular weights, bias, and an activation function. The essential operation of a neural network perceptron [29] can be given by the following Eq. (17.2).

$$X(n) = 1/(1 + \exp(-(W(n) \times f(n) + b(n))))$$ (17.2)

where n = number of features (f) dimensions, W is the weight, and b is the bias. A sigmoid activation function is used for the equation. The weights and bias are adjusted through beach-propagation through multiple iterations using an optimizer. The stochastic gradient descent (SGD) algorithm is a good optimizer with superior results [11]. Eventually, a neural network will get adjusted based on data distribution, but NN applies the steepest gradient descent. It can lead to a slow convergence rate and often yields suboptimal solutions. The linear-regression model (LM) algorithm overcame the problem, which provides faster convergence and better evaluation. LM algorithm converges faster; as a result, it starts memorizing the data, which will cost the model to generalize [31]. The loss function is essential as it needs to be optimized. Some famous loss functions for classifiers are cross-entropy. $1 = -\Sigma chat_i \log c_i$ and for regression are l2 loss $1 = \Sigma^n_1 (y - y_p)^2/n$ and l1 loss $1 = \Sigma^n_1 |y - y_p|/n$. Another adaptation to the NN is the extreme learning machine (ELM) algorithm [12]. ELM algorithm is used as it has a three-layer neural network with a sigmoid activation function. It prevents early scheduling, stopping, learning epochs, local minima traps, and early convergence. It is fast with a much better generalization technique. Since it uses a nonlinear activation function, it helps well in nonlinear data. After initializing the regularization term (C) and the number of hidden layers (L), the ELM value is calculated while training to optimize the network. A better neural network is a convolution neural network (CNN) that theoretically applies the convolution operation and deals with data to understand hidden relations. They work great with images and retrieving hidden signatures [11]. A basic intuition behind convolution filters is a matrix less than the size of the input data. A filter has values or weights drawn from a distribution, and then it is padded over the input data. A scalar product or element-wise dot product is performed among the padded region of the image and the filter. The single value is derived by adding the result, and padding values create a new matrix called a feature map [32]. A nonlinear activation function is used after this to handle nonlinearity. A pooling layer is added to it to reduce the dimension of the feature map. When data flows through these layers, the parameters are updated. Sometimes, these updated parameters make data passing through the network too big or too small, referred to as internal covariance shift. A particular layer known as batch normalization helps overcome this problem [33, 34]. Alzheimer's disease diagnosis framework contains a dense block containing dense layers with a skipping connection used to add information for better results. A transition block contains bath normalization, ReLu layer, 1×1 convolution, and polling layer [11]. We discuss some convolution networks here; Let-Net-5 is one small network used to benchmark other networks [32]. Radial basis function (RBF) network, or probabilistic neural network, is one of the platform's most important ANNs. A single layer hidden unit known as the radial bias function is injected between feed-forward neural networks for this type of network. There are three layers in this network the first layer is the input layer, the middle layer contains a Gaussian activation function, and the third layer is the output layer. When an input value is passed to the network, the network gives out a distribution of values and the linear sum of all the values is taken as an output from the last layer of the network [23, 29]. Unlike the traditional method that finds optimal control parameters in minimizing the loss of the regression via cross-validation, several multiple sparse regression models are better.

To transform the input vector into a matrix form, the learned sparse regression models with variable values for the parameter control were exploited as the output level representation learners. The output level representation matrix is then fed into a CNN to make proper clinical decisions. From a pattern classification standpoint, this CNN is an assembled classifier that finds the relation of different sparse regression models. A deep ensemble sparse regression network contains three layers, a convolution layer pooling layer, and a fully connected (FC) layer, and the optimization is done through SGD. The deep ensemble sparse regression network works better than single sparse regression [28]. Optimized multiple models are necessary to support standard ANN during training using genetic data. Data collected from numerous sources for disease prediction helps detect disease more clearly, so various models are created in the rare disease auxiliary diagnosis system (RDAD) algorithm for perfect analysis. The phenotype-based occasional disease similarity (PICS) algorithm used curated rare disease-phenotype associations as inclusion data and four variable disease similarity methods as the classifiers. As inclusion data to phenotype-gene-based rare disease similarity phosphoglyceric acids (PGAS) algorithm, curated phenotype-gene associations and rare disease-phenotype associations are used. A two-variable disease similarity method as a classifier is used.

Curated rare disease-phenotype associations are included in phenotype-based machine learning (CPML) models and curated and text-mined phenotype-based machine learning (APML) models. While the CPML and APML used six different ML algorithms, APML also used text-mined rare disease-phenotype associations as inclusion data [13]. The PCIS uses most informative common ancestor, cosine similarity, Tanimoto coefficient, and probability score as classifiers. The PGAS uses cosine similarity and Tanimoto coefficient as classifiers. The CPML and APML use logistic regression (LR), KNN, RF, extra trees (ETs), Naïve bias (NB), and deep neural networks (DNNs) as main classifiers. Their score is calculated using the Bayesian averaging score function to analyze. The four PICS, PGAS, APML, and CPML, build the RDAD algorithm.

Another way to deal with genetic data is through copy number change; it is a genetic variation in the human genome. This would result in the development and progression of cancers. Bayesian hidden Markov model is used along with Gaussian mixture clustering to identify copy number change [18]. In the first-order Markov process, the probabilities of the states depend only on the current state. Hence, the next stage of the system is dependent on the current state and is entirely independent of the previous conditions of the system [18]. Probabilistic parameters of a hidden Markov model are explained as follows: x and s = represent the states; a = represents the state transition probabilities; y = represents the possible observations; b = represents the output probabilities. The equation gives the likelihood of an observed sequence (Eq. (17.3)).

$$P(Y) = \sum_{x} P(Y/X)p(X) \tag{17.3}$$

As clustering algorithms are widely used for categorizing data based on features. A massive adaptation to clustering algorithms is Gaussian mixture clustering. To optimize the parameters, the expectation–maximization (EM) algorithm is used. The EM algorithm has two levels of optimization, the expectation step or E step and the maximization step or M step. The formulation of EM [18] is given as Eqs. (17.4)–(17.6).

$$\text{E step: } P(w_j/x_k, \Lambda_t) = P\left(x_k/w_j, \mu_i^{(t)}\sigma^2\right) p_i^{(t)} / \sum_{k} P\left(x_k/w_j, \mu_i^{(t)}\sigma^2\right) p_i^{(t)} \tag{17.4}$$

$$\text{M step: } \mu_i^{(t+1)} = \sum_{k} P(w_j/x_k, \Lambda_t)x_k / \sum_{k} P(w_j/x_k, \Lambda_t) \tag{17.5}$$

$$p_i^{(t+1)} = \sum_{k} P(w_j/x_k, \Lambda_t)/R \tag{17.6}$$

where w is the weights, μ is mean, σ is variance, Λ_0 is = $\{\mu^0{}_1, \mu^0{}_2, \ldots \mu^0{}_k\}$, and R is the number of parameters. For generating posterior samples, Hamilton Monte Carlo is used. The μ and σ are computed in stochastic forward-backward sampling algorithms and the Metropolis–Hastings method. EM clustering calculates the changes in the data's change-point indices. Markov chain Monte Carlo (MCMC) is used to examine each change point identified from the Gaussian mixture models (GM) clustering, classified into focal aberrations, outliers, transition points, complete chromosomal changes, and amplifications. The artificial immune recognition system (AIRS) algorithm was developed to deal with the immune-based system, including the following steps. It contains normalizing data, then memory cell identification and architecture review board (ARB) generation. Resources are allocated among ARB according to simulation level, and the determination of candidate memory cells is done in the next step. Adding candidate memory cells to the memory cell pool if some criteria are reached is done in the memory cell introduction step and thus releasing memory cells as a final step of the training [21].

The chapter is divided into different sections: Section 17.2 discusses performances as parameters used for healthcare monitoring systems objectives, and the motivation of this chapter is formulated in Section 17.3. The existing ML/DL techniques for healthcare monitoring and disease diagnosis are discussed in Section 17.4. Section 17.5 discusses the proposed model/methods for using ML/DL for healthcare monitor systems. Experimental results and discussion is shown in Section 17.6. The conclusion is concluded in Section 17.7, and the future scope is discussed in Section 17.8.

17.2 Performance Analysis Parameters

Measuring the performance of a model is essential and can be carried out by calculating the accuracy (ACC) (Eq. (17.7)) of the model. Sensitivity (SEN) (Eq. (17.8)) and specificity (SPEC) (Eq. (17.9)) is also a necessary factor for the diagnosis of the best model [4].

$$\text{Accuracy} = \Sigma((TP + TN)/(TP + FP + TN + FN)) \times 100)\% \text{ from 1 to } N \tag{17.7}$$

$$\text{Sensitivity} = (\Sigma((TP)/(TP + FN)) \times 100)\% \text{ from 1 to } N \tag{17.8}$$

$$\text{Specificity} = (\Sigma((TN)/(FP + TN)) \times 100)\% \text{ from 1 to } N \tag{17.9}$$

When dealing with medical research, Kaplan–Meier survival analysis (KM-plot) measures the survival time after treatment. An essential advantage of the KM-plot is that the method can account for censored data. A treatment log rank can be used to determine the differences in survival between groups [8].

Inference for multiple ML models is necessary. Ensemble ML is a field of ML where numerous models are used rather than making decisions from one model [4, 11, 20, 28]. It is used to find the best model from the sample for inference and training multiple models. It sometimes averages out the prediction from each model to give the best prediction; it may not be accurate. Still, it provides a diverse solution across multiple data, which deals with uncertainty. Bootstrap aggregating is a method where each model in the ensemble vote is assigned equal weight. Model variance is carried out by selecting a subset of the data for training. Random forest uses it as an inbuilt function for training multiple decision trees. Boosting is carried out in AdaBoost, which involves incremental models based on previously misclassified data [20].

Another popular ensemble learning is the bucket of models. Buckets of models have variable parameters, and only a single parameter set is best. So, the model that acquires the highest score is selected. Cross-validation is one of the famous bucket of models of ensemble learning. One of the most used is k-fold cross-validation, which selects the best model with optimal hyperparameters [2, 5, 7, 11, 15, 19, 20, 29]. Another variation of 10-fold is stratified 10-fold cross-validation [3]. Another ensemble modeling is Bayesian model averaging, where instead of making multiple models, weights are drawn from a posterior probability distribution in a model for a given data [18].

Another vital area in ML is localization. Apart from just classification, the location is important, especially in medical cases like localization of skin patches, tumor cells, or stones in abdomen organs. Landmark-based multi-instance learning is a way to detect a patch in the whole image. Pre-processing of photographs is required apart from data augmentation. Here, we discuss the context of Alzheimer's disease detection. For landmark discovery in the pictures, statistical testing is performed on active directory (AD) and numerical control (NC) groups. As a result, a *P*-value map in the template space is obtained, whose local minima can be defined as the location of discriminative landmarks in the temperate area. Finally, these landmarks are directly projected into linearly aligned training images using their respective deformation fields. For classification, convolutional neural networks are used. Here, L parallel sub-CNN architecture contains six convolutional layers, two fully-connected (FC) layers, and max-polling. The rectified linear unit (ReLU) activation function in convolution [16]. Another localization and region of interest implementation are done in [25]. An automatic segmentation algorithm is used to calculate the parameters of the iris, which is based on the circular Hough transform. A modified Consumer Health Technologies (CHT) algorithm was used to calculate the radius and center coordinates of the inner and outer boundaries of the iris. Rubber sheet normalization extracted the iris into a fixed rectangular representation. This algorithm also normalizes other factors like the visual angle of the camera, angular inconsistencies, and illumination level. The rectangular output is invariant to the dimensional irregularities, orientation, position, and camera angle of the pupil and iris within the eye's image. The region of interest was cropped from the normalized image through standard methods. First-order statistics features were calculated to describe the intensity distribution within the rectangular region of interest (ROI). The gray level distribution of the ROI can be analyzed using first-order distribution. Texture features help us know the position of the gray level in the image. Features while calculated from decent gray level co-occurrence matrix. Wavelet features and transformation show great potential in various fields like matching, biomedical application, and telecommunication. The Fisher score discriminator, *T*-test, and chi-square test for feature selection. Several algorithms, like binary tree model, SVM, adaptive boosting model, generalized linear model, neural network, and random forest, have been used as classifiers with a 10-fold cross-validation technique [26].

17.3 Objectives and Motivation

When developing a product, new algorithms are made for accurate prediction, but we start considering the algorithm's reliability when it comes to deployment. A new algorithm is reliable based on its accuracy to unseen data, but accuracy is not a metric to which we become reliable. Repeatability increases the reliability of a model's accuracy, which is why ensemble learning is essential. But repeatability requires enormous amounts of data. For a product, data is derived slowly, and the algorithm is updated to be more accurate over the years, but it may or may not work in the early stages. This could lead to a disaster in telemedicine, so having confidence in an algorithm is essential. So, this is the approach we use for prediction results and trust that an ensemble runs over a singular data. Higher the confidence, the higher the consideration of taking the prediction into account.

17.4 Existing ML/DL Techniques for Healthcare Monitoring and Disease Diagnosis

A detailed comparison of the algorithms used across several healthcare and disease datasets is given in Table 17.1. Datasets are from UCI repositories. Other algorithms that have significance are shown in [6]. These algorithms are deployed across devices and have been updated regularly.

Table 17.1 Comparison of technology for disease detection and diagnosis [1–3, 5, 7, 11–23, 25–28, 42–44].

Sr. No.	Disease	Resource of data	Algorithm	Accuracy	Specificity	Sensitivity	Others
1	CAD	UCI	KNN+SOM+PCA+ FUZZY SVM	0.968	0.943	0.96	Na
2	CAD	UCI	Hotdeck+SOM+PCA+ FUZZY SVM	0.9568	0.933	0.95	Na
3	CAD	UCI	SOM+PCA+SVM	0.9449	0.914	0.94	Na
4	CAD	UCI	PCA+SVM	0.8976	0.8571	0.9	Na
5	CAD	UCI	SOM+SVM	0.8661	0.8131	0.86	Na
6	CAD	UCI	SVM	0.8268	0.7642	0.83	Na
7	CAD	UCI	KNN+SOM+PCA+ FUZZY SVM	0.978	0.9697	0.969	Na
8	CAD	UCI	Hotdeck+SOM+PCA+ FUZZY SVM	0.958	0.9697	0.935	Na
9	CAD	UCI	SOM+PCA+SVM	0.9388	0.9697	0.9	Na
10	CAD	UCI	PCA+SVM	0.895	0.8529	0.85	Na
11	CAD	UCI Statlog	SOM+SVM	0.8723	0.8235	0.82	Na
12	CAD	UCI Statlog	SVM	0.851	0.7641	0.79	Na
13	CAD	Z-Alizadeh Sani	SVM	0.6977	Na	Na	AUC: 0.8090
14	CAD	Z-Alizadeh Sani	CHAID	0.8062	Na	Na	AUC: 0.8230
15	CAD	Z-Alizadeh Sani	C5.0	0.8217	Na	Na	AUC: 0.83
16	CAD	Z-Alizadeh Sani	Random forest	0.9147	Na	Na	AUC: 0.9670
17	CAD	Z-Alizadeh Sani	SVC+GA	0.92	Na	Na	$F1$: 0.906
18	CAD	Z-Alizadeh Sani	nuSVM+GA	0.93	Na	Na	$F1$: 0.903
19	CAD	Z-Alizadeh Sani	LinSVM+GA	0.92	Na	Na	$F1$: 0.910
20	CAD	Z-Alizadeh Sani	SVC+PSO	0.904	Na	Na	$F1$: 0.903
21	CAD	Z-Alizadeh Sani	nuSVM+PSO	0.911	Na	Na	$F1$: 0.898
22	CAD	Z-Alizadeh Sani	LinSVM+PSO	0.89	Na	Na	$F1$: 0.870
23	Hepatitis	UCI	ANN+LM	0.9187	Na	Na	Na
24	Hepatitis	UCI	LFDA+SVM	0.967	0.8571	1	Na
25	Lung cancer	NCI carry	MAS5+PCA+ANN	0.83	Na	Na	Na
26	Alzheimer's disease	OASIS	ConvNN+BN+ MP+NN	0.88	Na	0.93	$F1$: 0.92
27	Heart disease	Cleveland	ELM+NN	0.8	Na	Na	Na
28	Rare Mendelian	RAMEDIS	RDAD:APML	Na	Na	0.82	$F1$: 0.89 PRE: 0.95
29	Rare Mendelian	RAMEDIS	RDAD:CPML	Na	Na	0.95	$F1$: 0.97 PRE: 0.99
30	Rare Mendelian	RAMEDIS	RDAD:PGAS	Na	Na	0.67	$F1$: 0.76 PRE: 0.96

(continued)

Table 17.1 (Continued)

Sr. No.	Disease	Resource of data	Algorithm	Accuracy	Specificity	Sensitivity	Others
31	Rare Mendelian	RAMEDIS	RDAD:PICS	Na	Na	0.85	F1: 0.91, PRE: 0.99
32	Liver	UCI	Artificial immune+GA	0.9812	0.9603	0.9888	F1: 0.98 PRE: 0.98
33	Alzheimer	ADNI-1, ADNI-2, MIRIAD	L parallel sub-CNN	0.91	0.93	0.88	F1: 0.8974
34	Heart	UCI	Fuzzy rule based	0.79	0.9	0.84	Na
35	Cancer	EMR	MCMC+GM+ EM+HMM	0.8022	Na	Na	Na
36	Diabetes	UCI Pima	J48	Na	Na	0.738	F1: 0.73 PRE: 0.735
37	Diabetes	UCI Pima	MLP	Na	Na	0.75	F1: 0.751 PRE: 0.75
38	Diabetes	UCI Pima	Hoeffding Tree	Na	Na	0.757	F1: 0.759 PRE: 0.757
39	Diabetes	UCI Pima	JRip	Na	Na	0.755	F1: 0.755 PRE: 0.755
40	Diabetes	UCI Pima	BayesNet	Na	Na	0.741	F1: 0.74 PRE: 0.741
41	Diabetes	UCI Pima	RF	Na	Na	0.754	F1: 0.75 PRE: 0.75
42	Heart	Cleveland	CFS+Rotation Forest	0.775	Na	Na	Na
43	Parkinson's	Parkinson's dataset	CFS+Rotation Forest	0.844	Na	Na	Na
44	Diabetes	Diabetes	CFS+Rotation Forest	0.721	Na	Na	Na
45	Thyroid	UCI thyroid	AIRS+Fuzzy weight pre-processing	0.85	Na	Na	Na
46	Thyroid	Sets	Rule-based system+PSO	0.85	Na	Na	Na
47	Breast cancer	UCI breast cancer	TVMOPSO+RBF	96.53	Na	Na	Na
48	Diabetes	UCI Pima	TVMOPSO+RBF	78.02	Na	Na	Na
49	Hepatitis	UCI Hepatitis	TVMOPSO+RBF	82.26	Na	Na	Na
50	Parkinson's disease	MRI	PCA+SVM	84.7	87.5	83.8	Na
51	Iris's eye	Iris	RF	0.8997	Na	Na	Na
52	Hepatitis	UCI	SVM+SA	0.96	Na	0.98	F1: 0.97 PRE: 0.91
53	Brain disease	ADNI-1, ADNI-2, MIRIAD	MOLR+DeepESRNet	0.9028	89.05	92.65	Na

Na = Not applicable.
AUC = Area under the curve.
CHAID = Chi-square automatic interaction detection.

17.5 Proposed Model/Methods for Healthcare Monitoring System Using ML/DL

Discussing different approaches to making intelligent algorithms for better results in any IoT device has led us to select some exceptionally better algorithms than the rest. But yearly we need several such reviews because of new methods and techniques. Our model behaves perfectly by looking at the model prediction to be the same over the number of runs. Another important factor in ML is that neural networks are a black box. Uncertainty of how the model behaves is significant. So, it is necessary to exploit the model with random noise. Vanilla neural networks use point estimates to predict the actual prediction, which is good, but we still move to ensemble learning. But it is computationally expensive as it creates several models with separate parameters. Instead of finding the best model, it is more favorable to average our prediction. As a result, we get an optimum model which is overall generalized. The Bayesian neural network assigns a distribution to solve this ensemble problem instead of taking point values for weights and bias. It is generally known as the prior $P(w)$, selected to a distribution close to the data distribution. A Gaussian prior or a mixture of Gaussian distributions is more famous for biological neural network (BNN) [35]. This eventually helps us to get a better understanding of neural network uncertainty.

A likelihood function is constructed for a given dataset D and is provided by $P(D|w) = \prod_i P(y^{(i)}|x^{(i)}, w)$, where w is the weight. Maximum likelihood estimation aims to find the total value of $P(D|w)$ for a particular set of weights. Multiplying the prior with the likelihood gives us the posterior distribution. A posterior distribution is shown to sample out, and the posterior distribution is adjusted with the primary using maximum a posteriori (MAP), which eventually maximizes the posterior overweight (w). But the function is computationally intractable, so the approximation needs to be carried out using variational inference, which can minimize the Kullback–Leibler divergence between $q(w|\theta)$ and true posterior [40], and it is given by Eq. (17.10)

$$F(D, \ominus) = \text{KL}[q(w|\theta) \| P(w)] - E_{q(w|\theta)}[\log P(D|w)] \tag{17.10}$$

Where is the negative log-likelihood of the prediction? This method is commonly known as variational inference. In [25], Bayes by backdrop was introduced for approximating $F(D, \ominus)$, which is less computationally expensive (Eq. (17.11)).

$$F(D, \theta) \approx 1/N \sum_{i=1}^{N} [\log q(w^{(i)}|\theta) - \log p(w^{(i)}) - \log p(D|w^{(i)})] \tag{17.11}$$

Much better techniques and better estimators were introduced in [37], where they deal with global uncertainty of weights translated into local delays, which are not dependent across samples. This new estimator is more computationally efficient as it samples noise is not directly but as a function of $f(\in)$, so global noise is converted into local noise. The use of machine code (MC) dropout helps us to regularize uncertainty. Another recent development in estimators is the usage of flipout. In [36], which is based on two assumptions about weight distribution q_θ: (i) Perturbations of different weights are independent, and (ii) Distribution will be symmetric around zero. Then the perturbation distribution does not vary with the scalar multiplication by a random sign matrix whose entries are ± 1. If ΔW be a perturbation distribution and E be a lucky sign matrix, then $\Delta W = \Delta W \times E$ is identically distributed to ΔW. The loss gradient computed using ΔW and ΔW is identically distributed. We have used flipout estimator as the NN with MC dropout and label smoothing for better results in prediction [39]. We use Monte Carlo sampling to sample out from the distribution and predict the probabilities and the median. It is closer to the maximum value in the distribution and much more stable [41]. The proposed block diagram of methodology is shown in Figure 17.1.

We consider using flip-out layers, dropout, and Monte Carlo simulation. We were able to predict the probability of an accurate label; we also created random data and tried to predict the likelihood of its brands. We find that accuracy and possibility don't respond to random noise, but the ensemble representation of models helps us determine random noise.

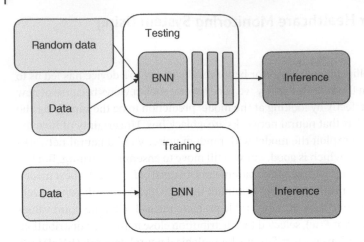

Figure 17.1 Block diagram of proposed system.

17.5.1 Case Study-I: Breast Cancer

The breast cancer dataset contains specific attributes:

- ID number: ID of the dataset
- Diagnosis (M = malignant, B = benign) – Type of diagnosis it is going through

The following real-value features are computed for each cell nucleus:

- Radius – mean of distances from the center to points on the perimeter
- Texture –standard deviation of gray-scale values
- Perimeter
- Area
- Smoothness –local variation in radius lengths
- Compactness –perimeter^2/area – 1.0
- Concavity –the severity of concave portions of the contour
- Concave points –number of concave portions of the contour
- Symmetry
- Fractal dimension – "coastline approximation" – 1

Flow-Chart Steps

- **Step 1:** Initialization of parameters w, θ of the network
- **Step 2:** Repeat steps 3–6 until convergence of parameters (w, θ)
- **Step 3:** Initialization of a minibatch of dataset $\mathbf{X^m}$
- **Step 4:** Initialization of weights $\mathbf{W}$
- **Step 5:** Calculate the gradient
- **Step 6:** Update the parameters concerning G

Inference-Monte Carlo

- **Step 1:** Repeat steps 3 and 4 until the given number of times
- **Step 2:** $\mathbf{W}$-initialization of parameters
- **Step 3:** Calculate all prediction distributions
- **Step 4:** Average the prediction for the final result.

17.6   Experimental Results and Discussion

Each data is normalized to the [−1, 1] range for feature transformation. This helps foster learning in neural networks as we have selected Gaussian prior. We use 250 Monte Carlo simulations for posterior sampling on breast cancer data containing two classes. When we use traditional neural networks, we get a high probability of random data (Figure 17.2).

The random data is normalized to get adjusted with accurate normalized data; we try to understand the overall response. When inference is carried out on BNN, we find reduced probabilities (Figure 17.3) but more importantly, we now get a probability distribution. Hence, we see the standard deviation in the probability distribution, which helps us understand the model response to random data (Figure 17.4).

We can see the distribution in Figure 17.4 has a significant variance, and their probability is in the middle, which helps to tell us about their uncertainty. We also use CNNs as it is a suitable feature extractor. We get a slightly better distribution with that in Figure 17.5.

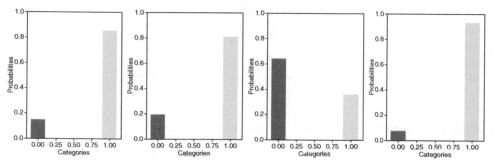

Figure 17.2   Categories versus probabilities using traditional NN of random data.

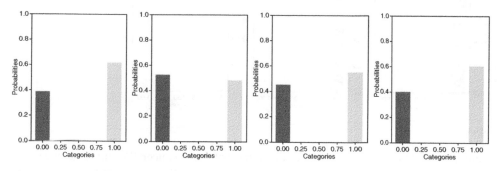

Figure 17.3   Probability of random data on BNN.

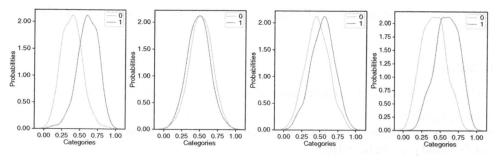

Figure 17.4   Probability distribution of random data using 20 binarized neural network machine code (BNNMC).

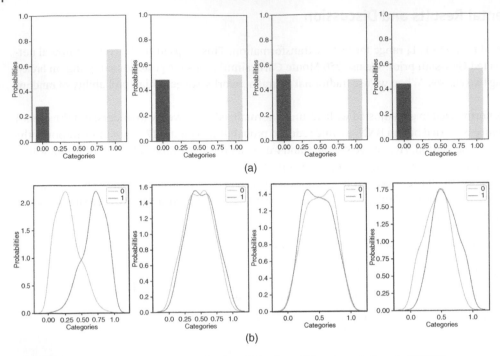

(a)

(b)

Figure 17.5 Probability (a) and its distribution (b) for four different predictions in BNNMC.

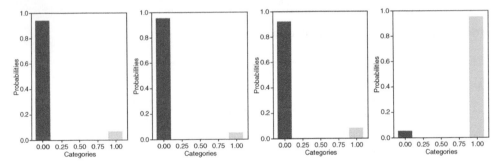

Figure 17.6 Probability using NN.

We have looked into random data. Now we look into the evaluation set of breast cancer. We almost get the same result in NN and BNN but what is worth it is we are now confident about our prediction. In Figure 17.6, we see only the probability achieved from NN, but in Figure 17.7, we can see the actual probability and the possible deviation in likelihood. We are now more confident about prediction as we can see the variance in the test set is less, so prediction is trustworthy.

17.6.1 Case Study-II: Diabetes

Pima Indians diabetes database contains diabetes conditions for patients. Some of its attributes are:

- Pregnancies = No of times pregnant
- Glucose = Plasma glucose concentration for a two hours in an oral glucose tolerance test
- Blood pressure = Diastolic blood pressure (mm Hg)
- Skin-thickness = Triceps skinfold-thickness (millimeter)
- Insulin = Two hour serum-insulin (muU/ml)
- BMI = Body-mass index (weight in kg/[height in m]²)

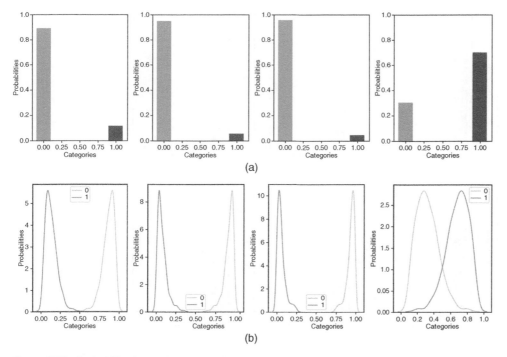

Figure 17.7 Probability (a) and its distribution (b) for four different predictions in BNNMC.

- Diabetes pedigree function = Diabetes pedigree function
- Age = Age of patient
- Outcome = 1 or 0

Flow-Chart Steps

- **Step 1:** Initialization of parameters w, θ of the network
- **Step 2:** Repeat steps 3–6 until convergence of parameters (w, θ)
- **Step 3:** Initialization of a minibatch of dataset $\mathbf{X^m}$
- **Step 4:** Initialization of weights $\mathbf{W}$
- **Step 5:** Calculate the gradient
- **Step 6:** Update the parameters concerning G

Inference-Monte Carlo

- **Step 1:** Repeat steps 3 and 4 until the given number of times
- **Step 2 : W**-initialization of parameters
- **Step 3:** Calculate all prediction distributions
- **Step 4:** Average the prediction for the final result.

Using BNN in diabetes, we achieved excellent results, but on diabetes, we didn't get too good results on NN in BNN, but since we used BNN, we could exploit the uncertainty. We used the same number of training for diabetes 100 iterations and 250 Monte Carlo samples for prediction. Since the number of training could increase the accuracy, but it can overfit the model, we traced loss and trained the network for 200 epochs. After that, we don't see a significant difference in loss. We got predictions from neural nets that are pretty uncertain, as shown in Figure 17.8.

On training a BNN, we found a massive uncertainty of probabilities across all samples. From Figure 17.9, no accurate predictions were made either by deterministic NN or BNN.

So, we tried to find the source of uncertainty by comparing datasets. We saw that the average correlation between all values in the diabetes dataset is less than that of breast cancer, which does not correctly help the neural network

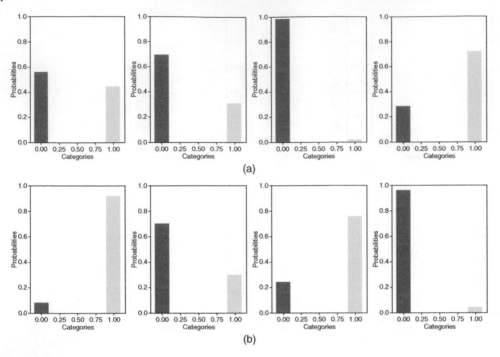

Figure 17.8 Random data (a) and Pima diabetes (b) with less confidence.

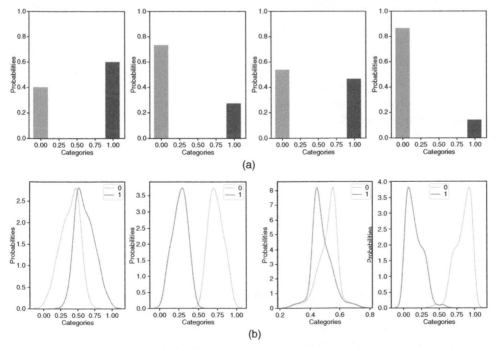

Figure 17.9 Probability estimate (a) and probability distribution (b).

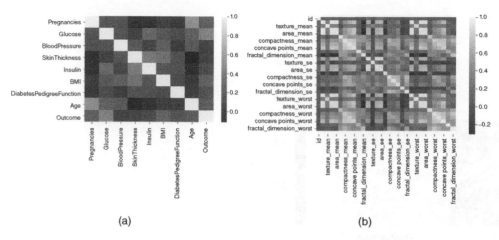

(a)   (b)

Figure 17.10 Correlation between samples of diabetes (a) and breast cancer (b).

estimate (Figure 17.10). Even the distribution of positive samples is 0.69, and the rest are harmful to diabetes, making it an imbalanced dataset. So, predictions on them are significantly less; we could increase that by overfitting. So, we get a low accuracy during the prediction. The confusion matrix also shows how each diabetes dataset's excellent data affect the prediction. The results comparison of proposed technology with existing technology is shown in Table 17.2.

The comparison between different techniques based on accuracy is shown in Figure 17.11.

The confusion matrix for breast cancer (above) and diabetes (below) is shown in Figure 17.12.

Table 17.2 Results comparison of proposed technology with existing technology.

Sr. No.	Techniques	Case	Accuracy (%)	No. of parameters
1	NN	1	98.88	4180
2	BNN+MC	1	98.86	4506
3	ConvBNN+MC	1	98.87	8506
4	NN	2	80.9	2722
5	BNN+MC	2	80.8	1098
6	ConvBNN+MC	2	80.8	5770

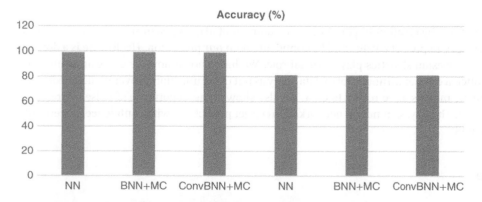

Figure 17.11 Comparison between different techniques based on the accuracy.

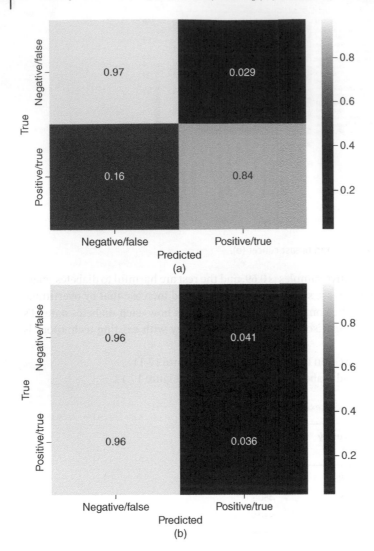

Figure 17.12 Confusion matrix for breast cancer (a) and diabetes (b).

17.7 Conclusions

We have shown the future of ML by creating importance in the domain of ML. Bayesian statistics is essential in healthcare, where it is necessary to predict a value and be confident about our prediction. Healthcare is a domain with a lot of uncertainty, so Bayesian statistics play a crucial role. We have shown how we can gain confidence in prediction, and we exploited a dataset without a lot of data analysis but using uncertainty over model prediction. This eventually helps us make models more robust to data. We also showed the importance of noise inference in finding a suitable model. Thus, the Bayesian neural network helps us get predictions with confidence, uncertainty analysis, and noise robustness.

17.8 Future Scope

Since Bayesian inference is new in its era, much research has been needed to increase model performance. One of its domains is the preliminary analysis, where selecting a prior is essential for faster training and best fit. Rather than using one initial, we can use several before analyzing massive dimension data. Another domain is causal inference, where the cause of uncertainty is deterministically found. This will help us find the source and eliminate them for better performance. Information entropy has also been looked at where we drop weights from the Bayesian model as it contains many parameters.

Another domain is implementing them in embedded devices that require a lot of computation, so they need to be quantized before deploying. IoT has enormous significance in data collection; regular update of the model is necessary through stored data. IoT is vulnerable to attacks, and a robust model will only help to segregate noise and factual data. So IoT has a lot of scope in using Bayesian statistics rather than point estimates.

References

1 Nilashi, M., Ahmadi, H., Manaf, A.A. et al. (2020). Coronary heart disease diagnosis through self-organizing map and fuzzy support vector machine with incremental updates. *International Journal of Fuzzy Systems* 22: 1376–1388. https://doi.org/10.1007/s40815-020-00828-7.

2 Joloudari, J.H., Joloudari, E.H., Saadatfar, H. et al. (2020). Coronary artery disease diagnosis; ranking the significant features using a random trees model. *International Journal of Environmental Research and Public Health* 17 (3): 731. https://doi.org/10.3390/ijerph1703073.

3 Abdar, M., Wojciech Książek, U., Acharya, R. et al. (2019). A new machine-learning technique for an accurate diagnosis of coronary artery disease. *Computer Methods and Programs in Biomedicine* 179: 104992, ISSN: 0169-2607, https://doi.org/10.1016/j.cmpb.2019.104992.

4 Al-Tashi, Q., Rais, H., and Abdulkadir, S.J. (2018). Hybrid swarm intelligence algorithms with ensemble-machine-learning for medical diagnosis. In: *2018 4th International Conference on Computer and Information Sciences*, Kuala Lumpur, 1–6. https://doi.org/10.1109/ICCOINS.2018.8510615.

5 Bascil, M.S. and Temurtas, F. (2011). A study on hepatitis disease diagnosis using multilayer neural network with Levenberg Marquardt training algorithm. *Journal of Medical Systems* 35: 433–436. https://doi.org/10.1007/s10916-009-9378-2.

6 Battineni, G., Sagaro, G.G., Chinatalapudi, N., and Amenta, F. (2020). Applications of machine learning predictive models in the chronic disease diagnosis. *Journal of Personalized Medicine* 10 (2): 21. https://doi.org/10.3390/jpm10020021. PMID: 32244292; PMCID: PMC7354442.

7 Chen, H.-L., Liu, D.-Y., Yang, B. et al. (2011). A new hybrid method based on local-Fisher discriminant-analysis and support vector machines for hepatitis disease diagnosis. *Expert Systems with Applications* 38 (9): 11796–11803, ISSN: 0957-4174, https://doi.org/10.1016/j.eswa.2011.03.066.

8 Chen, Y.C., Ke, W.C., and Chiu, H.W. (2014). Risk classification of cancer survival using ANN with gene expression data from multiple laboratories. *Computers in Biology and Medicine* 48: 1–7, https://doi.org/10.1016/j.compbiomed.2014.02.006. Epub February 22 2014. PMID: 24631783.

9 Chui, K.T., Alhalabi, W., Pang, S.S.H. et al. (2017). Disease diagnosis in smart healthcare: innovation, technologies and applications. *Sustainability* 9: 2309. https://doi.org/10.3390/su9122309.

10 Demir, E., Chaussalet, T., Adeyemi, S., and Toffa, S. (2012). Profiling hospitals based on emergency readmission: a multilevel transition modelling approach. *Computer Methods and Programs in Biomedicine* 108 (2): 487–499, ISSN: 0169-2607, https://doi.org/10.1016/j.cmpb.2011.03.003.

11 Islam, J. and Zhang, Y. (2018). Brain MRI analysis for Alzheimer's disease diagnosis using an ensemble system of deep convolutional neural networks. *Brain Informatics* 5 (2): 2. https://doi.org/10.1186/s40708-018-0080-3. PMID: 29881892; PMCID: PMC6170939.

12 Ismaeel, S., Miri, A., and Chourishi, D. (2015). Using the extreme learning machine (ELM) technique for heart disease diagnosis. In: *2015 IEEE Canada International Humanitarian Technology Conference*, Ottawa, ON, 1–3. https://doi.org/10.1109/IHTC.2015.7238043.

13 Jia, J., Wang, R., An, Z. et al. (2018). RDAD: a machine learning system to support phenotype-based rare disease diagnosis. *Frontiers in Genetics* 9: 587. https://doi.org/10.3389/fgene.2018.00587.

14 Kononenko, I. (2001). Machine learning for medical diagnosis: history, state of the art and perspective. *Artificial Intelligence in Medicine* 23 (1): 89–109. https://doi.org/10.1016/s0933-3657(01)00077-x. PMID: 11470218.

15 Liang, C. and Peng, L. (2013). An automated diagnosis system of liver disease using artificial immune and genetic algorithms. *Journal of Medical Systems* 37 (2): 9932. https://doi.org/10.1007/s10916-013-9932-9. Epub 1 March 2013. PMID: 23456763.

16 Liu, M., Zhang, J., Adeli, E., and Shen, D. (2018). Landmark-based deep multi-instance learning for brain disease diagnosis. *Medical Image Analysis* 43: 157–168, ISSN: 1361-8415, https://doi.org/10.1016/j.media.2017.10.005.

17 Malmir, B., Amini, M.H., and Chang, S.I. (2017). A medical decision support system for disease diagnosis under uncertainty. *Expert Systems with Applications* 88: 95–108. https://doi.org/10.1016/j.eswa.2017.06.031.

18 Manogaran, G., Vijayakumar, V., Varatharajan, R. et al. (2018). Machine learning-based big-data processing framework for cancer-diagnosis using hidden Markov model and GM clustering. *Wireless Personal Communications* 102: 2099–2116. https://doi.org/10.1007/s11277-017-5044-z.

19 Mercaldo, F., Nardone, V., and Santone, A. (2017). Diabetes mellitus affected patients classification and diagnosis through machine learning techniques. *Procedia Computer Science* 112: 2519–2528, ISSN: 1877-0509, https://doi.org/10.1016/j.procs.2017.08.193.

20 Ozcift, A. and Gulten, A. (2011). Classifier ensemble construction with rotation forest to improve medical diagnosis performance of machine learning algorithms. *Computer Methods and Programs in Biomedicine* 104 (3): 443–451. https://doi.org/10.1016/j.cmpb.2011.03.018. Epub 30 April 2011. PMID: 21531475.

21 Polat, K., Şahan, S., and Güneş, S. (2007). A novel hybrid method based on artificial immune recognition system (AIRS) with fuzzy weighted pre-processing for thyroid disease diagnosis. *Expert Systems with Applications* 32 (4): 1141–1147. https://doi.org/10.1016/j.eswa.2006.02.007.

22 Prasad, V., Rao, T.S., and Babu, M.S.P. (2016). Thyroid disease diagnosis via hybrid architecture composing rough datasets theory and machine learning algorithms. *Soft Computing* 20: 1179–1189. https://doi.org/10.1007/s00500-014-1581-5.

23 Qasem, S.N. and Shamsuddin, S.M. (2011). Radial basis function network based on time-variant multi-objective particle swarm optimization for medical diseases diagnosis. *Applied Soft Computing* 11 (1): 1427–1438, ISSN: 1568-4946, https://doi.org/10.1016/j.asoc.2010.04.014.

24 Sajda, P. (2006). Machine-learning for detection and diagnosis of disease. *Annual Review of Biomedical Engineering* 8: 537–565. https://doi.org/10.1146/annurev.bioeng.8.061505.095802. PMID: 16834566.

25 Salvatore, C., Cerasa, A., Castiglioni, I. et al. (2014). Machine learning on brain MRI data for differential diagnosis of Parkinson's disease and Progressive Supranuclear Palsy. *Journal of Neuroscience Methods* 222: 230–237. https://doi.org/10.1016/j.jneumeth.2013.11.016. Epub 26 November 2013. PMID: 24286700.

26 Samant, P. and Agarwal, R. (2018). Machine learning techniques for medical diagnosis of diabetes using iris images. *Computer Methods and Programs in Biomedicine* 157: 121–128, ISSN: 0169-2607,https://doi.org/10.1016/j.cmpb.2018.01.004.

27 Sartakhti, J.S., Zangooei, M.H., and Mozafari, K. (2012). Hepatitis disease diagnosis using a novel hybrid method based on a support vector machine and simulated annealing (SVM-SA). *Computer Methods and Programs in Biomedicine* 108 (2): 570–579. https://doi.org/10.1016/j.cmpb.2011.08.003.

28 Suk, H.I., Lee, S.W., Shen, D., and Alzheimer's Disease Neuroimaging Initiative (2017). Deep ensemble learning of sparse regression models for brain disease diagnosis. *Medical Image Analysis* 37: 101–113. https://doi.org/10.1016/j.media.2017.01.008. Epub 24 January 2017. PMID: 28167394; PMCID: PMC5808465.

29 Temurtas, H., Yumusak, N., and Temurtas, F. (2009). A comparative study on diabetes disease diagnosis using neural networks. *Expert Systems with Applications* 36 (4): 8610–8615. https://doi.org/10.1016/j.eswa.2008.10.032.

30 Yin, H. and Jha, N.K. (2017). A health decision support system for disease diagnosis based on wearable medical sensors and machine learning ensembles. *IEEE Transactions on Multi-Scale Computing Systems* 3 (4): 228–241. https://doi.org/10.1109/TMSCS.2017.2710194.

31 Lakshmi Narayanan, B., Pritzel, A., and Blundell, C. Simple and scalable predictive uncertainty estimation using deep ensembles. In: *Advances in Neural Information Processing Systems*, 6402–6413.

32 Lecun, Y., Bottou, L., Bengio, Y., and Haffner, P. (1998). Gradient-based learning applied to document recognition. *Proceedings of the IEEE* 86 (11): 2278–2324. https://doi.org/10.1109/5.726791.

33 Ioffe, S. and Szegedy, C. (2014). Batch normalization: accelerating deep network training by reducing internal covariate shift.

34 Bowman, S.R., Vilnis, L., Vinyals, O. et al. (2015). Generating sentences from a continuous-space. arXiv preprint arXiv:1511.06349.

35 Blundell, C., Cornebise, J., Kavukcuoglu, K., and Wierstra, D. (2015). Weight uncertainty in neural network. In: *Proceedings of the 32nd International Conference on ML*, vol. PMLR 37, 1613–1622.

36 Wen, Y., Vicol, P., Ba, J. et al. (2018). Flipout: efficient pseudo-independent weight perturbations on mini-batches. arXiv preprint arXiv:1803.04386.

37 Kingma, D.P., Salimans, T., and Welling, M. (2015). Variational dropout and the local reparameterization trick. In: *Advances in Neural Information Processing Systems*, 2575–2583.

38 Kingma, D.P. and Welling, M. (2013). Auto-encoding variational Bayes. arXiv preprint arXiv:1312.6114.

39 Hinz, T., Fisher, M., Wang, O. et al. (2020). Improved techniques for training single-image GANs. arXiv preprint arXiv:2003.11512.

40 Shridhar, K., Laumann, F., and Liwicki, M. (2019). A comprehensive guide to Bayesian convolutional-neural-network with variational inference. arXiv preprint arXiv:1901.02731.

41 Papadopoulos, C.E. and Yeung, H. (2001). Uncertainty estimation and Monte Carlo simulation method. *Flow Measurement and Instrumentation* 12 (4): 291–298.

42 Kumar, K., Bhaumik, S., and Tripathi, S.L. (2021). Health monitoring system. In: *Electronic Device and Circuits Design Challenges to Implement Biomedical Applications*, 461–480. Elsevier. ISBN: 978-0-323-85172 https://doi.org/10.1016/B978-0-323-85172-5.00021-6.

43 Singh, T. and Tripathi, S.L. (2021). Design of a 16-bit 500-MS/s SAR ADC for low power application. In: *Electronic Device and Circuits Design Challenges to Implement Biomedical Applications*, 257–273. Elsevier. ISBN: 978-0-323-85172-5 https://doi.org/10.1016/B978-0-323-85172-5.00021-6.

44 Kumar, K., Sharma, A., and Tripathi, S.L. (2021). Sensors and their application. In: *Electronic Device and Circuits Design Challenge to Implement Biomedical Applications*, 177–195. Elsevier. ISBN: 978–0–323-85172-5 https://doi.org/10.1016/B978-0-323-85172-5.00021-6.

18

Usage of AI and Wearable IoT Devices for Healthcare Data: A Study

Swarup Nandi[1], Madhusudhan Mishra[2], and Swanirbhar Majumder[1]

[1]*Department of Information Technology, Tripura University, Agartala, Tripura, India*
[2]*Department of ECE, NERIST, Nirjuli, Arunachal Pradesh, India*

18.1 Introduction

Recently we have seen the increasing part in wearing sensors, and presently we found numerous devices that are readily find in a commercial manner [1–3] for individual hygiene care, suitability, and mobility awareness. Wearable, hearable (in-ear wearing devices) and nearable (devices that interact with wearables) are built with the help of the idea of internet of things (IoTs), which provides the reliable process to manage oncoming hygiene care, and the benefits of the healthy lifestyles [4, 5]. This recitation starts along with the smartphone; it is currently represented with an extensive uninvited and thorough-going technology. Recently almost all wearable and nearable devices are decorated, including several kinds of high-degree complex sensors. These high-degree complex sensors are controlled by the systematic computational analytics are being investigated to implement the range of operations of truly bearable medical laboratories. The stocking integration of these assessments in smartphone applications provides permission to locate the facts to be released within due time, improving the user idea in representation support living continuity. The primary adoption, simple usage, and reliability of smartphones simplify user trust in various valuable applications that accommodate the dependence in the area of self-physiological realization and fitness controlling [6]. Wearable process has proposed attitudes through the most effective ability on fitness and hygiene- care craft [4, 7, 8].

The main motive of this chapter is to familiarize with different types of wearable-internet of things (W-IoT) devices that are widely used in medical science in recent times and provide a brief description of those wearable devices and the differences between them.

18.2 Literature Review

In recent time, we find that many IoT processes have been proposed for IoT medical-care facilities and co-operate contemporary applications. Wang et al. has been proposed a compound traffic norm balanced IoT method for medical-care facility [9]. Xu et al. has been designed a structure-based data accessing method (UDA-IoT)/system for the sake of medical-care facility intimation-profound uses [4]. A confirmation of medical support process that sounds with equal and IoT techniques has been proposed by Kolici et al. The proposed system consists of a smart box that is used to monitor the patient's condition. In addition, it carried out many experiments to elaborate the complete system with different stages [9]. With the increasing demand, a web real-time communication (WebRTC) and IoT device tunneling process that is widely used for hospitals has been introduced by Sandholm et al.

This system fully counts on stopping original parts of the WebRTC, is JSEP, and also uses endemic network portals that incorporate into a multiple-operation signal from multiple accompanying streams effectively except pricking any WebRTC traffic [10]. An achievement and preparation of bio-medical info with the help of IoT has been introduced by Antonovici et al. This system is placed by an Android app whose purpose is to collect and gather the fact and controlled systolic blood pressure, diastolic blood pressure and heart rate using the e-sphygmomanometer, which is controlled by through bluetooth. The mentioned process provides the chance of transferred healthcare info with the support of smartphones. Firstly data will be estimated with common values and if any fault has been found, then patient is notified. During the emergency period, the system provides emergency service for the patients and doctors will be notified automatically. The "text-to-speech" engine is supported by the patients with vision impairment, patient difficulties with diabetes, hypertension, or obesity. This engine allows the facts to be transferred as a typical string to the device [11]. A real-time internet application deals out with stream of conditions for medical IoT and has been proposed by Krishnan et al. In this system, the raw fact will be automatically plucked locally and transmitted to the server if the patient is out of Wi-Fi area, or even server is absent, and when the patient comes back in range of connectivity, the system automatically notified the patients about their health data [12]. An algorithm that is helpful for the detection of echocardiogram signal resolution, as well as arrhythmia detection which is fully supported by an IoT-based, attached wearable medical system has been proposed by Azariadi et al. It is compatible with 24-hours endless governance containing a Galileo board applied to equipment of the model [13].

A virtual reality safety framework is set up for IoT-based personal medical devices (PMDs) has been proposed by Mohan that qualify to improve mobility of the patient. In addition, it was naturalized to the better controlling ability of the patient's situation while they are moving. He represents the privacy impendence and drawbacks of PMD IoT that makes accosting these impendence challenging. In addition, they used some primary ways of solution in order to resolve these impendence challenging [14]. A cloud-computing-based mini healthcare system has been proposed by Yeh et al. that had been focused on refuge monitoring tricks for lightweight IoT devices. They handled all of the potential security issues as well as the issues of the cloud mutuality. From the experimental outcomes, we saw that the designed plan is hopeful for a cloud-based individual health information platform (PHI) [15]. A secured lightly built validating along with prime dedication set of rules for first to last exchange of data in the sake of awkward devices in IoT-enabled bound supported living process has been introduced by Porambage et al. They used delegate-based access for assigning the heavily computational operations as a co-helping device of more powerful devices used in medical sensors. The experimental outcomes are hopeful and related to the real-world applications [16].

A bearable e-travel aid special for person with vision problems has been proposed by Yelamarthi and Laubhan. It has been utilizing the ultrasonic range finders that find the distance between obstacles and the user, mounted on the belt of the user, the implemented devices fully support finding barriers placed before the user, and provide dedicated seamanship guidance with a bluetooth headphone [19]. A depth sensor-based navigation system have proposed by Yelamarthi et al. [17, 18], which is used to find out barriers in front of the user with elevated appropriateness, it also notified the people with vibrotactile feedback that is designed in the user hand gloves.

All of the above-mentioned systems are not provided with 100% accurate outcomes/results. There we find all of the systems have some minor limitations on them. Greeting these limitations, in this chapter, we represent an overview idea about the functionalities, features, and comparison among different IoT-based wearable devices related to the healthcare system. Some earlier works have shown the overview of existing related approach. It indicates the amount of focus in the diverged direction as reported in Refs. [25–27].

18.3 AI-Based Wearable Devices

Wearable technology related to healthcare system includes different types of computerized devices, being wired by patients, such as fit bits and smart watches, which are well designed for gathering the user's healthcare information and exercise. It can be observed that the use of wearable devices increased from 9% to 33% during four years (2014–2018) by patients of America [20]. The improvement of wearable technology with an increasing requirement from patient's to monitoring their self-body condition that is tremendous helpful to the healthcare system, also influence to construct of numerous wearable devices, such as fit bits, smart watches, and wearable monitors.

The observance of wearable devices by patients has reached near about four times within four years (2014–2018) that has been found from an observation conducted by Accenture states [23]. The practice of device has just 9% in 2014, and it can rapidly grow to 33% in 2018. From the above survey, it should give us a clear idea that these sophisticated, smart gadgets will be reached to the highest peak of medical services.

Advancement of extreme current technologies like big data, machinelearning (ML), cloud computing (CC), and artificial intelligence (AI), contribute to the W-IoT technology, as a result, it can lead to savings about $200 billion money globally in the medical-care system.

Along with abundant characteristics and range of operations of wearable devices includes outpatients controlling, tracking and storing information, influencing regular medical checkups and the way of life patterns, track out long-lasting conditions, and changing the environment of hospitality system. In total, this system/technology can become amazing assistance for medical-care system as well as individuals due to their simple use and working accuracy.

Most of the different types of wearable devices are developed using various IoT technologies like cloud-assisted agent-based smart environment (CASE), improved Bayesian convolution network (IBCN), and EDL are as follows:

18.3.1 Cloud-Assisted Agent-Based Smart Environment

The CASE artifice provides an easy way to designing and developing effective use of hospitality applications. CASE artifice is specifically applied in the situation where complicated mortal activities, such as walking and sleeping need to point out from the unanalyzed data of sensors. With the CASE artifice, we should construct applications related to the smart hospitality context. The CASE architecture is very effective in monitoring mortal functions for recognizing critical or dangerous situations.

18.3.2 Improved Bayesian Convolution Network

The IBCN permits to deliver everyone and every knowledgeable process to copy information via either basic telecommunications technique or lower ability back diffraction informing with cloud facilitation. IBCN consists of, an adjustment of the pattern's dormant fickle is planned and the shape are uprooted using fold layers, the achievement of the W-IoT has been uplifted by assembling a volatile auto-encoder with a grade sound genuine classifier. In addition, the IBCN provides assistance to locate the privacy consignment.

18.3.3 EDL

The main approach of deep learning is widely used in wearable devices for finding physiological activity recognition or human-movement analysis. Some ideas of the deep-learning approach in wearable sensor are, firstly, the deep-learning approach is the achievement of unanalyzed data from the wearable sensors. After the achievement of data, the data go for pre-processing task such as filtering, removing noising, normalization, and all data synchronization. After preprocessing, the pre-processed data go for feature extraction like in time or frequency domain and the selection of informative features for deep-learning training/testing. This feature extraction, such as expert opinion, patient metadata, and study metadata are considered or used as the input of training/testing deep- learning algorithms, and these will be used as the decision making of detection and prediction for medical care.

Recently, we have discussed most used top IoT-based wearable devices related to healthcare system. These wearable devices are as follows:

- **Women's Health Focused AVA Sensor:** To provide better monitoring of women's health, especially during the period cycles, the AVA sensor is an extremely effective wearable device for them. This device is worn by women only at night. It provides monitoring on women's fertility, pregnancy, and their total health condition. In addition, AVA sensor helps to token on women's weight, amount of sleep required, and stress levels during their pregnancy cycle. A clinical study conducted by the University Hospital of Zurich shows this wearable tool is well known for its appropriateness, it indicates a mean of 5.3 days of fertility during every cycle with 89% of accuracy.

- **AliveCor – Personal EKG:** For providing better monitoring of human health, such as atrial fibrillation, worsening of heart failure, rapid heartbeat, and the rhythm of a beating heart, it is widely and easily used bearable healthcare device. It can be easily handled and kept in our pocket, which is very helpful for emergency purposes and saves our lives. It is an FDA-cleared medical device that catches high-quality of EKG recordings and from that recordings it tracks out atrial fibrillation, worsening heart failure, rapid heartbeat, and the rhythm of a beating heart within 30 seconds and the information/data are directly send to the dedicated smartphone. The data have been used by the meditative doctors for the current investigation. To collect the data/recordings, patients should place their thumbs/fingers on the sensors, and they will be receiving the information in few seconds.

- **TempTraq – Controlling Sensor for Child:** To minimize the irritation of the use thermometers, especially for sick children, TempTraq is the supreme replacement of thermometers. It provides monitoring on the temperature of babies and children, especially during their illness without making any defenses when they take rest. TempTraq consists of a bluetooth sensors, and it is designed with a light and soft patch. Differ from traditionally used thermometers, TempTraq patch can freely be set under the arm of the child. The parents can easily get their governance over their baby's temperature during different time intervals. TempTraq is highly accurate and get the measures of temperature range from 87.0 to 109.3 °F. In addition, it should fulfill all the requirements and standards of ASTM E1112-00 as an ideal digital clinical thermometer.

- **BioScarf:** BioScarf is nothing but fashionable or up-to-date wearable clothing item, alternatively used in place of air pollution masks. It can be used by patients as well as other people to protect themselves from pollutant air and from potential respiratory health issues. Inside the BioScarf, an N95 filter is placed, as a result, it strains or removes more than 99.75% of pollutants transported by air. As an example, it can keep us away from from pet-irritation, farina, fume, and other contaminants, otherwise, we may suffer from many diseases like tonsillitis, common cold, fever, lung infection, and phthisis at bay. It is better than buying more masks regularly, and it can save money.

- **Blinq – Wearable Rings:** A fitness tracking wearable healthcare device available with an alternative of category and functionality is a Blinq wearable ring. It is available in different styles and features like publicity along with light emitting diodes, efficiency tracking option, and SOS distress feature setup by the Montreal-based startup. This smart Blinq ring is effectively used as an alternative of smart watches and also comes in a limited cost. This is water-resistant wearable healthcare device, which supports about 150 Android/iOS applications with 48 hours lasting capacity of battery and is available in different sizes.

- **SmartSleep Wearable:** For getting adequate amount of sleep in smart way, Philips SmartSleep soft headband is the latest that provides assistance to individuals to recognize the amount of their sleep, it also gets the clinically prescribed solution for their sleep. The soft headband consists of sensors. This sleep analyst controls so smoothly on the user who has suffered from less sleep. This headband is very helpful for people who do not get enough time for sleep or sleep too late/early by controlling their sleep cycles. This wearable device produces an audio tone, as a result, it elevates the depth of sleep as well as the duration of REM sleep. The dedicated app helps to collect the data, such as the metrics related to sleep, along with offering guidance and tips for a good night's rest. A combination of all these supreme features helps it to be considered an effective healthcare device.

- **BioPatch:** For monitoring heart-beat rate, echocardiogram, beat-rate inconstancy, and respiration rate, the currently developed BioPatch is very effective for healthcare. It is produced by SEER that can be easily used or placed on the left chest of the patient's/user's body. It consists of multiple-sensing algorithms for calculating the health data, such as heart-beat rate, echocardiogram, beat-rate inconstancy, and respiration rate. After collecting the data, it will send this data to the cloud-assisted platform, and this data is used to analyze by the assigning pathologist.

- **Smart Glasses:** The world-class smart glasses are built by the global smart eyewear technology in the year 2018. Their estimation of the demand for smart glasses is about 5847 USD this year. This number will increase up to USD 123 124 million approximately by 2027. For fulfillment of the demand, the Netherlands-based firm has taken the necessary steps for this. Their planning is the innovation of Minuet technology, where all health-communication platforms were developed in one mobile should present in this system containing most illuminated doctrine contact lenses for video sharing.. After the development of this device, we could find some massive changes made in telehealth. The name of most effected system is Vuzix's smart glasses that help to send secured videos and communicate with others.

- **Smart-Hearing Aids:** The long-term discovery of wearable technology is hearing aids. Today, various types of smart-hearing aids are available that have different types of sensors. Among those of hearing aids Starkey Hearing Technologies produced such futuristic wearable device for persons with hearing difficulties, in addition, this device accurately filters noise and points out only on the necessary source of sounds. The name of the product launched by the factory is Livio AI, which accurately filters the noise and points out only on necessary source of sound. In addition, it will reduce irritation of ringing of the ears of the patient. Wearable sensors and artificial intelligence support wearable device that must keep track on necessary health data, such as physical activities like running, walking, and are built to measure heart-beat rate in the future.

- **Wireless-Patient-Monitoring:** "Remote-patient monitoring" is the most precedence-based service in healthcare system that helps to reduce labor and save time and money for the patient as well as doctors. In real situation, few wearable devices containing most recent technology gives relief from the difficulties of the healthcare professionals. As an example, we can take "leaf healthcare," which provides an identical resolution and wireless-patient monitoring technology named "leaf patient monitoring system" that remotely control on patient's health issues. In addition, it gives relief from many pressure-related injuries of the patients.

- **Wearable-Fitness Trackers:** "Wearable-fitness trackers" is the simple and highest effective composition of wearable technology. Among all of the "wearable-fitness trackers" wristbands [22] that clipped along different types of sensors are modern ones. It helps to monitor on patient's physical activity as well as their heart rate. It comes with different health and fitness recommendation service that is fully analyzed by the smartphone apps.

- **Smart-Health Watch:** In recent time, smart watches are considered as a healthcare device because presently it was developed with clinically validate healthcare tools. In the year 2017, Apple had launched an app named "Apple heart study" that monitor on the patient's heart rates and alert for experiencing inconstancy in heat rate. In addition, they also launched another app named "movement disorder API," which is very helpful for scholars and researchers to collect information about the Parkinson's disease. Smart watch provides services to the users, such as read instructions, transmit messages, and makes cellular calls. It also gives few tricks of benefits of health as well as fitness trackers.

- **Wearable ECG Monitors:** Co-operative service of remote health monitoring is ECG monitoring. For monitoring the ECG of a patient remotely than wearable ECG monitors is very effective for it. It becomes inbuilt with smart watch that helps to capture the electrocardiograms of the patient. Recently a report was published by "Business Insider" that shows "Withings" is a very popular moveable ECG monitor in 2019 that is widely used to measure an electrocardiogram of the patients swiftly. In addition, it automatically sends the reading to the assigning doctor. It also detects the atrial fibrillation, pace, distance, elevation, and self-acting controlling for walking, running, swimming, and biking of the patient's condition.

- **Wearable Blood Pressure Monitors:** For tracking blood pressure "Omron Healthcare" launched a wearable device named "HeartGuide" in the year 2019, which is the first considered as the first wearable, for measuring the pressure of the blood in the circulatory system. This is nothing but an emblematic smart watch also an oscillometric blood pressure monitor used to monitor the pressure of the blood in circulatory system, regular activity, such as number of steps taken, total distance covered by the user, and amount of calories been used and required. The capability of the HeartGuide is to store about 100 readings in its storage space and transfer the data to the assigning mobile app. HeartGuide comprises the capability to store, track, and share healthcare data with their physician, and it provides a better idea about the effect of personal habits on blood pressure.

- **Biosensors:** Biosensor is the recently discovered wearable healthcare device that is totally different from wristbands and smart watches launched by Philips. Biosensor launched by Philips is coated with a sticky substance that allows the user to move around during the collection of data. It is used to monitor heart-beat rate, inhalation and exhalation rate, and body temperature of the patient's. Augusta University Medical Center conducted a survey that shows Biosensor reduced about 89% of patients/users can save their lives during a preventable cardiac or respiratory arrest. It is helpful for reducing labor and saves time and money.

18.4 Activities of Wearable Devices in Healthcare System

18.4.1 Women's Health focused AVA Sensor

It provides monitoring on women's fertility, pregnancy, and their total health condition. In addition, AVA sensor helps to token on women's weight, amount of sleep required, and stress levels during their pregnancy cycle. A clinical study conducted by the University Hospital of Zurich shows this wearable tool is well known for its appropriateness, it indicates a mean of 5.3 days of fertility during every cycle with 89% of accuracy.

Figure 18.1 shows the components used in AVA sensors and their activities are as follows:

Temperature Sensors: Takes the skin temperature at the wrist.
Accelerometer: Measures movement and determines sleep stage.
Photoplethysmograph: Captures heart rate variability, pulse rate, breathing rate, and skin perfusion.

The data flow of AVA sensor is shown in Fig. 18.2.

18.4.2 AliveCor – Personal EKG

It is an FDA-cleared medical device that catch high-quality EKG recordings and from those recordings it tracks out atrial fibrillation, worsening of heart failure, rapid heartbeat, and the rhythm of a beating heart within

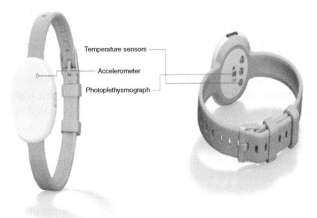

Temperature sensors
Accelerometer
Photoplethysmograph

Figure 18.1 Sensor combination inside AVA sensor. Source: Ava Security.

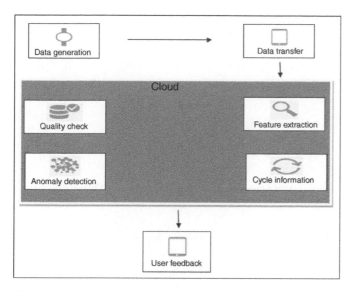

Figure 18.2 Dataflow of AVA sensor. Source: AliveCor, Inc.

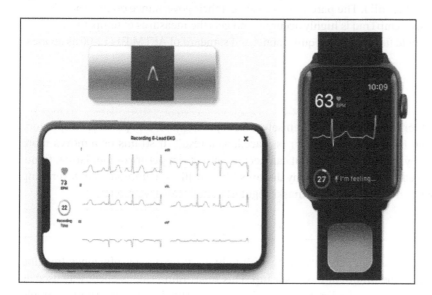

Figure 18.3 AliveCor – personal EKG.

30 seconds, and the information/data are directly send to the dedicated smart phone. The data have been used by the meditative doctors for current investigation. To collect the data/recordings patients should place their thumbs/fingers on the sensors, and they will be receiving the information in few seconds. The AliveCor – Personal EKG is shown in Fig. 18.3.

18.4.3 TempTraq

Temperature controlling sensor for child: It provides monitoring of temperature of babies and children, especially during their illness without making any defenses when they take rest (Figure 18.4). TempTraq consists of a blue-tooth sensors, and it is designed with a light and soft patch. Differ from traditionally used thermometers, TempTraq

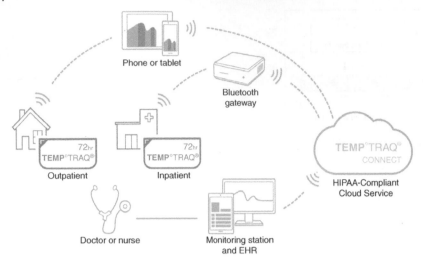

Figure 18.4 TempTraq. Source: G95, Inc.

patch can freely be set under the arm of the child. The parents can easily get their governance on their baby's temperature during different time intervals. TempTraq is highly accurate and gets the measures of temperature range from 87.0 to 109.3 °F. In addition, it should fulfill all the requirements and standard of ASTM E1112-00 as an ideal digital clinical thermometer.

18.4.4 BioScarf

It can be used by patients as well as other people to protect themselves from pollutant air and potential respiratory health issues (Figure 18.5). Inside the BioScarf an N95 filter is placed, as a result, it strains or removes more than 99.75% of pollutants transported by air. As an example, it can keep us away from pet-irritation, farina, fume, and other contaminants, otherwise, we may suffer from many diseases like tonsillitis, common cold, fever, lung infection, and phthisis at bay. It is better than buying more masks regularly, and it can save money.

18.4.5 Blinq – Wearable Rings

It is available in different styles and features like publicity along with light emitting diodes, efficiency tracking option, and SOS distress feature setup by the Montreal-based startup. This smart Blinq ring is effectively used as an alternative of smart watches and also comes at a limited cost (Figure 18.6). This is water-resistant wearable

Figure 18.5 BioScarf. Source: G95, Inc.

Figure 18.6 Blinq-wearable rings. Source: Koninklijke Philips N.V.

healthcare device that supports about 150 Android/iOS applications with 48 hours lasting capacity of battery and available in different sizes.

18.4.6 SmartSleep Wearable

This sleep analyst controls so smoothly on the user who have suffered from less sleep. This headband is very helpful for people who do not get enough time for sleep or sleep too late/early by controlling their sleep cycles (Figure 18.7). This wearable device produces an audio tone, as a result, it elevates the depth of sleep as well as duration of REM sleep. The dedicated app helps to collect the data, such as the metrics related to sleep, along with offering guidance and tips for a good night's rest. A combination of all these supreme features helps it to be considered as an effective healthcare device.

18.4.7 BioPatch

It is produced by SEER that can be easily used or placed on the left chest of the patient's/user's body (Figure 18.8). It consists of multiple-sensing algorithms for calculating the health data, such as heart-beat rate, echocardiogram, beat-rate inconstancy, and respiration rate. After collecting the data, it will send this data to the cloud-assisted platform, and this data is used to analyze by the assigning pathologist.

Figure 18.7 Wearable sleep headband. Source: Seers Technology.

Figure 18.8 BioPatches for bone regeneration. Source: Vuzix Corporation.

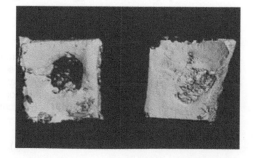

18.4.8 Smart Glasses

The Minuet technology where all health communication platforms were developed in one mobile should present in this system containing most illuminated doctrine contact lenses for video sharing (Figure 18.9). After development of this device, we could find some massive changes made in telehealth. The name of most effected system is Vuzix's smart glasses that helps to send secured videos and communicate with others.

18.4.9 Smart Hearing Aids

Starkey Hearing Technologies produced such futuristic wearable device for persons with hearing difficulties, in addition, this device accurately filter noise and point out only on the necessary source of sounds (Figure 18.10).

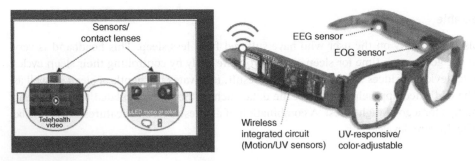

Figure 18.9 Smart glasses. Source: Vuzix Corporation.

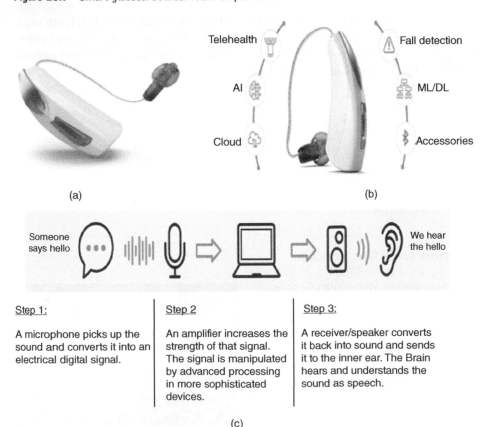

Figure 18.10 Smart hearing aids: (a) hearing aids, (b) technology used in Livio hearing aids, and (c) dataflow diagram of Livio hearing aids. Source: Smith+Nephew, Inc.

The name of the product launched by the factory is Livio AI that accurately filters the noise and points out only on the necessary source of sound. In addition, it will reduce irritation of ringing of the ears of the patient. Wearable sensors and artificial intelligence support wearable device must keep track of necessary health data, such as physical activities like running, walking, and are built to measure heart-beat rate in the future.

18.4.10 Wireless Patient Monitoring

"Remote patient monitoring" is the most precedence-based service in healthcare system that helps to reduce labor, save time and money for the patient as well as doctors (Figure 18.11). In real situation, few wearable devices contain most recent technology gives relief from the difficulties of the healthcare professionals. As an example, we can take "leaf healthcare" that provides identical resolution and wireless patient monitoring technology named as "leaf patient monitoring system" that remotely control on patient's health issues. In addition, it gives relief from many pressure-related injuries of the patients.

18.4.11 Wearable Fitness Tracker

Wristbands that are clipped along different types of sensors are modern one. It helps to monitor on patient's physical activity as well as their heart rate (Figure 18.12). It comes with different health and fitness recommendation service that is fully analyzed by the smartphone apps.

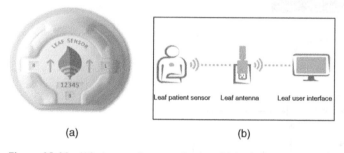

(a) (b)

Figure 18.11 Wireless patient monitoring: (a) leaf sensor. Source: Smith+Nephew, Inc., (b) data flow of leaf sensor.

Figure 18.12 Wearable fitness tracker: (a) wristbands, (b) algorithm of wearable fitness tracker. Source: Apple Inc.

(a)

(b)

18.4.12 Smart Health Watch

"Apple heart study" monitors on the patient's heart rates and alerts for experiencing inconstancy in heat rate. In addition, they also launched another app named "movement disorder API," which is very helpful for scholars and researchers to collect information about the Parkinson's disease. Smart watches provide services to the users, such as read instruction, transmit messages, makes cellular calls. It also gives few tricks of benefits of health as well as fitness trackers (Figure 18.13).

18.4.13 Wearable ECG Monitors

It becomes inbuilt with smart watch that helps to capture the electrocardiograms of the patient. Recently a report was published by "Business Insider" that shows "Withings" is a very popular moveable ECG monitor 2019 that

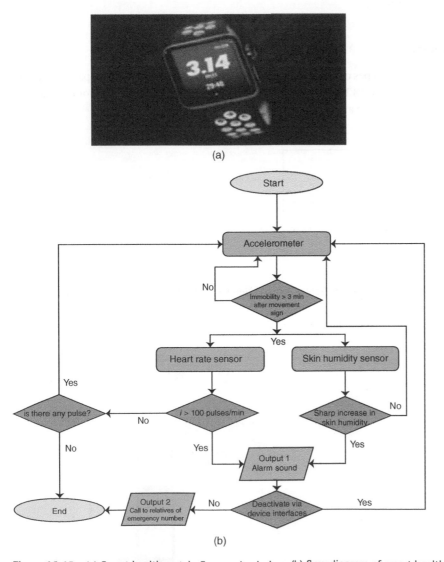

Figure 18.13 (a) Smart health watch. Source: Apple Inc., (b) flow diagram of smart health watch.

Figure 18.14   ECG monitors. Source: Withings.

is widely used to measure an electrocardiogram of the patients swiftly (Figure 18.14). In addition, it automatically sends the reading to the assigning doctor. It also detects the atrial fibrillation, pace, distance, elevation, and self-acting controlling for walking, running, swimming, and biking of the patient's condition.

18.4.14   Wearable Blood Pressure Monitors

This is nothing but an emblematic smart watch also an oscillometric blood pressure monitor used to monitor the pressure of the blood in circulatory system, regular activity, such as the number of steps taken, total distance covered by the user, and amount of calories has been used and required (Figure 18.15). The capability of the HeartGuide is to store about 100 readings in its storage space and transfer the data to the assigning mobile app. HeartGuide comprises the capability to store, track, and share the healthcare data with their physician, and it provide a better idea about the effect of personal habits on blood pressure.

18.4.15   Biosensors

It is used to monitor heart-beat rate, inhalation and exhalation rate, and body temperature of the patient's. Augusta University Medical Center conducted a survey that shows biosensor reduced about 89% of patient/user and save their lives during the preventable cardiac or respiratory arrest (Figure 18.16). It is helpful for reducing labor and saves time and money. In the field of personal healthcare, viable biosensor are becoming pretty popular, especially, self-monitoring of blood glucose.

The comparison among the premium wearable devices is shown in Table 18.1.

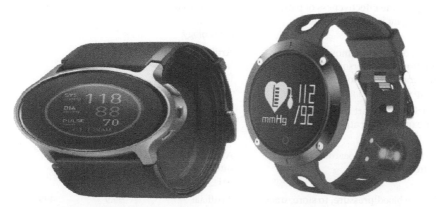

Figure 18.15   Blood pressure monitor. Source: Koninklijke Philips N.V.

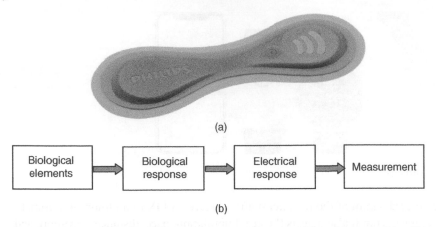

(a)

| Biological elements | → | Biological response | → | Electrical response | → | Measurement |

(b)

Figure 18.16 (a) Biosensor. Source: Koninklijke Philips N.V., (b) working principle of biosensor.

Table 18.1 Comparison among the premium wearable devices.

Sr. No.	Wearable devices	Purpose	Technology used	Accuracy (in %)	Currency (in US $)
1	AVA sensor	Women fertility tracking	Artificial intelligence and IoT	90	249
2	AliveCor personal EKG	Atrial fibrillation or normal heart rhythm	Machine learning	97	149
3	TempTraq	Senses, records and sends alerts of child's temperature to dedicated mobile app	Artificial intelligence and IoT	99	72
4	Sleep headband	Measures EEG that correlated with that of the PSG, also measure biomarkers reliably throughout the night	Artificial intelligence and IoT	86	400
5	Smart hearing aid	Accurately purify the unwanted source of sound and focuses on needed origin of sound	Artificial intelligence and IoT	96.70	3500
6	Leaf sensor	Measure the effectiveness of patients turning for injury prevention	FDA-cleared medical technology	97	189
7	Fitness trackers	Track orientation, movement and rotation of the person. Collects data and converts it into steps, calories, sleep quality and general activity	Artificial intelligence and IoT	93	99
8	Smart health watch	Track user's heart rhythm and alert for experiencing atrial fibrillation	Artificial intelligence and IoT	93	89
9	Wearable ECG monitors	To measure electrocardiogram of patient's heart	Artificial intelligence and IoT	96	89
10	Wearable blood pressure monitor	Taking blood pressure, to store, track and share data, tracks steps, distances and calories burned and quality of sleep	Artificial intelligence	99.9	499

18.5 Barriers to Wearable's Adoption

18.5.1 Cost

Due to increasing numbers of user in the market globally, along with the cheap price of wearable products, the acceptance of wearable tools/devices are coming out as the mainstream consumer product should expectedly increase.

18.5.2 Designation

Most wearable devices are designed with an attractive and imitative omnipresent style and appearance able goods like smart watches, bracelets, and glasses are the effective thought of the users are looking out and style of the tools.

18.5.3 Absence of Initiative Use Case

There are so many drawbacks found in the wearable devices, the most common barrier is that it could limit the smart watch acceptance, which is present in all kinds of wearable devices globally that user still do not follow in every situation.

18.5.4 Lack of a Killer App

Still there are not adequate applications that have been set up for really compelling on wrist-worn devices. Fragmentation is the most effective cause for the absence of powerful applications along with a suitable environment.

18.5.5 Limited Functionality

Most of the devices, such as fitness bands, smart watches, and wristbands contain limited functions like tracking breath rate, blood pressure monitoring related information, and transferring the analyzed data out into the dedicated smart phone and tablet.

An observation conducted by Nielsen contains over 4000 participants responded that the popularity of the uses of wearable devices is rapidly increased and up to 15% of the consumers who were well known about the term "wearable IoT devices" was recently used but its high cost is the main obstacle.

18.6 Wearable Devices Consumers

According to HRI/CIS wearable consumer survey, by Stistica.com in 2014, 70% of consumers were aware of wearable devices at that time. Of all the wearable-IoT devices, the most effective and popular device is fitness bands that have consumed with 61% of respondents using them after the number of consumers of smart watches (45%) and mobile-health devices (17%). As per HRI/CIS wearable consumer survey, 2014 fitness bands remain consumers' top wearable pick as shown in Figure 18.17.

From Figure 18.17, it is observed that

- 45% wearable customers wish to use fitness band.
- 35% wearable customers want to utilize smart watches.
- 20% wearable customers desire to consume smart clothing.
- 19% wearable customers expect to employ smart glasses.
- 13% wearable customers need to make use of people tracking devices.

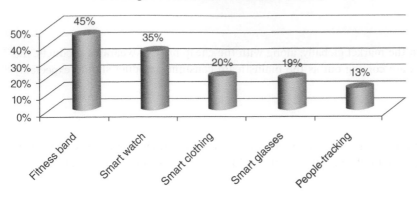

Figure 18.17   Percentage of consumers of wearable devices. Source: HRI/CIS wearable consumer survey 2014.

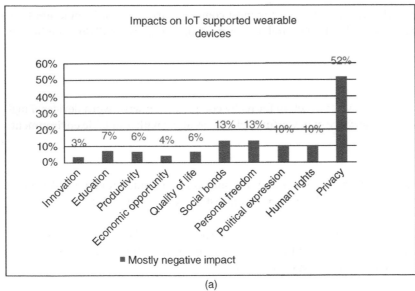

(a)

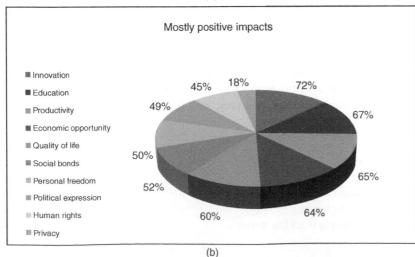

(b)

Figure 18.18   (a) Impacts on IoT supported wearable devices, (b) impacts on IoT supported wearable devices. Source: Microsoft, Stistica.com.

Based on a survey of 12 002 internet users in 12 countries done by Microsoft, Figure 18.18a,b indicate the loss of privacy is only the downside to the rise of personal technology that impacts on IoT-supported wearable devices.

Consumers hope wearables help them exercise smarter, pull together their medical information, and eat better. The information that the consumers desire to obtain from the wearable technology is shown in Figure 18.19.

Predictable worldwide wearable device shipments (in million units) with respect to the global wearable market within the five years span is shown in Figure 18.20.

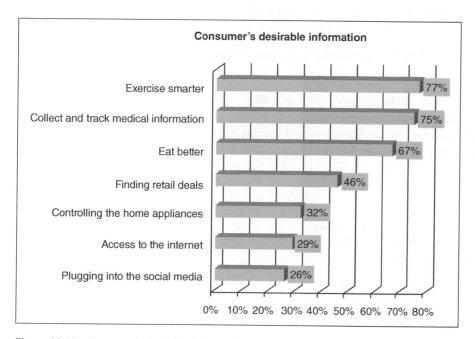

Figure 18.19 Consumer's desirable information.

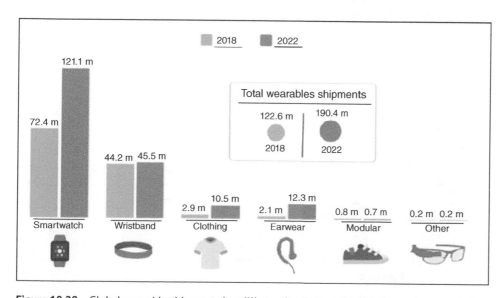

Figure 18.20 Global wearable shipments in million units. Source: IDC, Stistica.com.

Estimated growth rate of wearable shipments per million per unit during the five years span (2018–2022) among the various IoT-based wearable devices is shown in Table 18.2.

Figures 18.21–18.24 shows the wearable consumers in India (in percentage [%]) as well as Global (in million units).

Figure 18.21 shows that

- Consumers whose age range between 25 and 34 years used 47.5% of the wearable devices out of 100%.
- 42.9% of wearable devices are used by male of age range 45–54 years.
- 41.2% of wearable devices are used by female of age range 45–54 years.

So it is cleared that middle-aged Indian people preferred most amount of wearable devices for their health issues. Figure 18.22 shows that

- Number of consumers gradually increased from year 2017 to 2020.
- Number of predicted consumers slowly decreased from 2020 to 2022.
- Most abundant number of consumers of wearable devices in India is 2020.
- Least number of consumers used wearable devices in India was 2017.

Table 18.2 Global growth rate.

Sr. No.	Wearable devices	2018	2022	Growth rate (in %)
1	Smartwatch	72.4	121.1	67.3
2	Wristband	44.2	45.5	2.9
3	Clothing	2.9	10.5	262.1
4	Ear-wear	2.1	12.3	485.7
5	Modular	0.8	0.7	−12.5
6	Other	0.2	0.2	00.0

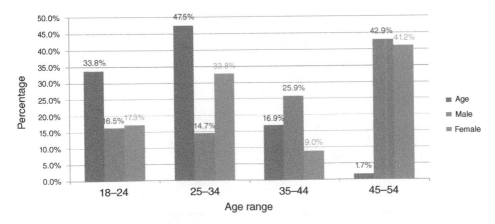

Figure 18.21 Consumers in India age wise. Source: Stistica.com.

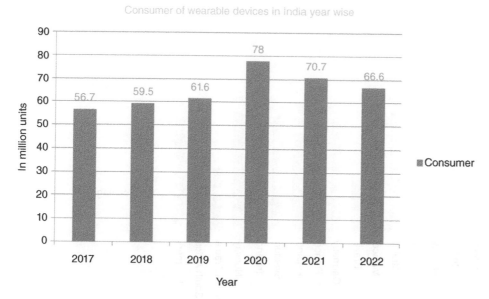

Figure 18.22   Consumer details in India year wise ($). Source: Stistica.com.

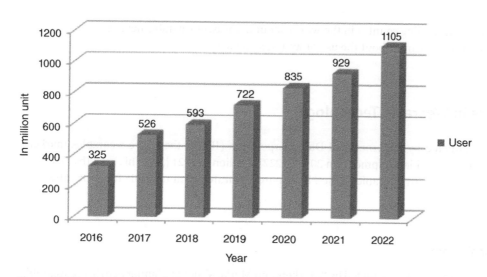

Figure 18.23   Consumer details in global year wise ($). Source: Stistica.com.

Figure 18.23 shows that

- Number of consumers of wearable devices is gradually increased globally.
- Most predicted number of consumers globally in the year 2022.
- Least number of consumers globally in the year 2016.

Figure 18.24 points out the percentage of global comparison of the consumer of wearable device. We are here to present the data of the world's top most 20 leading countries.

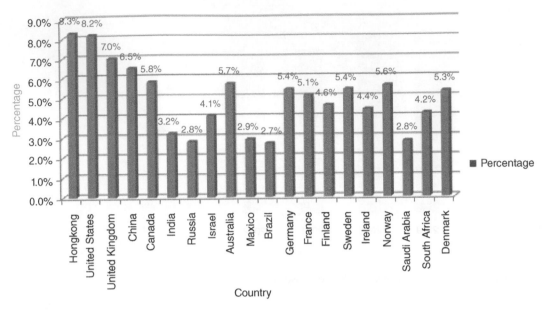

Figure 18.24 Percentage of global comparison of consumer of wearable device. Source: Stistica.com.

- It shows that Hong Kong is the leading country in the world about the uses of wearable devices.
- Brazil has the least position in the world about the use of wearable device.

18.7 Recent Trends in Wearable Technology

On the basis of a survey conducted by "International Data Corporation (IDC)," the entire wearables market's expected growth rate is from 113.2 million shipments in 2017 to 222.3 million in 2021 [24]. This growth consists of the development and design of modern wearables which have purely featured with multiple functionalities and measure health data.

18.7.1 Wearables in Healthcare

On the basis of the market research study conducted by "Markets and Markets" held in January 2018, the wearable market will expectedly grow 14.41 USD billion by 2022 from 6.22 USD billion in 2017.

Wearables in healthcare include the following:

- Blood pressure monitors
- Electrocardiogram monitors
- Defibrillators
- Drug-delivery products
- Insulin pumps
- Smart glasses

18.7.2   Wearables in Industry

- Industrial wearables like smart glasses, gloves, head-mounted displays, cameras, audio devices, and sensors equipped with clothes are adopting most of the industrial sectors in our modern world currently.
- Industrial wearables include
  - Smart PPE (personal protective equipment): It provides safety during potentially dangerous work to the workers.
  - Smart watches and glasses: These provides smart, fast, and effective contact with the worker to each other during low and loud visibility surroundings.
  - Exoskeletons
  - Body-worn terminals

18.7.3   Wearables in Robotics

- Wearable robot also known as bionic robot or exoskeleton is a special type of device that help to make the person's motion more softly in hard working environment.
- Application of wearable robots includes in post-surgery of medical science; auto-mobiles industries help workers for carrying heavier loads, industrial automotive, civil engineering, and in defense.
- Wearable exoskeleton is used to reduce the strain of worker's body in automotive industry.
- Exoskeleton/wearable robot is designed to clear the heap of broken building parts.

18.7.4   Wearables in Defense

- Wearable in Defense includes exoskeletons, smart tactical vests, and body-worn terminals.
- Most of the wearables are used in aerospace and defense for getting instant access to mission-critical information, increasing coordination, and increasing the acceleration and vitality of decision making.
- A soldier carries equipment weighing anywhere between 40 and 80 kg using this wearable technology-based smart clothing.
- Smart clothing's wireless communication capabilities enable soldiers' location, safety, and potential hazards to be monitored with greater accuracy.
- The adoption of the ARA ARC4 system helps to overcome the soldier's to manage the traditional way of finding the location on map.
- Hands-free wearable is very helpful for increasing the autonomy of the soldier.
- To enhance the soldier's mission, the innovative power and energy system is a unique wearable system.
- For enhancing the management of soldier, they adopt a wearable solar- harvesting system.

18.7.5   Wearables in Sport

- Wearable in sport includes activity monitors, fitness and heart rate monitors, foot pods and pedometers, sleep sensors, smart glasses, clothing, and watches.
- The main wearable e-textiles have embedded capacitive, resistive, and optical sensors allowing the textile to sense touch, strain, pressure, temperature, and humidity.
- For the USA Olympic and Paralympics teams, Ralph Lauren designed water-repellent jackets equipped with a button on a slender battery pack that athletes could push easily, even wearing gloves, to get a jolt of warmth.

18.7.6 Wearable in CPS

- Nowadays, wearable devices play important roles in the cyber-physical systems (CPSs) for measuring the different kinds of health issues. In remote healthcare monitoring system, these wearable devices are utilized for measuring the health parameters that are effectively used for the diagnosis of the patient health condition.
- Interoperability, scalability, lightweight security for relevance levels, and reliability of wearable devices are successfully employed throughout the design phase of a health-monitoring system [21].
- In CPS, wearable sensors are geared up with low-power wireless transponders with rechargeable batteries having sovereign properties for any kind of test or other diagnostic procedures.

18.8 Conclusion

In this work, a thorough study on the design and usage of AI and wearable IoT devices for healthcare data has been done. With the advances of Industry 4.0 technology, a specialized CPSs on healthcare area has developed with the advances of artificial intelligence, deep and machine learning, IoT, and the 5G technologies. The on-device AI solutions of wearable devices help transmission of vital body parameters to the healthcare system connected to the CPS for both assisted and controlled applications. The various proposed overviews of the working and description about the different types of IoT-based wearable devices with or without on-device AI are widely used in medical science for monitoring critical patients that help the doctors as well as patients to save their time and money. The primary motive of the work presented here is to increase the awareness of the common people about the uses of these wearable devices. We hope that if the people (both healthy and those who are not) start to use them, they will increase their knowledge for tracking their health issues in simple and easy way.

References

1 Tresp V, Overhage JM, Bundschus M, Rabizadeh S, Fasching PA, Yu S. Going digital: a survey on digitalization and large-scale data analytics in healthcare. *Proceedings of the IEEE* 2016 104 (11): 2180–2206.

2 Kirk, S. (2014). The wearables revolution: is standardization a help or a hindrance?: mainstream technology or just a passing phase? *IEEE Consumer Electronics Magazine* 3 (4): 45–50.

3 Framingham M. IDC (2019). IDC Reports Strong Growth in the Worldwide Wearables Market, Led by Holiday Shipments of Smartwatches, Wrist Bands, and Ear-Worn Devices.

4 Xu, B., Xu, L.D., Cai, H. et al. (2014). Ubiquitous data accessing method in IoT-based information system for emergency medical services. *IEEE Transactions on Industrial Informatics* 10 (2): 1578–1586.

5 Scott, C.D. and Smalley, R.E. (2003). Diagnostic ultrasound: principles and instruments. *Journal of Nanoscience and Nanotechnology* 3 (2): 75–80.

6 Yelamarthi, K. and Laubhan, K. (2015). Space perception and navigation assistance for the visually impaired using depth sensor and haptic feedback. *International Journal of Engineering Research & Innovations* 7 (1): 56–62.

7 Nalwa, H.S. (ed.) (2003). *Magnetic Nanostructures*. Los Angeles: American Scientific Publishers.

8 H.V. Jansen, N.R. Tas and J.W. Berenschot, "*Encyclopedia of Nanoscience and Nanotechnology*", Edited H.S. Nalwa, American Scientific Publishers, Los Angeles, vol. 5, (2004), pp. 163–275.

9 Wang, X., Wang, J.T., Zhang, X., and Song, J. (2013). A multiple communication standards compatible IoT system for medical usage. In: *IEEE Faible Tension Faible Consommation (FTFC)*, Paris, 1–4.

10 Kolici, V., Spaho, E., Matsuo, K. et al. (2014). Implementation of a medical support system considering P2P and IoT technologies. In: *Eighth International Conference on Complex, Intelligent and Software Intensive Systems*, Birmingham, 101–106.

11 Sandholm, T., Magnusson, B., and Johnsson, B.A. (2014). An on-demand WebRTC and IoT device tunneling service for hospitals. In: *International Conference on Future Internet of Things and Cloud*, Barcelona, 53–60.

12 Antonovici, D.A., Chiuchisan, I., Geman, O., and Tomegea, A. (2014). Acquisition and management of biomedical data using internet of things concepts. In: *International Symposium on Fundamentals of Electrical Engineering*, Bucharest, 1–4.

13 Krishnan, B., Sai, S.S., and Mohanthy, S.B. (2015). Real time internet application with distributed flow environment for medical IoT. In: *International Conference on Green Computing and Internet of Things*, Noida, 832–837.

14 Azariadi, D., Tsoutsouras, V., Xydis, S., and Soudris, D. (2016). ECG signal analysis and arrhythmia detection on IoT wearable medical devices. In: *5th International Conference on Modern Circuits and Systems Technologies*, Thessaloniki, 1–4.

15 Mohan, A. (2014). Cyber security for personal medical devices internet of things. In: *IEEE International Conference on Distributed Computing in Sensor Systems*, Marina Del Rey, CA, 372–374.

16 Yeh, L.Y., Chiang, P.Y., Tsai, Y.L., and Huang, J.L. (2015). Cloud-based fine-grained health information access control framework for lightweight IoT devices with dynamic auditing and attribute revocation. *IEEE Transactions on Cloud Computing* 6: 532–544.

17 Porambage, P., Braeken, A., Gurtov, A. et al. (2015). Secure end-to-end communication for constrained devices in IoT-enabled ambient assisted living systems. In: *IEEE 2nd World Forum on Internet of Things*, Milan, 711–714.

18 Laubhan, K., Trent, M., Root, B. et al. (2016). A wearable portable electronic travel aid for the blind. *IEEE International Conference on Electrical, Electronics, and Optimization Techniques*.

19 Yelamarthi, K., DeJong, B.P., and Laubhan, B.P. (2014). A kinect-based vibrotactile feedback system to assist the visually impaired. *IEEE Midwest Symposium on Circuits and Systems*.

20 Kimura, J. and Shibasaki, H. (1995). Recent advances in clinical neurophysiology. *Proceedings of the 10th International Congress of EMG and Clinical Neurophysiology*, Kyoto, Japan, (15–19 October).

21 Quarto, A., Soldo, D., Gemmano, S. et al. (2017). IoT and CPS applications based on wearable devices. A case study: monitoring of elderly and infirm patients. *2017 IEEE Workshop on Environmental, Energy, and Structural Monitoring Systems (EESMS)* (July 2017).

22 Larsen, C.E., Trip, R. and Johnson, C.R. (1995). Methods for procedures related to the electrophysiology of the heart. U.S. Patent 5,529,067, filed 19 August 1994 and issued 25 June 1995.

23 https://fitbit.com/in/home (accessed 4 February 2021).

24 Brossard J., Jean-Marie, B. (2021). Emerging Trends In Wearable Technology Across Several Markets, Trend paper, Reimaging connectivity together, fischer connectors®.

25 Alazab, M. and Tang, M. (2019). *Deep Learning Applications for Cyber Security*. Springer. ISBN: 978-3-030-13056-5.

26 Asghar, M.Z., Subhan, F., Ahmad, H. et al. (2020). *Senti-eSystem: A Sentiment-Based eSystem-Using Hybridized Fuzzy and Deep Neural Network for Measuring Customer Satisfaction*. Wiley Online Library https://doi.org/10.1002/spe.2853.

27 Bhattacharya S., Somayaji K. R. S., Gadekallu T. R., Alazab M. and Maddikunta R. K. P., " *A Review of Deep learning for Future Smart Cities*", Wiley (2020), https://doi.org/10.1002/itl2.187.

19

Impact of IoT in Biomedical Applications Using Machine and Deep Learning

Rehab A. Rayan[1], Imran Zafar[2], Husam Rajab[3], Muhammad Asim M. Zubair[4], Mudasir Maqbool[5], and Samrina Hussain[6]

[1]*Department of Epidemiology, High Institute of Public Health, Alexandria University, Alexandria, Egypt*
[2]*Department of Bioinformatics and Computational Biology, Virtual University of Pakistan, Lahore, Punjab, Pakistan*
[3]*Department of Telecommunications and Media Informatics, Budapest University of Technology and Economics, Budapest, Hungary*
[4]*Department of Pharmaceutics, The Islamia University of Bahawalpur, Pakistan*
[5]*Department of Pharmaceutical Sciences, University of Kashmir, Hazratbal, Srinagar, India*
[6]*Department of Drug Design and Pharmacology, University of Copenhagen, Denmark*

19.1 Introduction

As we move into the 2020s, more devices than ever are connected to the internet, and this will continue. Accordingly, more than 21 billion devices will be connected to the internet around the world by 2020, which is five times as many as there were four years ago [1]. The internet of things (IoT) is, at its most basic, a network that connects items that can be used to identify them to the internet. This lets them send, store, and collect information. IoT can be defined in terms of healthcare as any device that can collect health data from people. This includes mobile phones, computers, wearables and smart bands, surgical devices, digital medications that are implanted in the body, and other portable devices that can measure health data and connect to the internet [2].

As IoT technology has grown, it has garnered attention in a number of health practices that aim to improve the health of the population as a whole [3]. In recent reviews, the many services, and uses of IoT in healthcare have been discussed viz mobile health (mHealth), eHealth, semantic devices, ambient assisted living, smartphones and wearable devices, and community-based healthcare [2, 4]. These solutions have been described in great detail and can be used for a wide range of single-condition and cluster-condition management purposes, such as letting health-care professionals monitor and track patient's condition from a distance, making it easier for people with chronic conditions to take care of themselves, helping to spot problems early, identifying symptoms and clinical diagnoses faster, and so on. These apps can help make better use of healthcare resources while still giving high-quality and low-cost care [5, 6].

IoT is a complicated network of "things" that each has a unique identifier and connects to a server that provides the right services. They can talk to each other and people in the real world by sharing relevant information from the real and virtual worlds. These things can react on their own to things that happen around them. Some of these processes can be started by people or by machines talking to each other. They can also provide services. IoT opportunities will soon change the way healthcare is done. This technology will be a big part of tele-monitoring patients in hospitals and, even more important, at home [6–8]. By capturing illnesses and hazardous situations early and helping people avoid them, remote patient monitoring is a wonderful way to improve the quality of healthcare and lower costs at the same time [9, 10].

Machine Learning Algorithms for Signal and Image Processing, First Edition.
Edited by Deepika Ghai, Suman Lata Tripathi, Sobhit Saxena, Manash Chanda, and Mamoun Alazab.
© 2023 The Institute of Electrical and Electronics Engineers, Inc. Published 2023 by John Wiley & Sons, Inc.

19.1.1 Artificial Intelligence and Machine Learning

Artificial intelligence (AI) is a field that includes machine learning (ML). The main goal of ML is to learn from past situations and patterns. Instead of just making code, big data is fed into a generic algorithm, and analysis is done using the available data. IoT and ML systems can quickly train a system to spot medical abnormalities by using simple data and big data. The accuracy of predictions is related to how much big data has been taught [11–13]. As a result, big data improves the predictive accuracy of ML algorithms used in healthcare prediction platforms. Professionals now have access to a vast array of biological data, including diagnostic metrics and assessments, socio-demographic factors, and diagnostic imaging technologies, because of advances in technology and research. Biomedical data is unbalanced and nonstationary, with a prominent level of complexity, due to the abundance of data and the veracity of certain circumstances [14]. In this situation, ML is still especially important to: (i) To help doctors, naturalists, and health experts use and process substantial amounts of medical information much better; (ii) To reduce the chances of medical mistakes; and (iii) make sure that predictive and therapeutic rules and procedures work well together. Deep learning (DL) and artificial neural networks (ANNs) are the two data-mining algorithms that are most often used for image processing and finding flaws. DL algorithms are used at all levels of medicine in biomedical fields, from genomics tasks like figuring out how genes are expressed to global health-management tasks like predicting population growth or the spread of a virus [15]. In recent years, ML systems have been used in healthcare increasingly. ML techniques are used by several clinical decision support systems to create enhanced learning models that can be used to improve health-care service applications [13, 16–18]. ANNs and support vector machines (SVMs) are two applications of ML in health. Such models are used to accurately diagnose the type of cancer in a number of cancer classification applications. These algorithms work by analyzing data from sensors and other sources to find a patient's clinical problems and patterns of behavior [19]. Like noticing changes in a patient's behavior, in their daily routine, in how they sleep, eat, drink, and digest, and in how they move around. Then, health-care apps and clinical decision-support systems can use these algorithms to suggest changes to a patient's lifestyle and routines, as well as to prescribe various kinds of specialized therapies and health-care programmers. This lets doctors make a caring plan to support patients make the necessary changes to their lives [13, 16, 17, 20, 21]. ML is used to make patient load prediction techniques, which makes it easy for hospitals to share information about how many patients they have. In a hospital, past data is used to predict how many patients will come so that the right plans can be made. IoT devices with built-in ML techniques are used to make a classifier that spots certain health incidents, like falls, in older people. The clustering algorithms could find forms of patient's unruly behavior and let health-care workers know about them. In the same way, IoT microchips are used to keep track of what a patient does every day through daily habit modeling. The information is used to find things that are different about older people. Many countries have created new skills and laws to make the most of the IoT in biomedical systems. Because of this, modern biomedical research is now more interesting to investigate. This study aims to give a full review of innovative studies in biomedical systems based on the IoT and a summary of how advanced studies in biomedical systems based on the IoT have grown over time [22].

This chapter introduces the basic concepts of IoT, ML, and DL. Then it discusses the history of DL and ML in Section 19.2. Next, it reviews ML and DL algorithms and classifications, emphasizing DL architectures in Section 19.3. Further, it presents applied ML and DL techniques in the biomedical field in Section 19.4. Moreover, it highlights some IoT-based ML and DL case studies in biomedical systems in Section 19.5. Finally, it concludes with opportunities and challenges along with future insights in Section 19.6.

19.2 History of DL and ML

ANN was inspired by biological systems in the 1960s when it was discovered that different cells in the visual cortex were active when cats looked at different objects. These tests showed that the eyes and cells in the visual cortex were connected and that the visual cortex processed data in layers. ANNs could copy the way we see things by connecting artificial neurons in layers that could help us find out what they are.

After the 1960s, ANN development stopped because it could not do enough because its structures were too shallow, and computers could not do enough computing. Thanks to improvements in computing and technology, effective back propagation (BP) of RNA made it possible for pattern recognition studies [23]. First, the categorizations were done with an ANN model in a neural net with BP. Then, the parameters were changed by comparing the predicted class labels to the real ones. Even though it helped to reduce mistakes by using gradient descent, BP seemed to only work for some types of ANN. BP, adaptive learning rate, momentum, quasi-Newton methods, least-square techniques, and conjugated gradients (CGs) methods were all suggested as ways to improve steeper gradients through learning. Because ANN is so complicated, other fundamental ML approaches, like SVM support machines, random forests, and k-nearest neighbor (k-NN) algorithms, have quickly caught up with it [24].

When an ANN has more hidden layers, it is much easier to figure out what its functions are. When RNA has deep and complicated roles, it often sticks to what is best locally or moves through a gradient of spread. When a gradient is passed backward through the layers, the slope quickly loses its steepness. This means that the weights of the layers closest to the entry do not change much. Data denoising and dimension reduction for data visualization are regarded as two of the most intriguing real-world uses for autoencoders. Autoencoders can learn data projections that are more intriguing than PCA or other fundamental methods if the dimensionality and sparsity restrictions are the right ones. The goal of autoencoders (AE), which are neural networks, is to replicate the inputs in the outputs. In order to produce the desired output, they first compress the input into a latent-space representation and then rebuild it. Layered pre-training gets around the gradient diffusion barrier and helps deep neural networks (DNNs) choose better weights. This keeps reconstructed data from reaching a local optimum, which is often caused by choosing random weights at the start [25]. Graphics processing units (GPUs) have given academics a new reason to be interested in DL. Deep understanding has become more popular in recent years because people are paying more attention and putting in more effort. It is used a lot in business right now. Deep belief networks (DBNs) and limited Boltzmann machine stacks (RBM) have been used to recognize audio and video and to process natural language. Convolutional neural networks (CNNs) have been used a lot in image recognition, image segmentation, video recognition, and natural language processing since they were first made to better mimic how animals see the world. In recurrent neural networks (RNNs), which are a type of ANN, artificial neural cells are linked to time steps and show dynamic activity. RNN has been used to handle sequential data in natural-language processing and document identification. In recent years, sparse EAs, stacked EAs (SAEs), and denoising EAs have become more popular as part of in-depth online training [26].

19.3 Methods of ML and DL Algorithms and Classification

ML can frequently better explain data than model systems, giving technological answers as well as a helpful benchmark. It may also be utilized to assist people in their learning. A DNN is an ANN having multiple layers between the input and output layers (DNN). Neurons, synapses, weights, biases, and functions are all core elements of ML, which exist in various forms and dimensions [27].

ML allows a machine to autonomously analyze and understand a set of inputs as an experience without the need for external help [13, 28]. The training and testing phases are crucial in the development of an effective ML model. The training phase (which requires a lot of studies) entails giving the system labeled or unlabeled inputs. The algorithm afterward keeps the training feeds in the feature space for future predictions to regard. Ultimately, the system is inputted an unlabeled input and should expect the proper result in the testing phase. Simply defined, ML predicts outcomes for unlabeled input by using known data in its feature space. As a result, a good ML model can predict outputs by referring to prior encounters. The precision of this model is determined by the correctness of the output and the training.

Advanced biological and medical technologies have provided us with explosive volumes of physical and physiological data, medical imagery, functional magnetic resonance imaging, genomic, and protein complexes are just a few examples of the ability to understand public health and illness is made more accessible by using this data [29]. CNNs-based DL approaches offer a lot of potential for extracting functions and predicting outcomes from massive datasets. In biomedicine, ML and DL techniques enhance clinical care by utilizing the enormous volume of medical information given by IoT technology [30]. Although these techniques have a lot of promise, they also have several shortcomings. ML is measured in three realms: image acquisition, computational linguistics of hospital data, and genetic data. These areas are concerned with diagnosing, discovering, and predicting outcomes [31]. A massive infrastructure for medical devices provides data because there is rarely any standard framework to use such health information efficiently. Health records are available in multiple forms, which could also store future formatting more challenging and raise distortion [32]. We study the history of ML and DL and the fundamentals of techniques and technology in biomedical applications.

ML algorithms use patterns and experiences to improve the efficiency of activity. ML could be divided into three types: supervised learning, unsupervised learning, and improved learning. Unsupervised learning, for example, focuses on finding similarities in datasets while pooling samples [33]. Supervised learning is concerned with identifying the best or most appropriate behavior to execute in a situation to maximize a reward, via putting it differently. Semi-supervised learning is a technique of learning that falls in between supervised and unsupervised. Here, the algorithm can operate with both labeled and unlabeled data. When the data given is scarce and learning interpretations are necessary, this collection of methods is extraordinarily successful [34].

DL is a field of ML that is still relatively young and quickly expanding. It represents large-scale data abstraction with deep multilayer neural networks (DNNs), which generate a sense of the data as pictures, audio, and text. DL includes two attributes: multiple layers of irregular processing elements, each with a supervised or unsupervised learning function. In the 1980s, ANNs were employed to establish the groundwork for DL, but it was not until 2006 that the true impact of DL was realized. DL has since been used in various fields, including automatic voice recognition, picture recognition, natural language processing, drug development, and bioinformatics [35].

19.3.1 Deep Learning Architectures

19.3.1.1 Auto Encoders

Auto encoders (AEs) take features from unlabeled data and set target values identical to the inputs, unlike regular ANNs. Given the input vector $\{X^1, X^2, X^3 \ldots X^n\}$, $X^{(i)} \in R^n$, the AE tries to learn the model and is given by Eq. (19.1):

$$h_{w,b}(x) = g(Wx + b) \approx x \tag{19.1}$$

where W and b are the model's parameters, g is the activation function (exact definition as in the following context), and $h_{w,b}$ is the hidden units. The AE achieves a reduction of data dimensionality comparable to principal component analysis when the number of hidden units, which reflects the dimension of features, is lower than the input dimension. An AE with a classifier in the last layer may also do classification tasks and pattern recognition [36].

19.3.1.2 Deep Multilayer Perceptron

The well-known Rumelhart neural network, invented in 1986 and taught using the training algorithm approach, was the forerunner to DL details mentioned in Figure 19.1. Simultaneously, the most concealed levels were at least two or three, each containing only a few units [37]. Due to the advancement of various methods for training big architectures, including GPU technologies, the NN scale can currently reach multiple concealed layers with a little over 650 000 neurons and 630 million learned elements (e.g. AlexNet).

19.3.1.3 Deep Auto-encoders

A one hidden layer MLP is a particular instance of an auto-encoder. The goal of an AE is to reproduce the input vector $x^\wedge = F(x)$, where x and x are the input and output vectors, respectively. In an AE, the input/output layers have the same number of units as the hidden layer, but the hidden layer has fewer units. By stacking many AEs, deep AEs may be created. DAEs are utilized in DL for feature extraction/reduction and pre-training elements for compound networks [38].

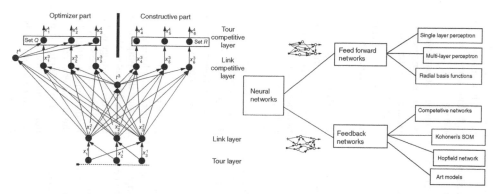

Figure 19.1 Most common neural network structures applied in biomedical: (optimized and forward) feedforward neural networks with various depths (one concealed layer, two concealed layers, and a deep structure with several concealed layers); (X_1, X_2, and X_n) an auto-encoder and a deep auto-encoder, respectively; referrals to a confined Boltzmann machine and a deep belief network, respectively; and an AlexNet, respectively.

19.3.1.4 Restricted Boltzmann Machine and Deep Belief Networks

A bounded Boltzmann machine is a dynamic probabilistic system that consists of a visible layer and a hidden layer with no linkages (RBM). In other words, the observational and functional applicators are represented by the visible and hidden layers, respectively. While the final model is a lower generic than a Boltzmann machine, it may teach you how to extract critical functions [39]. In RBM, an energy-derived model, the energy throughout the combined arrangement of transparent and concealed storage units is expressed as a Kohonen dynamical system. This energy role allocates a probability to every pair of seen and concealed vectors in the modeled network. RBM typically replicated the probability of input data or the combination of data input and target classes as a combined distribution. In DL, RBM, like EA, may perform the characteristics of a complex network. A deep belief network is a stack of RBMs, with the concealed conditions of every RBM serving as training data for a second RBM. Consequently, each RBM recognizes pattern expressions at the lowest level and learns to code patterns independent monitoring [40].

19.3.1.5 Convolutional Neural Networks

CNNs portray multidimensional networks that include input data, such as two-dimensional pictures with three datasets. They are motivated via the visual cortex's neurobiological model in which cells are liable to limited visual field areas. The multilayer perceptron and the clustering layer are the three types of layers in CNN design, as mentioned in Figure 19.2. Several neural maps, also known as function maps or filters, make up a precipitation layer. In contrast to a highly centralized repository, every neuron in a function map is just linked to a small patch

Figure 19.2 The multilayer perceptron and the clustering layer of the CNN design.

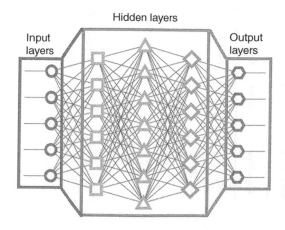

Table 19.1 Review of the examined ML and DL algorithms.

Algorithm	Scope	Learning type	Uses	Applied technique	Pros	Cons
Support vector machine (SVM) [47]	ML	Supervised	Binary classification, nonlinear classification	Decision boundary, soft margin, kernel trick	SVM is very successful in three-dimensional spaces. The fundamental strength of SVM rests in its application of the kernel's strategy.	Selecting the appropriate hyperplane and kernels approach is difficult.
Random forest [48]	ML	Supervised	Classification, regression	Bagging	In the context of random forest, the correlation coefficients among decision trees (DTs) are weaker.	Random forest performs badly when dealing with high-dimensional, data sparsity.
Naïve Bayes (NB) [17]	ML	Supervised	Probabilistic classification	Continuous variables (maximum likelihood)	Data scanning in which each attribute is checked individually. Acquiring simple data from each characteristic helps to improve the accuracy of the premises.	A modest quantity of instructional analysis is required. This function simply computes the parameters' eccentricities for each categorization.
Gradient boosted decision trees [49]	ML	Supervised	Classification, regression	Strong pre pruning	GBDT is used to investigate and progressively improve forecasting accuracy.	In GBDT training takes an extended period and requires constant varied modification. It struggles with sparse, high-dimensional data.
Decision trees (DTs) [50]	ML	Supervised	Prediction, classification	Continuous target variable (reduction in variance) Categorical target variable (Gini impurity)	The design is straightforward. It is applicable toward both discrete and continuous characteristics. Synthesis of content is low to non-existent.	The skewed dataset and interference influence the testing dataset, making it more expensive and using more RAM.
K-nearest neighbor (K-NN) [51]	ML	Supervised	Classification, regression	Continuous variables (Euclidean distance) Categorical variables (Hamming distance)	K-NN decided to use a stochastic technique. Simple to understand. It is simple to implement. There is no need for formal education. Changes may be easily implemented by integrating them into the current collection of labeled data.	K-NN takes an exceptionally long time to compute the comparison across databases. The main disadvantage is that the numbers are skewed. Productivity is affected by the classifier chosen (K value). We must use homogenous features since knowledge may be forgotten.

Convolutional neural network (CNN) [52]	DL	Spatial features	Facial recognition systems Analyzing and parsing through documents Smart cities (traffic cameras, for example) Recommendation systems, among other use cases	Convolution operation	Acquire the parameters dynamically without saying anything explicitly. These conduct a search and users retrieve useful information from data streams.	CNN is unable to handle regionally fluctuating Gaussian noise, which is common in fuzzy pictures.
Recurrent neural networks (RNNs) [53]	DL	RNN is created by implementing a repeating constraint on an ANN's hidden layers. Extension of recurrent neural networks (RNNs)	Time Series data Text data Audio data	Vanishing gradient	It captures the chronological data provided in the inputs, such as the dependency of terms in this case, while making assumptions. Distribute the parameters over a number of time steps. This reduces the number of features to train as well as the computational complexity.	All forms of neural network models are affected by the disappearing and expanding potential problems.
Long short-term memory (LSTMs) Networks [54].	DL	In sequential logistic regression, acquiring configuration dependence is crucial.	Common applications are cursive recognition and production, knowledge representation and interpretation, speech acoustic modeling, voice recognition, protein conformational predictions, and interactive multimedia information processing.	Long-term contextual connections may be represented without having to cope with the minimization concerns that beset the main, resulting in a greater performance	Provide a diverse model parameter, such as data for training as well as input and output preconceptions. Backpropagation Through Time, and LSTMs lower the difficulty of maintaining individual weighting to one output.	Require many resources and time to be trained and become ready for real-world applications. They respond identically to a feed-forward neural system as a measure of the initialization of several unrelated weights.

(continued)

Table 19.1 (Continued)

Algorithm	Scope	Learning type	Uses	Applied technique	Pros	Cons
Stacked auto-encoders [55].	DL	Unsupervised pre-training. Supervised fine-tuning	A study on P300 Segment Identification and Tracking of 3D Spine Modelling in Adolescent Idiopathic Scoliosis was published.	Random over-sampling	Reducing the dimensionality of the data we are using, as well as the learning time for your cases. The compactness and speed in coding using backpropagation.	Processing time, hyperparameter tuning, and model validation before you even start building the real model. Prone to overfitting, though this can be mitigated via regularization.
Deep Boltzmann machine (DBM) [56]	DL	Markov random field with multiple layers of hidden random variables	Application in linguistics, robotics, computer vision, and artificial intelligence.	Backpropagation method	An appropriate selection of interactions between visible and hidden units can lead to more tractable versions of the model. Can capture many layers of complex representations of input data, and they are appropriate for unsupervised learning since they can be trained on unlabeled data.	The high computational cost of inference, which is almost prohibitive when it comes to joint optimization in sizable datasets.
Deep belief networks (DBN) [57]	DL	Unsupervised, probabilistic generative model	Aircraft engine fault diagnosis	Used in a feed-forward neural network and fine-tuned to optimize discrimination	Show a higher classification capability to multiple features of input patterns hierarchically. With the pre-trained RBM. Training time is short on GPU-powered machines. Exactly accurate compared to a shallow net.	The approximate inference procedure is limited to a single bottom-up pass. Ignoring top down influences on the inference process are that the mode can fail to account for uncertainty when interpreting ambiguous sensory inputs.

of neurons in the preceding layer, named receptive field. Following that, various precipitation filters are used to corrupt the input data, progressively changing the convolution layers [41].

The precipitation filters employ the same settings in each tiny region of the picture, reducing hyperparameters in the model. A clustering layer leverages the image's stationing feature to decrease variance and capture essential functions by taking the average, max, or other parameters for the functionalities at many sites on the function maps. A CNN is made up of numerous unions and grouped layers that allow for the learning of increasingly abstract data. In the final layers of a CNN, a totally linked classification is utilized to categories the information retrieved by earlier aggregated and precipitated levels [42]. The most often used CNNs in ML applications are AlexNet, Clarifai, VGG, and GoogleNet.

19.3.2   Findings of Applied ML and DL Techniques

ML is a connected field of AI. ML is a system based on AI algorithms that analyzes and interprets a collection of inputs to determine knowledge without the need for human participation [28]. Classifiers and training models are two essential components to develop an effective training model for learning approaches. During the training phase, the machine may receive scored or unacknowledged inputs. After that, in the function area, the system creates such instruction signals to represent forecasts. The logistics network appears to have used an unidentifiable input during the validation method to predict the correct results. The employment of diverse ML techniques in transportation, for instance, proposes to suffice the challenges of rising travel requirements, safety concerns, energy consumption, radiations, and environmental degradation [43]. As a result, an effective learning algorithm may anticipate outcomes by highlighting its prior beliefs and viewpoints. A model's efficiency is determined by its output reliability and model design. Researchers have previously built and deployed several well-known ML algorithms for categorization.

The modern development in AI and large medical datasets has driven significant interest in developing DL algorithms that would more quickly and accurately distinguish diagnostic tests than subjective evaluation and other traditional methods. DL is a subgroup of ML, which performs excellent power and versatility by representing the world as a nested scale of notions [44–46]. We examined ML and DL algorithms and summarized them in Table 19.1 presenting the advantages and disadvantages along with techniques applied and deployment.

19.4   ML and DL Applications in Biomedicine

The field of medication gets advantages due to the structure of IoT that helps in the coordination of IoT technology and cloud computing. The situation additionally spreads out conventions for spread of the patient's information from various sensors and clinical tools to a specified medical care organization. The geography of IoT is a plan of various segments of an IoT medical care framework/network that are soundly associated with the medical care climate [58]. An IoT system consists of three main components such as the publisher, broker, and subscriber.

An organization of associated sensors and other clinical gadgets that might work separately or in a group to record a patient's fundamental data. The data might be temperature, a saturation of oxygen blood pressure, heart rate, EMG, electrocardiograph (ECG), electroencephalogram (EEG), etc. [2]. These pieces of information can be sent consistently through an organization to a specialist. The broker is liable for the preparation and storage of obtained information in the cloud. In the end, the patient's data can be observed and taken through a phone, PC, tablet, etc. the distributor can deal with this information and give criticism after getting the perception of any humiliation or physiological peculiarity in the patient's ailment. The IoT adopts distinct parts into an amalgam grid, where a reason for the existence of every segment on the IoT organization and cloud in the medical services organization is committed [59].

Table 19.2 Review of the examined ML and DL applications.

Application	Description	Algorithm	Solved problems	Obstacles	Future insights
ECG monitoring [62]	ECG stands for electrocardiogram, and it is the movement of heart due to the depolarization and repolarization of atria and ventricles.	Deep learning based on bio potential chip to assemble virtuous eminence ECG data	Used for the initial detection of heart aberrations with the help of ECG monitoring.	There is a problem of power consumption linked with an ECG monitoring system that is wearable.	The ECG and accelerometer data of elderly patients will be real-time monitored using this system.
Diagnosis of disease [63]	The interventions that need to be attempted are determined by this. It is applied to determine the hazard elements related to the condition, also the signs and symptoms, to enhance accuracy and detection proficiency	Convolutional neural networks, and back propagation networks	Diseases are diagnosed using this that includes deep-learning systems and support vector machines. Diagnosis of wrong patients can become the cause of incongruous mediations and adversarial consequences.	The sound legislation that defines the use of ML in healthcare. Obtaining well-annotated data for supervised learning is challenging	To support the correct and timely diagnosis of diseases, the ML will be assimilated into medical records
Medical imaging [64]	ML has several applications in the field of medical imaging. Medical imaging is the process and method for the diagnosis and treatment by creating images of body parts. The health professional then manually examines these images to detect the abnormalities.	ANNs (artificial neural networks) and CNNs (convolutional neural networks).	In manual imaging, the use of ML resolves the issues of efficiency and accuracy.	Time-consuming and prone to errors. Also, the outputs of deep learning techniques are difficult to explain logically. Great dependency on the volume of training sets and the quality.	To improve patient-centeredness by improving the quality of training datasets.
Behavioral modification or treatment [65]	Encompasses helping a patient change undesirable behavior. The amalgamation of ML into behavioral alteration programs can be helpful for the determination of what works.	Several reasoning and machine learning techniques, containing natural language administering	The incompetence for the evidence that to be synthesized and delivered on behavioral changes needs of user involvements and perspective to recover the expediency of evidence.	The deficiency of a change in behavior intrusion awareness system consists of ontology, process and assets for marking reports, a computerized annotator, machine learning and intellectual algorithms and user interface.	The consumption of substantiation from ML programs for the guidance of change in behavior mediations.
Clinical trial studies [5]	There is an urge to advance ML algorithms accomplished for ceaseless gaining from medical information	Methods of DL	The trouble of drawing bits of knowledge from huge measures of clinical information utilizing human abilities.	The issue of using reflective learning models to composite clinical datasets. Preparing datasets that are well categorized marks are required.	The proceeded assortment of preparing datasets for advanced relevance of deep learning in clinical exploration preliminaries.

Smart electronic health records [66]	The consideration of ML in digital wellbeing records makes shrewd frameworks with the ability to achieve a finding of diseases, movement expectation, and hazard assessment.	DL, usual language dispensation, and administered ML	The electronic well-being records stock clinical information yet does not help dynamic.	Planning information before they are taken care of into a machine learning algorithm remnant, a thought-provoking job. Furthermore, it is troublesome to consolidate patient-explicit elements in ML models.	Far-reaching selection of shrewd electronic wellbeing records, so it can help the administration.
Epidemic outbreak prediction [67]	Scrutiny of disease can be beneficial from ML as it considers the forecast of epidemics, henceforth, empowering the execution of proper protections.	DNN (deep neural network), LSTM (long short-term memory) learning, and ARIMA (autoregressive integrated moving average)	The trouble of getting ready for and managing irresistible ailments because of the deficiency of information.	Low precision of prescient models. The trial indicating constraints to exploit with ML models	To predict a variety of transferable diseases, the usage of prognostic models
Diagnostic and prognostic models for COVID-19 [68]	This study scrutinized the forecast simulations for COVID-19 and concluded that they are not well-designed.	DL models	There is a need to review forecast simulations	The established simulations are challenging.	Assemble quality and high-volume datasets to train COVID-19 prediction models.
Personalized care [69]	ML algorithms run an opportunity for contribution of person-centered care.	Deep neural networks (DNN), deep learning (DL)	The incapability to offer adapted care despite the mounting amassing of data.	It is essential for the sustained accumulation of elevated-quality training datasets.	Making frameworks that could be coordinated into digital wellbeing records to advance customized medication.

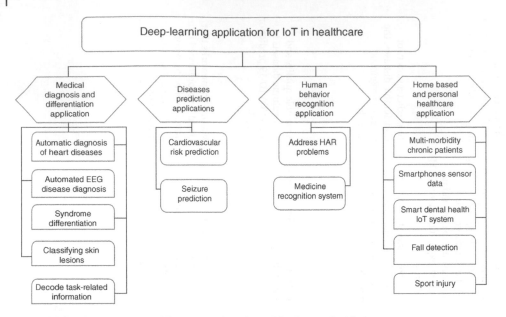

Figure 19.3 Categorization of DL applications for IoT in the medical industry.

Since the connection of the devices for IoT relies upon the interest and application of the medical services, it is difficult to recommend a comprehensive design for IoT. Various underlying changes have been embraced in the past for an IoT framework [60]. It is pivotal to drill down completely related exercises identified with the ideal wellbeing application while planning another IoT-based medical services' framework for continuous patient observing. The accomplishment of the IoT framework relies upon how it is fulfilling the prerequisites of medical care suppliers. Every disease needs devastating methods of medical services exercises; the topology should keep the clinical guidelines and steps in the diagnosis system [61]. Distinguishing the right issue to be addressed utilizing ML is the initial phase in building an ML product. Even though a model can be made to deliver understanding, it can affect patient contemplation. Useful applications are listed below (Table 19.2).

Figure 19.3 shows a thorough categorization of DL applications for IoT in the medical industry, which includes applications for medical forecasting, disparity, personal health-care programs, disease prediction technologies, and disease human behavior tracking apps [70]. We looked at publications that dealt with DL applications in the medical field in complicated subjects such as smartphone anomaly detection, chronic multi-morbidity, learning the patient's biological signal, perceptive dental health IoT systems, case identification models, injuries in sports, and nutritional monitoring systems in the health and lifestyle and home automation systems section [71].

19.5 Discussions of IoT-Based ML and DL Case Studies in Biomedical Systems

The internet of health things (IoHT) concept was investigated. By collecting and combining data on critical signs required in hospitals, the exchange of objects leads to regulating patients' physical conditions. IoT includes four steps: collection, storage, processing, and presentation. IoT has two advantages: it can eliminate service interruptions and efficiently distribute limited resources [72]. The number and intricacy of the data collected in this study are the study's principal flaws, making it a challenging chore for caretakers.

Tuli et al. anticipated HealthFog, a DL and IoT-based solution for automatic heart disease diagnosis. HealthFog delivers a light mist and effective administration of data from numerous IoT devices relevant to heart disease patients. The benefits of a new pattern for fog and edge equipment include energy-saving options and low-latency data processing systems [73]. In terms of fog's flaws in medical applications, it is crucial to understand the response and latency and how difficult it is to make the best use of the Quality of Service (QoS) characteristics in real-time fog.

Sarraf and Tofighi clarified that DL technologies have enabled neoteric technological improvements in the EEC's automated disease detection and diagnosis systems. The excellent efficiency of EEC deciphering due to the autonomous extraction feature is one of the study's positive findings. In addition, assessing your CEE can reveal unique health concerns [74]. Moreover, finding EEC pathology datasets will be difficult because although some are available online, most are small and inadequate for DL models.

Cerebral vascular accident (CVA), also known as cerebrovascular accident, is a condition in which specific brain areas cease to function due to ischemia or blood clots. This issue can be resolved with prompt diagnosis. To diagnose a stroke, CT scans and MRIs are often employed [75]. IoT frameworks can also be used to identify strokes based on CT images distributed by CNNs to determine whether the brain is robust, the stroke is ischemic, or a hemorrhage causes the stroke. The advantages of implementing IoT in healthcare include regions that are less reliant on humans, resulting in fewer human errors. The proposed architecture cannot be used in other medical imaging. Hence, this study's limitation is that it must expand.

Faust et al. established using pulse signals, the researcher created a DL model centered on long short-term memory (LSTM) to detect atrial fibrillation (AF) episodes. With tagged heart rate signal data from Physio Net's Atrial Fibrillation Database, evaluated the DL LSTM-based system in 20 individuals. The enactment of ML methods is more limited than that of the DL model. Additionally, it can extrapolate the knowledge derived from a restricted data set to a more extensive collection. The problem with this study is that it ignores the crucial concept of education [76].

Differentiation of the thymus is a fundamental aspect of Chinese medicine, used to treat infectious fevers. In old classical Chinese medicine, it is difficult to distinguish between infectious febrile disorders and their complications. Through convergence, DL is a promising approach for determining computer-assisted conditions from contagious fever. The proposal is for a stacked automatic encoder with an adaptive abandon function [77]. The strengths of this study are considered to prevent over-assembly and increase classification accuracy, however, deficiency in many clinical instances, infectious fever that it cannot distinguish.

Bray et al. investigated BP learning models for computer-assisted lung cancer diagnostic and therapeutic applications. Lung cancer is a severe concern to humanity today. Lung tumors can be benign or cancerous, and many patients have both. Deep enhancement models can detect lung cancers and provide a reliable outcome. To cure lung cancer utilizing deep reinforcement learning models, the most challenging part is developing an appropriate function to update the Q value of each metric [78].

Ma and Tavares stated that melanoma is a dangerous kind of skin cancer that is more prone to spread. There are three main types of melanocytic lesions: common nevi, atypical nevi, and melanoma. The skin lesions in this study are classified using an IoT-based approach. To obtain images, the suggested method used CNN models on the ImageNet dataset. The benefits of this strategy can be used in a variety of situations and are convenient [79].

Respecting internet access: A good connection must connect to the LINDA API (application programming interface) and send photos. ConvNets learn to employ a wide range of power at high alpha, beta, and gamma frequencies, according to Schirrmeister et al., who also gave unique methods for visualizing functions. Furthermore, the research depicts the ConvNets design process, which decodes the information and is linked to a task derived from a smooth EEG without artisan functions [80]. One of the benefits is being able to provide edge learning and scalability for massive amounts of data. ConvNets have the flaw of displaying erroneous predictions and need training data.

19.6 Opportunities and Challenges

Biomedical systems have had substantial expansion in recent years, contributing significantly to income and employment. Few years back, diseases and irregularities in the human body could be diagnosed only through a physical examination in a hospital. During the duration of therapy, most patients were required to stay in the hospital [81]. As a result, biomedical prices have risen and biomedical institutions in rural and isolated areas have been overburdened. The technical advancements accumulated over the past have now enabled the diagnosis of different ailments and the surveillance of health using tiny, embedded sensors in wristbands.

Furthermore, development has revolutionized the biomedical system from a hospital-centered to a patient-centered approach. Various diagnostic tests, for example, can be conducted at home without the assistance of a biomedical specialist (such as measuring pO_2, blood–glucose level, and blood pressure). Advanced computing technologies also can be used to send clinical data to biomedical centers from remote locations. The accessibility of biological resources has a qualitative part in the usage of various communication services in conjunction with advancements in technology (e.g. big data analytics, cloud computing, mobile computers, IoT, wireless sensing, and ML) [82].

In recent decades, advances in high-throughput technologies have resulted in a considerable rise in biomedical data, including genetic sequences, protein structures, and medical imaging. Efficient and effective computation techniques are necessary to store, analyze, and comprehend this flow of massive biological data. DL-based algorithmic frameworks emphasize these challenging difficulties.

IoT has not only increased independence but has also broadened human capabilities to interact with the outside world. IoT is a significant means of global communication by employing futuristic algorithms and protocols. It connects many items to the internet, including wireless sensors, home appliances, and electrical devices [83]. The IoT is gaining popularity due to its advantages of improved accuracy, reduced cost, and the capacity to better forecast events in the future [84]. Increased knowledge of computing technologies, software, apps, quick access to wireless technology, the modernization of mobile, and a broader digital economy have all benefited the IoT's rapid development [83]. Sensors, actuators, and other IoT devices have been combined with some further physical gadgets to exchange and monitor data via communication decorum, such as IEEE 802.11 Wi-Fi, Zigbee and bluetooth. Sensors incorporated or transportable in the human body are often used in biomedical applications to capture physiological data from the body of the patient, such as ECG, EEG, pressure frequency, and temperature [85].

Furthermore, environmental conditions such as humidity, temperature, time, and date can be captured. Such data assist in drawing relevant and accurate extrapolations about a patient's health condition. Since much information is acquired from a wide range of sources, storage capacity, and accessibility are especially vital in the IoT system (sensors, mobile phones, email, software, and applications). Physicians and other authorized individuals have access to the data collected by the above-mentioned sensors. Interacting with biomedical practitioners via the cloud/server allows for faster patient diagnosis and, if necessary, medical intervention.

For a practical and secure transfer, coordination among users, patients, and the communication unit is preserved [86]. Most IoT systems have a user experience that serves as a console for caregivers and allows them to regulate, visualize, and be concerned about their data. Many studies on the advancement of the IoT system in biological superintend, security, control, and integrity have been discovered in the literature [87]. These accomplishments demonstrate the value of IoT in the biomedical field and its promising future. However, maintaining the quality of the service matrix, which includes the integrity, security, cost, dependability, and availability of the information exchange, is crucial in designing an IoT device [88].

Computer-based intelligence, and explicitly DL, upholds the formation of dynamic IoT frameworks both on the plan of the correspondence foundation and on the investigation of information [89].

In any case, the huge measure of information gathered by IoT frameworks considers the neural networks' preparation that can oversee the exhibition and security of the framework. Such points require added consideration from scholastic and modern specialists. It is true that QoS does not just affect the clients' experience yet, additionally assumes an urgent part in the administration of crisis of basic situations. The devices are required to have the option to recognize between the high-need, crisis-related traffic, and ordinary traffic. Solid prioritization of the network traffic and utilization of systems administration devices to satisfy certain QoS guidelines is still, an open matter for a large part, that DL can explain with significant outcomes [90].

Besides, DL bids should be examined in the entire correspondence stack, from an actual level where sign aggregation, encoding, and disaggregation of channels might be concluded because of AI models, conversely with customary unequivocal improvement approach, at the level of application. Additionally, the organization layer requires novel methods for choosing and adjusting steering calculations that might be absolutely or unconscious of the organizations geography and state [91].

Without a doubt, the IoT networks development is frequently uncontrolled, and the circumstances of certain regions are exceptionally unpredictable. Simulated intelligence will help in the plan or initiative-taking responsive systems focused on guaranteeing the base QoS even in the event of irregularities. IoT security frameworks are another area that will attract several examination endeavors before exceedingly long [92].

Scientific literature has received some interesting contributions due to the utilization of DL in IoT networks. Still, abundant area for enhancements and additional directions needs to be investigated, comprising the conservation of classification, and versatility to assaults that destabilize portions of the correspondence foundation (e.g. sticking assaults, etc.).

The biggest region for upcoming placement of DL in IoT is the undeniable level applications, i.e. the examination of IoT gathered information [93]. The developed range of IoT devices considers a substantial number of contextual analyses pointed toward making our social orders more comprehensive, secure, and, as a rule, vivid. In the following several years, one critical difficulty that many societies must confront is the population aging and the resulting distributed support and monitoring that vulnerable, many elders, and no autonomous citizens will require.

Dependable and secure IoT structures furnished with AI are the central progression to disseminate observing of individuals needing consideration, identification of peculiarities (e.g. circulatory strain issues, falls strokes), and giving data to a better life (e.g. calculation of person on foot courses without architectural obstruction). The high transmission capacity needed by interactive media streams and the QoS severe prerequisites of these types of uses present significant issues to the current IoT foundations, and advancement, both from the hypothetical and practical are necessary [94].

19.6.1 Future Insights

In many cases, our health-care services are now more expensive than at any time before, and many patients are compelled to be hospitalized for the length of their treatment. These difficulties can be overcome by using technologies that can remotely monitor patients. IoT technologies will cut the cost of health-care services by gathering real-time medical data from patients and passing it to health-care providers. This will allow for the cure of health problems before they become critical.

In the coming years, IoT technology will be extensively used in healthcare. The health-care industry is continually in the search for innovative methods to deliver services while reducing costs and quality improvement; consequently, this sector's dependency on IoT technology will continue to grow [95–97]. Patients are more able to follow self-care principles using such technologies, resulting in improved cost-effectiveness of health services, better self-management, and increased patient satisfaction.

IoT-based solutions can also be used for remote surveillance of physiological status in patients needing constant supervision. Rapprochement of multiple IoT designs has also recently allowed the development of intelligent health-care systems. IoT-driven solutions can be advantageous in developing a consistent system by the interconnection of varied objects to gain a comprehensive picture of the patient's health status [98–101].

19.6.2 Conclusions

The biomedical field is among the greatest tangled regarding the degree of responsibility and rigorous commands that makes it a relevant and indispensable field for creativity. Biomedical systems have had substantial expansion in recent years, contributing significantly to income and employment. The technical advancements accumulated over the past have now enabled the diagnosis of different ailments and the surveillance of health using tiny, embedded sensors in wristbands. This development has revolutionized the biomedical system from a hospital-centered to a patient-centered approach. Moreover, advances in high-throughput technologies have resulted in a considerable rise in biomedical data, including genetic sequences, protein structures, and medical imaging. Efficient and effective computation techniques are necessary to store, analyze, and comprehend this flow of massive biological data. DL-based algorithmic frameworks emphasize these challenging difficulties.

The IoT has presented a realm of opportunities in the medical field and might be the answer to several dilemmas. Introducing the IoT has not only increased independence but has also broadened human capabilities to interact with the outside world. IoT is a significant means of global communication by employing futuristic algorithms and protocols. Using the medical IoT would produce great possibilities for telemedicine, distant supervision of patients' status, among other applications. This might be with the sides of ML and DL frameworks. Computer-based intelligence, and explicitly DL, upholds the formation of dynamic IoT frameworks both on the plan of the correspondence foundation and on the investigation of information. However, the development of IoT networks is frequently uncontrolled, and the conditions of certain regions are exceptionally erratic. The security of IoT frameworks is another field that will draw many examination endeavors before exceedingly long.

The utilization of DL in IoT networks has led to some interesting contributions in the scientific literature. Still, abundant space for enhancements and further directions still needs to be investigated, including the conservation of classification, and the versatility to assaults that destabilize portions of the correspondence foundation (e.g. sticking assaults, etc.). In this chapter, we reviewed the commonly dominant ML and DL algorithms, named some ML and DL applications in the biomedical domain, and examined IoT-based ML and DL applications in the medical system.

References

1 Kelly, J.T., Campbell, K.L., Gong, E., and Scuffham, P. (2020). The internet of things: impact and implications for health care delivery. *Journal of Medical Internet Research* 22: e20135. https://doi.org/10.2196/20135.

2 Dang, L.M., Piran, M.J., Han, D. et al. (2019). A survey on internet of things and cloud computing for healthcare. *Electronics* 8: 768. https://doi.org/10.3390/electronics8070768.

3 Saarikko, T., Westergren, U.H., and Blomquist, T. (2017). The internet of things: are you ready for what's coming? *Business Horizons* 60: 667–676. https://doi.org/10.1016/j.bushor.2017.05.010.

4 Nazir, S., Ali, Y., Ullah, N., and García-Magariño, I. (2019). Internet of things for healthcare using effects of mobile computing: a systematic literature review. *Wireless Communications and Mobile Computing* 2019: e5931315. https://doi.org/10.1155/2019/5931315.

5 Yin, Y., Zeng, Y., Chen, X., and Fan, Y. (2016). The internet of things in healthcare: an overview. *Journal of Industrial Information Integration* 1: 3–13. https://doi.org/10.1016/j.jii.2016.03.004.

6 Jara, A.J., Zamora-Izquierdo, M.A., and Skarmeta, A.F. (2013). Interconnection framework for mHealth and remote monitoring based on the internet of things. *IEEE Journal on Selected Areas in Communications* 31: 47–65.

7 Tundis, A., Kaleem, H., and Mühlhäuser, M. (2020). Detecting and tracking criminals in the real world through an IoT-based system. *Sensors* 20: 3795. https://doi.org/10.3390/s20133795.

8 Trinugroho, D. and Baptista, Y. (2014). Information integration platform for patient-centric healthcare Services: design, prototype and dependability aspects. *Future Internet* 6: 126–154. https://doi.org/10.3390/fi6010126.

9 Ahmadi, H., Arji, G., Shahmoradi, L. et al. (2019). The application of internet of things in healthcare: a systematic literature review and classification. *Universal Access in the Information Society* 18: 837–869. https://doi.org/10.1007/s10209-018-0618-4.

10 Taylor, R., Baron, D., and Schmidt, D. (2015). The world in 2025 – predictions for the next ten years. In: *2015 10th International Microsystems, Packaging, Assembly and Circuits Technology Conference IMPACT*, 192–195. Taiwan: IEEE https://doi.org/10.1109/IMPACT.2015.7365193.

11 Hung, C.-Y., Chen, W.-C., Lai, P.-T. et al. (2017). Comparing deep neural networks and other machine learning algorithms for stroke prediction in a large-scale population-based electronic medical claims database. In: *2017 39th Annual International Conference of the IEEE Engineering in Medicine and Biology Society (EMBC)*, 3110–3113. South Korea: IEEE https://doi.org/10.1109/EMBC.2017.8037515.

12 Ngiam, K.Y. and Khor, I.W. (2019). Big data and machine learning algorithms for health-care delivery. *The Lancet Oncology* 20: e262–e273. https://doi.org/10.1016/S1470-2045(19)30149-4.

13 Al Dhahiri, A., Alrashed, B., and Hussain, W. (2021). Trends in using IoT with machine learning in health prediction system. *Forecast* 3: 181–206. https://doi.org/10.3390/forecast3010012.

14 Yue, L., Tian, D., Chen, W. et al. (2020). Deep learning for heterogeneous medical data analysis. *World Wide Web* 23: 2715–2737. https://doi.org/10.1007/s11280-019-00764-z.

15 Mahmud, M., Kaiser, M.S., McGinnity, T.M., and Hussain, A. (2021). Deep learning in mining biological data. *Cognitive Computation* 13: 1–33. https://doi.org/10.1007/s12559-020-09773-x.

16 Alharbi, R. and Almawashi, H. (2019). The privacy requirements for wearable IoT devices in healthcare domain. In: *2019 7th International Conference on Future Internet of Things and Cloud Workshops (FiCloudW)*, 18–25. Turkey: IEEE https://doi.org/10.1109/FiCloudW.2019.00017.

17 Kaur, G. and Oberoi, A. (2020). Novel approach for brain tumor detection based on Naïve Bayes classification. In: *Data Management, Analytics and Innovation* (ed. N. Sharma, A. Chakrabarti and V.E. Balas), 451–462. Singapore: Springer https://doi.org/10.1007/978-981-32-9949-8_31.

18 Ibrahim, N., Akhir, N.S.M., and Hassan, F.H. (2017). Predictive analysis effectiveness in determining the epidemic disease infected area. *AIP Conference Proceedings* 1891: 020064. https://doi.org/10.1063/1.5005397.

19 Huang G, Hu J, He Y. et al. (2021). Machine learning for electronic design automation: A survey. *ACM Transactions on Design Automation of Electronic Systems* 26: 1–46.

20 Raeesi Vanani, I. and Amirhosseini, M. (2021). IoT-based diseases prediction and diagnosis system for healthcare. In: *Internet of Things for Healthcare Technologies* (ed. C. Chakraborty, A. Banerjee, M.H. Kolekar, et al.), 21–48. Singapore: Springer https://doi.org/10.1007/978-981-15-4112-4_2.

21 Deshields, T.L., Wells-Di Gregorio, S., Flowers, S.R. et al. (2021). Addressing distress management challenges: recommendations from the consensus panel of the American Psychosocial Oncology Society and the Association of Oncology Social Work. *CA: A Cancer Journal for Clinicians* https://doi.org/10.3322/caac.21672.

22 Issac, A.C. and Baral, R. (2020). A trustworthy network or a technologically disguised scam: A bibliomorphological analysis of bitcoin and blockchain literature. *Global Knowledge, Memory and Communication* 69: 443–460. https://doi.org/10.1108/GKMC-06-2019-0072.

23 Orhorhoro, E.K., Ebunilo, P., and Sadjere, G. (2017). Development of a predictive model for biogas yield using artificial neural networks (ANNs) approach 2017. https://www.semanticscholar.org/paper/Development-of-a-Predictive-Model-for-Biogas-Yield-Orhorhoro-Ebunilo/d65df521b9feceb147a01ca0949f318463bce7a9 (accessed 5 August 2021).

24 Khan, G.M. (2018). *Evolution of Artificial Neural Development: In Search of Learning Genes*. Springer International Publishing https://doi.org/10.1007/978-3-319-67466-7.

25 Zhang, L., Tan, J., Han, D., and Zhu, H. (2017). From machine learning to deep learning: progress in machine intelligence for rational drug discovery. *Drug Discovery Today* 22: 1680–1685. https://doi.org/10.1016/j.drudis.2017.08.010.

26 Zhou, X., Li, Y., and Liang, W. (2021). CNN-RNN based intelligent recommendation for online medical pre-diagnosis support. *IEEE/ACM Transactions on Computational Biology and Bioinformatics* 18: 912–921. https://doi.org/10.1109/TCBB.2020.2994780.

27 Zendesk (2022). Deep learning vs. machine learning: what's the difference? https://www.zendesk.com/blog/machine-learning-and-deep-learning (accessed 5 August 2021).

28 Dike, H.U., Zhou, Y., Devarasetty, K.K., and Wu, Q. (2018). Unsupervised learning based on artificial neural network: a review. In: *2018 IEEE International Conference on Cyborg and Bionic Systems (CBS)*, 322–327. China: IEEE https://doi.org/10.1109/CBS.2018.8612259.

29 Holzinger, A., Haibe-Kains, B., and Jurisica, I. (2019). Why imaging data alone is not enough: AI-based integration of imaging, omics, and clinical data. *European Journal of Nuclear Medicine and Molecular Imaging* 46: 2722–2730. https://doi.org/10.1007/s00259-019-04382-9.

30 Srivastava, A., Jain, S., Miranda, R. et al. (2021). Deep learning based respiratory sound analysis for detection of chronic obstructive pulmonary disease. *PeerJ Computer Science* 7: e369. https://doi.org/10.7717/peerj-cs.369.

31 Myszczynska, M.A., Ojamies, P.N., Lacoste, A.M.B. et al. (2020). Applications of machine learning to diagnosis and treatment of neurodegenerative diseases. *Nature Reviews Neurology* 16: 440–456. https://doi.org/10.1038/s41582-020-0377-8.

32 Bote-Curiel, L., Muñoz-Romero, S., Guerrero-Curieses, A., and Rojo-Álvarez, J.L. (2019). Deep learning and big data in healthcare: a double review for critical beginners. *Applied Sciences* 9: 2331. https://doi.org/10.3390/app9112331.

33 Bao, W., Lianju, N., and Yue, K. (2019). Integration of unsupervised and supervised machine learning algorithms for credit risk assessment. *Expert Systems with Applications* 128: 301–315. https://doi.org/10.1016/j.eswa.2019.02.033.

34 Lai, W.-S., Huang, J.-B., and Yang, M.-H. (2017). Semi-supervised learning for optical flow with generative adversarial networks. In: *Advances in Neural Information Processing Systems*, vol. 30. Curran Associates, Inc.

35 Lee, J.H., Shin, J., and Realff, M.J. (2018). Machine learning: overview of the recent progresses and implications for the process systems engineering field. *Computers and Chemical Engineering* 114: 111–121. https://doi.org/10.1016/j.compchemeng.2017.10.008.

36 Won, Y.-S. and Bhasin, S. (2021). On use of deep learning for side channel evaluation of black box hardware AES engine. In: *Industrial Networks and Intelligent Systems*. (ed. N.S. Vo, V.-P. Hoang and Q.-T. Vien), 185–194. Cham: Springer International Publishing https://doi.org/10.1007/978-3-030-77424-0_15.

37 Yao, X., Wang, X., Wang, S.-H., and Zhang, Y.-D. (2020). A comprehensive survey on convolutional neural networks in medical image analysis. *Multimedia Tools and Applications* https://doi.org/10.1007/s11042-020-09634-7.

38 Shao, H., Jiang, H., Lin, Y., and Li, X. (2018). A novel method for intelligent fault diagnosis of rolling bearings using ensemble deep auto-encoders. *Mechanical Systems and Signal Processing* 102: 278–297. https://doi.org/10.1016/j.ymssp.2017.09.026.

39 Liu, R., Rong, Z., Jiang, B. et al. (2018). Soft sensor of 4-CBA concentration using deep belief networks with continuous restricted Boltzmann machine. In: *2018 5th IEEE International Conference on Cloud Computing and Intelligence Systems (CCIS)*, 421–424. China: IEEE https://doi.org/10.1109/CCIS.2018.8691166.

40 Li, Z., Wang, Y., and Wang, K. (2020). A data-driven method based on deep belief networks for backlash error prediction in machining centres. *Journal of Intelligent Manufacturing* 31: 1693–1705.

41 Yamashita, R., Nishio, M., Do, R.K.G., and Togashi, K. (2018). Convolutional neural networks: an overview and application in radiology. *Insights Into Imaging* 9: 611–629. https://doi.org/10.1007/s13244-018-0639-9.

42 Ulku, I. and Akagündüz, E. (2022). A survey on deep learning-based architectures for semantic segmentation on 2d images. *Applied Artificial Intelligence* 11: 1–45.

43 Abduljabbar, R., Dia, H., Liyanage, S., and Bagloee, S.A. (2019). Applications of artificial intelligence in transport: an overview. *Sustainability* 11: 189. https://doi.org/10.3390/su11010189.

44 Asaoka, R., Murata, H., Hirasawa, K. et al. (2019). Using deep learning and transfer learning to accurately diagnose rarly-onset glaucoma from macular optical coherence tomography images. *American Journal of Ophthalmology* 198: 136–145. https://doi.org/10.1016/j.ajo.2018.10.007.

45 Shibata, N., Tanito, M., Mitsuhashi, K. et al. (2018). Development of a deep residual learning algorithm to screen for glaucoma from fundus photography. *Scientific Reports* 8: 146–165. https://doi.org/10.1038/s41598-018-33013-w.

46 Hood, D.C. and De Moraes, C.G. (2018). Efficacy of a deep learning system for detecting glaucomatous optic neuropathy based on colour fundus photographs. *Ophthalmology* 125: 1207–1208. https://doi.org/10.1016/j.ophtha.2018.04.020.

47 Zhou, X., Zhang, X., and Wang, B. (2016). Online support vector machine: a survey. In: *Harmony Search Algorithm* (ed. J.H. Kim and Z.W. Geem), 269–278. Berlin, Heidelberg: Springer 10.1007/978-3-662-47926-1_26.

48 Khan, Z., Gul, A., Perperoglou, A. et al. (2020). Ensemble of optimal trees, random forest and random projection ensemble classification. *Advances in Data Analysis and Classification* 14: 97–116. https://doi.org/10.1007/s11634-019-00364-9.

49 Shadkani, S., Abbaspour, A., Samadianfard, S. et al. (2021). Comparative study of multilayer perceptron-stochastic gradient descent and gradient boosted trees for predicting daily suspended sediment load: The case study of the Mississippi River, U.S. *International Journal of Sediment Research* 36: 512–523. https://doi.org/10.1016/j.ijsrc.2020.10.001.

50 Sen, P.C., Hajra, M., and Ghosh, M. (2020). Supervised classification algorithms In machine learning: a survey and review. In: *Emerging Technology in Modelling and Graphics* (ed. J.K. Mandal and D. Bhattacharya), 99–111. Singapore: Springer https://doi.org/10.1007/978-981-13-7403-6_11.

51 Hussain, W. and Sohaib, O. (2019). Analysing cloud QoS prediction approaches and its control parameters: considering overall accuracy and freshness of a dataset. *IEEE Access* 7: 82649–82671. https://doi.org/10.1109/ACCESS.2019.2923706.

52 Li, Z., Liu, F., Yang, W. et al. (2021). A survey of convolutional neural networks: analysis, applications, and prospects. *IEEE Transactions on Neural Networks and Learning Systems* https://doi.org/10.1109/TNNLS.2021.3084827.

53 Alkhodari, M. and Fraiwan, L. (2021). Convolutional and recurrent neural networks for the detection of valvular heart diseases in phonocardiogram recordings. *Computer Methods and Programs in Biomedicine* 200: 105940. https://doi.org/10.1016/j.cmpb.2021.105940.

54 Lindemann, B., Müller, T., Vietz, H. et al. (2021). A survey on long short-term memory networks for time series prediction. *Procedia CIRP* 99: 650–655. https://doi.org/10.1016/j.procir.2021.03.088.

55 Azarbik, M. and Sarlak, M. (2020). Real-time transient stability assessment using stacked auto-encoders. *COMPEL - The International Journal for Computation and Mathematics in Electrical and Electronic Engineering* 39: 971–990. https://doi.org/10.1108/COMPEL-12-2019-0477.

56 Gm, H., Gourisaria, M.K., Pandey, M., and Rautaray, S.S. (2020). A comprehensive survey and analysis of generative models in machine learning. *Computer Science Review* 38: 100285. https://doi.org/10.1016/j.cosrev.2020.100285.

57 Sohn, I. (2021). Deep belief network based intrusion detection techniques: a survey. *Expert Systems with Applications* 167: 114170. https://doi.org/10.1016/j.eswa.2020.114170.

58 Oryema, B., Kim, H.-S., Li, W., and Park, J.T. (2017). Design and implementation of an interoperable messaging system for IoT healthcare services. In: *2017 14th IEEE Annual Consumer Communications & Networking Conference (CCNC)*, 45–52. USA: IEEE https://doi.org/10.1109/CCNC.2017.7983080.

59 Birje, M.N. and Hanji, S.S. (2020). Internet of things based distributed healthcare systems: a review. *Journal of Data, Information and Management* 2: 149–165. https://doi.org/10.1007/s42488-020-00027-x.

60 Ahad, A., Tahir, M., and Yau, K.-L.A. (2019). 5G-based smart healthcare network: architecture, taxonomy, challenges and future research directions. *IEEE Access* 7: 100747–100762. https://doi.org/10.1109/ACCESS.2019.2930628.

61 Kadhim, K.T., Alsahlani, A.M., Wadi, S.M., and Kadhum, H.T. (2020). An overview of patient's health status monitoring system based on internet of things (IoT). *Wireless Personal Communications* 114: 2235–2262. https://doi.org/10.1007/s11277-020-07474-0.

62 Tekeste, T., Saleh, H., Mohammad, B., and Ismail, M. (2019). Ultra-low power QRS detection and ECG compression architecture for IoT healthcare devices. *IEEE Transactions on Circuits and Systems I: Regular Papers* 66: 669–679. https://doi.org/10.1109/TCSI.2018.2867746.

63 Xu, J., Xue, K., and Zhang, K. (2019). Status and future trends of clinical diagnoses via image-based deep learning. *Theranostics* 9: 7556–7565. https://doi.org/10.7150/thno.38065.

64 Kim, M., Yun, J., Cho, Y. et al. (2019). Deep learning in medical imaging. *Neurospine* 16: 657–668. https://doi.org/10.14245/ns.1938396.198.

65 Michie, S., Thomas, J., Johnston, M. et al. (2017). The Human Behaviour-Change Project: harnessing the power of artificial intelligence and machine learning for evidence synthesis and interpretation. *Implementation Science* 12: 121. https://doi.org/10.1186/s13012-017-0641-5.

66 Shah, P., Kendall, F., Khozin, S. et al. (2019). Artificial intelligence and machine learning in clinical development: a translational perspective. *npj Digital Medicine* 2: 1–5. https://doi.org/10.1038/s41746-019-0148-3.

67 Lin, W.-C., Chen, J.S., Chiang, M.F., and Hribar, M.R. (2020). Applications of artificial intelligence to electronic health record data in ophthalmology. *Translational Vision Science & Technology* 9: 13. https://doi.org/10.1167/tvst.9.2.13.

68 Wynants, L., Calster, B.V., Collins, G.S. et al. (2020). Prediction models for diagnosis and prognosis of covid-19: systematic review and critical appraisal. *BMJ* 369: m1328. https://doi.org/10.1136/bmj.m1328.

69 Ahmed, Z., Mohamed, K., Zeeshan, S., and Dong, X. (2020). Artificial intelligence with multi-functional machine learning platform development for better healthcare and precision medicine. *Database: The Journal of Biological Databases and Curation* 2020: baaa010. https://doi.org/10.1093/database/baaa010.

70 Atitallah, S.B., Driss, M., Boulila, W., and Ghézala, H.B. (2020). Leveraging deep learning and IoT big data analytics to support the smart cities development: review and future directions. *Computer Science Review* 38: 100303. https://doi.org/10.1016/j.cosrev.2020.100303.

71 El-Sappagh, S., Ali, F., El-Masri, S. et al. (2019). Mobile health technologies for diabetes mellitus: current state and future challenges. *IEEE Access* 7: 21917–21947. https://doi.org/10.1109/ACCESS.2018.2881001.

72 Da Costa, C.A., Pasluosta, C.F., Eskofier, B. et al. (2018). Internet of health things: toward intelligent vital signs monitoring in hospital wards. *Artificial Intelligence in Medicine* 89: 61–69. https://doi.org/10.1016/j.artmed.2018.05.005.

73 Tuli, S., Basumatary, N., Gill, S.S. et al. (2020). HealthFog: an ensemble deep learning based Smart Healthcare System for Automatic Diagnosis of Heart Diseases in integrated IoT and fog computing environments. *Future Generation Computer Systems* 104: 187–200. https://doi.org/10.1016/j.future.2019.10.043.

74 Sarraf, S. and Tofighi, G. (2016). Classification of Alzheimer's disease using fMRI data and deep learning convolutional neural networks. arXiv preprint arXiv:1603.08631.

75 Kamnitsas, K., Ledig, C., Newcombe, V.F.J. et al. (2017). Efficient multi-scale 3D CNN with fully connected CRF for accurate brain lesion segmentation. *Medical Image Analysis* 36: 61–78. https://doi.org/10.1016/j.media.2016.10.004.

76 Faust, O., Shenfield, A., Kareem, M. et al. (2018). Automated detection of atrial fibrillation using a long short-term memory network with RR interval signals. *Computers in Biology and Medicine* 102: 327–335. https://doi.org/10.1016/j.compbiomed.2018.07.001.

77 Ren, J.-L., Zhang, A.-H., and Wang, X.-J. (2020). Traditional Chinese medicine for COVID-19 treatment. *Pharmacological Research* 155: 104743. https://doi.org/10.1016/j.phrs.2020.104743.

78 Bray, F., Ferlay, J., Soerjomataram, I. et al. (2018). Global cancer statistics 2018: GLOBOCAN estimates of incidence and mortality worldwide for 36 cancers in 185 countries. *CA: A Cancer Journal for Clinicians* 68: 394–424. https://doi.org/10.3322/caac.21492.

79 Ma, Z. and Tavares, J.M.R.S. (2017). Effective features to classify skin lesions in dermoscopic images. *Expert Systems with Applications* 84: 92–101. https://doi.org/10.1016/j.eswa.2017.05.003.

80 Schirrmeister, R.T., Springenberg, J.T., Fiederer, L.D.J. et al. (2017). Deep learning with convolutional neural networks for EEG decoding and visualisation. *Human Brain Mapping* 38: 5391–5420. https://doi.org/10.1002/hbm.23730.

81 Mumpe-Mwanja, D., Barlow-Mosha, L., Williamson, D. et al. (2019). A hospital-based birth defects surveillance system in Kampala, Uganda. *BMC Pregnancy and Childbirth* 19: 372. https://doi.org/10.1186/s12884-019-2542-x.

82 Verma, P. and Fatima, S. (2020). Smart healthcare applications and real-time analytics through edge computing. In: *Internet of Things Use Cases for the Healthcare Industry* (ed. P. Raj, J.M. Chatterjee, A. Kumar and B. Balamurugan), 241–270. Cham: Springer International Publishing https://doi.org/10.1007/978-3-030-37526-3_11.

83 Pradhan, B., Bhattacharyya, S., and Pal, K. (2021). IoT-based applications in healthcare devices. *Journal of Healthcare Engineering* 2021: e6632599. https://doi.org/10.1155/2021/6632599.

84 Perera, C., Zaslavsky, A., Christen, P., and Georgakopoulos, D. (2014). Context aware computing for the internet of things: a survey. *IEEE Communication Surveys and Tutorials* 16: 414–454. https://doi.org/10.1109/SURV.2013.042313.00197.

85 Jan, S.U., Ali, S., Abbasi, I.A. et al. (2021). Secure patient authentication framework in the healthcare system using wireless medical sensor networks. *Journal of Healthcare Engineering* 2021: e9954089. https://doi.org/10.1155/2021/9954089.

86 Banerjee, A., Chakraborty, C., Kumar, A., and Biswas, D. (2020). Chapter 5 - Emerging trends in IoT and big data analytics for biomedical and health care technologies. In: *Handbook of Data Science Approaches for Biomedical Engineering* (ed. V.E. Balas, V.K. Solanki, R. Kumar and M. Khari), 121–152. Academic Press https://doi.org/10.1016/B978-0-12-818318-2.00005-2.

87 Farahani, B., Firouzi, F., Chang, V. et al. (2018). Towards fog-driven IoT eHealth: promises and challenges of IoT in medicine and healthcare. *Future Generation Computer Systems* 78: 659–676. https://doi.org/10.1016/j.future.2017.04.036.

88 Aghdam, Z.N., Rahmani, A.M., and Hosseinzadeh, M. (2021). The role of the internet of things in healthcare: future trends and challenges. *Computer Methods and Programs in Biomedicine* 199: 105903. https://doi.org/10.1016/j.cmpb.2020.105903.

89 Haider, S.A., Adil, M.N., and Zhao, M. (2020). Optimization of secure wireless communications for IoT networks in the presence of eavesdroppers. *Computer Communications* 154: 119–128. https://doi.org/10.1016/j.comcom.2020.02.027.

90 Chen, Y., Song, B., Du, X., and Guizani, N. (2020). The enhancement of catenary image with low visibility based on multi-feature fusion network in railway industry. *Computer Communications* 152: https://doi.org/10.1016/j.comcom.2020.01.040.

91 Irshad, O., Khan, M.U.G., Iqbal, R. et al. (2020). Performance optimization of IoT based biological systems using deep learning. *Computer Communications* 155: 24–31. https://doi.org/10.1016/j.comcom.2020.02.059.

92 Ali, R., Kim, B., Kim, S.W. et al. (2020). (ReLBT): a reinforcement learning-enabled listen before talk mechanism for LTE-LAA and Wi-Fi coexistence in IoT. *Computer Communications* 150: 498–505. https://doi.org/10.1016/j.comcom.2019.11.055.

93 Khan, S., Alvi, A.N., Javed, M.A. et al. (2021). An efficient medium access control protocol for RF energy harvesting based IoT devices. *Computer Communications* 171: 28–38. https://doi.org/10.1016/j.comcom.2021.02.011.

94 Grande, E. and Beltrán, M. (2020). Edge-centric delegation of authorization for constrained devices in the internet of things. *Computer Communications* 160: 464–474. https://doi.org/10.1016/j.comcom.2020.06.029.

95 Anurag, M.S.R., Rahmani, A.-M., Westerlund, T. et al. (2014). Pervasive health monitoring based on Internet of Things: Two case studies. In: *2014 4th International Conference on Wireless Mobile Communication and Healthcare-Transforming Healthcare Through Innovations in Mobile and Wireless Technologies (MOBIHEALTH)*, 275–278. Greece: IEEE https://doi.org/10.1109/MOBIHEALTH.2014.7015964.

96 Thimbleby, H. (2013). Technology and the future of healthcare. *Journal of Public Health Research* 2: e28. https://doi.org/10.4081/jphr.2013.e28.

97 Kulkarni, A. and Sathe, S. (2014). Healthcare applications of the internet of things: a review. *International Journal of Computer Science and Information Technologies* 5 (5): 6229–6232.

98 Uckelmann, D., Harrison, M., and Michahelles, F. (2011). An architectural approach towards the future internet of things. In: *Architecting the Internet of Things* (ed. D. Uckelmann, M. Harrison and F. Michahelles), 1–24. Berlin, Heidelberg: Springer https://doi.org/10.1007/978-3-642-19157-2_1.

99 Distefano, S., Bruneo, D., Longo, F. et al. (2017). Hospitalised patient monitoring and early treatment using IoT and cloud. *BioNanoScience* 7: 382–385. https://doi.org/10.1007/s12668-016-0335-5.

100 Chan, W.M., Zhao, Y., and Tsui, K.L. (2017). Implementation of electronic health monitoring systems at the community level in Hong Kong. In: *Smart Health* (ed. H. Chen, D.D. Zeng, E. Karahanna and I. Bardhan), 94–103. Cham: Springer International Publishing https://doi.org/10.1007/978-3-319-67964-8_9.

101 Segura, A.S. (2016). The internet of things: business applications, technology acceptance, and future prospects, Doctoral Thesis. Universität Würzburg.

20

Wireless Communications Using Machine Learning and Deep Learning

Himanshu Priyadarshi[1], Kulwant Singh[2], and Ashish Shrivastava[3]

[1]*Department of Electrical Engineering, Manipal University Jaipur, Jaipur, India*
[2]*Department of Electronics and Communication Engineering, Manipal University Jaipur, Jaipur, India*
[3]*Faculty of Engineering and Technology, Shri Vishwakarma Skill University, Gurgaon, India*

20.1 Introduction

Genuine leaders across the board have repeatedly emphasized the need of technology to go green. Wireless communication pervades the life of the modern-age citizens in many ways (refer Figure 20.1). Hence, the need of wireless-communication systems to go green can hardly be overemphasized. Seamless, secure, synergistic, and affordable communication is vital for any civilization to thrive.

Wireless communication implemented by networks comprised of energy harvesting devices has been hailed to bring many disruptive changes in wireless communications: (i) self-adequacy and sustainability in terms of energy requirements-sufficient, (ii) perpetuality of operation, (iii) mitigation of conventional energy dependencies as well as carbon cost, (iv) unbridled mobility, and (v) extension of wireless networks to difficult-to-deploy avenues, such as geographically challenging topographies, energy-derived rural areas of third world countries, human bodies, and structures.

The implementation of energy harvesting has brought newer dimensions to wireless communication scenario due to the intermittent and stochastic nature of energy availability. This has mandated the need to revisit the model of wireless-communication system at the transport layer, network layer, data link layer, and physical layer (Figure 20.2). The aspects pertaining to scheduling and optimization of energy harvesting communications at the networking and medium access layer have been comparatively well explored [1]. Such attempts have been directed toward obtaining the maximum throughput by optimization of power transmitted over time; nonetheless, the achievement of information-theoretic asymptotes in the physical layer rate-power correlation is subjected to the luxuries of batteries' energy density and time buffer between energy harvests.

Inspired by Shannon's information theory meant for communications in which the average power is constrained; another cross-section of recent literature aims to demystify the information-theoretic capacities and formulate the information theory for energy harvesting wireless communications. This information-theoretic capacity of energy harvesting links has been explored with smaller-scale dynamics, with energy harvests being considered at the channel utilization level. In-depth analysis of algorithms and transmission schemes at the physical layer has been instrumental for translating the objectives into outcomes.

Exploratory efforts related to the intelligent reflecting surface (IRS) have gained momentum in wireless communications toward configuring the characteristics of channel propagation for favoring information transmission [1]. An IRS is made up of a large matrix of passive (in contrast to active elements like batteries) scattering elements. Each scattering element is capable of inducing individual phase change to the incoming communication signals by exerting control over its operating state. The cumulative coordinated phase control of all scattering elements has

Machine Learning Algorithms for Signal and Image Processing, First Edition.
Edited by Deepika Ghai, Suman Lata Tripathi, Sobhit Saxena, Manash Chanda, and Mamoun Alazab.
© 2023 The Institute of Electrical and Electronics Engineers, Inc. Published 2023 by John Wiley & Sons, Inc.

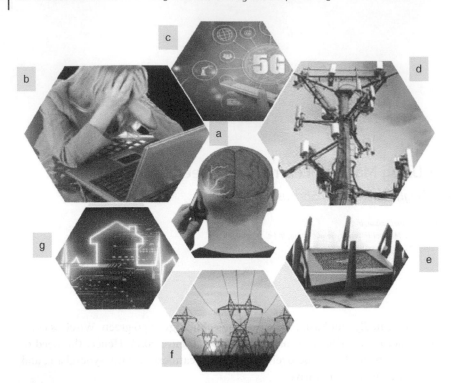

Figure 20.1 (a) cellular communication may affect the human brain; (b) working on laptops for prolonged durations causes increased stress; (c) with fifth generation of communication devices, humanity is surrounded by internet of things ; (d) for improved communication, more number of cell-phone towers are required, previously with lesser users lesser number of towers were required, but with increased cell-phone users; towers are being installed in close proximity to human residences with their effect on human health still being debated; (e) apart from devices and towers, routers and access points also have exposure effects on human health; (f) transmission towers also affect human health apart from being fatal on being directly contacted; (g) however human existence has literally become unseparable from communication devices. Source: (a) Adobestock. (c) nirutft/Adobe Stock. (d) CounterDarkness/Pixabay. (e) Proxima Studio/Adobe Stock. (f) TebNad/Adobe Stock. (g) Who is Danny/Adobe Stock.

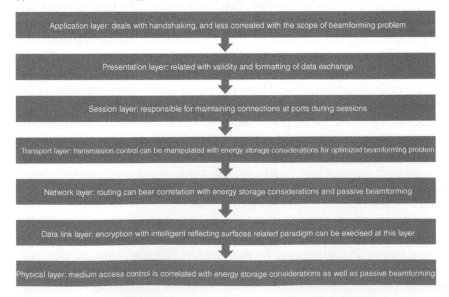

Figure 20.2 Open-system interconnection model for communication systems with energy harvesting considerations for optimized beamforming.

been termed as passive beamforming. By dint of passive beamforming, the orientation and strength of the signals can be purposefully tuned to manifest desirable channel conditions. Thus, the channel environment can be transformed into a smart space with the capability of being optimized in terms of its configuration for enhancing the performance of the network.

The objectives of IRS implementation toward power-transmission performance are to improve the spectral efficiency as well as the energy efficiency. IRS implementation can also be utilized for better security, wireless-energy exchange, vehicular communications, and mobile-edge computing. The typical formulation of the performance maximization of IRS-aided wireless-communication systems is done as a combined optimization problem of the passive and active beam-formation strategies [2–6]. The non-convexity characteristic of the problem structure forces the solution to be carved around the alternating optimization (AO) chassis, which unfortunately leads to sub-optimal solutions. In every iterative step of the alternating optimization framework, convex approximation (also known as semi-definite relaxation) is generally used to optimize either among the active or passive beamforming.

Implementation of IRS's passive beamforming has two pertinent challenges to grapple with: (i) high space and time complexity for algorithms to coordinate the passive beamforming with enormity of the number of scattering elements; (ii) the dependency on the awareness of channel information, which becomes more difficult for passive beamforming in contrast to active beamforming (where the communication is powered solely by batteries or equivalent active elements). Matter-of-factly, this led to optimization problem formulation for collaboration between active and passive beamforming. Convex approximations have been utilized for the solution to such optimization problems, with convergence obtained at a sub-optimal solution.

Being a self-learning (also known as heuristic) approach, the performance depreciation of the alternating optimization method lacks exactitude and faces difficulty in characterization. Any pliable formulation of beamforming optimization is matter-of-factly based on not-so-exact system modeling, which might be an over-simplification (assumptions like continuity and exactitude of phase control, perfect information of the channel) of the real-life system. Approximation to the original problem formulation might deviate the solution further astray from the optimum; hence model-based optimization approach might not yield anything better than the lower performance bound of the actual problem.

It would be unrealistic to invoke traditional algorithms for solving the optimization problem of collaboration between active and passive beamforming, due to the enormity of the scattering elements in the IRS. The system model becomes very complicated with the increase in the scattering elements, and hence it is not advisable to work toward obtaining traditional algorithm, particularly with the facility of machine learning (ML) (which includes deep learning). ML has been hailed to give reasonable success for complicated and difficult-to-decode system models.

ML can be utilized for performance premium in IRS-facilitated wireless-communication networks with a huge number of scattering elements in the IRS. The purpose of utilizing a suitable machine-learning algorithm (MLA) for collaborative beamforming is to achieve robustness defying the vagaries of channel dynamics and lack of channel information knowledge. In the subsequent section to follow, we have deliberated upon the IRS-enabled wireless networks, underscoring the performance dividends and the contending issues. Subsequently, an overview of various use cases of MLAs in IRS-enabled wireless networks has been presented. In comparison to the traditional optimization algorithms, MLAs provide better malleability and robustness against problems of channel dynamics, lack of channel information knowledge, and imprecise system modeling in a constantly changing radio ambience. Herein, we would also underscore some practical issues in deploying MLAs, stemming due to the necessity of a huge corpus of data samples required for offline training and the slow speed of convergence during online learning. The fundamental scheme is to bifurcate the decision parameters of a complicated control problem into two sub-divisions. The first part may be searched in the outer-loop MLA, while the second part is optimized instantaneously by efficiently solving an approximation problem. The verification by dint of numerical simulations establishes the following: (i) accelerates the quest for optimal solutions by utilizing the attributes of control structure; and (ii) simplifies the action space complexity of the MLA.

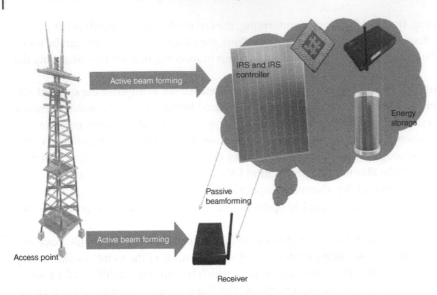

Figure 20.3 Simplified schematic of IRS in a wireless-communication system.

Fortunately, MLAs are more robust against the stochastic-system models and have been utilized for IRS's phase control. Deep-neural network (DNN) has been employed to: (i) correlate the receiver's location with the IRS's optimal phase configuration for a complicated indoor ambience; (ii) provide real-time prognostics for the IRS's passive beamforming with lesser time and space complexity; (iii) improve the signal-to-noise ratio based on ambience interactivity using policy-gradient method; (iv) enhance secrecy against eavesdropping; and (v) to adjust channel loading as per receiver's feedback [7–12].

20.1.1 IRS-Enabled Wireless-Communication Systems

The architectural tunability of IRS's is dependent on its tuneable chips inbuilt into the IRS matrix [13]. As depicted in Figure 20.3, an individual tuneable chip is governed by the IRS controller to modulate the phase shift of the unit scattering element. The IRS controller is capable of having communion with the external radio frequency (RF) transceivers about its reconfiguration settings. The role of IRS in a wireless-communication scenario along with its challenges has been presented in the following.

20.2 Contributions of Intelligent Reflecting Surfaces (IRS) in Wireless-Communication Systems

20.2.1 IRS As Signal Reflector in Wireless-Communication System

The communication signal in the direct link existing between the radio frequency transceivers can be superposed constructively with its reflected versions at the receiver or manipulated destructively to the effect of information spillage reduction to uncalled-for receivers. It can be observed that this mechanism supports high-data transmission rate, without any active power overhead. A higher data rate implies lesser transmission delays in the network. Investment in an increasing number of scattering elements has established to incentivize the performance of the wireless-communication system [14]. Larger dimension IRS can be divided into smaller flexible-to-be configured sub-matrices, to have multiplicity of reflected versions of the original signal. IRSs can work in unison for safe navigation of the reflected signals to improperly conditioned receivers.

20.2.2 IRS As Signal Transmitter in Wireless-Communication System

Exercising control over the IRS's phase shifts in the temporal domain empowers the signal emanating from the IRS to exhibit variegated radiation patterns and thus may be utilized to carry and encrypt confidential information. It might be speculatively suggested that this is general case of the traditional backscatter communications [15], wherein the reflecting states of the antenna are modulated by altering the load impedance, and thus varying the information bits. IRS-based backscatter communications, so to say, offer greater flexibility as IRS is capable of providing more peculiar reflection paradigms, faster data rates, lesser transmission delays, and larger transmission distances for information communication.

20.2.3 IRS As Signal Receiver in Wireless-Communication System

The passive reflecting units of IRS may be considered as individual receivers for multiplexed data transmissions, provided IRS is equipped with the signal-processing support. There configurability feature of IRS can accomplish commendable capacity gain by mitigating the interference among various users. The big-size IRS structure may be fragmented into smaller units, manipulating the signals received from different users at an individual basis. The IRS may also be gainfully utilized as a matrix of sensors to track the location of the mobile devices based on fluctuations on the radio-frequency power sensed on the IRS surface.

20.3 Merits of IRS-Aided Wireless-Communication Systems for Performance Enhancement

20.3.1 Enhancement in the Channel Capacity

The collaborative beamforming approach for capacity enhancement by dint of perfect IRS may be decoupled in two parts. The objective of passive beamforming is to maximize the equivalent channel gain by a gainful combination of the direct as well as the reflected signals, whereas active beamforming can be aligned with the equivalent channel. Considering a more pragmatic implementation, the active beamforming is utilized to transfer power for sustaining the IRS's operations by dint of radio-frequency energy harvesting. Since IRS comprised of passive-circuit elements has more economic implementation, larger dimension IRS may be utilized to boost the channel conditions as well as the incoming signal strength.

20.3.2 Savings on the Transmit Power of Base Station in IRS-Aided Wireless-Communication System

Improvement in channel condition tacitly implies that the transmitting network device can economize the power required for signal transmission without compromising on the quality of service to the clients. This has certainly led to the evolution of a more energy-efficient paradigm in wireless communications. In accordance with the power scaling law mentioned in [14], the transmit power associated with the base station of an IRS-enabled system is inversely proportional to the number of scattering elements in the IRS, this power-saving becomes even more conspicuous for users situated away from the base station, and also for IRS having low phase resolution.

20.3.3 Protection Against Eavesdropping and High Confidentiality Rate

Assurance of confidentiality is of utmost importance in mission-critical communication, and IRS may be utilized to snub attempts at wireless eavesdropping by curbing the information leakage to any illicit entity. According to revelations of the simulation-based studies of some researchers [5], it is more pragmatic to obtain enhanced confidentiality rate along with energy efficiency by the deployment of higher dimensions of IRS matrix, in contrast to increasing the dimensions of the active transmitter's antenna array. Protection for the legally subscribed user, against eavesdropping by an unauthenticated user can be obtained by curbing unwanted conditions of propagation, through proper leveraging of IRS matrix.

20.4 Issues in Collaboration Between Active and Passive Beamforming

The combined optimization problem has been reportedly [5, 16] solved by invoking the methodology of alternating optimization (AO). The AO approach decouples the phase control component of IRS and the transmit beamforming into two sub-categories. Semi-definite programming with high computational burden has to be run for optimizing the beamforming strategy with semi-definite relaxation. Such methods driven by optimization are meritorious in the sense that they can sometimes even yield standard acceptable solutions, and the performance metrics are generally predictable.

20.4.1 Overhead on Algorithms Due to Time and Space Complexity

Increasing the number of scattering elements provides much more flexibility available at the IRS's phase control, even with finite-phase resolution or hardware limitations. The overhead incurred in getting the flexibility of IRS's phase control comes with the computational burden associated with the semi-definite programming required to tackle the convex optimization formulation for getting the right balance between active and passive beamforming. More time and space complexity associated with the algorithm of formulation naturally translates to challenging overhead in training as well as increased power requirement for channel estimation, with the upscaling of IRS size.

20.4.2 Lack of Channel Information

The exchange of information and estimation of channel necessitates the use of highly sophisticated and sprightly protocols. Moreover, such protocols must be executed within one time cycle of synchronism to ensure valid handshaking with regard to the channel information, particularly in a dynamic ambience. It is vital that real-time implementation of the joint beamforming optimization must exhibit robustness to manage the errors, omissions, and uncertainties with respect to the information about the channel.

20.4.3 Simplifying Assumptions Lead to Unrealistic System Modeling

The assumption of perfect availability of channel information is an oversimplifying assumption for the joint beamforming problem formulation; this causes the model to be oversimplified, which is rendered further unrealistic due to approximations of continuity and exactitude for the phase control of IRS. This leads to sub-optimal solution with uncertainty in practical performance. Hence, it is naturally expected that methods based solely on optimization can provide at most the extrema of the intended problem.

These issues have catalyzed the adoption of model-less ML approaches to solve the joint beamforming problem for IRS-aided systems.

20.5 Scope of Machine Learning for IRS-Enabled Wireless-Communication Systems

MLAs are classified as supervised or unsupervised based on the question whether the data used to train and test the MLAs is labeled or unlabeled, respectively.

While supervised MLAs provide a decision boundary or surface for segregating the labeled data, whereas unsupervised MLAs give a regression function for continuous mapping between input and output. The concept of thread-based programming has tremendously increased the computational capability as well as speed, and hence the metrics of ML can be tremendously improved by deploying DNNs for extracting the tacit characteristics of the data to be trained, although these features may be deeply entrenched within the framework, as the weight of different neurons in the DNN may be iteratively adapted. The concept of learning from experience and interaction is an intrinsic part of all types learning and intelligence cultivation in all domains. What we term as experiential learning in the domain of pedagogy is the most outcome-based approach based on interaction, and captures the essence of reinforcement learning (RL) among MLAs. RL-based MLAs involves decision making regarding how to correlate the ambient situations to action being taken, using a numerical reward attribute. The RL-based MLAs might get crippled with instability while dealing complicated systems which have humongous size of state and action space. To overcome these issues, DNNs and deep Q-neural nets are utilized for more efficient approximation of the parameters associated with reinforcement learning, and the approach is generally known as deep-reinforcement learning (DRL).

One remarkable DRL approach, namely policy gradient deep-reinforcement learning, bifurcates the interaction based learning into two DNNs: actor DNN and the critic DNN. The actor DNN determines the optimal policy for decision-making based on the value function evaluated by the critic DNN. Deep learning fits naturally in the scheme of things for IRS because huge number of scattering elements in the IRS matrix will be generating huge dataset anyway.

20.5.1 Pathway Assessment for Communication Channel and Signal Diagnostics

In order to estimate the channel for a particular training time quantum, a discrete sequence of pre-conceived pilot is sent at the transmitter, and subsequently the expected response of the channel, as well as the response observed at the receiver is utilized for channel estimation. In case the experimenter wants to train the machine for supervised learning, pilot and the expected version of the channel response may be used as the labeled data. With the enormity of dataset inherently associated with the humongous IRS matrix, the DNNs can suitably handle the dataset for better training.

Researchers [17, 18] have also investigated the possibility of using convolutional neural network for channel probation for the IRS-enabled wireless-communication system. In their studies they have used simulation-based on input signal, expected channel outputs, and phase variations; after training the convolutional neural net rigorously, it can be used for real-time predictions, with cost function on DNNs being manipulated to increase the affine value between the transmitted and received vectors.

20.5.2 Machine Learning for Passive Beamforming

The way DNNs have been fashioned, makes it very congenial for these to be rigorously trained to retrieve highly-complex mapping (from the ambient conditions like channel response as well as the spatial coordinates of the receiver) for arriving at the optimal solution for passive beamforming. The tracking of spatial coordinates of the receiver with respect to the optimal phase configuration of IRS can be done to strengthen the intensity of the received signal [8]. The labeled dataset for the unsupervised learning is derived from IRS's phase vector in its optimal form. The rigorously-trained DNN may be subsequently utilized in online mode to forecast the optimal phase vector associated with IRS with reference to the spatial coordinates of the subscriber.

The utility of ML based on unsupervised approach is to get highest attainable value of signal-to-noise ratio from the receiver vantage point [19] in the context of optimal passive beamforming for IRS. The cost function for the regression in the unsupervised learning is formulated by negating the expected value of the signal-to-noise ratio from the receiver's vantage point. The DNN has to be trained to minimize this cost function to be employed for regression. Invoking Markov's property, one can obtain more malleability with DRL making it even more competent under alien and dynamic ambience. A state signal capable of retaining all the relevant information can be said to have the Markov property (one may refer [20] for knowing more about Markov states, which is an introductory text on reinforcement learning). The Markovian basis of reinforcement learning, wherein the machine is trained to remember all the relevant states in a compact way, is very congenial for maximizing the signal-to-noise ratio for based on agent-ambience interaction wherein the relevant states are decided depending on the numerical reward for corresponding actions. Deep deterministic policy gradient algorithm has been used very effectively [21, 22] to optimize the continuous phase vector for ensuring the best-possible signal-to-noise ratio at the receiver end of the communication channel, with much lesser time complexity.

20.5.3 Prevention of Denial of Service Attacks and Stealth in Communications

Generally, the literature emphasizes about the security aspects of communication, at the application layer; however, by employing the frameworks based on DRL, the security of the physical layer can also be bolstered to counteract stealth as malignant nodes trying to decrease the signal-to-noise ratio for denial of service to bonafide subscribers. In order to prevent stealth in communication, IRS-enabled wireless-communication systems increase the confidentiality rate by suppressing the information leakage, in comparison to the bonafide exchange of information in the channel among legal subscribers. The numerical reward signal in DRL has been formulated as the deviation of the rate of confidentiality with reference to a pre-set baseline known as penalty [23], to provide a solution for the problems related to denial of service attacks and stealth.

In comparison to algorithms based on optimization, the ML-based algorithms in IRS-enabled systems have exhibited more malleability and robustness against constantly changing ambience, stochastic nature of information, and modeling inaccuracies. This cannot lead us to assume that the practical implementation of ML-based algorithms will not be having challenges. Two formidable challenges in the practical implementation have been reported to be the following: (i) during online training the convergence speed is undesirably slow; (ii) a huge amount of data samples has to be mustered for offline training of the ML-based algorithm. Researchers employing DNN training [8, 17–19] adopted either of the following approaches: (i) procure a sufficiently large dataset or (ii) randomly generate an adequately big dataset from simulation tools. Approach (i) has data accessibility issues as people working in this area might not be willing to share; whereas the dataset from approach (ii) is based on simplified modeling, which might have an inherent bias when it comes to online prognostics. The solutions based on reinforcement learning as well as DRL are capable of making decisions from the very beginning [21–24], and hence are less vulnerable to bias, however, they suffer with slow convergence issues percolating through the ambience. Table 20.1 summarizes the contribution of ML-based approaches for IRS-enabled communication systems.

Table 20.1 Contribution of machine learning-based approaches for IRS-enabled communication systems.

Reference no.	Machine-learning algorithm(s)	Target	Outcome
[17]	Supervised learning with convolutional neural nets	Probing the communication channel	Error mitigation, and robustness enhancement
[8]	Supervised learning with deep-neural nets	Data rate improvement	Accomplished the upper extremum with an exactitude of channel knowledge
[19]	Unsupervised learning with deep-neural nets	Signal-to-noise ratio of receiver	Better performance with semi-definite programming
[21]	Deep-deterministic policy gradient algorithm	Signal-to-noise ratio of receiver	Accomplished near-optimal value of the signal-to-noise ratio with lower time complexity
[22]	Deep Q-networks	Confidentiality rate	Better quality of service, and rate of confidentiality

20.6 Summary

Manipulating the phase associated with communication signals reflected from intelligently configured surfaces (nevertheless passive from an energy point of view) can be utilized gainfully to make communication systems and network ecologically green, affordable as well as accessible on an even greater scale. To this end, researchers have been importing training algorithms from the domain of ML, and have been reporting their results in terms of meaningful communication channel metrics. Going further, one can achieve even better if the training resilience of ML techniques can be gainfully blended with the efficiency of model-based optimization methods.

Acknowledgment

We are very grateful to our parents, teachers and friends, who invested their precious time in the thoughtful discussions.

References

1 Gunduz, D., Stamatiou, K., Michelusi, N., and Zorzi, M. (2014). Designing intelligent energy harvesting communication systems. *IEEE Communications Magazine* 52 (1): 210–216. https://doi.org/10.1109/MCOM.2014 .6710085.

2 Wu, Q. and Zhang, R. (2018). Intelligent reflecting surface enhanced wireless network: joint active and passive beamforming design. In: *2018 IEEE Global Communications Conference (GLOBECOM)*, 1–6. https://doi.org/10 .1109/GLOCOM.2018.8647620.

3 Lyu, B., Hoang, D.T., Gong, S., and Yang, Z. (2020). Intelligent reflecting surface assisted wireless powered communication networks. In: *2020 IEEE Wireless Communications and Networking Conference Workshops (WCNCW)*, 1–6. https://doi.org/10.1109/WCNCW48565.2020.9124775.

4 Jiang, T. and Shi, Y. (2019). Over-the-air computation via intelligent reflecting surfaces. In: *2019 IEEE Global Communications Conference (GLOBECOM)*, 1–6. https://doi.org/10.1109/GLOBECOM38437.2019.9013643.

5 Yu, X., Xu, D., and Schober, R. (2019). Enabling secure wireless communications via intelligent reflecting surfaces. In: *2019 IEEE Global Communications Conference (GLOBECOM)*, 1–6. https://doi.org/10.1109/ GLOBECOM38437.2019.9014322.

6 Yu, X., Xu, D., Ng, D.W.K., and Schober, R. (2020). Power-efficient resource allocation for multiuser MISO systems via intelligent reflecting surfaces. In: *GLOBECOM 2020 – 2020 IEEE Global Communications Conference*, 1–6. https://doi.org/10.1109/GLOBECOM42002.2020.9348054.

7 Huang, C., Alexandropoulos, G.C., Yuen, C., and Debbah, M. (2019). Indoor signal focusing with deep learning designed reconfigurable intelligent surfaces. In: *2019 IEEE 20th International Workshop on Signal Processing Advances in Wireless Communications (SPAWC)*, 1–5. https://doi.org/10.1109/SPAWC.2019.8815412.

8 Gao, J., Zhong, C., Chen, X. et al. (2020). Unsupervised learning for passive beamforming. *IEEE Communications Letters* 24 (5): 1052–1056. https://doi.org/10.1109/LCOMM.2020.2965532.

9 Taha, A., Alrabeiah, M., and Alkhateeb, A. (2019). Deep learning for large intelligent surfaces in millimeter wave and massive MIMO systems. In: *2019 IEEE Global Communications Conference (GLOBECOM)*, 1–6. https://doi.org/10.1109/GLOBECOM38437.2019.9013256.

10 Song, Y., Khandaker, M.R.A., Tariq, F. et al. (2021). Truly intelligent reflecting surface-aided secure communication using deep learning. In: *2021 IEEE 93rd Vehicular Technology Conference (VTC2021-Spring)*, 1–6. https://doi.org/10.1109/VTC2021-Spring51267.2021.9448826.

11 Feng, K., Wang, Q., Li, X., and Wen, C. (2020). Deep reinforcement learning based intelligent reflecting surface optimization for MISO communication systems. *IEEE Wireless Communications Letters* 9 (5): 745–749. https://doi.org/10.1109/LWC.2020.2969167.

12 Yang, H., Xiong, Z., Zhao, J. et al. (2020). Deep reinforcement learning based intelligent reflecting surface for secure wireless communications. arXiv preprint arXiv:2002.12271.

13 Yang, H., Xiong, Z., Zhao, J. et al. (2021). Deep reinforcement learning-based intelligent reflecting surface for secure wireless communications. *IEEE Transactions on Wireless Communications* 20 (1): 375–388. https://doi.org/10.1109/TWC.2020.3024860.

14 Wu, Q. and Zhang, R. (2019). Intelligent reflecting surface enhanced wireless network via joint active and passive beamforming. *IEEE Transactions on Wireless Communications* 18 (11): 5394–5409. https://doi.org/10.1109/TWC.2019.2936025.

15 Park, S.Y. and In Kim, D. (2020). Intelligent reflecting surface-aided phase-shift backscatter communication. In: *2020 14th International Conference on Ubiquitous Information Management and Communication (IMCOM)*, 1–5. https://doi.org/10.1109/IMCOM48794.2020.9001811.

16 Zou, Y., Gong, S., Xu, J. et al. (2020). Wireless powered intelligent reflecting surfaces for enhancing wireless communications. *IEEE Transactions on Vehicular Technology* 69 (10): 12369–12373. https://doi.org/10.1109/TVT.2020.3011942.

17 Elbir, A.M., Papazafeiropoulos, A., Kourtessis, P., and Chatzinotas, S. (2020). Deep channel learning for large intelligent surfaces aided mm-wave massive MIMO systems. *IEEE Wireless Communications Letters* 9 (9): 1447–1451. https://doi.org/10.1109/LWC.2020.2993699.

18 Khan, S. and Shin, S.Y. (2019). Deep-learning-aided detection for reconfigurable intelligent surfaces. arXiv: 1910.09136.

19 Lee, G., Jung, M., Kasgari, A.T.Z. et al. (2020). Deep reinforcement learning for energy-efficient networking with reconfigurable intelligent surfaces. In: *ICC 2020 – 2020 IEEE International Conference on Communications (ICC)*, 1–6. https://doi.org/10.1109/ICC40277.2020.9149380.

20 Ozel, O., Tutuncuoglu, K., Ulukus, S., and Yener, A. (2015). Fundamental limits of energy harvesting communications. *IEEE Communications Magazine* 53 (4): 126–132. https://doi.org/10.1109/MCOM.2015.7081085.

21 Sutton, R.S. and Barto, A.G. (2018). *Reinforcement Learning: An Introduction*. MIT Press.

22 Gong, S., J. Lin, B. Ding, D. Niyato, D. I. Kim et al. (2020). Optimization-driven machine learning for intelligent reflecting surfaces assisted wireless networks. arXiv: 2008.12938.

23 Lin, J. Y. Zout, X. Dong, S. Gong, et al. (2020). Optimization-driven deep reinforcement learning for robust beamforming in IRS-assisted wireless communications. arXiv: 2005.11885.

24 Huang, C., Mo, R., and Yuen, C. (2020). Reconfigurable intelligent surface assisted multiuser MISO systems exploiting deep reinforcement learning. *IEEE Journal on Selected Areas in Communications* 38 (8): 1839–1850. https://doi.org/10.1109/JSAC.2020.3000835.

21

Applications of Machine Learning and Deep Learning in Smart Agriculture

Ranganathan Krishnamoorthy[1], Ranganathan Thiagarajan[2], Shanmugam Padmapriya[3], Indiran Mohan[6], Sundaram Arun[5], and Thangaraju Dineshkumar[4]

[1] *Centre for nonlinear Systems, Chennai Institute of Technology, Chennai, India*
[2] *Department of Information Technology, Prathyusha Engineering College, Chennai, India*
[3] *Department of Computer Science and Engineering, Loyola Institute of Technology, Chennai, India*
[4] *Department of Electronics and Communication Engineering, Kongunadu College of Engineering and Technology, Trichy, India*
[5] *Department of Electronics and Communication Engineering, Jerusalem College of Engineering, Chennai, India*
[6] *Department of Computer science and Engineering, Prathyusha Engineering College, Chennai, India*

21.1 Introduction

Agriculture is the backbone of India, which is the non-technical sector and need a better way of implementation. A farmer generally uses the traditional method in farming, which can cause a loss in yield due to the manual labor. The mutation over the crops can regain the growth in the minerals in the soil. It enhances the ability to cultivate the crop. This crop mutation uses the leftover type of mineral to grow the crops. To improve soil fertility constantly, the mutation is implemented. It analyzes the yield and identifies the crop yield by checking the unfit soil yield, i.e. the one which is unfit to grow a crop. These machine learning (ML) and deep learning (DL) can predict the soil moisture, crop mutation, and climatic factors, which can cause the loss in crop yield. As the prior intervention takes place, the crop yield loss gets reduced as it reduces the loss when compared to the traditional method. Precision smart farming is used to improve the demand for food and increase sustainable productivity over the growth in crops. Precision smart farming aims to reduce crop loss by using the decision-making process, which is an alternative method of traditional farming. Based upon the crop selection, pH level, and soil moisture, the growth of the crop is improvised. In modernized agriculture, it seeks out a controlled-type environment such as it locates the specified areas to grow certain crops. To improve the productivity in the crop, different optimized precise management is enhanced. Agricultural machines with the internet-of-things (IoTs) sensors identify the dataset which are normalized to attain the precise results.

21.1.1 Major Contributions of Smart Agriculture

Smart agriculture is one of the modernized approaches which is used nowadays to reduce the time and cost of efficiency. These ensure the agriculture smart management which results as per the specific conditions. Some of the high-intensity plantings and huge productivity, these DL and ML approaches can be used. Batch-wise production of the different crop mutations can be achieved. Locusts that can damage the agricultural crops can be minimized by segmentation of the image of the crop yield. In this modern era, it has a great potential to enhance the sustainability and massive productivity in crops. The agricultural productivity and quality of the crop are increased by reducing the farmer's time and cost-efficiency. It gathers the data by collecting it from the sensor and analyzing it by checking the pH rate, soil quality, and minerals effectively. Image segmentation analyzes the precision agriculture for identifying and monitoring the diseases prediction. In India, the agricultural sector contributes 17.5% of GDP with a total of 45% of labor in manual. Since they follow the traditional method of farming, it consumes more time and complexity in cultivating the yields. Irrigation for crop cultivation can be analyzed using the DL and ML which reduces the complexity with minimal cost. It checks and analyzes the best crop selectively and chooses the best yield for production using the prediction by choosing the specific region in agriculture. The meteorological climatic factors include the climatic weather condition, average temperature, max–min temperature, and state of humidity. They have been impacted to check the crop yield prediction. Based upon the land type such as dry land or wet land, it categorizes accordingly. This dataset compares the image and structures along with the pattern to attain the soil structural composition. As the soil structural composition is essential in analyzing the shape of the soil. Based upon the soil structure, the cultivation of the crop is enhanced. With soil texture, moisture, and soil consistency, the growth of the crops improve. Some of the data are complex and large, which is difficult to analyze new information. Some of the useful information which is extracted from the sensors is used to attain accurate results. In this, the aim of this chapter is to provide the applications of deep-learning and machine-learning techniques in the smart agriculture field. The way of using this technique in agriculture is been investigated using several research communities [1]. In this field, some of the different techniques are improvised to attain a smart way of reducing the cost in farming and reducing the complexity. This chapter describes the concept of ML and DL architecture, challenges, computation methods, and applications in various fields of application. The main aim is to provide knowledge on the concept of deep learning, smart cultivation, computation methods, and application of ML and DL in smart agriculture.

The chapter is discussed as follows: Section 21.2 indicates the concept of machine learning and its types in ML algorithms which is to categorize the characteristics in agriculture. Section 21.3 indicates the concept of deep learning and its types. It defines deep learning and the way input is parameterized based upon their characteristics. Section 21.4 represents that agriculture is refined in a smart way using the digital technology. Section 21.5 represents the computational methods used in the farming. In Sections 21.6 and 21.7, it represents the security issues and application domain used in agriculture, respectively. Section 21.8 indicates the different case studies used in smart agriculture, and Section 21.9 indicates the agro city detailed explanation. In Section 21.10, prediction of analyzing different parameters in agriculture are analyzed using deep-learning and machine-learning techniques. The results and discussions are discussed in Section 21.11 and conclusions are concluded in Section 21.12.

21.2 Concept of Machine Learning

This section provides a conceptual view of smart agriculture, which defines the precise way of crop yielding and prediction. Machine learning is used to analyze the data with automation in the model building with minimal human intervention.

ML is the study of the algorithm which is used to improve the use of data [2, 3]. Figure 21.1 represents the basic concept of machine learning. Machine-learning algorithms build the model using the sample data with the training data using prediction and decision approach. ML can be used in different applications to analyze complex and large data. In this, the subset of machine learning is closely related to computational statistics. It uses methods, theory, and computational models to deliver accuracy over results. Machine learning was initiated by Arthur Samuel. He is an American IBM worker who is a pioneer in gaming and artificial intelligence (AI). Later in 1973, pattern recognition was developed by Duda and Hart. Machine learning performs the tasks without the explicit program by using certain tasks [4–6].

21.2.1 Types of Machine Learning

A machine-learning algorithm allocates the decision autonomously without any use of external support. Deep learning is the subset of machine learning which is from the same family. This trains up the model from a large

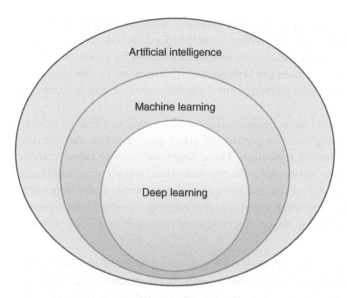

Figure 21.1 Basic concept of machine learning.

set of datasets using the artificial neural network (ANN) to correlate the input. This program extracts and collects the information and achieves based upon the knowledge of the data. It builds the framework to predict the data using the decisions. Machine learning is one of the sophisticated techniques which are used to discover data with unknown regularities along with the patterns of the data. ML allows in designing the computer to learn the way of structures. This regulates the performance and determines the computational analysis using statistics to analyze the patterns. ML is also called computational learning which statistically solves complex problems. Using the structural patterns, climate prediction, weed prediction, and disease identification can be analyzed. Large datasets are derived from established insights using the relative type difficulty analysis in the different environments. In machine-learning algorithms, there are three different types which are explained as follows: (i) un-supervised learning, (ii) supervised learning, and (iii) semi-supervised learning.

21.2.1.1 Supervised Learning

Supervised learning predicts the set of variables from the independent variables. This set of variables generates the function which amps the input into the desired set of output. It uses the trained process which achieves the desired accuracy as results. It maps the function by generating the desired output. It analyzes the behavior of the input by mapping the vector into the varied class by analyzing the input and output function. Large-scale data are sampled from the available data which are expensive in approach. Data which are analyzed are in the input variables with the output target values. It maps the input and learns how to target from input and output. Supervised learning holds sufficient data with pre-determined classifications. It follows the set of rules in instances such as training set. It minimizes the network classifier by determining the attributes to solve the classified approach. It consists of a pre-determined approach to analyze the trained datasets. These datasets have a sufficient approach in the implementation of the structured patterns by underlying the classified data. Mainly to analyze the classification approach, the supervised approach is implemented. It recognizes the determined set of classification with precision and accuracy. This makes the computer understand the classification system which has been created in the system. Every instance of the classified approach is predetermined, and it easily works out with the agent. Some of the examples in supervised learning includes regression and classification approach.

Regression Regression is one of the examples of the supervised learning approach, which is used to predict the continuous set of values. It sets out the goals by plotting the best-fitted line to analyze the data. It uses the metrics to evaluate the data using the trained regression model. It validates the continuous value using the prediction. These techniques use the model to analyze the relationships between the variable and its outcomes. In this regression method, it uses the statistical method to model the relationships of both predicted variables and independent variables.

These determine the data by finding the correlated data between the variables and continuous output. Regression plots the graph with the variables by using the data points and graph plot. It points out the data points and targets the data points in the minimum line of regression. These determine the one target variable with the series of independent variables. It learns the curve plot with the real-valued output which indicates the predicted new data. It is a type of predictive modeling that analyzes the relation between the dependent variable and other variables within the dataset. These target variables consist of continuous values which predict only the continuous set of values with the best-fit plot curve. Regression consists of two types as follows: (i) linear regression, and (ii) logistic regression

(i) **Linear Regression:** It is used to estimate the original set of values such as the constant variable. These values begin the correlation among the independent and dependent variables. Figure 21.2 represents the relationship between those variables using linear regression technique. It fits out the best line called the regression line using the characterized linear equation. Linear regression is characterized to attain the real set of continuous values. It analyzes the dependent variable and independent variable. It minimizes the sum of the squared set of difference distance between the regression line and data points. It uses the scalar responses by modeling the relationship using the linear predictor functions. It has two different types of linear regressions: (i) simple and (ii) multiple linear regressions.

The simple linear regression uses one set of explanatory variables in this case. For one such case, both the dependent and independent variable are used. If the multivariate systems of linear regression are predicted than the single variable, then it is termed multiple linear regressions. It focuses on the conditional property using the given values of predictors.

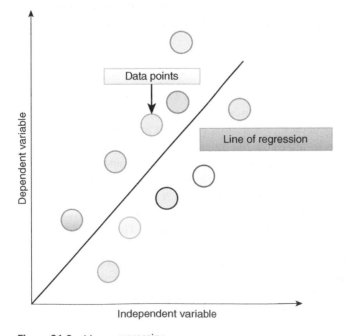

Figure 21.2 Linear regression.

Figure 21.3 Logistic regression.

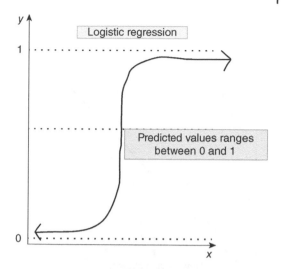

(ii) **Logistic Regression:** Logistic regression is one of the types of classification algorithms which is used to predict the probability of the targeted variable. These variables use only the possible classes which predict in a logistic manner. This model predicts the function by using the categorical department using the set of independent variables. The logistic regression predicts the output using the dependent variable. It uses the statistical model, which follows the conditional probability. It uses the predicted output such as $Y = 0$ or 1, which is the cost function as shown in Figure 21.3. Logistic regression is used to classify the predictive analysis with the probability. The cost function ranges from 0 to 1 used as a core in the methods. This logistic function uses sigmoid function, which populates the growth of the capacity. It is used to predict the probability over the target variable. It is one of the simplest methods to predict agriculture disease prediction and weed detection etc.

Classification   Classification is used to categorize the set of data into the defined class with the structured and unstructured forms. This process predicts the data in the target form, labeled form, and category form. In the supervised learning technique, classification is one of the types to understand the program by classifying the new observations into the number of classes or groups. It follows the statistics in which the computer program understands the data which are given into the observations and performs the process by categorizing it. The categorized data are used to perform the structured data and unstructured data according to their observations. It predicts the classes based upon the data points. Using the predictive modeling, the task can be predicted by using the mapping function where the input variables are changed into a discrete set of variables. These algorithms are used to map the input data by using specific categorical data. It sets out the data in a classified manner according to the way of detecting and recognizing it. It identifies the dataset by identifying similar data within it. It mainly predicts the output of the categorical data.

(i) **K-Nearest Neighbor:** *K*-nearest neighbor (KNN) is used for classification and regression issues. These categorize in both ways by identifying the similar nearest neighbors as shown in Figure 21.4. It gathers the available cases and easily classifies the majority state of the *K*-neighbors. It assigns the best class and defines the distance using the functions such as Murkowski, Euclidean, Manhattan, and Hamming distance, respectively. These are used to assign the continuous functions by categorizing the variables using the distance functions. KNN performs the performance modeling by mapping the information by gaining access to it. KNN maps the data since it is simple to use. It implements the data, which is classified using the classification algorithm. It imputes the missing values and re-sampled the known datasets. The KNN algorithm is used for the prediction of the crop yield, which includes the soil moisture, temperature conditions, and crop suggestions in agriculture.

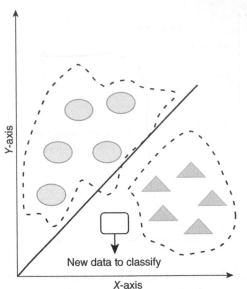

Figure 21.4   *K*-nearest neighbor.

(ii) **Decision Tree:** One among the supervised learning algorithm is decision tree, which are used to classify the category and continuous dependent variables. It defines the structured tree-like structures by determining the variables in the distinct groups based upon their different attributes. It classifies based upon the attributes and parameters in a defined manner. Decision tree algorithm is a tree-like structure, which is like the flowchart-like structure with the internal node. It represents the class label based upon their computation models and branches. Figure 21.5 represents the way of defining the crop yield using the decision tree approach. They define the decision node and leaf node to represent the different identity. Decision tree is

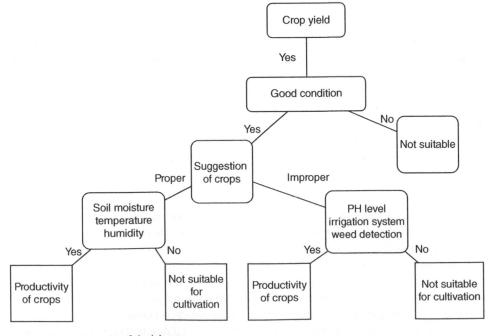

Figure 21.5   Example of decision tree.

used to predict the class label, where it initiates from the root of the label. It uses the decision tree tool to construct the outcomes of the data classification. It is a tree-structured classification technique that uses the internal nodes. Each branch represents the decision rules where the leaf nodes are considered as the output of those decisions. This tree analysis consists of spanning of a number of different leaf constructions. It identifies the way by splitting the dataset based upon the different conditions. It is a non-parametric supervised learning method that is used for the regression and classification. In this algorithm, it uses the decision tree to classify the leaf node and root node. In the decision-tree algorithm, each leaf node is labeled within the class where it uses the subordinate feature decision nodes as the input feature. In this, each leaf node is classified using a specific class with a particular probability distribution. Each leaf signifies the dataset by classifying the tree with the probability distribution. Each decision is well-constructed with the skewed level of subsets within the classes.

(iii) **Random Forest:** It is used for classifying the collection of objects. A random forest is the ensemble of decision trees that are classified as the object-based upon attributes. Each tree is classified using the majority of the votes in that class. It includes the training and test set to split up the tree-based upon the majority of votes as shown in Figure 21.6. It uses input variables that are specified in each node and identifies which tree has the greatest extent toward it. Random forest is a machine-learning technique which combines the classifiers to produce the necessary solution to complex problems. It consists of many decision trees which are generated using the trained set of bagging or Bootstrap aggregation algorithm. Random forest establishes the general outcome of predictions over the decision trees by taking the average of the output from various decision trees. Bagging is a type of meta algorithm which improves the accuracy level in the machine-learning algorithms. As the number of trees increases, the prediction over the outcome also increases. Random forest reduces the limitations over the use of the decision-tree algorithm. It increases precision and reduces over fitting of the dataset. It can be used in both the classification and regression techniques. Random forest is one of the popular machines learning algorithms which are a type of supervised technique. It uses the ensemble learning for using multiple classifiers to solve a complex problem and to increase the performance of the model. It has a number of decision trees where subsets of the given dataset are included which improves the average predictive accuracy of the dataset. Rather than using the decision-tree algorithm, a random forest easily predicts each tree based upon the majority of votes of the prediction. As the number of trees increases, the accuracy of the over-fitting problem gets reduced. It predicts the correct output by predicting the classes of the dataset. It features out the variable by predicting the accurate results. Low correlations are detected as predicts from each tree. It utilizes ensemble learning which classifies the many classifications by constructing many decision trees.

(iv) **Support Vector Machine:** Support vector machine (SVM) is a classification method to plot the data item in n-dimensional space. It features out the data by using the particular coordinate. Those coordinates are also called support vectors which are taken from the splits of the data. From the different classified groups of data, the splits of data are taken. The closest point between the two groups is calculated rather than the farthest one. Figure 21.7 represents the way the classes are defined. The SVM performs the classified approach by creating the n-dimensional hyperplane which splits the data into different categories using optimal decision. SVM prototypes are related to the neural networks which use the sigmoid kernel function and sub-divides into two different layers. The support vector model is considered as the alternative for the classical multi-layer preceptor. It classifies the weight of the network using the quadratic programming problem by using the linear constraints. The SVM predictor variable is called an attribute. It transforms these attributes into a defined hyperplane called as a feature. The way of representing the suitable task is known as feature selection. These features are defined in a row of values called a vector. They are separated by the optical hyperplane by the cluster of vector. The SVM margins find out the best line in between the support vectors which is maximized. The SVM can be able to generalize to different classes where the set of labeled data is provided using the trained set of the algorithm.

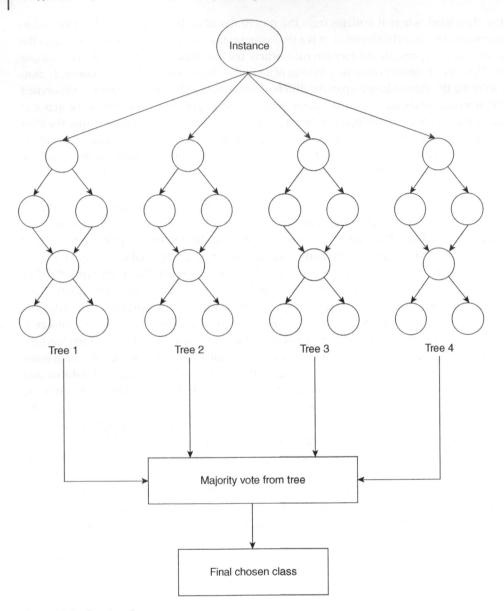

Figure 21.6   Random forest.

One of the main functions of SVM is to analyze the hyperplane to predict and distinguish between two classes. SVM is a linear type of model which can solve linear and nonlinear problems. It creates a line or a hyperplane that separates the dataset into classes. It is a type of predictive analysis used in data-classification algorithm, which designs new different data elements to the labeled set of categories. The main tasks are to form the decision boundary that can easily segregate the n-dimensional space into the different classes, where it is easily predicted a new data point into the correct analysis of the category. SVM chooses the support vectors creating their hyperplane. The best decision boundary or vector is called the hyperplane. The features can be two or three according to their dimensional plane. The hyperplane generally uses maximum margin and maximum distance between the data points. SVM represents different classes in the hyperplane in terms of multi-dimensional space. The main goal is to divide each dataset into different classes and to find the

Figure 21.7 Support vector machine.

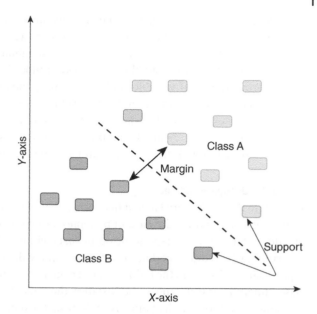

maximum marginal hyperplane. Maximizing each distance of margin gives the reinforcement of the data points which are being classified for further purpose.

(v) **Bayes Classifier:** The naive Bayes classification technique is used to predict the assumptions of the predicted class. It assumes the particular feature which relates to the presence of the other feature as shown in Figure 21.8. The naive Bayes model is easy and can be used in large datasets to build the data. A feature

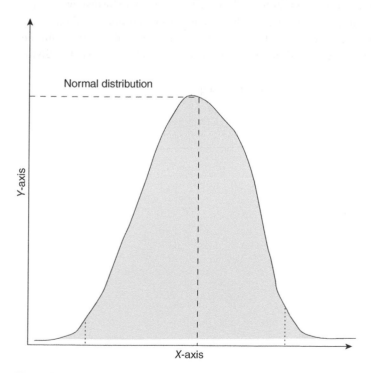

Figure 21.8 Naïve Bayes method.

depends upon each other on data and its properties. It uses outperform the classified approach in an efficient way. The classification algorithm is a fast way of machine learning for prediction purpose. It means the probabilistic classifier, which predicts the probability of the object. It predicts the probability of the different classes on the various attributes. It uses multiple classifications and requires less state of training data. It categorizes the input variables rather than using the numerical variables. A high-dimensional training dataset is used in text classification. It predicts the large volumes of data into good results in the sentimental analysis. It checks the conditional equality of the trained dataset. It uses the particular feature of the class using the class variable. A small amount of trained data estimates based upon the parameters and variances of the variables. It is a discrete and continuous type of attribute that performs well with real-life examples.

21.2.1.2 Unsupervised Learning

It is where the models are labeled in a format manner. It compares the data by using the structure and similarities in the data. It identifies the dataset by representing the data in a compressed manner. It analyzes the input data and doesn't have the corresponding output data within it. Unsupervised algorithm is one type of machine-learning algorithm, which draws certain interferences in the datasets. It uses k-means and clustering to cluster up similar type of data. It discovers the hidden type of data that are grouped together without the help of human intervention. Based upon the interferences and distance, the k-means and clustering are used to analyze the data. It deselects the data by discovering the patterns underlying the data. It checks the information and gathers the relevant sorts of information to identify the patterns. Unlabeled data are categorized using the unsupervised machine-learning approach, which uses the k-means to cluster up the data. These k-means algorithm clusters up the data by partitioning the data points which tend to be the nearest. Uncategorized and unlabeled data are categorized using unsupervised learning techniques.

Clustering It is an unsupervised machine-learning approach that is used to cluster up the data by discovering the interpretation of the data in groups or clusters. It groups up the unlabeled dataset by defining the data points using the possible similarities within the data as represented in Figure 21.9. It compares and identifies similar data by analyzing the patterns of size and structural composition. Clustering is a method that is used to divide

Figure 21.9 Clustering.

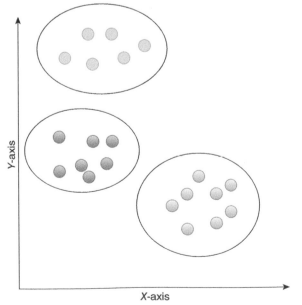

Y-axis

X-axis

the data points according to a similar data type. It identifies the dissimilar type of data from the unlabeled group of data. Homogeneous groups and heterogeneous groups are divided by the way data are grouped. It finds out useful information from the unknown properties. It groups up the data and structures according to the way of classification. Grouping up the unlabeled data is termed clustering. It consists of data points with high and low similarities. The main goal of the clustering is to divide the set of data clusters into the same clusters where the dissimilar are clustered accordingly. It identifies the data with similar metrics and checks the pattern recognition, retrieval of information, and image processing.

21.2.1.3 Reinforcement Learning

It is the type of machine learning which is suitable for performing actions that can maximize the reward. It maximizes the reward by using a particular solution. It identifies the possible behavior in specific situations. It enables the agent to provide the interaction over the environment using the trial and error method which solves complex problems. Reinforcement learning rewards the desired outcomes by using the behavior of the undesirable ones. Agent can be able to perceive the dataset and interpret it within the environment. It uses the trial and error method to identify whether the situation is compromised or not. It is one of the subsets of machine learning which trains up the models to make a sequence of decision models. It either uses the rewards or other penalty under the actions. It tries to maximize the total reward to the environment.

Q-learning   Q-learning is one of the types of reinforcement machine-learning technique. It is used to learn the value of action in a particular environment, such as some of the stochastic transition problems. Q-learning attains the maximum reward without even any updating. Rather than using the new action and reward, it selects the same policy. Q-learning is the value-based learning approach that is used to find the optimal action policy by the q-policy. This type of learning is a free type that is value-based where it informs the action of an agent should be done.

Markov Decision Model   Markov decision process is a framework that is used to describe the environment of reinforcement learning. Using this method, the agent and environment partly interact with the discrete-time process. Markov decision process is the stochastic discrete process that forecasts the value under the decision process. It predicts the value within the current state by randomly selecting the variable. The Markov decision process solely computes both the transition property and transition probability. This probability is that the agent as in Figure 21.10 can move from one state to another by the use of the state transition property. It uses the discrete-time stochastic control technique which computationally forecasts the current activity of the agent and environment. It is the extension of the Markov chains which use actions and rewards. This decision process reduces the overall probability of the Markov chain. Markov process is generally the memory-less type of random process. It uses the sequence of the random state such as s [1], s [2], s [3], and s [n] using the Markov property. Rewards are the statistical values that each agent receives from various actions at some state (s) in the environment. A policy defines what actions needed to be performed in a particular state. Reinforcement learning is a type of experience where the agent determines the change in policy.

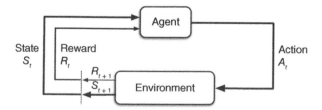

Figure 21.10   Markov decision process via reinforcement learning.

21.3 Concept of Deep Learning

Deep learning is a subset of machine learning and AI which imitates the working of the human brain by creating patterns and process data for the decision-making process. Deep learning is said to be the subset of machine learning where it labels the data which is in unstructured or unlabeled form as shown in Figure 21.11. In the digital era, deep learning has been evolved for the explosion of data in different forms. Enormous amount of data are easily accessible and shared in the cloud computing format using deep learning. Data are normally unstructured and irrelevant in format. The maximum amount of data that is in unstructured form has to be changed to the structured form. Deep learning plays an important role in data science which consists of statistics and predictive type of modeling. Structures that are modeled are emphasized using deep learning. Deep learning consists of classes of machine-learning algorithms that extract the multiple layers progressively featuring their raw input. Image processing and other identification edges can be easily attained by using deep learning. Deep learning is also called deep-structured learning, as it represents the data in a structured format. It is used in different fields of applications which include speech recognition, computer intelligence, computer vision, biomedical informatics, and image processing analysis. They attain accurate result when compared to the human brain. ANN is used for information processing and other distributed communication nodes within the biological system. Neural network tends to be static and dynamic in representing the structure. Deep learning uses multiple layers within the network. It is a modernized version of variation where the unbounded number of layers is within the bounded size. A neural network simulates the behavior of the human brain according to its ability. Each single layer can make predictions approximately. Some of the additional hidden layers help to optimize and refine the accuracy output. It consists of multiple layers of interconnected nodes which are built from the previous layer to optimize and refine the prediction. The input and the output layer within the deep neural network is the visible layer. The input layer consists of data processing and the output layer consists of final production. The deep neural network is a multi-hidden network player that layout the single hidden layer preceptor from it. It reduces the complexity of feature engineering and improves the accuracy by connecting all the nodes. It transforms the data and extracts features from the data using the different levels of abstractions. One of the essential parts of deep learning is its pattern recognition, which does not depend upon extended programming. It is a type of iterative-learning method, which exposes huge dataset.

21.3.1 Types of Deep Learning

Deep learning is characterized based upon their way of networks are communicated among each other. In deep learning, there are different types to emphasize deep-learning techniques. Some of the types of deep learning techniques are as follows:

- Convolutional neural network (CNN)
- Recurrent neural network (RNN)

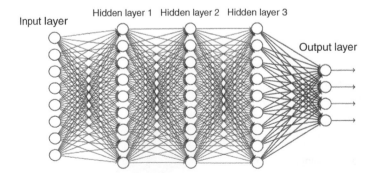

Figure 21.11   Deep learning.

- Generative adversarial network (GAN)
- Multilayer perception (MLP)
- Radial basis function network (RBFN)
- Long short term memory (LSTM) network
- Self-organizing map (SOM)

21.3.1.1 Back Propagation

The back propagation (BP) algorithm is one of the best algorithms which have the fundamental building blocks within the neural networks. Figure 21.12 represents the way of propagations by minimizing the errors. It effectively trains the model by using the neural network which performs the backward pass using the model parameters. The model parameters are adjusted in the backward pass. Training and optimal layer of the network are implied using the neural network. Back propagation is used as a mathematical tool to improve the accuracy over the predictions of the data. ANN uses back propagation to compute the gradient descent along with the respective set of weights. It consists of a way of propagating the total loss that the neural network calculates the error function. It is the function of the neural network. It has the access to the impact in reducing the error and to recognize the speech and image processing given variable. To reduce the error and predict the accuracy over the nodes, back propagation is used. The objective of the back propagation is to maintain the multilayer feed-forward over the neural networks. It is a method of fine-tuning that weights the neural network-based upon the error rate. It calculates the error by calculating the gradient loss function with the weights of the networks.

21.3.1.2 CNN in Agriculture

The CNN consists of multiple convolutional layers which results in providing speech recognition, NLP, and image processing. The CNN structure is composed of convolutional layers based upon the extraction of features. Fully connected functions are classified using CNN neural networks. Some of the BP neural networks feature the map using the networks with specific values. Firstly, it converts the signal into features and targets the features. CNN has the strongest capability in processing the image, as it classifies the crop structure and selection. To classify the crop, yield prediction, best quality crop, and disease detection, CNN with AI is implemented. Based upon the plant structure and soil composition, the yield training iterations are analyzed. Performance tunics are improvised with the fine-tuning. Using the image recognition of the plant and soil, the yield prediction can be categorized.

21.3.1.3 RNN in Agriculture

RNN uses temporal information, which is time series based. It follows semantic information and temporal information as a breakthrough. RNN is actually a variant of ANN which uses current input of the network from the

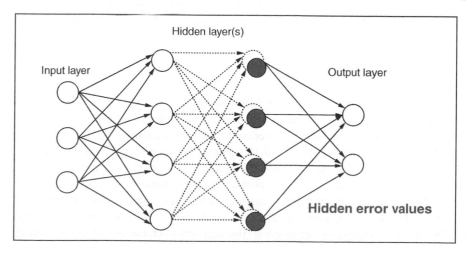

Figure 21.12 Back propagation.

previous output. Each network is gathered from the previous information and applied to calculate the current network output. ANN is treated similarly to BP neural network where the output is used as the input of the next network. RNN can solve the problem using time series in a theoretical manner. As it has long-time series, it is difficult to solve the problems due to the length of the information. Some of the gradients can disappear or explode due to the length of the information. In the RNN, LSTM network is said to be the improvement of RNN, which is mainly used to solve the time series of problems and long intervals. It selectively chooses a state of the moment from the RNN. The RNN process time-series data and it is used in many other applications such as plant phenotype recognition, leaf index estimation, climate prediction, and soil moisture and temperature estimation report. One of the challenging tasks in agriculture land covers classification. Using the key point of RNN, land classes are predicted. RNN models are rational and stable in rules which use multi-class cases. Plant phenotype is estimated using the RNN approach by static observations. LSTM sequentially built a model which features the extraction and output in an accurate manner. RNN-based classifier is used to map the coverage of crop cultivation using mono temporal models. The model performed accuracy of about 99% with fivefold cross-validation dataset. RNN consists of the two stacked LSTM layers which are connected with a dense layer surcharge activation layer and a repeat layer. RNN model can provide predictions with a low error state. One of the main mechanisms of improvement is that the sequence model is used to overcome the bias which occurs in the typical location or at a time period.

21.3.1.4 GAN in Agriculture

A GAN is a new way of technique which is a combination of BP existing algorithms. It sets out the noise which learns the distribution of the real data and generates new data. This generation model captures the distribution of the data and binary classifies the discrimination model. GAN is a different neural network that can enrich the datasets. It solves the feature loss problems and reduces the down sampling process. A feature gets inaccurate and loses its value due to image compression. To avoid such scenarios, the GAN is implemented. GAN is also used in many different applications. GAN identifies the similarity between the empirical images and synthetic images. These images increase the qualitative results to improve the feature translation.

21.4 Smart Agriculture

Smart agriculture uses different technologies to improve the field in the agriculture. It increases the quality and quantity of the crop cultivation. Sustainable increase in the agricultural productivity develops the adaptability over the agricultural farming. Continuous monitor the crop yield using different sensors improves the precise accuracy and its environmental conditions.

21.4.1 Smart Farming

Smart farming decreases the ecological footprint over the agricultural farming. These farming can make agriculture profitable to the farmers. It decreases the resources by reducing the money and labor for farming. Traditional method is avoided as it increases the reliability of reducing the explicit risks [7–10]. It optimally forecasts the weather conditions and climatic data of the yield to cultivate the crops in a modernized way. Farming-related data are stored in the cloud server which manages the concept of smart farming using advanced technology as shown in Figure 21.13. Different technologies are used to attain the data of the natural resources which are collected, gathered, and processed to attain future outcomes. IoT, cloud applications, and big data are used to track the data of crop yield. It tracks, monitor, automate, and analyze all the operations of the farming in a smart way [11]. Due to the usage of smart farming, the demand over the crop yield is getting increased.

Figure 21.13 Smart farming.

21.4.2 Precision Farming

Precision farming [12–14] uses the IoT-based approaches which control and predict higher accuracy in agriculture. As shown in Figure 21.14, it precisely determines the need in agriculture with higher computational accuracy at a low cost. One of the big differences between the traditional approach and the precision farming approach is that it allows the algorithms to effectively predict the crops in agriculture. The quality and the quantity of the crops are

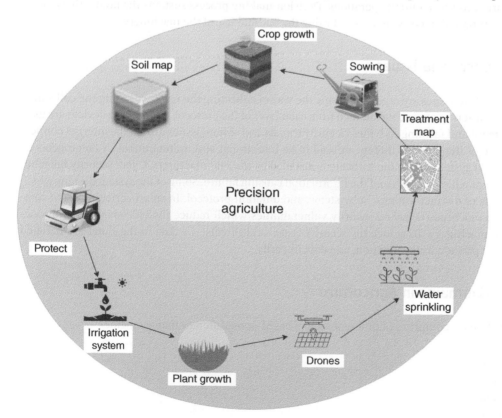

Figure 21.14 Precision agriculture.

significantly increased as it uses the modernized approach in [15]. It continuously monitors the farming using the practical frameworks, which state the condition of the crop and yield management. It improves the quality using the sophisticated techniques which are used. Precision farming uses the smart way in agriculture such as it uses different technologies such as IoT, location positioning systems, sensors, and AI on farming. Many different smart technologies such as smart irrigation systems, plant and crop growth, sowing methodology, water drone sprinkling, and treatment weed map are predicted using the precision farming methodology [16–19]. Precision agriculture is used to improve crop management using the way of response, observe, and measure the variability in the crops. To reduce the environmental risks, predictive modeling is constructed. It benefits the way of monitoring the soil, physiological parameters of the soil using sensors, and potential hazards. It improves the accountability, efficiency, productivity, quantity, profitability of the agricultural farming. Precision farming increases the productivity of the crops by reducing the diseases and other factors which affect agriculture.

21.5 Computation Methods

The recent development in the smart automation system provides efficient work performance and improved capacity [20–22]. The labor cost and usage of the materials are comparatively lesser than the traditional approach. Using the automation systems, a better way of control of quality can be enhanced. Based upon the agricultural fleet parameters, the different tasks and optimization techniques are developed. Some of the important tasks are managed using the sowing; fertilizer, harvesting, and sprinkling are monitored. Physical optimization plays a vital key to increase the efficiency and productivity in the crops. The environmental factors and biological factors are two different criteria to be analyzed. Optimization and dimension capability focus on the operation planning where it attains an integral part of the agricultural operations. Decision-making process sustains the productivity of the agricultural system. The production costs are reduced using the optimal use of the machinery.

21.6 Security Aspects and Issues

The security smart agriculture issues are organized using the way of collecting the resources using the collecting, processing, storing, and gathering of the data attained from each layer of the protocol. Some of the devices are used to monitor the environment and farming activities. Security threats and several issues during the internet connectivity are analyzed. Some of the security threats are said to be intentional and unintentional. A heterogeneous device comprises the smart system where the software is installed by the manufacturers. Cyber security breaches can compromise the system which can mislead the information causing failure. Some of the system designs which consider compatibility are of distinct features, subsystems, and different protocol. In smart agriculture, the communication between the machines is having security vulnerabilities which reduce the efficiency of the devices used in the agricultural machinery. To reduce the security issues and threats, the data in the cloud are handled using access control, data privacy, authentication, and data integrity.

21.7 Application Domains in Agriculture

The smart agricultural domain comprises of different sectors based upon different management.

- ➢ Irrigation management
  - Specific crop irrigation
  - Soil moisture sensors
- ➢ Farming management
  - Cyclic growth of crops
  - Specific climatic conditions

Figure 21.15 ML and DL applications in agriculture.

➢ Livestock farming management
  • Tracking and monitoring of the animal health
  • Bio tags usage
➢ Farming assistance
  • Issues solved by experts
  • Experts assistance and solutions
➢ Crop management
  • Protection of crops
  • Usage of water alert need a mechanism

Using the above different management systems as shown in Figure 21.15, the agricultural aspects over the use of smart agriculture are improvised. Each management in the smart agriculture is used to improve the quality and quantity of the productivity in the growth of the crops. ML and DL applications in the agriculture illustrate the way these techniques are been used in this sector.

21.8 Case Study

In the smart agriculture, there are various challenges that are faced and new technologies are improved to develop the aspects of smart agriculture. One of the case studies of the smart way of agriculture is the Aarav Unmanned System. It is developed by a startup-based company in Bangalore that mainly focused upon the representation of agricultural domains. It specializes in representing the image in a three-dimensional way. It is comparatively precise and faster than other drones. It is mainly used in the agricultural sector for productivity purposes. Crop farm is the other technique that was also initiated by a startup that deals with the food supply chain of organic vegetables. Agricultural supply chain faces challenges as it is immersive in structure. To increase the productivity and reach the consumer's side, the online platform is used. Farmers directly send the food supply to the online and offline retail platform. Due to this latest technology, the shortage over the food chain will be reduced. CropIn technology is another method that provides an innovative way of digitization in agriculture.

21.9 Agro Smart City

Agro-smart city indicates the modernized approach to agriculture used in the urban farming to cultivate food, practice, and distribute food products. It includes animal husbandry; horticulture and aquaculture are included in smart agriculture. It is the practice of cultivation, processing, and distributing farming in urbanized areas. In urban farming, affordable and small space is needed to cultivate the crops. It provides the essential commodities for growing food within the smaller space. Urban agriculture reflects the various levels of both in social and economic development. This modernized agriculture improves food productivity, safety, and security.

21.10 Concept of Application of ML and DL in Smart Agriculture

In agriculture, machine learning is used to derive the performance of the learning process. These methodologies are used to perform a particular task. Machine learning consists of data with different parameters and attributes. Based upon the different features, the datasets are characterized. Performance metrics calculate the datasets of each variable and identify similar patterns. The machine-learning algorithms use the mathematical and statistical models to compute the trained dataset. Using the corresponding input, the data are mapped into the output. It controls the fertility of the soil, monitor of crops, and soil texture using deep-learning algorithms [23]. To save the crop from weeds and other prediction, ML algorithms are used. The data which are generated from the various sensors are conditionally gathered and analyzed with the trained dataset using the machine-learning algorithms. Using these algorithms, fast way of result-oriented decisions and quick solutions are easily implemented.

21.10.1 Prediction of Plant Disease

The plant disease diagnosis plays an important role as it determines the standard of growing in the yield. Using the trained set of models, the datasets of the plant leaves are determined using the fully convolutional datasets. It identifies predefined plant disease images which are already generated inside the dataset. Based upon the richness of the grains, nutritious value and yield moisture can be estimated. Alexnet, image net, and Google net models are extracted for identifying the datasets. Based upon the computational methods, the plant disease can be identified. Both the machine-learning algorithm and deep-learning algorithm can estimate the plant disease by pre-determined datasets of the plant. Using the best algorithm, the performance of the predictability can be improved. According to the plant disease classification model, as shown in Figure 21.16, pathogens play an important role in causing diseases in the plant. A viral disease is an infectious disease that can extend from one plant to another plant either by pest or contact. Ascomycetes and basidiomycetes are fungi mostly accountable for disease in the plant. *Oomycetes* and *Phytomyxea* are fungal-like organisms that enclose a vicious pathogen in the plant. The spots which are caused by the blights can extend quickly in plants.

Figure 21.16 Plant disease classification. Source: Alinsa/Adobe Stock.

21.10.2 Locust Prediction

The machine learning and DNN approach identify the locality of the locust by deriving the specific data of the reproduction state of the locust. The harm of the locusts is enormous because the scalability is high. A desert locust usually occurs in the warmth-gradient areas, plagues that concern the North African countries. The locust assault has to be predicted prior to their creation of swarms to begin up the circumstances before threatening in an efficient way. Locust lifecycle is surveyed where it typically takes a number of weeks to potentially assault the yield. Locust normally follows a communal set of movements such as their own actions, way of network, and its own rules. Machine-learning algorithm can be used for identifying locust attack. High forecast in risky regions with cost-effective is been implemented using a machine-learning algorithm approach which is to diminish the threat over agricultural regions. An immature locust does not fully grow, and it can damage other locusts. A plentiful amount of plant life gets nourished by swarms. To govern this, pest chemicals are used to avoid the organization of locust eggs and swarm. Locust pests can cause harm to the economy as well as food scarcity. The control of a normal locust swarm is enormous in destruction.

21.10.3 Plant Classification

A plant plays a significant role in human life. Due to deforestation, climatic conditions, and other natural calamities, the growth of the plants is getting decreased. Numerous numbers of plants are vanishing due to natural calamities. Plants produce organic products, which is helpful to humans. Using deep-learning and machine-learning techniques, the leaves classification can be determined. Image classification is used to attain the leaves condition. Based upon the plant's size, shape, texture, and color, the plant's condition can be determined. The anatomy and taxonomy of the leaves are classified for the plant classification. Leaf analysis detects the automatic plant classification of leaves. Based upon the plant texture as shown in Figure 21.17, leaves shape, veins of leaf, and color of the leaf, the plant classification can be determined. By using the scientific methodology, the extraction of features is computed and calculated by using the classification algorithm. Some of the machine-learning algorithms classify and extract the outline of the leaf using predefined features. Based upon the descriptions of the leaf, the shape, vein, and color of the plant, the texture of the leaf is classified. The color structure of the leaves is determined using the RGB structure complexity. CNN and the probability neural network employ the classification of the image in the 2×2 structure, which achieves the best performance and high-level test accuracy. To determine the structure of the leaves, the edge of the object with the shape measurement is followed. Different algorithms are experimented to determine the leaves and plant classification based upon the samples. Various datasets are modeled using the different strategies of the plant arrangement. A characteristic of machine-learning and deep-learning approach is the features of the plant classification using the trained dataset [24]. High dimensionality is used to train the dataset in an effective way. The plant identification is determined using the pre-processing technique, image segmentation, classification, and extraction of the plant dataset. Deep-learning and machine-learning extract the feature which provides the in-depth information of the image. This automated plant classification helps the farmers and botanists to identify the plant.

21.10.4 Livestock Farming

Livestock production is intensified to enhance the productivity of the animal. It influences the perception over sustainability, food security, and safety. In livestock production, the monitoring of the welfare of production and animal welfare are determined. Using the precision livestock farming technique, animal health can be recognized using the real-time monitoring. These roles improve livestock farming using the support decision-making process. It relies on sensors, gyroscopes, and cameras to identify livestock farming agriculture. These models define biological awareness by determining the predictions and suggestions for farming.

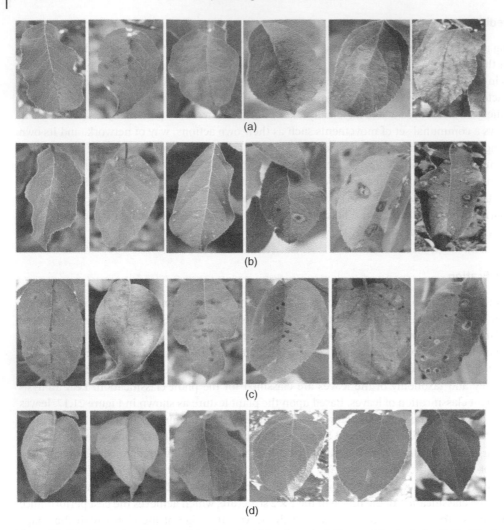

Figure 21.17 Plant classification a) Apple cedar b) Apple scab c) Diseased Apple Cedar d) Healthy Apple Cedar. Source: Thapa, Ranjita et al., 2020/John Wiley & Sons.

21.10.5 Smart Irrigation System

A smart irrigation system constitutes the consumption of water for the crops in agriculture. Effective water management is needed to improve the purpose of conserving water to accomplish sustainable crop productivity. Water savings are improvised to enhance the improvement in the agricultural sector. The feasibility and effectiveness of using the irrigation approach depend upon the topological factors, soil properties, and soil texture. These factors enhance by accomplishing water savings and optimization in the yield. Monitoring of the crop growth, climatic conditions, groundwater level management are utilized using the image spatial associatively. This efficient irrigation ensures the sustainability of water and weather prediction. Irrigation is based upon the climatic conditions and ground-level management of the soil. An automatic irrigation system is used to improve the water level of the soil and soil moisture content as shown in Figure 21.18. A machine-learning algorithm improves the high-quality research to enhance the experimental results. Using machine-learning and deep-learning techniques, the analysis over the soil moisture and control over the water irrigation are determined. The smart irrigation system enhances the soil moisture and climatic conditions are used to determine the landscape irrigation side. This process maximizes the efficiency of irrigation by minimizing the wastage of water, plant health, and the soil quality.

Figure 21.18 Smart irrigation in agriculture. Source: Jim/Adobe Stock.

A different sensor controls the usage of the irrigation system by accurately determining the moisture content of the soil.

21.10.6 Pest Control Prediction

Pest controls are utilized to reduce the pathogens on the plants. A machine-learning algorithm predicts the prediction of diseases and crop pests. Some of the plant pathogens and plant-feeding insects can affect the growth of the agricultural ecosystems. The pest controls are controlled by the dynamic way of concerning the dynamics in the pest and disease control model. This pest management can reduce crop yield loss due to the weed in the plants. Image classification visually characterizes the observable disease patterns in the plant. Pest control treatment improves the controlled environment in the plant pesticide. A strategic way of control over the insect and weeds can be controlled using the pest control methodology. In an agricultural crop, managing the crop, spaces are handled to prevent pests. Pest control is required to monitor, identify, and take action over the control of the pesticide environment as shown in Figure 21.19.

21.10.7 Soil Management

Soil management is the better way to manage the soil. It enhances the nutrient supply, imbalance in water supply, and erosion of soil. It gives precise information regarding the soil consistency, land degradation, and nutrient imbalance in the soil. Soil properties [25–27] entail with the soil texture, organic matter compounds, and nutrients content using the sampling of the soil. Low sensors and mapping sensors can provide an effective solution for the soil spatial environment. Degradation of soil, climatic condition, and water from sustainable land management generate the benefit over the multiple ecosystems. These use policy support which reduces the conversion of land and adapts to sustainable practices. This management consists of a large number of strategies to protect the soil resources and reduce the soil erosion shown in Figure 21.20. Agricultural soil can affect the entire ecosystem as

Figure 21.19 Pest control treatment. Source: belyjmishka/Adobe Stock.

Figure 21.20 Soil management system. Source: Okea/Adobe Stock.

it has a domineering effect on the vegetation crops. To improvise the quality and increase the quality of the crop input, soil management has to be predicted.

21.10.8 Crop Quality and Its Management

Crop quality is based upon the assessment of the quality of the crop by soil conditions and climatic conditions. Crop characteristics are determined to attain the structural composition of the crop yield prediction. High accuracy of performance in crop quality is achieved using smart farming techniques. The harvesting of the crop in smart agriculture gives a detailed description about the growth of the crops in the yield. One of the best ways of cultivation is selective harvesting. It harvests quality crops with the management practices to develop the quality and its structure. Each crop deviates according to its texture, structure, color and size composition. A machine-learning algorithm combines with the image segmentation to achieve better, improved results [28]. In the machine-learning approach, the classification of the plant and disease prediction is attained using supervised machine-learning algorithms and other different techniques. It reduces the huge loss for the farmers as it priory predicts the yield loss. The way of preserving the crop and managing it for the hybrid way of cultivation can improve the productivity over the growth of crops in agriculture [29–33].

21.10.9 Weed Protection

Weeds usually occur in the larger agricultural field as it spreads massively within a short span of time. It affects the growth of the crops as it is destroyed by the weeds. Prolific seed production and longevity in the seeds reduce the growth of the weeds in the plants. To reduce crop destruction, the weeds have to be controlled using weed control using some mechanical tasks, such as using the herbicides treatment to improve the productivity of the crops. Due to the usage of weed control herbicides, the resistant durability gets increased, which causes the control over the weed. Spraying of herbicides in specific zones is the best solution to reduce crop disease in agriculture. Both the ML and DL techniques extract the features and analyze the dataset with the associated samples to identify the specific zones where the herbicides have to be sprayed. As it has higher accuracy in performance, the prediction over the weed zones is accurate in the state.

21.10.10 Yield Prediction

Crop yield prediction is one of the important tasks as it decides whether the cultivation land is sustainable or not. There are many different approaches to analyze the sustainable land which is needed for cultivation. In the modern agriculture, crop yield prediction plays a challenging task as it emphasizes the crops cultivation. It gives an accurate model to compute the decision approach in managing the cultivation yield. In the yield prediction, there are some characteristics such as genotype and phenotype that are achieved to attain the crop mutations. Both the ML and DL techniques simulate the overall performance in the prediction of the yield. Based upon the conditions, the yield predictions can be analyzed to get the maximum range of the field. The crop yield rate influences the good quality of seeds based upon the favorable and unfavorable conditions. These machine-learning techniques improve the yield rate of crops cultivation. Based upon the soil textural composition, weather condition, and soil type, the growth of the crops is increased. The traditional method uses only single spaces, and in modern agriculture, multiple data models are predicted. Crop yield and quality of the crops are emphasized using machine-learning algorithms. By the use of different issues such as climatic conditions, weather forecast, temperature, and other factors, the yield prediction can be improvised.

21.11 Results and Discussion

Deep-learning and machine-learning applications detect and solve problems that emerge during agricultural crop growth. Smart agriculture has been used to regulate crop monitoring and agricultural pests in order to increase crop output. Machine learning improves the iterative process by discovering information based on patterns and their associations. ML can improve the vision over image quality and categorize the various crop yields. ML and DL can forecast soil moisture, crop mutation, and climatic factors that can lead to agricultural yield loss. A sustainable increase in agricultural production fosters flexibility in agricultural farming. Convolutional sigma deep-learning algorithm agricultural machines equipped with IoT sensors detect the dataset, which is then normalized to get exact results. This chapter elaborates on the many applications of deep learning and machine learning in smart agriculture and precision agriculture. In conclusion, it is said that precision agriculture greatly improves agricultural crop productivity. Smart farming approaches achieve high precision in crop quality performance. Both ML and DL methods are utilized in agriculture for a variety of purposes, including plant disease categorization, locust prediction, soil quality, crop quality, irrigation management, and many more.

21.12   Conclusion

Different sensors are used to gather data of the soil moisture, structure, climatic conditions, temperature, humidity, and water level of the soil. These data are collected using various sensors. These sensors collect the data within the cloud using big data technology. The gathered data are compared with the trained dataset using the classification and regression techniques in the machine-learning algorithms. Deep-learning algorithm convolutional segments the pre-defined image with the followed image. Using the image segmentation process, the data are computationally classified in the matrix form. To improve the productivity of the crop, the monitoring of the crop and agricultural pests had to be controlled using smart agriculture. A machine-learning algorithm combines with image segmentation to achieve better, improved results. In the machine-learning approach, the classification of the plant and disease prediction is attained using supervised machine-learning algorithms and other different techniques. Deep learning is considered one of the types of machine learning which explores the advantages of smart farming. Due to the usage of smart farming, the demand for crop yield is getting increased. Sustainable increase in the agricultural productivity develops the adaptability of the agricultural farming. The agricultural productivity and quality of the crop are increased by reducing the farmer's time and cost-efficiency. Irrigation for crop cultivation can be analyzed using the DL and ML, which reduces the complexity with minimal cost. Agricultural machines with the IoT sensors identify the dataset which is normalized to attain the precise results. Machine learning enhances the iterative process by discovering knowledge based on patterns and their associations.

References

1 Ayaz, M., Ammad-Uddin, M., Sharif, Z. et al. (2019). Internet-of-things (IoT)-based smart agriculture: toward making the fields talk. *IEEE Access* 7: 129551–129583.

2 Alazab, M., and Tang, M. (Eds.). (2019). Deep Learning Applications for Cyber Security. Advanced Sciences and Technologies for Security Applications. https://doi.org/10.1007/978-3-030-13057-2.

3 A Review on Deep Learning for Future Smart Cities. https://onlinelibrary.wiley.com/doi/full/10.1002/itl2.187.

4 Sciforce (2020). Smart farming: the future of agriculture. https://www.iotforall.com/smartfarming-future-of-agriculture/ (accessed 2 December 2019).

5 Schuttelaar-partners.com (2020). Smart farming is key for the future of agriculture. https://www.schuttelaar-partners.com/news/2017/smart-farming-is-key-for-thefuture-of-agriculture (accessed 14 December 2019).

6 Varghese, R. and Sharma, S. (2018). Affordable smart farming using IoT and machine learning. In: *Proceedings of the 2nd International Conference on Intelligent Computing and Control Systems (ICICCS), Madurai, India*, 645–650.

7 Goodfellow, I., Bengio, Y., and Courville, A. (2016). *Deep Learning*, 1–15. Cambridge, MA, USA: MIT Press.

8 Raschka, S. and Mirjalili, V. (2017). *Machine Learning and Deep Learning with Python, Scikit-Learn and Tensor-Flow*, 2–6. Birmingham, UK: Packt Publishing.

9 Fyfe, C. (2000). Artificial neural networks and information theory. PhD dissertation. Department of Computing and Information System, University of Paisley, Paisley, UK.

10 Deng, L. and Yu, D. (2014). Deep learning: methods and applications. *Foundations and Trends in Signal Processing* 7 (3–4): 197–387.

11 Heaton, J. (2015). Artificial intelligence for humans. In: *Neural Networks and Deep Learning*, vol. 3, 165–180. Chesterfield, MO, USA: Heaton Research.

12 Graupe, D. (2007). *Principles of Artificial Neural Networks*, 1–12. Chicago, IL, USA: World Scientific.

13 Kröse, B. and Smagt, P. (1996). *An Introduction to Neural Networks*, 15–20. Amsterdam, The Netherlands: The University of Amsterdam.

14 Freeman, J.A. and Skapura, D.M. (1991). *Neural Networks Algorithms, Applications, and Programming Techniques*, 18–50. Reading, MA, USA: Addison-Wesley.

15 Emir, Ş. (2013). Classification performance comparison of artificial neural networks and support vector machines methods: an empirical study on predicting stock market index movement direction. PhD dissertation. Institute of Social Sciences, Istanbul University, Istanbul, Turkey.

16 Sharma, A. (2017). Understanding activation functions in deep learning. https://www.learnopencv.com/understanding-activation-functions-in-deep-learning/ (accessed 1 December 2019).

17 Bengio, Y. (2009). Learning deep architectures for AI. In: *Foundations and Trends in Machine Learning*, vol. 2, no. 1, 1–127. Boston, MA, USA: Now.

18 Gulli, A. and Pal, S. (2017). *Deep Learning With Keras*, 68–90. Birmingham, UK: Packt Publishing Ltd.

19 Toussaint, M. (2016). Introduction to optimization. Second order optimization methods. https://ipvs.informatik.uni-stuttgart.de/mlr/marc/teaching/13-Optimization/04-secondOrderOpt.pdf (accessed 1 October 2019).

20 Patterson, J. and Gibson, A. (2017). *Deep Learning: A Practitioner's Approach*, 102–121. Newton, MA, USA: O'Reilly Media.

21 Despois, J. (2017). Memorizing is not Learning!—6 Tricks to prevent overfitting in machine learning. https://ai-odyssey.com/2018/03/22/memorizing-is-not-learning%E2%80%8A-%E2%80%8A6-tricks-to-prevent-overfitting-in-machinelearning/ (accessed 1 March 2019).

22 Ganguly, K. (2017). *Learning Generative Adversarial Networks*, 25–41. Birmingham, UK: Packt Publishing.

23 Lu, J., Hu, J., Zhao, G. et al. (2017). An in-field automatic wheat disease diagnosis system. *Computers and Electronics in Agriculture* 142: 369–379.

24 Fuentes, A., Yoon, S., Kim, S., and Park, D. (2017). A robust deep-learning based detector for real-time tomato plant diseases and pests recognition. *Sensors* 17 (9): 2022.

25 Kerkech, M., Hafiane, A., and Canals, R. (2018). Deep leaning approach with colorimetric spaces and vegetation indices for vine diseases detection in UAV images. *Computers and Electronics in Agriculture* 155: 237–243.

26 Hu, G., Wu, H., Zhang, Y., and Wan, M. (2019). A low shot learning method for tea leaf's disease identification. *Computers and Electronics in Agriculture* 163, Art No.: 104852.

27 Coulibaly, S., Kamsu-Foguem, B., Kamissoko, D., and Traore, D. (2019). Deep neural networks with transfer learning in millet crop images. *Computers in Industry* 108: 115–120.

28 Cruz, A., Ampatzidis, Y., Pierro, R. et al. (2019). Detection of grapevine yellows symptoms in *Vitis vinifera* L. With artificial intelligence. *Computers and Electronics in Agriculture* 157: 63–76.

29 Picon, A., Alvarez-Gila, A., Seitz, M. et al. (2019). Deep convolutional neural networks for mobile capture device-based crop disease classification in the wild. *Computers and Electronics in Agriculture* 161: 280–290.

30 Grinblat, G.L., Uzal, L.C., Larese, M.G., and Granitto, P.M. (2016). Deep learning for plant identification using vein morphological patterns. *Computers and Electronics in Agriculture* 127: 418–424.

31 Rahnemoonfar, M. and Sheppard, C. (2017). Deep count: fruit counting based on deep simulated learning. *Sensors* 17 (4): 905.

32 Ampatzidis, Y., Partel, V., Meyering, B., and Albrecht, U. (2019). Citrus rootstock evaluation utilizing UAV-based remote sensing and artificial intelligence. *Computers and Electronics in Agriculture* 164: Art No.: 104900.

33 Singh, G. et al. (2021). Bio inspired optimization algorithms for machine learning in agriculture applications. In: *Smart Agriculture: Emerging Pedagogies of Deep Learning, Machine Learning and Internet of Things*. AAP, CRC Taylor and Francis.

22

Structural Damage Prediction from Earthquakes Using Deep Learning

Shagun Sharma[1], Ghanapriya Singh[1], Smita Kaloni[2], Ranjeet P. Rai[1], and Sidhant Yadav[1]

[1]*Department of Electronics Engineering, National Institute of Technology, Uttarakhand, India*
[2]*Department of Civil Engineering, National Institute of Technology, Uttarakhand, India*

22.1 Introduction

Structural damage detection (SDD) has been an area of interest in recent years to propose techniques that could detect anomalies in the structures, susceptible to damages caused due to vibrations affecting their stiffness, stability, and reducing their life [1]. SDD or structural health monitoring (SHM), however, is a general term often referring to the analysis methods present in engineering that would detect changes in the integrity of a structure without physically measuring it, also known as condition monitoring (CM). These methods include the acquisition of response data of the structure and extracting features from it to observe relevant damage. The techniques invented so far involve the application of machine-learning algorithms to automate this process of monitoring the structure's health, which requires manually selecting the relevant features, demanding expertise in the domain and, thus, failing to recognize underlying patterns [2]. This seemingly impossible task of selecting the relevant features that define the formation of anomalies in the structure and accesses any pattern underlying that might go unnoticed can be dealt with the application of deep learning techniques instead of prevailing machine-learning techniques [3–5]. Therefore, acquiring response data is the most important step in SHM, since this can directly influence the performance of the study. Although there are prevalent methods of acquiring or generating data experimentally (e.g. experimental observations from an accelerometer network) or computationally (using certain numerical techniques like finite element method), the challenge lies within the fact that the data acquired should represent the overall population of possible scenarios and uncertainties. Experimentally collected data might not be sampled to represent the vast possibilities of scenarios in real-world, while computationally collected data could carry differences from the actual data when the complexity of the study increases. Since these issues have been there for long, lab-scaled models were used to collect data for study. They can be modeled in a controlled environment and could be suitably simulated with external applications to represent real-life scenarios. The issue with them was producing enough sets of data to process relevant details from it.

The aspect of this study is to accumulate relevant data on damaged and undamaged structures in order to proceed with the training of the networks. This chapter aims to study structures with hinge properties. To fulfill the demand for data, a G+20 building is designed, consisting of a ground floor, 19 upper floors, and one terrace. G+20 buildings are suitable for such studies considering the fact that such buildings are designed to handle gravity loads, and improve the ductility of the structure without compromising the stiffness and resistivity to displacements in the structure.

SDD methods involve two main steps: first is feature extraction and the other one is classification. In the first step, the acceleration signals are extracted. The second step involves building a classifier and training it on the

Machine Learning Algorithms for Signal and Image Processing, First Edition.
Edited by Deepika Ghai, Suman Lata Tripathi, Sobhit Saxena, Manash Chanda, and Mamoun Alazab.

extracted features. This chapter involves using a G+20 building as the structure and detecting the damage using a neural network. The neural network is trained to recognize the acceleration responses of the structure whose elements have different degrees of damage or no damage. The trained neural network will then have the capability of recognizing the severity of the individual member and whole structural damage. The training of neural network can be thought of as an optimization process to find a set of network weights that minimize the cost function. This works nearly the same for 1D convolutional neural networks (CNNs) and long short-term memories (LSTMs). The approach deviates slightly to build a classification model using 2D CNNs as the type of data fed into this network differs from theirs. The acceleration signals extracted are transformed into spectrograms and fed into 2D CNNs. A spectrogram is a visual way of observing the strength of signals in a particular waveform over time at various frequencies. The main objective is to test the conventional approaches of using 1D CNNs against newer networks, such as LSTMs that perform better than CNNs in many vibrational-based classification applications, and using 2D CNNs as well, which have been rigorously used in the image-processing sectors and have been performing well in the classification applications, to build a ground-level framework that would imply the advantages and disadvantages of using these networks in a domain relevant to the one discussed in this chapter.

22.2 Literature Review

SHM has been a reliable solution to improve the reliability, life cycle, and durability of a structure [6]. With the latest innovations in sensors, high-speed cloud computation, and internet facilities, SHM has improved drastically to provide better insights into civil structures [2, 7, 8]. In recent couple of years, SHM has relied heavily on computer vision-based technologies that include sensors to diagnose a problem in 2D or 3D domain for example, displacement, stress, strain and cracks [1, 9] by collecting vibrational time series data and converting it into two-dimensional image vectors. With developments in sensors and computer vision-based technologies, the diagnosing of damages to the structure has improved drastically, but the solutions adopted till now are not globally applicable. This is because the vision-based technologies rely on the visual form of data, e.g. images, videos, and the variability in such data due to environmental noises challenges the robustness of the proposed solution [5, 10]. SHM is a multi-disciplinary domain and to analyze the faults and damages in any structure requires a considerable expertise in every domain, diagnosing faults in multiple aspects of the structure and finding underlying patterns that might not be directly observable [3, 11–13]. Also, a lack of cooperative mechanism is observed due to the multi-disciplinary nature of SHM often involving failure in identifying a problem.

SHM methods can be broadly classified into "global health monitoring (GHM) methods" and "local health monitoring (LHM) methods." Currently, the state-of-the art techniques in GHM can be used to detect whether the structure is damaged or not so that further analysis can be kicked into analyze the extent of damage and the locations in the structure. In this, techniques like ultrasonic waveguides and eddy currents are used to measure the stress and locate the cracks in the structure. These fall under the LHM methods. However, these techniques are quite expensive in terms of time and money and may get hindered at places where physical access is constrained [14].

Since last couple of years, a similar approach is being applied and tested on timber-based structures as well where the buildings are equipped with sensors and accelerometers to monitor their performance along with other new systems, processing the data to monitor the health of the structure and analyzing any damage that might affect the health of the structure. A key point to note in studies like this is that most of the structures considered in the study are single-storied buildings and the handful amount of multistoried buildings that were considered were also low-rise buildings, pointing to the fact that the results obtained from these would not be applicable to the high-rise buildings that are the most sought-after structures in engineering [15]. Thus, the methodology of this study if applied to the high-rise buildings at a similar scale could provide beneficiary insights into the damage detection and monitoring of such structures.

Most GHM methodologies are generally rooted from finding shifts in the resonant frequencies of the structure. However, these changes could also be built up due to environmental changes often affecting the dynamic nature of the structure. Imaging- and pattern-recognition methods, which are often used in USA, comprise reflected or absorbed lights from the cracks in the structure [16]. However, since we need a threshold for the gray levels in the images of these cracks, it is difficult to generalize this technique as the gray levels match significantly in images and regions, thus, preventing it to evolve into a more robust solution [17].

Also, the prevalent vision-based technologies require well-operating sophisticated sensors and hardware with computational efficiencies and battery power to be able to detect faults at remote locations [18]. Another drawback of these is the usage of physical laboratory models to accumulate relevant data for the study [19]. The circumstances generated in a laboratory might not always resemble the actual cases, and the work might fail to justify in case of a new scenario. Often real-time automation of data collections and solution deployment becomes challenging due to the presence of noise in the structure's environment leading to either incorrect detection or negligence of any anomaly that might be present [20–22]. Even though the application of SHM ranges far beyond than just civil structures and includes applications in military, aerospace, etc. a generic approach to identify and proceed to solve the issues in these sectors has not been proposed yet [2, 23–27]. Multiple works in the domain of SHM involve conflicts regarding the techniques to collect relevant data, choosing a suitable algorithm to solve the issue, and finding a most efficient and promising technology among the available choices to eradicate the issue. Numerous studies have used machine-learning algorithms to solve this issue [23, 28–32]; however, these algorithms require manual selection of features that will be used to identify the problem and since this requires an insight into the structure from different domains, an expertise in each one of them is required to achieve state-of-the art results.

22.3 Proposed Methodology

After accumulating the dataset, the main task is to extract the important features of the structure after the hinges started to form. As mentioned earlier, the deep-learning techniques prove to be a better alternative over manual extraction of features for machine-learning algorithms. The proposed methodology involves training of the following:

- CNNs
- LSTM networks

Over the vibrational data, these networks have been used widely in various vibration-based domains. The process of data generation, segregating it according to the joints, splitting into sets, passing into a network for training, calculating the probabilities of damage, and the average probability of damage (PoD_{avg}) for the monitored structure is illustrated in Figure 22.1.

First, the data recorded for different axes (X- and Y-axes) were appended to form a sequence of 8002 time stamps (4001 time stamps each) as described above. These appended sequences were now segregated into training and validation sets, containing 60 and 16 sequences of vibrational data each for the 440 joints of the structure. Thus, a total of 26 400 data points were available for the training, and 7040 data points were available for validating the neural network.

The data sequences were appended and all the different earthquake scenarios were pooled into a single database to make sure that the splitting of the sequences into training and validation sets is randomly shuffled, therefore, avoiding the chance of any bias that might be present due to the different sets of values of "g" used to simulate different earthquake scenarios.

The data after segregation (as training and validation sets) are fed into neural networks. Vibrational data from each joint are fed into a network that is reserved for that particular joint. In this way, 440 different networks are trained for the 440 joints present in the structure designed for this study, and the results (probability of damage)

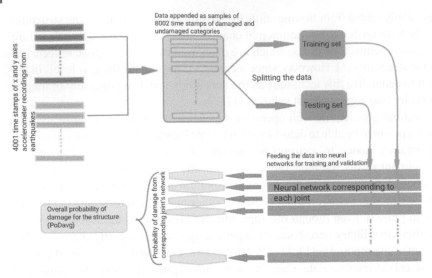

Figure 22.1 Flowchart of proposed methodology.

from each network are averaged to calculate the "average probability of damage (PoD_{avg})" for the structure (Eq. (22.1)).

$$PoD_{avg} = \frac{PoD_1 + PoD_2 + PoD_3 + \cdots + PoD_n}{n} \tag{22.1}$$

where PoD_{avg} is a measure to show the probability of damage of the structure. The higher the value, the greater is the probability of the structure to be anticipated as a damaged one.

The observed drawback in the approaches discussed in the chapter is that slightly large amount of data is needed to efficiently train a model as compared to previous works done in this domain and to achieve that, multiple simulations have to be done using varying "*g*" acceleration values as impulses to the structure requiring efficient computational hardware.

22.3.1 Deep-Learning Models

Another aspect of this study was to compare the performance of the neural networks that are being used to achieve state-of-the-art performances in multiple frequency-based vibrational data and conclude their advantages and disadvantages in a context relevant to the theme of this chapter.

Three different neural networks have been used in this study that are:

(a) **1D CNNs:** 1D CNNs are very effective for deriving features from a fixed-length segment of the overall dataset, where it is not so important where the feature is located in the segment.
(b) **2D CNNs:** 2D CNNs are very effective for deriving features from a two-dimensional data matrix, like an image. Therefore, the vibrational data are fed in the form of spectrograms to the network.
(c) **LSTMs:** LSTM networks are a special kind of recurrent neural networks (RNNs), capable of learning long-term dependencies in a frequency-based vibrational data and, therefore, has been used in this study.

22.4 Proposed Methodology for Deep-Learning Models

22.4.1 One-Dimensional Convolutional Neural Network (1D CNNs)

The proposed methodology requires the use of acceleration measurement from an undamaged and a damaged structure to train a neural network for damage detection purposes. The data of acceleration from each joint are used to train the neural network for those joints which will further help in predicting the damage in structure. The output is proposed in a score out of 100 (PoD$_{avg}$), which reflects the damage of structure. In this application of 1D CNNs, it is found that the ReLu activation function provided the necessary nonlinear transformations to learn. In this study, different sizes of layers are considered and all neural networks have the same activation function in CNN's layers and output layer. The hidden layer has ReLu (positive linear) function, while the output layer has sigmoid (sigm) function. Stochastic gradient descent optimizer "Adam" has been used that can handle sparse gradients on noisy problems. The layers of the neural network used in the chapter are described in Figure 22.2.

After segregating the database into training and validation sets (as described above), the data were now reshaped as an array of shapes (number of samples, 4001, 2). The purpose of choosing this shape was the fact that our data comprised readings from two different axes (X and Y); hence, we fed the data into the 1D CNN as two channels of 4001 readings each to extract features from both the axes simultaneously, which is denoted by the third argument of the shape parameter. Therefore, a total of 440 CNNs were trained for 440 joints. Once the training was over, the predictions from each joint's CNN were evaluated, and average was taken to get the probability of damage for the overall building (PoD$_{avg}$).

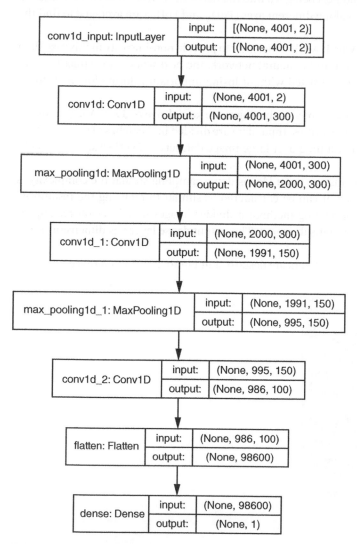

Figure 22.2 1D CNN layers.

22.4.2 Two-Dimensional Convolutional Neural Network (2D CNNs)

In 2D CNN network, first the dataset is produced which is in the form of spectrograms that will be used as the input for training and validation purposes. The spectrograms are generated from the same vibrational data that were collected from the designed structure. The only difference is in the way the input is fed into the network. Each sequence of shape 4001×2 (as discussed in Section 22.3) is transformed into a spectrogram as shown in Figure 22.3.

The images produced possess a resolution of width and height of 432 and 288 pixels, respectively, having an approximate size of 33 KB. These spectrograms' generation is the first step for 2D CNN. The spectrogram datasets are divided into two axes datasets. which are X- and Y-axes datasets.

Each axis consists of spectrograms of 440 joints and those joints are further classified under damaged and undamaged categories. Under every joint, there are damaged and undamaged cases of both training and validation dataset. In training dataset, there are 32 spectrograms in the damaged category, and 28 spectrograms in the undamaged category. In the validation dataset, there are 5 spectrograms in damaged and 11 spectrograms in undamaged cases. The input shape of the 2D-CNN is $432 \times 288 \times 3$. Here, $432 \times 288 \times 3$ represents the dimension of the image, whereas 3 represents the number of channels being fed into the network (RGB channels of the image). Later the predictions of CNN training of 440 joints along different axes (PoD_X and PoD_Y) are averaged to find the overall PoD_{avg} of the structure.

The input layer has "ReLu" activation function which works efficiently to provide non-linearity in this case than other activation functions. After the inputs are given to the neural network, the next work is to extract features from the given input images. Multiple features are extracted without losing important information, and this is done by filters as shown in Figure 22.4.

The 64 filters were used during the feature extraction process having same kernel size of 3×3. A larger kernel size would become unable to extract some important features. Input data are divided into batches of size 4 to avoid the resource exhaustion error encountered during training over large image datasets in 2D CNNs, which might vary with the training hardware's specifications.

Once the input is done, the next process is pooling. Its main idea is down sampling so that the complexity is reduced for further layers. In the image processing, it can be considered as similar to reducing the resolution. Max-pooling is one of the most common types of pooling method; it divides images in sub-region rectangles, and it only provides the maximum value output of that sub-region. The output dense layer has dimension 1 and activation function "sigmoid" is used.

Different layers used for this network and their parameters are described in Figure 22.5 for reference.

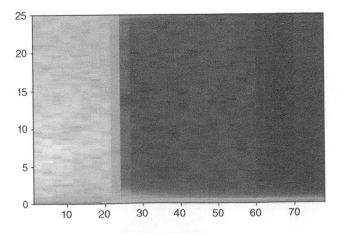

Figure 22.3 Spectrogram for 2D CNN.

Figure 22.4 Feature extraction using filters in 2D CNN.

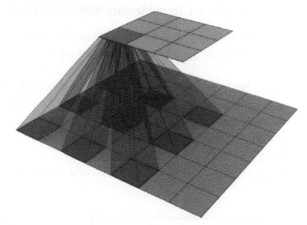

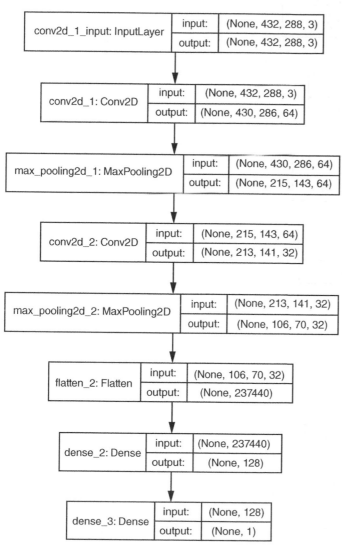

Figure 22.5 2D CNN layers.

22.4.3 Long Short-Term Memory Network (LSTMs)

These networks have loops in them that provide the property of persistence compared to other traditional networks and could be beneficial in extracting the repetitive features in the vibrational data to detect any dependencies in them.

LSTMs perform well when the frequency-based vibrational data are sequential. The activation function "ReLu" is used in the hidden layer, and the activation function softmax is used in "output" layer. The "softmax" function converts the numbers into a probability distribution at the output layer and the sum of probabilities is one only. The optimizer "Adam" is used. The "Adam" optimizer is the replacement for stochastic gradient descent for network training.

Structure of the LSTM network used and the associated parameters are described in Figure 22.6.

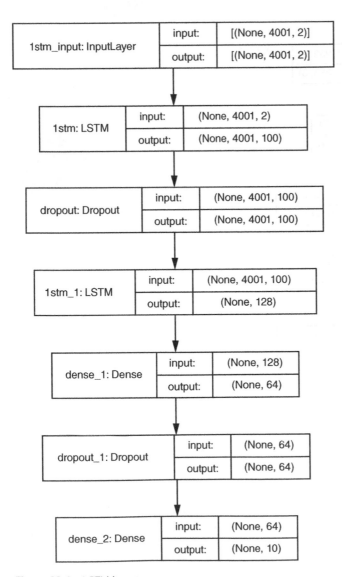

Figure 22.6 LSTM layers.

22.5 Experimental Results and Discussions

22.5.1 Dataset

To accumulate relevant data of damaged and undamaged scenarios, a G+20 building was designed and analysis was done using SAP2000 to study the structures with hinge properties [33]. These hinges show that the damage is about to start in the structure, and by using these data, we categorize the building as damaged or undamaged [34]. Different scales of acceleration are used to simulate earthquakes, such as 0.1g, 0.3g, 0.5g, 0.7g, 1g, 2g, and 3g. The vibrational data from accelerometer were recorded for 80 seconds at a frequency of 50 HZ producing 4001 time stamps of data. This sequence was repeated in the same manner for X- and Y-axes, respectively. Thus the final data for a particular acceleration value of "g" contained 8002 time stamps of vibrational data (4001 time stamps of data for X- and Y-axes, respectively) recorded at 50 HZ frequency for 440 joints of the G+20 building designed for this study.

The 11 different earthquakes' scenarios were simulated by scaling the acceleration responses of earthquakes, each containing different sets of "g" accelerations (from the values provided above) that were provided to the structure to analyze the formation of hinges in the joints and record the vibrational data from them to produce the required dataset.

The first set of data is recorded by measuring the vibration response of the joints when structure is in undamaged condition after the input is fed to the structure, while the second set is recorded when the structure is in damaged condition after the input is fed. The data, after being labeled as undamaged or damaged, are then provided with labels for the same categories to be fed to a neural network.

Secondarily, the aspect of this study was damage detection of building using deep-learning techniques. The damage detection is based on the formation of different types of hinges during the different magnitudes of earthquakes' acceleration. As the different scaled earthquakes' accelerations have been used, such as 0.1g, 0.3g, 0.5g, 0.7g, 1g, 2g, and 3g, it can be noticed that hinges had formed at a different location in different scaled earthquakes, so it is very difficult to estimate a proper threshold acceleration values after which the joint started hinge formation in the designed structure.

22.5.2 Results and Discussions

Following results have been observed from this study as shown in Table 22.1.

The **number of layers** column signifies the number of times the respective neural network's layer is used in the model defined for a particular joint of the structure in the study.

The above results display the PoD_{avg} value, stated in the **efficiency of prediction** column. These are the average values of probability of the damage occurring in all the joints of the designed structure, once the impulses are

Table 22.1 Experimental results.

Neural network	Number of layers	Axis of accelerometer	Efficiency of prediction (%)
1D CNN	2 Layered	X- and Y-axes	77
	3 Layered		79
	4 Layered		73
LSTM	2 Layered	X- and Y-axes	68
2D CNN	2 Layered	X-axis	72
	2 Layered	Y-axis	69

applied and the hinges are formed in the joint. Thus, more the efficiency of prediction, the more precisely it gives the PoD_{avg}, hence classifying whether the structure is damaged or not.

Axis of accelerometer signifies the direction along which the impulses are applied to the structural joints.

22.6 Conclusion

Results from the study show that 1D CNNs perform better than all the other neural networks that have been used in the study. Although LSTMs perform better than 1D CNNs in many vibration-based domains, less efficiency of LSTM as compared to 1D CNN is observed because LSTMs work better on the sequential dataset than 1D CNN indicating that the dataset provided is not sequential in nature.

Additionally, data generation for 2D CNNs (spectrogram's generation) requires significantly powerful hardware in order to compensate the time required for applying these networks into industrial applications.

1D CNNs, which feed on frequency-based vibrational data directly from the accelerometers, do not require such additional hardware and are relatively easy to be applied to industrial applications or devices with minimal computational efficiency and battery power, thus making them a suitable choice for any such application in future relevant to the context of this study.

This study hints at a correlation with the unidentified potential of the newly developed neural network techniques to quantify parameters that could be neglected while handpicking features for a machine-learning algorithm. Although the results obtained are not the state of the art, but considering the fact that this is an initial attempt to pave the path of upcoming efforts in the area, they look promising enough to demonstrate the possibility of actually being able to obtain an algorithm that would be acceptable globally for any civil structure. The techniques discussed in this study prove to be useful with the fact that they clear out some assumptions around the types of networks that could be used and give an insight into the possible ways of collecting the data and processing them to achieve the results.

Moreover, it tries to demonstrate the constraints associated with the hardware resources required to train the models and tune them to achieve the best possible results.

There are still a lot of aspects of this study where further experimentation could bring in newer insights and conclusions.

References

1 Doebling, S.W., Farrar, C.R., and Prime, M.B. (1998). A summary review of vibration-based damage identification methods. *Shock and Vibration Digest* 30: 91–105. https://doi.org/10.1177/058310249803000201.

2 Farrar, C.R. and Lieven, N.A.J. (2007). Damage prognosis: the future of structural health monitoring. *Philosophical Transactions of the Royal Society A - Mathematical Physical and Engineering Sciences* 365: 623–632. https://doi.org/10.1098/rsta.2006.1927.

3 Avci, O., Abdeljaber, O., Kiranyaz, S. et al. (2021). A review of vibration-based damage detection in civil structures: from traditional methods to machine learning and deep learning applications. *Mechanical Systems and Signal Processing* 147: 107077. https://doi.org/10.1016/j.ymssp.2020.107077.

4 Avci, O. and Abdeljaber, O. (2016). Self-organizing maps for structural damage detection: a novel unsupervised vibration-based algorithm. *Journal of Performance of Constructed Facilities* 30: 04015043. https://doi.org/10.1061/(asce)cf.1943-5509.0000801.

5 Abdeljaber, O., Avci, O., Kiranyaz, M.S. et al. (2018). 1-D CNNs for structural damage detection: verification on a structural health monitoring benchmark data. *Neurocomputing* 275: 1308–1317. https://doi.org/10.1016/j.neucom.2017.09.069.

6 Santos, A., Figueiredo, E., Silva, M.F.M. et al. (2016). Machine learning algorithms for damage detection: kernel-based approaches. *Journal of Sound and Vibration* 363: 584–599. https://doi.org/10.1016/j.jsv.2015.11.008.

7 Farrar, C.R. and Worden, K. (2007). An introduction to structural health monitoring. *Philosophical Transactions of the Royal Society A - Mathematical Physical and Engineering Sciences* 365: 303–315. https://doi.org/10.1098/rsta.2006.1928.

8 Worden, K., Farrar, C.R., Manson, G., and Park, G. (2007). The fundamental axioms of structural health monitoring. *Philosophical Transactions of the Royal Society A - Mathematical Physical and Engineering Sciences* 463: 1639–1664. https://doi.org/10.1098/rspa.2007.1834.

9 Sohn, H. and Farrar, C.R. (2001). Damage diagnosis using time series analysis of vibration signals. *Smart Materials and Structures* 10: 446.

10 Abdeljaber, O., Avci, O., Kiranyaz, S. et al. (2017). Real-time vibration-based structural damage detection using one-dimensional convolutional neural networks. *Journal of Sound and Vibration* 388: 154–170. https://doi.org/10.1016/j.jsv.2016.10.043.

11 Alvandi, A. and Cremona, C. (2006). Assessment of vibration-based damage identification techniques. *Journal of Sound and Vibration* 292: 179–202. https://doi.org/10.1016/j.jsv.2005.07.036.

12 Bijalwan, V., Semwal, V.B., Singh, G., and Crespo, R.G. (2021). Heterogeneous computing model for post-injury walking pattern restoration and postural stability rehabilitation exercise recognition. *Expert Systems* e12706. https://doi.org/10.1111/exsy.12706.

13 Kudva, J.N., Munir, N., and Tan, P.W. (1992). Vibration-based damage identification methods: a review and comparative study. *Smart Materials and Structures* 1: 108. https://doi.org/10.1088/0964-1726/1/2/002.

14 Chang, P.C., Flatau, A., and Liu, S.C. (2003). Health monitoring of civil infrastructure. *Structural Health Monitoring* 2: 257. https://doi.org/10.1177/1475921703036169N.

15 Riggio, M. and Dilmaghani, M. (2019). Structural health monitoring of timber buildings: a literature survey. *Building Research and Information* https://doi.org/10.1080/09613218.2019.1681253.

16 Agarwal, A., Sondhi, K.C., and Singh, G. (2021). Transfer learning: survey and classification. In: *Smart Innovations in Communication and Computational Sciences*, 145–155. Singapore: Springer.

17 Kanagaraj, N., Hicks, D., Goyal, A. et al. (2021). Deep learning using computer vision in self driving cars for lane and traffic sign detection. *International Journal of Systems Assurance Engineering and Management* 1–15. https://doi.org/10.1007/s13198-021-01127-6.

18 Kaloni, S., Singh, G., and Tiwari, P. (2021). Nonparametric damage detection and localization model of framed civil structure based on local gravitation clustering analysis. *Journal of Building Engineering* 103339, 2352-7102, https://doi.org/10.1016/j.jobe.2021.103339.

19 Brownjohn, J.M.W. (2007). Structural health monitoring of civil infrastructure. *Philosophical Transactions of the Royal Society A - Mathematical Physical and Engineering Sciences* 365: 589–622. https://doi.org/10.1098/rsta.2006.1925.

20 Figueiredo, E. and Santos, A. (2018). Machine learning algorithms for damage detection. In: *Vibration-Based Techniques for Damage Detection and Localization in Engineering Structures*, 1–40. https://doi.org/10.1142/9781786344977_0001.

21 Singh, G., Chowdhary, M., Kumar, A., and Bahl, R. (2019). A probabilistic framework for base level context awareness of a mobile or wearable device user. In: *2019 IEEE 8th Global Conference on Consumer Electronics (GCCE)*, 217–218. IEEE https://doi.org/10.1109/GCCE46687.2019.9015237.

22 Singh, G., Singh, R.K., Saha, R., and Agarwal, N. (2020). IWT based iris recognition for image authentication. *Procedia Computer Science* 171: 1868–1876.

23 Hou, R. and Xia, Y. (2021). Review on the new development of vibration-based damage identification for civil engineering structures: 2010–2019. *Journal of Sound and Vibration* 491: 115741. https://doi.org/10.1016/j.jsv.2020.115741.

24 Singh, G., Chowdhary, M., Kumar, A., and Bahl, R. (2020). A personalized classifier for human motion activities with semi-supervised learning. *IEEE Transactions on Consumer Electronics* 66 (4): 346–355. https://doi.org/10.1109/TCE.2020.3036277.

25 Nagarajaiah, S. and Basu, B. (2009). Output only modal identification and structural damage detection using time frequency & wavelet techniques. *Earthquake Engineering and Engineering Vibration* 8: 583–605. https://doi.org/10.1007/s11803-009-9120-6.

26 Kaloni, S. and Shrikhande, M. (2017). Damage detection in a structural system via blind source separation. *Proceedings of 16th World Conference in Earthquake Engineering*, Santiago, Chile (9–13 January).

27 Kaloni, S. and Shrikhande, M. (2019). Estimation of normal modes via SYNCHROSQUEEZED transform. *ISET Journal of Earthquake Technology* 56 (N): 1–10.

28 Kumar, H.S., Srinivasa Pai, P., Sriram, N.S., and Vijay, G.S. (2013). ANN based evaluation of performance of wavelet transform for condition monitoring of rolling element bearing. *Procedia Engineering* 64: 805–814. https://doi.org/10.1016/j.proeng.2013.09.156.

29 Singh, G. and Rawat, T. (2013). Color image enhancement by linear transformations solving out of gamut problem. *International Journal of Computer Applications* 67 (14): 28–32.

30 Singh, P., Singh, R.K., and Singh, G. (2018). An efficient iris recognition system using integer wavelet transform. In: *2018 2nd International Conference on Trends in Electronics and Informatics (ICOEI)*, 1029–1034. IEEE https://doi.org/10.1109/ICOEI.2018.8553796.

31 Singh, R.K., Saha, R., Pal, P.K., and Singh, G. (2018). Novel feature extraction algorithm using DWT and temporal statistical techniques for word dependent speaker's recognition. In: *2018 Fourth International Conference on Research in Computational Intelligence and Communication Networks (ICRCICN)*, 130–134. IEEE https://doi.org/10.1109/ICRCICN.2018.8718681.

32 Chhillar, S., Singh, G., Singh, A., and Saini, V.K. (2019). Quantitative analysis of pulmonary emphysema by congregating statistical features. In: *2019 3rd International Conference on Recent Developments in Control, Automation & Power Engineering (RDCAPE)*, 329–333. IEEE https://doi.org/10.1109/RDCAPE47089.2019.8979081.

33 SAP2000 (2016). *CSI Analysis Reference Manual*. Computers & Structures, Inc.

34 FEMA-356 (2000). *Pre-Standard and Commentary for the Seismic Rehabilitation of Buildings*. Federal Emergency Management Agency.

23

Machine-Learning and Deep-Learning Techniques in Social Sciences

Hutashan V. Bhagat and Manminder Singh

Department of Computer Science and Engineering, Sant Longowal Institute of Engineering and Technology, Longowal, Sangrur, India

23.1 Introduction

The term "Social Sciences" is a collection of several pedagogical disciplines that are especially dedicated to examining society. More precisely, "Social Sciences" is a branch of science that figure out how an individual or group of individuals interact, behave, and develop as a culture within a society. Social Scientists combine Social Science, Computer Science (Machine Learning), and Mathematics (Statistics) in order to get new insights into the social behavior. As a result, social phenomenon is now modeled, simulated, and analyzed by using artificial intelligence (AI) of the systems. Traditional statistical techniques are used as a base for the formation of complex machine-learning models. Based on the learning mechanisms, a machine can adopt supervised, unsupervised, or reinforcement learning [1]. Depending upon the complexity of the problems and available data, these learning mechanisms can be used accordingly. This section discusses the basic terminology of the machine learning followed by an intuitive explanation of the machine-learning models that are proved to be fruitful in the field of social sciences. Moreover, various applications of the machine learning [2, 3] that are capable to generate solutions even for the complex problems [4, 5] are also highlighted in this section.

23.1.1 Machine Learning

Humans, being as a top of the creation, evolve day to day from their experiences. In 1959, the author Samuel tried to mimic the functioning of a human brain into a computer system in order to make a model that could defeat him in a checker game [6]. After enough training of the model, the computer intelligence overtook the human intelligence in playing checkers game. Today, machines are becoming so intelligent that they learned how to drive and park a vehicle (autonomous vehicles), intelligent robots that reduce the human efforts in the field of machinery, to make predictions of weather and natural disasters, predict diseases and their cure in human body, etc. From the last two decades, machine learning has evolved even with more accurate and efficient models that now a day's giving fruitful response in distinct areas of applications. An operational definition of machine learning was given by Tom Mitchell in 1997 stated: "A computer program is said to learn from experience E with respect to some class of tasks T and performance measure P, if its performance at tasks in T, as measured by P, improves with experience E" [7]. It is a task-oriented definition that considers machine learning as a tool to improve the performance of the system under consideration. In simple words, machine learning can be defined as a way of creating a suitable model (as shown in Figure 23.1) based on the given problem, then, training that model with the available data to get learn and further test it to get the results.

Machine Learning Algorithms for Signal and Image Processing, First Edition.
Edited by Deepika Ghai, Suman Lata Tripathi, Sobhit Saxena, Manash Chanda, and Mamoun Alazab.
© 2023 The Institute of Electrical and Electronics Engineers, Inc. Published 2023 by John Wiley & Sons, Inc.

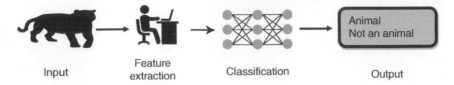

| Input | Feature extraction | Classification | Output |

Figure 23.1 Machine-learning model.

Below given are some of the important applications of the machine learning that come across in our day-to-day life.

- **Diseases Detection and Diagnosis:** Machine-learning models are capable of learning from the previous existing examples and identifying hard-to-discern patterns from complex, noisy, and redundant datasets. This ability of machine-learning models makes them suited for the detection as well as diagnosis of diseases. Recently, diseases like cancer (breast cancer, liver cancer, lung cancer, etc.) can easily be prognoses and predicted using machine-learning models [8].
- **Recommendation Systems:** As the name suggests, "Recommendation System" is a system that is designed in a manner to suggest users a new set of choices based on their existing patterns of exploration. Most of the e-commerce firms (Amazon, Flipkart, Alibaba, etc.) use recommendation systems to suggest user's new things based on their interest. In the world of entertainment, Netflix, SonyLiv, Hotstar, etc. are the best online movies applications that make an individual playlist for every user based on their choices [9].
- **Biometrics:** Security is the most important concern of every organization. Authentication by means of password-only is not sufficient as the password can be easily hacked or stolen. Machine learning lets an organization to use biometric authentication to secure their information as well as assets from illegitimate users. Commonly used biometric authentication techniques are finger-prints scanning, face scanning, retina scanning, etc. [10] are the latest techniques based on machine learning for security.
- **Fraud Detection:** Financial frauds are considered to be an ever-growing menace having ill-consequences to the society. Machine learning, with its imperative techniques, helps in detecting such frauds efficiently. Credit-card frauds are the most occurring frauds nowadays because almost every society in this world has moved toward the cashless payments. Fraud transactions can be done either by the consent of the card holder and the card issuing authority by issuing false identity to an external card or by the use of stolen credit card [11]. Medicare frauds are also very common which results in increased health-care cost [12]. Machine-learning techniques help in identifying the false credit card transactions as well as invalid medical claims to protect the financial loss.
- **Autonomous Machines:** Today, human beings are surrounded by several machines that can work even without human interaction. Such machines are autonomous machines and they are trained in such a manner that they can work endlessly and efficiently. Industry 4.0 is an era of such machines that is totally based on the artificial intelligence. For instance, an autonomous car can drive and park itself; an autonomous truck can be used for the mining purposes [13].
- **Traffic Prediction:** Ancient Greek Philosopher Theophrastus stated: "Time is the most valuable thing a man can spend." Machine learning made it possible for a man to save time while visiting from one place to another. Machine-learning models can take the real-time location of the user and the average time of some past days for the destination place and predict the best route (less traffic and less time) for the user [14].
- **Speech Recognition:** The speech recognition system is a system that takes the human voice as an input and converts it into text as an output. Such systems are also known as Computer Speech Recognition systems. Common examples of such systems are: "Google Assistant," which is present in almost every smart phone, "Alexa" developed by Amazon, "Siri" and "Cortana." All the above examples are based on voice recognition technology

that follows the voice instructions. Voice recognition systems are developed by training the system with different voices having different pitch or accent as an input and simple text as an output of the system [15].

- **Automatic Language Translation:** We are living in a world having different countries and each country has its native language or a country, like India, can have different languages itself. In order, that a language of one region will not become a barrier for another region, a framework based on neural machine-learning techniques known as neural machine translation is used [16]. The commonly used application is "Google Neural Machine Translation," [17] which translates the input text to our native language.

23.1.2 Deep Learning

Deep learning is defined as the branch of artificial intelligence, which is based on the ideology of the working of neurons in the human brain. It includes several statistical methods of machine learning in order to learn the hidden feature hierarchies based on artificial neural networks (ANNs) [18]. In Figure 23.2, in order to identify the object, an image is given as an input to the deep-learning model. The model then itself extracts the features of the image that are further processed by different hidden layers of the neural networks in order to predict the input image is an image of an animal.

23.1.3 Social Data Analysis

Social scientists need to analyze the social data in order to take care of the dependent, independent, and useful features present within the data. So, analysis can be categorized into four basic types, which are given below:

- **Descriptive Analysis:** Descriptive analysis is the first step for a social scientist to analyze the data. It uses various statistical methods in order to know the distribution of the data, helps to remove redundancy and outliers, and identifies the correlation among different variables [19].
- **Detective Analysis:** Detective analysis includes identifying the emerging anomalies, associative patterns, or events hidden within the data. Social scientists make use of this data for predictive analysis [20].
- **Predictive Analysis:** The processed information that social scientists fetch during the descriptive and detective analysis is now used to predict future events. Some typical examples are early detection of tsunamis can help in the evacuation of the coastal areas to save lives, early detection of some viral disease pandemics, rise or fall of share prices in the share markets, etc. [21].
- **Causal Analysis:** In causal analysis, design experimentation is involved in order to identify the cause and effect to influence the outcome of the experiment. For example: Did the use of insecticide causes the crops to grow faster? All such causal questions can be better answered when social scientists use designed experiments to collect data [22].

Before heading toward the various machine-learning models that will be useful for social scientists, let us first discuss the various steps involved in the machine-learning process.

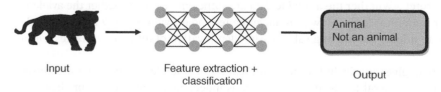

Input   Feature extraction +   Output
            classification

Figure 23.2   Deep-learning model.

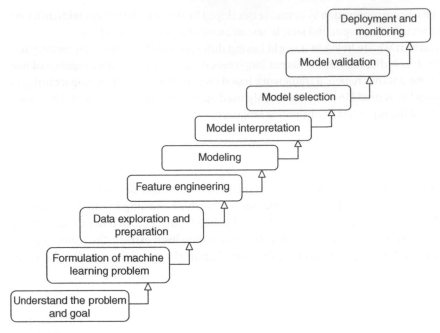

Figure 23.3 Machine-learning process.

23.1.4 Machine-Learning Process

Social scientists can think of solving problems using machine-learning models only when there are large data-driven problems. Figure 23.3 shows the step-wise roadmap that a social data scientist should follow in order to solve any real-world problem using machine learning.

1. **Understand the Problem and Goal:** A problem is a situation when a system is not able to give the desired output or meet the expectations of a given task. Problems generally lead to a vague set of goals, such as increasing the market value of a product, improving medical facilities, and how a variable "X" influences the outcome "Y,". A knowledgeable team should be there who have concrete information regarding the problem and are capable enough to formulate the metrics that need to be improved.
2. **Formulation of Machine-Learning Problem:** Identify your problem first, whether it is a binary classification problem or a one-dimensional regression problem? Is the expected outcome is particular class label for a given input in case of predictive modeling problems or the outcome is a real or continuous value in case of regression problems? Having sufficient knowledge regarding different tasks that can be done using machine learning, one can easily find the methods suitable for a particular problem.
3. **Data Exploration and Preparation:** Exploration of data is an important step in the machine-learning process that lets social scientists to interpret – Whether the available data are sufficient or not? Whether the available data are in categorical form or numerical form? How to remove the redundant values? How to filter out the outliers within the data? How to impute the missing values? And identification of different variables within the dataset.
4. **Feature Engineering:** Feature engineering includes the identification of suitable features that are closely related to the outcome. Features in general terms are also known as independent variables or predictors or factors. The more precisely and more accurately the selection of features is done, the more accurate will be your outcome.
5. **Modeling:** Once the problem formulation and feature selection is done, a method needs to be chosen from a suite of methods available in the machine-learning libraries. It is not always true that a method that is suitable

to solve the problem "A" can also solve the problem "B" and sometimes a chain of different methods need to be applied to get the desired outcome for complex problems. An empirical approach needs to apply in order to validate which method works best for a particular problem from a suite of methods.

6. **Model Interpretation:** After the modeling phase, we come out with several machine-learning models. Model interpretation requires identifying the various flaws present in the models, i.e. what are the errors? What are the features that are found to be important? Which features need to be eliminated? Which independent variable was found to be important? All such queries need to resolve in the model interpretation phase.

7. **Model Selection:** Selection of a suitable machine-learning model for a particular problem from the available set of machine-learning models is done by taking into consideration distinct evaluation metrics. In terms of machine learning, this is known as model validation, which is performed by using historical data.

8. **Model Validation:** The validation performed in the model selection phase is completely different from the validation performed in model validation phase. In this phase, validation is done with respect to the new data instead of the historical data, as well as design experimentation trials are carried out to validate the model.

9. **Deployment and Monitoring:** A machine-learning model that comes out from field trials as well as validated over historical data are considered to be the best model to put into practice for real-world problems. Model monitoring is required because a model that is suitable now for real-world problems might not be able to perform well or need to be changed after some period because the world is not static, and there might be several other factors get involved for which the model has not been trained.

All the above steps that we have discussed are considered to be critical steps. A social scientist needs to go thoroughly through all these steps in order to solve a problem using machine-learning model.

23.1.5 Machine-Learning Terminology

The basic terminology that social scientists must know while solving problems using machine-learning models is given below [23].

- **Dataset:** A dataset is an organized and well-defined set of data, which is in a tabular form, i.e. data are well distributed in the form of rows and tables. Machine-learning models usually learn from these datasets.
- **Instance:** An instance in a dataset is generally a data point or we may call it an observation or an example. It is represented as a row in a dataset.
- **Features:** Feature in machine-learning is defined as the independent variable that is given as an input to a machine model in order to predict or classify the unknown variable. Features are also known as "attributes," "predictors," or "explanatory" variables.
- **Labels:** Labels are the dependent variables that a machine learns to predict. It is also known as response variable or the outcome.
- **Learning:** Learning in terms of machine learning is simply training a model using historical data in order to predict future outcomes. It is also known as "fitting," "building," or an "estimating function" of a model.
- **Underfitting:** Underfitting refers to a model that is less trained and has very low performance in terms of accuracy. Such models are usually trained over a limited set of features, or some irrelevant features were used to train a model that is further unable to predict the outcome.
- **Overfitting:** Overfitting refers to a model that is trained to such an extent that it includes the noise and other details of the training data that will impact the overall performance of the model in a negative manner [24].
- **Regularization:** In order to avoid a model to get overfitted, some additional constraints are formulated and applied during the training phase of the model. For example, L_1 regularization and L_2 regularization are commonly used in machine-learning models based on logistic regression to avoid overfitting.

In this section, we briefly discussed the basic concept of machine-learning and deep-learning models, a step-wise roadmap representing the machine-learning process that social data scientists must follow to solve problems utilizing machine-learning models, and the basic terminology to structure the real-world problems in context of machine-learning and deep-learning techniques. In Section 23.2, we discuss the distinct learning mechanisms that can be utilized to train a machine-learning or deep-learning model depending upon the availability of the data as well as the complexity of the problem. Further, the applications of machine-learning and deep- learning techniques in the field of "Economics," "Social Health," "Linguistic," "Human Resource Management" (HRM), "Behavior Analysis," "Marketing," "Inventory Control," and "Education" are discussed in this chapter.

23.2 Machine-Learning and Deep-Learning Techniques

In Section 23.1, we have discussed the various steps in machine-learning process and some basic terminology used in machine learning. Different learning mechanisms have different techniques associated. There are basically three major types of learning mechanism: supervised learning, unsupervised learning, and reinforcement learning. Depending upon the complexity of the problem and available data, a model is trained with one of the above learning mechanisms. Let us discuss each learning mechanism separately followed by different techniques.

23.2.1 Supervised Learning Techniques

In supervised learning mechanism, as the name itself suggests that training is provided to the model in a supervised manner, i.e. given pairs of inputs and corresponding outputs are provided so that the model could learn to produce the desired output for any new input. Such models, without any human intervention, can create an output from an input that has never been given before. Machine-learning models that learn from the input–output traits are known as supervised learning models.

Various techniques that come under supervised machine learning are given below:

- **Bayesian Networks** [25] use directed acyclic graphs in order to determine the features. This technique is not useful for high-dimensional datasets.
- **Naïve Bayes** [26] is a Bayesian network with class conditional independence and hence, takes less computation time during training.
- **Logistic Regression** [27] is a statistical model that uses a logistic curve to fit in a dataset. This technique is only helpful when predicted variables are dichotomous.
- **Decision Trees (DTs)** [28] is an outlier-independent technique that can handle linearly inseparable data as well as the feature interactions within the dataset. The commonly used algorithms are: ID3 [29], C4.5 [30], C5.0 [31], and CART [32].
- **Random Forests** [27] makes use of various decision trees in order to train the model and returns the majority decision of all decision trees.
- **Support Vector Machines (SVM)** [33] is a complex technique but it has a high accuracy rate. It is commonly used in text classification problems and is very efficient for high-dimensional datasets also.
- **K-Nearest Neighbor (KNN)** [34] is a nonparametric technique that uses a class of the nearest neighbors (previously labeled points). It is considered to be a lazy-learning mechanism and has low efficiency.
- **Neural Networks (NNs)** [35] are based on the neural structure of the human brain and are used when there is a nonlinear relationship among the variables. Some commonly used NN classifiers are multi-layer perceptron (MLP), back propagation neural network (BPN), complementary neural network (CMPNN) [36], etc.
- **Discriminant Analysis** [37] includes two techniques: linear discriminant analysis, where variables are combined in a linear fashion, and quadratic discriminant analysis (QDA), where variables are combined in a quadratic manner. LDA technique is mostly used dimension reduction technique for large datasets.

23.2.2 Unsupervised Learning Techniques

In unsupervised learning mechanism, there are no explicit labels known to the model. The model has to find the label itself by finding the hidden structure and relationship among the data items. The commonly used unsupervised learning techniques are given below:

- **Clustering** [38] is an unsupervised learning mechanism that groups together more similar data points to form a cluster. Every cluster is a set of similar data points that are having minimum inter-cluster similarities and maximum intra-cluster similarities. Commonly used clustering algorithms are K-means [39], K-medoids [40], K-modes [41], partition around medoids (PAMs) [42], clustering large applications (CLARAs) [42], clustering large applications based on randomized search (CLARANS) [43], fuzzy C-means (FCM) [44], etc.
- **Hierarchical Clustering** [45] where, data items are decomposed in a hierarchical manner depending on the measure of proximity, i.e. how close the data items are in the dataset. Proximities can be measured from the intermediate nodes in the hierarchy. The datasets are represented using the dendrogram in which each data item is a leaf node. The decomposition can be performed in two ways: agglomerative and divisive. An agglomerative clustering [46], which is a bottom-up approach, starts with one object for each cluster and recursively merges two or more of the most appropriate clusters. However, divisive clustering, which is a top-down approach, starts with the dataset as one cluster and recursively splits the most appropriate cluster and the process continues until some stopping criterion is reached (mostly the number of k clusters). The major drawback of this method is that it is not possible to move to the previous step once a merge or split step is performed. Some of the well-known algorithms under this category are BIRCH [47], CURE [48], ROCK [49], Chameleon [50], Echidna [51], etc.
- **Principal Components Analysis** [52] is a method used for reducing the dimensions of large datasets in order to make calculations easier and avoid overfitting. The principal components analysis (PCA) algorithm is based on variance and covariance within the datasets.
- **Independent Component Analysis** [53] is used to analyze the multivariate data. The main goal of independent component analysis (ICA) is to find the linear representation of non-normal data in order to have independent components. FastICA [54] is the most commonly used algorithm.
- **Anomaly Detection** [55] is a technique to identify the rare, irregular, or irrelevant observations within the datasets that results in an unexpected pattern. DBSCAN [56], Isolation Forest [57], Boxplots [58], etc. are the commonly used anomaly detection algorithms.

23.2.3 Reinforcement Learning Techniques

The training given to a machine-learning model in order to generate a sequence of decisions is known as reinforcement learning. In a potentially complex and uncertain environment, the agents (reinforcement learning algorithms) are trained in such a way that always tend to increase the total reward. A system generally employs a trial-and-error approach to reach a solution to a given problem. The model either gets a reward or penalty for every action it performed. It is up to the agents to perform task in such a way that every given solution should approach to maximize the reward. Reinforcement learning can be used, for example, in training a model that can control a self-driving car. Preparing a simulation environment for such learning mechanism is a quite complex task. The algorithms under reinforcement learning mechanism are SARSA-Lambda [59], Deep Q-Network (DQN) [60], Soft Actor-Critic (SAC) [61], Deep Deterministic Policy Gradient (DDPG) [62], etc.

23.2.4 Deep-Learning Techniques

Deep learning, sometimes also referred to as deep-structured learning or differential programming, is a technique based on the ANNs with more than one hidden layer that results in a large and complex network. Convolutional neural network (CNN) [63], recurrent neural networks (RNNs) [64], stacked auto-encoders [65], deep Boltzmann machine (DBM) [66], long short-term memory networks (LSTMs) [67], deep belief networks (DBNs) [68].

23.3 Social Sciences Applications Using Machine-Learning and Deep-Learning Techniques

Social science is the branch of science that encompasses a wide set of pedagogical disciplines, including "Economics," "Social Health," "Linguistic," "HRM," "Behavior Analysis," "Marketing," "Inventory Control," and "Education." Social scientists make use of machine- learning models in order to solve distinct problems in distinct fields of social sciences. Applications of machine-learning and deep-learning techniques in distinct areas of Social Sciences are discussed below.

23.3.1 Education

We are now in an era of Education 4.0 [69] where education is not limited to classrooms and students. Today, we are so equipped with digital resources that a whiteboard and marker are replaced by graphical computer presentations to get into a deep insight of every given concept. Numerous of online platforms ("Udemy," "WizIQ," "Ruzuku," "Educadium," "LearnWorlds," "Thinkific," etc.) are available for online courses. Machine learning in the field of education has shown its application in:

- **Grading or Evaluation of Students:** "Biasing" is the toughest problem in education institutions. Machine learning helps to remove human biasing while grading students. Models based on supervised learning use text classification to evaluate students and predict the final grades for each course [70]. E-learning platforms nowadays are getting more popularity as there are no direct interactions between a tutor and a student. Moreover, the online courses are less expensive and, hence, the dropout rate of students is also very high. It is very difficult to evaluate the students on such platforms. In Ref. [71], the authors use data generated by TEL (Technology Enhanced Learning) system known as DEEDS (stands for "Digital Electronics Education and Design Suite") to evaluate the students on basis of exercises designed at different difficulty levels. The machine-learning techniques ANNs, SVM, decision trees, and naïve Bayes classifiers are used to generate results.

- **Analyze and Predict Student Performance:** Several web-based online learning environments use logged data of the students in order to study the students' performance. MOOCs (Massive Open Online Courses) [72] and LMSs (Learning Management Systems) [73] are the two world-wide using learning environment that offers free education as well as numerous university courses. In Ref. [74], authors use correlation-based feature selection (CBFS) for the input features (gender, attendance, board marks, income) to predict the performance of the students using MLP, 1-nearest neighborhood (1-NN), C4.5, and sequential minimal optimization (SMO) techniques. SMO shows the best results among the other techniques. Another ANN approach takes grades, the period of study, and school scores as the input features [75] to predict the students' performance. The proposed technique has achieved an accuracy of 85%. Authors in Ref. [76] use both SVM and multilinear regression models to predict the performance of students for engineering courses. The course grades of all semesters are considered as the input features with exam scores as the corresponding output variable of the model. SVM technique is suitable for predicting the individual performance of student, whereas multilinear regression model suitably predicts the performance of overall courses of the students. In Ref. [77], a machine-learning model based on Bayesian knowledge trace is used to predict the performance of the student while playing a computer game. However, the model is inefficient in predicting the hidden patterns of the students.

- **Testing Students:** Machine learning is not limited in predicting the performance of students, but also, provides models for their assessment. A feedback system can be generated to inform teachers, students, and their parents about the learning of the student, their progress, and what could be the supportive measures to increase the performance of the students. In Ref. [78], the authors developed an e-learning-based model to act as a warning system for the students. The system uses the average login time of students, course login time, and delay in access to their assignments. It is observed that the C4.5 technique has achieved an accuracy of 93%, whereas, the CART technique has achieved an accuracy of 94%. A self-learning model based on machine-learning algorithm known as XGBoost [79] is proposed. The model identifies the students at-risk for the upcoming courses. In Ref. [80], the authors developed a similar model using the linear regression (LR) technique to get the early prediction of the students at-risk for MOOC study courses. A comparative analysis is performed by the authors in [81] using five different machine- learning models in order to identify the students at-risk. The key features used in this study are grades, attendance of the students, quizzes, weekly assignments, offline exam courses, mathematical activities, and project milestones. Among all other techniques (SVMs, DTs, ANNs, and logistic regression), naïve Bayes classifier outperforms with an accuracy of 85%.

- **Improving Students Retention:** The term "student retention" also referred to as "persistence" is a measure that is used to evaluate the school performance and helps to improve the gray areas within a school or a university. Sometimes, it also affects the various important segments of evaluation (ranking, reputation, financial, etc.) of a school or university. An analytical system [82] based on machine learning that provide a real-time support from teachers to their students to solve retention problems. The system considers behavioral characteristics, previous academic history, student efforts, and student grades as inputs. In Ref. [83], the authors use two classifiers: J-48 and J-Rip, to identify the students who have not continued the last orientation stage and such models are efficient enough in providing information to teachers to curb the problem of student attrition. In Ref. [84], the authors proposed a model known as learning analytics model (LAM) to provide personalized learning and support services. Self-organizing maps (SOMs) are used to provide visual analytics of the data. LAM model proved to be efficient in increasing student retention rates. Another model [85] is based on qualitative and quantitative analysis to identify the various factors that lead to an increase in retention rates. The model is based on neural networks (NNs) to predict the retention rate of the students for first-year students of various disciplines. A case study [86] to find out the reasons for student dropout for data provided by "Catholic University of the North" using different machine-learning techniques. The authors use Bayesian belief network, decision trees, and neural networks techniques on a dataset containing information of 89 056 students with 11 attributes. Results have shown that the socio-economic factors (loans, scholarships, etc.) have a great impact on the retention rate.

23.3.2 Economics

Till mid-1980s, economist uses causal inference to make predictions that require human interactions [87]. However, such techniques were inefficient in making prediction. The inclusion of machine-learning models helps economists to work efficiently even with big datasets and to solve complex problems over the last few decades. Machine learning has a number of applications for policy problems in economics. In Ref. [88], the authors have concluded that machine-learning model are the key components for decisions as well as for policy problems. The authors use machine-learning models in real-world problems, e.g. in order to take a decision whether to do an operation or hip replacement for an elderly patient? In Ref. [89], the authors use the machine-learning models based on random forests and logistic regression techniques that analyze the mobile data of the user to predict the loan repayment and credit score. In Ref. [90], the authors have studied the racial discrepancies in stop-and-frisk tactics by analyzing continuously for five years over the three million stops in New York City to find out the probability that a person has a weapon or not. The random forests technique of machine learning helps to depict that the whites are more susceptible to having weapons like knife or guns than blacks. In Ref. [91], a machine-learning model using v-support vector regression (v-SVR) technique is developed to improve the services in urban cities. Further, Google Street View images are used to find out the poverty, safest places, and the value of a house in New York City. Another machine-learning model is based on the random forests technique, which is used to give early warnings of the upcoming recession in the country [92]. The authors used some important explanatory variables in the market while forecasting. In Ref. [93], the authors used a machine-learning model to find an efficient way of pattern recognition on economic data. A period of 10 years from 2007 to 2016 is used to collect 2719 experimental samples of various companies and 470 financial indicators (features) are used to train and test the model using KNN and SVM techniques of machine learning to obtain an efficient recognition model. The overall accuracy of the proposed model is 95%. In Ref. [94], the authors proposed a machine-learning model that can accurately predict the economic turning point in the real world. It sometimes becomes a notorious task for economists to predict whether a new phase of the economy has begun or not. The proposed model is based on random forests and boosting techniques to develop several models that can accurately predict the economic growth cycle turning points in the real world. The average outcome of all these models is then taken into consideration to make the final decision. In Ref. [95], the authors used the similar approach used in Ref. [94]. A multivariate time series model is developed to estimate the cyclic components in the real-world financial sectors (bank credits, GDP, the value of a house) in order to find the relation between the financial cycles and business cycles in Taiwan City. A machine-learning model based on random forests and boosting techniques is developed. The model concludes that the length of both businesses, as well as financial cycles, is much shorter than in industrialized economies. In Ref. [96], the authors proposed a comparative approach using different machine-learning models to improve the empirical understanding about asset pricing by identifying the asset risk premium and other estimation errors. Boosted regression trees (BRTs) and neural networks (NNs) techniques are the best performing machine-learning models. In Ref. [97], the authors mainly focus on the problems of high dimensionality ("curse of dimensionality") and irregularities in the economic datasets. A framework is proposed that is capable enough to find a solution for stochastic economic models with irregularly shaped datasets. Both the Bayesian Gaussian mixture (BGM) model and Gaussian regression are combined to learn with irregularly shaped ergodic sets. In Ref. [98], the authors modified the commonly used numerical technique (Black–Scholes Option Pricing Model [BSOPM]) to find the price of the stocks based on the buying price (price of the call option) and selling price (price of the put option) to predict the stock price. The new technique, named as modified BSOPM, includes two more parameters: "strike price" and "time of expiration" in order to evaluate the stock price in the frontier markets using decision trees (DTs), neural networks (NNs), and ensemble learning techniques of machine learning. In Ref. [99], a comparative analysis of the traditional time series prediction models with machine-learning models for predicting stock prices is performed. The major goal of an investor while investing in the stock market is to minimize the losses and maximize the gains in order to accumulate overall profit. Because of the volatile nature of the stock market, it is hard for investors to raise their profits to certain goals. The authors concluded that the NARX model (nonlinear autoregressive exogenous), which is based on RNNs technique of machine learning, is suitable for short-term forecasts. In Ref. [100], an integrated machine-learning model is proposed that is based on the integration of three learning algorithms: Genetic algorithm, probabilistic-SVM, and AdaBoost, to evaluate the current trends in the stock market as well as to predict the future value of the stock price. The model is tested for 20 shares selected from the SZSE, a Chinese stock exchange, whereas 16 shares from NASDAQ, an American stock exchange. The simulated results showed that the proposed integrated model makes the better classification and generalization to achieve better profitability as compared to the traditional time-series models.

23.3.3 Marketing

Digital marketing is the backbone of marketing 4.0, which integrates online and offline interactions between different companies and their customers. Companies are now paying much attention to Big Data Analytics in order to get deep data insights for better customer engagement. Machine-learning models learn from the previous choices of the customers and make personal recommendations to assist customers with what they are looking for during online shopping. It also provides customer with various alternatives for a particular item in order to ensure the availability of the product. ChatBots that act as a human virtual assistant are able to provide assistance as well as advice to every new customer visiting the website [101]. The potential of a machine-learning model to "learn on the go" makes it a backbone of the marketing 4.0. The applications of machine learning in various aspects of marketing are as shown:

- **Social Media Analysis:** It is true to say "The more you know your customers, the more you earn from your business." A marketing analyst should have fruitful information regarding their customers (their requirements, interests, products, geographical location, sex, etc.) in order to make qualified decisions [102]. Digitalization of the markets helps marketing analysts to get real-time information regarding their customers. The customers manually enter their choices and interests that can be recorded to provide more valuable suggestions without spending much effort and time using distinct machine-learning models [103]. A marketing analyst can collect customer data from the customer's reviews and surveys, by analyzing the social media contents of customers using various techniques, such as text mining, sentiment analysis and social sensing. These data are then analyzed to make qualified decisions regarding personal recommendations to customers [104]. Fuzzy-SVM learning model are the most efficient models in such cases.

- **Products Launch and Purchase:** The traditional way of launching a product requires a lot of time and effort where most of the companies were reliant on manually done door-to-door social surveys. Nowadays, the e-commerce giants like Amazon, Flipkart, Alibaba, etc. provide companies with a well-established platform to launch their new products. An apparel industry can take inspiration from social media sites (Instagram) by following the top trending dress designs in the markets and take decisions for launching new products [105]. To predict what a customer will buy next is an important question for online retail and marketing entrepreneurs. Machine-learning models latent Dirichlet allocation (LDA) [106] and mixtures of Dirichlet-multinomials (MDMs) [107] are the most promising models that are capable enough to handle large datasets and predict customers' future purchase.

- **Advertisement:** Machine learning played a significant role in digital advertisements also. There are numerous machine-learning techniques for a data analyst to enhance online advertising. The authors in Ref. [108] categorized the online advertisement machine-learning techniques into two categories: user-centric machine-learning techniques and content-centric machine-learning techniques. The user-centric techniques mainly target the behavior and user-profiling of the user. The behavioral characteristics of a user can be obtained from various sources such as search engines, visited pager, or websites, previous purchase from the browsing history, links that are accessed, etc. [109]. The user profiling is a technique that is used to identify the hidden pattern in the user's behavior in order to get to know what a user likes and whatnot. Such systems are known as recommendation systems [110] and are truly modeled using machine-learning techniques. The content-centric approach includes contextual advertising and real-time bidding techniques. The contextual advertising technique provides the user personalized advertising environment on the third-party web pages based on the previous patterns of the user [111]. In order to know the user patterns, web pages are analyzed in real-time to get a collection of keywords and deliver the closely matched advertisement to the user. The real-time bidding technique is based on the real-time machine-learning model that makes decisions based on the current patterns of the user (searches, link clicks, etc.) in order to determine suitable advertisements at that particular instant [112]. Sequential pattern mining (SPM) technique for the extraction of meaning full keywords and then meaningful language patterns from online broadcasting content [113]. The authors then used these language patterns to identify the real-time suggestions. The click-through rate (CTR) is a measure that is used to find the number of times the user clicks the advertisement. In Ref. [114], the authors used two approaches in order to find the CTR. A deep-learning model known as DBNLR, which is a combination of two techniques DBN and logistic regression, is used to find the probable value of the predicted CTR. In Ref. [115], the authors use an attention mechanism based on deep neural networks to cluster similar value pairs (users having similar patterns). An attention stacked autoencoder (ASAE) model is used to predict the advertising CTR. In Ref. [116], the authors proposed a technique to identify the attributes (gender, age, and other personality attributes) of the users that are not mentioned. Three major techniques: SVM, LR, and naïve Bayes are used. In Ref. [117], the authors proposed a technique known as follow the Regularized Factorized Leader (FTRL) for the prediction of advertising CTR. The technique has a higher convergence rate than other factorizing machines. In Ref. [118], the authors proposed a framework consisting of four modules (user intention, sentiments, target advertisement, and term expansion) to predict the best advertisement for a user. The proposed BCCA framework is an integration of contextual advertisement with text mining in order to identify the predicted advertisement and rank them as per their relevancy. Another response prediction framework based on logistic regression technique is proposed in Ref. [119] in order to have simple, scalable, and efficient display advertisements for the users. The proposed framework proved to be efficient, as it consumes less memory for scaling large samples and parameters.

23.3.4 Miscellaneous Applications

Machine learning has shown its applications in HRM. Business analysts come out with several machine-learning models that provide assistance in efficient utilization of human resources in big organizations. A conceptual framework known as artificial intelligence based human resource management (AIHRM) is proposed in [120]. The AIHRM framework is based on the six important traits of HRM: HRM strategy and planning, human resource recruitment, human resource training and development, human resource performance management, employee salary evaluation, and relationship management. In Ref. [121], the authors have shown how big organizations make use of machine-learning models in Green Human Resource Management (GHRM). Several processes, such as screening of employees, employee engagement, and career development can be deployed using machine-learning models without making much use of the resources.

Machine learning has also shown its application in the field of criminology. A probabilistic model, which is capable enough to predict the crime rates, provides the generic demographic factors incorporated with the crime is proposed in Ref. [122]. The Markov chain Monte Carlo (MCMC) learning model and Bayesian linear regression (BLR) technique are used to build the whole framework. Another machine-learning model based on the random forests algorithm is designed to predict future crime at micro places [123]. The model is able to forecast the future robberies in Dallas and find out the demographic factors associated with them. The proposed model outperforms the existing kernel density and risk terrain model (RTM).

Machine learning makes use of the social media contents in order to analyze social behavior, community threats, etc. within a community. A similar approach is used by the authors in Ref. [124] for sentiment analysis of Czech social media. Maximum entropy and SVMs machine learning classifiers are used in the proposed work. Social networking giants: Facebook, Instagram, and Twitter have attracted the whole world population in the recent few years [125]. Sentiments are usually considered to be positive, negative, or neutral sentiments. A feature-based model known as maximum entropy is a widely used sentiment analysis model that does not consider independent assumptions [126]. A famous supervised learning-based classification model known as SVM is performed efficiently and accurately as compared to other state-of-the-art techniques (stochastic gradient descent [127], MLP [128], SailAil Sentiment Analyzer [129], and random forests [130]).

Being a part of a multilingual world, machine learning comes up with advent models that prominently remove all language barriers. No matter which part of the world you are in, machine-learning models provide you with real-time language translation. Nowadays, there are various speech recognition models that are based on the CNN and RNN techniques of deep learning [131]. There are intelligent devices that can follow the voice command of a user to perform any particular task (for example, "Google Assistant" in smart phones, "Alexa," etc.). A model based on Gaussian mixture model (GMM) and universal background model is proposed in Ref. [132] that are used to identify the age of the speaker (child, young, adults, etc.) from voice. Such models can be used in age-restricted areas, content selection, etc. The model is able to give an accuracy of 74%. Another speech recognition model that intelligently recognizes the voice of male or female is proposed in Ref. [133]. The model gives simple binary outputs (male or female) and hence shows great accuracy. In Ref. [134], the authors proposed a speech emotion recognition model that uses deep neural networks (DNNs) to extract important features required to identify the emotions. A single-layer neural network model known as the extreme learning machine (ELM) is used to identify the utterance-based emotions from the features that are extracted from raw data. The proposed model shows high accuracy and 10 times faster training time than SVMs models.

23.4 Conclusion

We are living in a social world where one society depends on another to fulfill its requirements. The advent of machine-learning models provides society with a convenient way of communication, marketing, management of resources, trading, etc. that helps society to work smoothly. At the beginning of this chapter, a brief explanation of machine-learning and deep-learning models is given followed by a stepwise roadmap of the machine-learning process that helps social data scientists to solve real-time and real-world problems efficiently. Different learning mechanisms have different techniques associated. Bayesian networks, naïve Bayes, logistic regression, decision trees, random forest, SVMs, KNN, neural networks, discriminant analysis, and so on are commonly used supervised learning techniques that are discussed with their applications. Unsupervised learning techniques, such as clustering, PCA, ICA, and anomaly detection can be used to find hidden structures in unlabeled datasets. For problems associated with uncertainty, reinforcement learning has the potential to generate a sequence of decisions. SARSA-Lambda, DQN, SAC, and DDPG are some of the important techniques of reinforcement learning. The DEEDS systems played an important role in evaluating student performance in the modern Education 4.0 system. LAM models are proved to be efficient in increasing student retention rates. BSOPM is an efficient model to predict stock prices. For an investor, the NARX model is the best machine-learning model for short-term forecasts. In the field of marketing, LDA and MDM models are capable enough to handle large datasets and predict customers' future purchase. The MCMC learning model and BLR technique collaborate to design a framework that is capable of predicting crime rates. This chapter highlights the various machine learning and deep-learning techniques, and literature has been reviewed that provides a deep insight into the applications of these models in Social Sciences. Moreover, this chapter provides a roadmap for a social scientist to solve problems using machine-learning models.

References

1 Qiu, J., Wu, Q., Ding, G. et al. (2016). A survey of machine learning for big data processing. *EURASIP Journal on Advances in Signal Processing* 2016 (1): 1–16. https://doi.org/10.1186/s13634-016-0355-x.

2 Beauchamp, N. (2017). Predicting and interpolating state-level polls using twitter textual data. *American Journal of Political Science* 61: 490–503. https://doi.org/10.1111/ajps.12274.

3 Bamman, D., Underwood, T., and Smith, N.A. (2014). A Bayesian mixed effects model of literary character. *Proceedings of the 52st Annual Meeting of the Association for Computational Linguistics (ACL'14)*, Baltimore, MD. https://doi.org/10.3115/v1/p14-1035.

4 Cranshaw, J., Schwartz, R., Hong, J.I., and Sadeh, N. (2012). The livehoods project: utilizing social media to understand the dynamics of a city. *Proceedings of the Sixth International AAAI Conference on Weblogs and Social Media* 6 (1): 58–65.

5 Hipp, J.R., Faris, R.W., and Boessen, A. (2012). Measuring 'neighborhood': constructing network neighborhoods. *Social Network* 34: 128–140. https://doi.org/10.1016/j.socnet.2011.05.002.

6 Samuel, A.L. (1959). Some studies in machine learning using the game of checkers. *IBM Journal of Research and Development* 3 (3): 210–229. https://doi.org/10.1147/rd.33.0210.

7 Mitchell, T.M. (1997). *Machine Learning*. McGraw-Hill.

8 Mccarthy, J.F., Marx, K.A., Hoffman, P.E. et al. (2004). Applications of machine learning and high-dimensional visualization in cancer detection, diagnosis, and management. *Annals of the New York Academy of Sciences* 1020 (1): 239–262. https://doi.org/10.1196/annals.1310.020.

9 Khanal, S.S., Prasad, P.W.C., Alsadoon, A., and Maag, A. (2019). A systematic review: machine learning based recommendation systems for e-learning. *Education and Information Technologies* 1–30. https://doi.org/10.1007/s10639-019-10063-9.

10 Bhanu, B. and Kumar, A. (ed.) (2017). *Deep Learning for Biometrics*. Switzerland: Springer.

11 Awoyemi, J.O., Adetunmbi, A.O., and Oluwadare, S.A. (2017). Credit card fraud detection using machine learning techniques: a comparative analysis. In: *2017 International Conference on Computing Networking and Informatics (ICCNI)*, 1–9. IEEE https://doi.org/10.1109/ICCNI.2017.8123782.

12 Bauder, R.A. and Khoshgoftaar, T.M. (2017). Medicare fraud detection using machine learning methods. In: *2017 16th IEEE International Conference on Machine Learning and Applications (ICMLA)*, 858–865. IEEE https://doi.org/10.1109/ICMLA.2017.00-48.

13 Hyder, Z., Siau, K., and Nah, F. (2019). Artificial intelligence, machine learning, and autonomous technologies in mining industry. *Journal of Database Management (JDM)* 30 (2): 67–79. https://doi.org/10.4018/JDM.2019040104.

14 Siddiqui, S.Y., Khan, M.A., Abbas, S., and Khan, F. (2020). Smart occupancy detection for road traffic parking using deep extreme learning machine. *Journal of King Saud University-Computer and Information Sciences*. https://doi.org/10.1016/j.jksuci.2020.01.016.

15 Padmanabhan, J. and Johnson Premkumar, M.J. (2015). Machine learning in automatic speech recognition: a survey. *IETE Technical Review* 32 (4): 240–251. https://doi.org/10.1080/02564602.2015.1010611.

16 Jiang, B., Song, Y., Wei, S. et al. (2014). Deep bottleneck features for spoken language identification. *PloS One* 9 (7): e100795. https://doi.org/10.1371/journal.pone.0100795.

17 Wu, Y., Schuster, M., Chen, Z. et al. (2016). Google's neural machine translation system: bridging the gap between human and machine translation. arXiv preprint arXiv:1609.08144.

18 Vargas, R., Mosavi, A., and Ruiz, R. (2017). Deep learning: a review. In: *Advances in Intelligent Systems and Computing*, 1–11.

19 Kaur, P., Stoltzfus, J., and Yellapu, V. (2018). Descriptive statistics. *International Journal of Academic Medicine* 4 (1): 60–63.

20 Heymann, H., King, E.S., and Hopfer, H. (2014). Classical descriptive analysis. In: *Novel Techniques in Sensory Characterization and Consumer Profiling*, 9–40. CRC Press.

21 Bowles, M. (2015). *Machine Learning in Python: Essential Techniques for Predictive Analysis*. Wiley.

22 Morgan, S.L. and Winship, C. (2015). *Counterfactuals and Causal Inference*. Cambridge University Press.

23 Mooney, S.J. and Pejaver, V. (2018). Big data in public health: terminology, machine learning, and privacy. *Annual Review of Public Health* 39: 95–112. https://doi.org/10.1146/annurev-publhealth-040617-014208.

24 Dietterich, T. (1995). Overfitting and undercomputing in machine learning. *ACM Computing Surveys (CSUR)* 27 (3): 326–327. https://doi.org/10.1145/212094.212114.

25 Franz, P. (2005). Bayesian network classifiers versus selective k-NN classifier. *Pattern Recognition* 38 (1): 1–10. https://doi.org/10.1016/j.patcog.2004.05.012.

26 Kuncheva, L.I. (2006). On the optimality of naive Bayes with dependent binary features. *Pattern Recognition Letters* 27 (7): 830–837. https://doi.org/10.1016/j.patrec.2005.12.001.

27 Lorena, A.C., Jacintho, L.F., Siqueira, M.F. et al. (2011). Comparing machine learning classifiers in potential distribution modelling. *Expert Systems with Applications* 38 (5): 5268–5275. https://doi.org/10.1016/j.eswa.2010.10.031.

28 Rokach, L. and Maimon, O. (2005). Top-down induction of decision trees classifiers-a survey. *IEEE Transactions on Systems, Man, and Cybernetics, Part C: Applications and Reviews* 35 (4): 476–487. https://doi.org/10.1109/TSMCC.2004.843247.

29 Peng, W., Chen, J., and Zhou, H. (2009). An implementation of ID3-decision tree learning algorithm. From web. arch. usyd. edu. au/wpeng/DecisionTree2. pdf Retrieved date: 13 May.

30 Sharma, S., Agrawal, J., and Sharma, S. (2013). Classification through machine learning technique: C4. 5 algorithm based on various entropies. *International Journal of Computer Applications* 82 (16): 0975–8887. https://doi.org/10.5120/14249-2444.

31 Pandya, R. and Pandya, J. (2015). C5. 0 algorithm to improved decision tree with feature selection and reduced error pruning. *International Journal of Computer Applications* 117 (16): 18–21.

32 Lawrence, R.L. and Wright, A. (2001). Rule-based classification systems using classification and regression tree (CART) analysis. *Photogrammetric Engineering and Remote Sensing* 67 (10): 1137–1142.

33 Campbell, C. and Ying, Y. (2011). Learning with support vector machines. *Synthesis Lectures on Artificial Intelligence and Machine Learning* 5 (1): 1–95. https://doi.org/10.2200/S00324ED1V01Y201102AIM010.

34 Peterson, L.E. (2009). K-nearest neighbor. *Scholarpedia* 4 (2): 1883. https://doi.org/10.4249/scholarpedia.1883.

35 Kotsiantis, S.B., Zaharakis, I., and Pintelas, P. (2007). Supervised machine learning: a review of classification techniques. *Emerging Artificial Intelligence Applications in Computer Engineering* 160 (1): 3–24.

36 Sudheer, K.P., Gosain, A.K., Mohana Rangan, D., and Saheb, S.M. (2002). Modelling evaporation using an artificial neural network algorithm. *Hydrological Processes* 16 (16): 3189–3202. https://doi.org/10.1002/hyp .1096.

37 Tharwat, A., Gaber, T., Ibrahim, A., and Hassanien, A.E. (2017). Linear discriminant analysis: a detailed tutorial. *AI communications* 30 (2): 169–190. https://doi.org/10.3233/AIC-170729.

38 Berry, M.W., Mohamed, A., and Yap, B.W. (ed.) (2019). *Supervised and Unsupervised Learning for Data Science*. Springer Nature.

39 Ralambondrainy, H. (1995). A conceptual version of the k-means algorithm. *Pattern Recognition Letters* 16 (11): 1147–1157.

40 Harikumar, S. and Surya, P.V. (2015). K-medoid clustering for heterogeneous datasets. *Procedia Computer Science* 70: 226–237. https://doi.org/10.1016/j.procs.2015.10.077.

41 Cao, F., Liang, J., Li, D. et al. (2012). A dissimilarity measure for the k-modes clustering algorithm. *Knowledge-Based Systems* 26: 120–127. https://doi.org/10.1016/j.knosys.2011.07.011.

42 Kaufman, L. and Rousseeuw, P.J. (2009). *Finding Groups in Data: An Introduction to Cluster Analysis*, vol. 344. Wiley.

43 Ng, R.T. and Han, J. (2002). CLARANS: a method for clustering objects for spatial data mining. *IEEE Transactions on Knowledge and Data Engineering* 14 (5): 1003–1016. https://doi.org/10.1109/TKDE.2002 .1033770.

44 Askari, S. Fuzzy C-Means clustering algorithm for data with unequal cluster sizes and contaminated with noise and outliers: Review and development. *Expert Systems with Applications* 165: 113856. https://doi.org/10 .1016/j.eswa.2020.113856.

45 Xu, J., Wang, G., and Deng, W. (2016). DenPEHC: density peak based efficient hierarchical clustering. *Information Sciences* 373: 200–218. https://doi.org/10.1016/j.ins.2016.08.086.

46 Wang, J., Zhu, C., Zhou, Y. et al. (2017). From partition-based clustering to density-based clustering: fast find clusters with diverse shapes and densities in spatial databases. *IEEE Access* 6: 1718–1729. https://doi.org/10 .1109/ACCESS.2017.2780109.

47 Zhang, T., Ramakrishnan, R., and Livny, M. (1996). BIRCH: an efficient data clustering method for very large databases. *ACM Sigmod Record* 25 (2): 103–114. https://doi.org/10.1145/235968.233324.

48 Guha, S., Rastogi, R., and Shim, K. (1998). CURE: an efficient clustering algorithm for large databases. *ACM Sigmod Record* 27 (2): 73–84. https://doi.org/10.1145/276305.276312.

49 Guha, S., Rastogi, R., and Shim, K. (2000). ROCK: a robust clustering algorithm for categorical attributes. *Information Systems* 25 (5): 345–366. https://doi.org/10.1016/S0306-4379(00)00022-3.

50 Karypis, G., Han, E.H., and Kumar, V. (1999). Chameleon: hierarchical clustering using dynamic modeling. *Computer* 32 (8): 68–75. https://doi.org/10.1109/2.781637.

51 Murtagh, F. and Contreras, P. (2012). Hierarchical clustering for finding symmetries and other patterns in massive, high dimensional datasets. In: *Data Mining: Foundations and Intelligent Paradigms*, 95–130. Berlin, Heidelberg: Springer https://doi.org/10.1007/978-3-642-23166-7_5.

52 Abdi, H. and Williams, L.J. (2010). Principal component analysis. *Wiley Interdisciplinary Reviews: Computational Statistics* 2 (4): 433–459. https://doi.org/10.1002/wics.101.

53 Comon, P. (1994). Independent component analysis, a new concept? *Signal Processing* 36 (3): 287–314. https://doi.org/10.1016/0165-1684(94)90029-9.

54 Oja, E. and Yuan, Z. (2006). The FastICA algorithm revisited: convergence analysis. *IEEE Transactions on Neural Networks* 17 (6): 1370–1381. https://doi.org/10.1109/TNN.2006.880980.

55 Chandola, V., Banerjee, A., and Kumar, V. (2009). Anomaly detection: a survey. *ACM Computing Surveys (CSUR)* 41 (3): 1–58. https://doi.org/10.1145/1541880.1541882.

56 Schubert, E., Sander, J., Ester, M. et al. (2017). DBSCAN revisited, revisited: why and how you should (still) use DBSCAN. *ACM Transactions on Database Systems (TODS)* 42 (3): 1–21. https://doi.org/10.1145/3068335.

57 Xu, D., Wang, Y., Meng, Y., and Zhang, Z. (2017). An improved data anomaly detection method based on isolation forest. In: *2017 10th International Symposium on Computational Intelligence and Design (ISCID)*, vol. 2, 287–291. IEEE https://doi.org/10.1109/ISCID.2017.202.

58 Dawson, R. (2011). How significant is a boxplot outlier? *Journal of Statistics Education* 19 (2) https://doi.org/10.1080/10691898.2011.11889610: 1–13.

59 Chen, S.L. and Wei, Y.M. (2008). Least-squares SARSA (lambda) algorithms for reinforcement learning. In: *2008 Fourth International Conference on Natural Computation*, vol. 2 (ed. 636), 632. IEEE https://doi.org/10.1109/ICNC.2008.694.

60 Fan, J., Wang, Z., Xie, Y., and Yang, Z. (2020). A theoretical analysis of deep Q-learning. In: *Learning for Dynamics and Control*, 486–489. PMLR.

61 Haarnoja, T., Zhou, A., Hartikainen, K. et al. (2018). Soft actor-critic algorithms and applications. arXiv preprint arXiv:1812.05905.

62 Li, S., Wu, Y., Cui, X. et al. (2019). Robust multi-agent reinforcement learning via minimax deep deterministic policy gradient. *Proceedings of the AAAI Conference on Artificial Intelligence* 33 (01): 4213–4220. https://doi.org/10.1609/aaai.v33i01.33014213.

63 Liu, T., Fang, S., Zhao, Y. et al. (2015). Implementation of training convolutional neural networks. arXiv preprint arXiv:1506.01195.

64 Medsker, L.R. and Jain, L.C. (2001). Recurrent neural networks. *Design and Applications* 5: 64–67.

65 Sun, Y., Xue, B., Zhang, M., and Yen, G.G. (2018). An experimental study on hyper-parameter optimization for stacked auto-encoders. In: *2018 IEEE Congress on Evolutionary Computation (CEC)*, 1–8. IEEE. doi: 10.1109/CEC.2018.8477921.

66 Salakhutdinov, R. and Hinton, G. (2009). Deep Boltzmann machines. In: *Artificial Intelligence and Statistics*, 448–455. PMLR.

67 Woodbridge, J., Anderson, H. S., Ahuja, A. et al. (2016). Predicting domain generation algorithms with long short-term memory networks. arXiv preprint arXiv:1611.00791.

68 Hinton, G.E. (2009). Deep belief networks. *Scholarpedia* 4 (5): 5947. https://doi.org/10.4249/scholarpedia.5947.

69 Ciolacu, M., Tehrani, A.F., Beer, R., and Popp, H. (2017). Education 4.0—fostering student's performance with machine learning methods. In: *2017 IEEE 23rd International Symposium for Design and Technology in Electronic Packaging (SIITME)*, 438–443. IEEE https://doi.org/10.1109/SIITME.2017.8259941.

70 Wu, J.Y., Hsiao, Y.C., and Nian, M.W. (2018). Using supervised machine learning on large-scale online forums to classify course-related Facebook messages in predicting learning achievement within the personal learning environment. *Interactive Learning Environments* 1–16. https://doi.org/10.1080/10494820.2018.1515085.

71 Hussain, M., Zhu, W., Zhang, W. et al. (2019). Using machine learning to predict student difficulties from learning session data. *Artificial Intelligence Review* 52 (1): 381–407. https://doi.org/10.1007/s10462-018-9620-8.

72 Imran, H., Hoang, Q., Chang, T.-W., and Kinshuk, G.S. (2014). A framework to provide personalization in learning management systems through a recommender system approach. In: *Intelligent Information and Database System. ACIIDS 2014*, Lecture Notes in Computer Science, vol. 8397, 271–280. https://doi.org/10.1007/978-3-319-05476-6_28.

73 Kloft, M., Stiehler, F., Zheng, Z., and Pinkwart, N. (2014). Predicting MOOC dropout over weaks using machine learning methods. In: *Proceeding of the EMNLP 2014 Workshop on Analysis of Large Scale Social Interaction in MOOCs*, 60–65.

74 Acharya, A. and Sinha, D. (2014). Early prediction of students performance using machine learning techniques. *International Journal of Computers and Applications* 107 (1): 37–43.

75 De Albuquerque, R.M., Bezerra, A.A., de Souza, D.A. et al. (2015). Using neural networks to predict the future performance of students. In: *IEEE International Symposium on Computers in Education (SIIE)*, vol. 2015, 109–113. https://doi.org/10.1109/SIIE.2015.7451658.

76 Huang, S. and Fang, N. (2013). Predicting student academic performance in an engineering dynamics course: a comparison of four types of predictive mathematical models. *Computers and Education* 61: 133–145. https://doi.org/10.1016/j.compedu.2012.08.015.

77 Käser, T., Hallinen, N.R., and Schwartz, D.L. (2017). Modeling exploration strategies to predict student performance within a learning environment and beyond. In: *17th International Conference on Learning Analytics and Knowledge 2017*, 31–40. https://doi.org/10.1145/3027385.3027422.

78 Hu, Y.-H., Lo, C.-L., and Shih, S.-P. (2014). Developing early warning systems to predict students online learning performance. *Comput Human Behav* 36: 469–478. https://doi.org/10.1016/j.chb.2014.04.002.

79 Hlosta, M., Zdrahal, Z., and Zendulka, J. (2017). Ouroboros: early identification of at-risk students without models based on legacy data. In: *7th International Conference on Learning Analytics & Knowledge (LAK'17)*, 6–15. https://doi.org/10.1145/3027385.3027449.

80 He, J., Bailey, J., Rubinstein, B.I.P. et al. (2015). Identifying at-risk students in massive open online courses. *Proceedings of the AAAI Conference on Artificial Intelligence* 29 (1): https://doi.org/10.1609/aaai.v29i1.947.

81 Marbouti, F., Diefes-Dux, H.A., and Madhavan, K. (2016). Models for early prediction of at-risk students in a course using standards-based grading. *Comput Educ* 103: 1–15. https://doi.org/10.1016/j.compedu.2016.09.005.

82 Arnold, K.E. and Pistilli, M.D. (2012). Course signals at Purdue: using learning analytics to increase student success. In: *2nd International Conference on Learning Analytics and Knowledge (LAK'12)*, 267–270. https://doi.org/10.1145/2330601.2330666.

83 Kai, S, Miguel, J, Andres, L et al. (2017). Predicting student retention from behavior in an online orientation course. *Proceedings of 10th International Conference on Educational Data Mining*, pp. 250–255.

84 De Freitas, S., Gibson, D., Du Plessis, C. et al. (2015). Foundations of dynamic learning analytics: using university student data to increase retention. *British Journal of Educational Technology* 46 (6): 1175–1188. https://doi.org/10.1111/bjet.12212.

85 Alkhasawneh, R. and Hargraves, R.H. (2014). Developing a hybrid model to predict student first year retention in STEM disciplines using machine learning techniques. *Journal of STEM Education: Innovations and Research* 15 (3): 35–42.

86 Miranda, M.A. and Guzma'n, J. (2017). Analysis of university student dropouts using data mining techniques. *University Education* 10 (3): 61–68.

87 Hamermesh, D.S. (2013). Six decades of top economics publishing: who and how? *Journal of Economic Literature* 51: 162–172. https://doi.org/10.1257/jel.51.1.162.

88 Kleinberg, J., Ludwig, J., Mullainathan, S., and Obermeyer, Z. (2015). Prediction policy problems. *American Economic Review* 105 (5): 491–495. https://doi.org/10.1257/aer.p20151023.

89 Bjorkegren, D., and D. Grissen. 2018. "Behavior revealed in mobile phone usage predicts loan repayment." Available at SSRN: https://ssrn.com/abstract=2611775

90 Goel, S., Rao, J.M., and Shroff, R. (2016). Precinct or prejudice? Understanding racial disparities in New York City's stop-and-frisk policy. *Annals of Applied Statistics* 10 (1): 365–394. https://doi.org/10.1214/15-AOAS897.

91 Glaeser, E.L., Kominers, S.D., Luca, M., and Naik, N. (2018). Big data and big cities: the promises and limitations of improved measures of urban life. *Economic Inquiry* 56 (1): 114–137. https://doi.org/10.1111/ecin.12364.

92 Nyman, R. and Ormerod, P. (2017). Predicting economic recessions using machine learning algorithms. arXiv preprint arXiv:1701.01428.

93 Wei, X., Chen, W., and Li, X. (2020). Exploring the financial indicators to improve the pattern recognition of economic data based on machine learning. *Neural Computing and Applications* 1–15. https://doi.org/10.1007/s00521-020-05094-0.

94 Raffinot, T., & Benoit, S. (2018). Investing through economic cycles with ensemble machine learning algorithms. Available at SSRN: https://ssrn.com/abstract=2785583.

95 Cheng, H.L. and Chen, N.K. (2021). A study of financial cycles and the macroeconomy in Taiwan. *Empirical Economics* 61 (4): 1749–1778.

96 Gu, S., Kelly, B., and Xiu, D. (2020). Empirical asset pricing via machine learning. *The Review of Financial Studies* 33 (5): 2223–2273. https://doi.org/10.1093/rfs/hhaa009.

97 Scheidegger, S. and Bilionis, I. (2019). Machine learning for high-dimensional dynamic stochastic economies. *Journal of Computational Science* 33: 68–82. https://doi.org/10.1016/j.jocs.2019.03.004.

98 Chowdhury, R., Mahdy, M.R.C., Alam, T.N. et al. (2020). Predicting the stock price of frontier markets using machine learning and modified Black–Scholes Option pricing model. *Physica A: Statistical Mechanics and its Applications* 124444. 10.1016/j.physa.2020.124444.

99 Hushani, P. (2019). Using autoregressive modelling and machine learning for stock market prediction and trading. In: *Third International Congress on Information and Communication Technology*, 767–774. Singapore: Springer https://doi.org/10.1007/978-981-13-1165-9_70.

100 Zhang, X.D., Li, A., and Pan, R. (2016). Stock trend prediction based on a new status box method and AdaBoost probabilistic support vector machine. *Applied Soft Computing* 49: 385–398. https://doi.org/10.1016/j.asoc.2016.08.026.

101 Miklosik, A., Kuchta, M., Evans, N., and Zak, S. (2019). Towards the adoption of machine learning-based analytical tools in digital marketing. *IEEE Access* 7: 85705–85718. https://doi.org/10.1109/ACCESS.2019.2924425.

102 Łukowski, W. (2017). The role of knowledge management in mobile marketing. *Marketing of Scientific and Research Organizations* 25 (3): 135–155. https://doi.org/10.14611/minib.25.09.2017.16.

103 Miklosik, A. and Evans, N. (2020). Impact of big data and machine learning on digital transformation in marketing: a literature review. *IEEE Access*. https://doi.org/10.1109/ACCESS.2020.2998754.

104 Fan, S., Lau, R.Y.K., and Zhao, J.L. (Mar. 2015). Demystifying big data analytics for business intelligence through the lens of marketing mix. *Big Data Research* 2 (1): 28–32. https://doi.org/10.1016/j.bdr.2015.02.006.

105 Campbell, C., Sands, S., Ferraro, C. et al. (2020). From data to action: how marketers can leverage AI. *Business Horizons* 63 (2): 227–243. https://doi.org/10.1016/j.bushor.2019.12.002.

106 Blei, D.M., Ng, A.Y., and Jordan, M.I. (2003). Latent Dirichlet allocation. *Journal of Machine Learning Research* 3: 993–1022.

107 Heimo, J. (2017). Marketer's guide to the algorithm–using machine learning to enhance marketing. *Journal of Marketing* 5: 8–616.

108 Choi, J.A. and Lim, K. (2020). Identifying machine learning techniques for classification of target advertising. *ICT Express* https://doi.org/10.1016/j.icte.2020.04.012.

109 Chen, Y., Kapralov, M., Canny, J., and Pavlov, D. (2009). Factor modeling for advertisement targeting. *Advances in Neural Information Processing Systems* 22: 324–332.

110 Addis, A., Armano, G., and Vargiu, E. (2009). Profiling users to perform contextual advertising. *Proceedings of the 10th Workshop dagli Oggetti agli Agenti (WOA 2009)* (Vol. 73).

111 Zhang, Y., Surendran, A.C., Platt, J.C., and Narasimhan, M. (2008). Learning from multi-topic web documents for contextual advertisement. In: *Proceedings of the 14th ACM SIGKDD International Conference on Knowledge Discovery and Data Mining*, 1051–1059. https://doi.org/10.1145/1401890.1402015.

112 Pepelyshev, A., Staroselskiy, Y., and Zhigljavsky, A. (2015). Adaptive targeting for online advertisement. In: *International Workshop on Machine Learning, Optimization and Big Data*, 240–251. Cham: Springer https://doi.org/10.1007/978-3-319-27926-8_21.

113 Li, H., Zhang, D., Hu, J. et al. (2007). Finding keyword from online broadcasting content for targeted advertising. In: *Proceedings of the 1st International Workshop on Data Mining and Audience Intelligence for Advertising*, 55–62. https://doi.org/10.1145/1348599.1348608.

114 Jiang, Z. (2016). Research on CTR prediction for contextual advertising based on deep architecture model. *Journal of Control Engineering and Applied Informatics* 18 (1): 11–19.

115 Wang, Q., Liu, F.A., Xing, S., and Zhao, X. (2018). A new approach for advertising CTR prediction based on deep neural network via attention mechanism. *Computational and Mathematical Methods in Medicine* 2018: https://doi.org/10.1155/2018/8056541.

116 Bin Tareaf, R., Berger, P., Hennig, P. et al. (2017). Identifying audience attributes: predicting age, gender and personality for enhanced article writing. In: *Proceedings of the 2017 International Conference on Cloud and Big Data Computing*, 79–88. https://doi.org/10.1145/3141128.3141129.

117 Ta, A.P. (2015). Factorization machines with follow-the-regularized-leader for CTR prediction in display advertising. In: *2015 IEEE International Conference on Big Data (Big Data)*, 2889–2891. IEEE https://doi.org/10.1109/BigData.2015.7364112.

118 Fan, T.K. and Chang, C.H. (2011). Blogger-centric contextual advertising. *Expert Systems with Applications* 38 (3): 1777–1788. https://doi.org/10.1016/j.eswa.2010.07.105.

119 Perlich, C., Dalessandro, B., Raeder, T. et al. (2014). Machine learning for targeted display advertising: transfer learning in action. *Machine Learning* 95 (1): 103–127. https://doi.org/10.1007/s10994-013-5375-2.

120 Jia, Q., Guo, Y., Li, R. et al. (2018). A conceptual artificial intelligence application framework in human resource management. In: *Proceedings of the International Conference on Electronic Business*, 106–114. https://aisel.aisnet.org/iceb2018/91.

121 Garg, V., Srivastav, S., and Gupta, A. (2018). Application of artificial intelligence for sustaining green human resource management. In: *2018 International Conference on Automation and Computational Engineering (ICACE)*, 113–116. IEEE https://doi.org/10.1109/ICACE.2018.8686988.

122 Marchant, R., Haan, S., Clancey, G., and Cripps, S. (2018). Applying machine learning to criminology: semi-parametric spatial-demographic Bayesian regression. *Security Informatics* 7 (1): 1–19. https://doi.org/10.1186/s13388-018-0030-x.

123 Wheeler, A.P. and Steenbeek, W. (2020). Mapping the risk terrain for crime using machine learning. *Journal of Quantitative Criminology* 1–36. https://doi.org/10.1007/s10940-020-09457-7.

124 Habernal, I., Ptáček, T., and Steinberger, J. (2013). Sentiment analysis in czech social media using supervised machine learning. In: *Proceedings of the 4th Workshop on Computational Approaches to Subjectivity, Sentiment and Social Media Analysis*, 65–74.

125 Ahmad, M., Aftab, S., Muhammad, S.S., and Ahmad, S. (2017). Machine learning techniques for sentiment analysis: a review. *International Journal of Multidisciplinary Sciences and Engineering* 8 (3): 27.

126 Wang, H., Can, D., Kazemzadeh, A. et al. (2012). A system for real-time twitter sentiment analysis of 2012 us presidential election cycle. In: *Proceedings of the ACL 2012 System Demonstrations*, 115–120.

127 Bifet, A. and Frank, E. (2010). Sentiment knowledge discovery in Twitter streaming data. In: *International Conference on Discovery Science*, 1–15. Berlin, Heidelberg: Springer https://doi.org/10.1007/978-3-642-16184-1_1.

128 Singh, P.K. and Husain, M.S. (2014). Methodological study of opinion mining and sentiment analysis techniques. *International Journal on Soft Computing* 5 (1): 11.

129 Ghiassi, M., Skinner, J., and Zimbra, D. (2013). Twitter brand sentiment analysis: a hybrid system using *n*-gram analysis and dynamic artificial neural network. *Expert Systems with Applications* 40 (16): 6266–6282. https://doi.org/10.1016/j.eswa.2013.05.057.

130 Breiman, L. (2001). Random forests. *Machine Learning* 45: 5–32. https://doi.org/10.1023/A:1010933404324.

131 Nassif, A.B., Shahin, I., Attili, I. et al. (2019). Speech recognition using deep neural networks: a systematic review. *IEEE Access* 7: 19143–19165. https://doi.org/10.1109/ACCESS.2019.2896880.

132 Bocklet, T., Maier, A., Bauer, J.G. et al. (2008). Age and gender recognition for telephone applications based on GMM supervectors and support vector machines. In: *2008 IEEE International Conference on Acoustics, Speech and Signal Processing*, 1605–1608. IEEE https://doi.org/10.1109/ICASSP.2008.4517932.

133 Vogt, T. and André, E. (2006, May). Improving automatic emotion recognition from speech via gender differentiation. In: *LREC*, 1123–1126.

134 Han, K., Yu, D., and Tashev, I. (2014). Speech emotion recognition using deep neural network and extreme learning machine. In: *Fifteenth Annual Conference of the International Speech Communication Association*.

24

Green Energy Using Machine and Deep Learning

R. Senthil Kumar[1], S. Saravanan[1], P. Pandiyan[2], K.P. Suresh[1], and P. Leninpugalhanthi[1]

[1]*Department of EEE, Sri Krishna College of Technology, Coimbatore, Tamil Nadu, India*
[2]*Department of EEE, KPR Institute of Engineering and Technology, Coimbatore, Tamil Nadu, India*

24.1 Introduction

Global electricity demand is expanding on a daily basis as a result of the rapid development of industries, commercial sectors, as well as the growth of the global population. Traditional electricity generators that rely on coal, fossil fuels, and oil are running out of fuel and will be unable to keep up with demand in the foreseeable future [1]. Thus, prior to exhausting non-renewable conventional energy sources, we must develop alternative energy sources to sustain industrial growth and the global economy. This challenge can be overcome by major renewable-energy sources or green energy, such as solar, wind, hydropower, geothermal, and biomass. Solar, wind, and hydropower are particularly important because of their abundant availability in nature, as well as the fact that they have no negative environmental consequences. As a result, these green-energy sources are the best alternative sources because they are pollution-free, low -cost, environmentally friendly, and less complicated. Figure 24.1 shows the advantages of green energy in terms of energy access, security and reliability, mitigation of environmental impacts, and economic development. Even though it has more advantages, efficiency-wise, it provides a low range due to a lack of maintenance and operation. It makes it more difficult for researchers working on renewable energy to analyze how to demonstrate the effectiveness of the system using emerging techniques. Furthermore, predictive maintenance and its challenging opportunities focus on making it more efficient and extending the life of energy generation.

Nowadays, machine learning (ML) and artificial intelligence (AI) play a significant contribution in the field of renewable-energy generation. Green-energy forecasting is the process of gathering both current and historical data on renewable energy in order to forecast the effectiveness of green-energy generation. Green energy includes solar energy, wind energy, and hydropower, which are mainly used to reduce climate effects, as well as to prevent global warming [2]. Hence, prediction is very essential, with different types of prediction methodologies that have been initiated to increase the generation in terms of energy predictions.

Renewable-energy sources get more benefits from machine-learning and deep-learning techniques in terms of analyzing data in a smart manner. The outcome of renewable energy is not constant due to climatic changes and the environment. To predict the amount of energy available from solar, wind, and hydropower based on every season as well as the exact time, machine learning provides recent, realistic data. This technique is also useful for determining the angle of solar panels, wind speed and direction, and the flow of water in a hydropower plant's pipes in order to intelligently increase the amount of generating power throughout the season. So, green-energy generation is one of the fields, which consist of more datasets, and researchers can implement machine learning to solve more challenging problems more easily. In this chapter, we will first go over the significance of green energy

Machine Learning Algorithms for Signal and Image Processing, First Edition.
Edited by Deepika Ghai, Suman Lata Tripathi, Sobhit Saxena, Manash Chanda, and Mamoun Alazab.

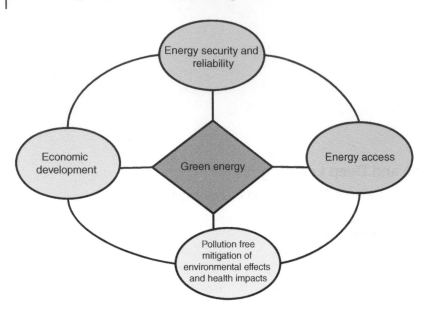

Figure 24.1 Advantages of green energy.

in detail using statistical data. Second, we discussed the types of machine-learning techniques that are used in green energy. Finally, we discussed deep-learning architectures for renewable energy with different methods.

24.1.1 Solar Energy

Solar energy is an abundant energy source among other renewable-energy resources viz. wind, biomass, hydro, etc. It is solar light energy that can be converted into electrical and thermal energy for use in a variety of domestic, industrial, and commercial applications. Solar energy is energy derived from the light and heat emitted by the sun's rays, which is a very clean and abundant renewable source, making it green energy. Global electricity demand has increased in recent years [2], but conventional sources such as oil and fossil fuels are limited, resulting in severe energy demand and global warming. These factors make researchers get into these crises and look for an alternative to green energy.

Earth receives an enormous amount of energy from the sun, which is almost 8000 times more than the total demand in the world [3]. The availability of the sun's light is approximately 300 days out of 365 days in a year, particularly in India. As a result, the intensity of the sunlight falls evenly on the earth's surface, regardless of location. Figure 24.2 depicts the key steps involved in harnessing solar energy.

(a) Photovoltaic (PV) cells
(b) Concentrating solar power

The sun's light is converted into electric energy as DC power by using PV cells. It is a type of harvesting that operates on the basis of PV effects. The solar cell consists of many layers of semiconductor materials, mainly p-type and n-type layers. It is possible to obtain PV cells in a variety of configurations, the most common of which are monocrystalline silicon cells, polycrystalline silicon cells, and thin-film technology. Concentrating solar power is nothing but the generation of thermal power generation.

24.1.1.1 Photovoltaic (PV) Cell

The solar PV cell is used to convert solar energy into electric energy with nonlinear output characteristics due to ambient environmental conditions, such as irradiation and climatic changes [4]. Practically, groups of PV cells are connected to get a higher amount of power for each and every application. When a group of PV cells is connected

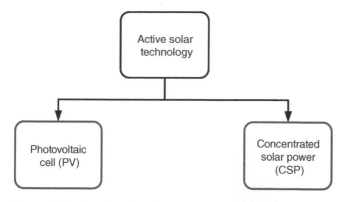

Figure 24.2 Classification of solar energy technology.

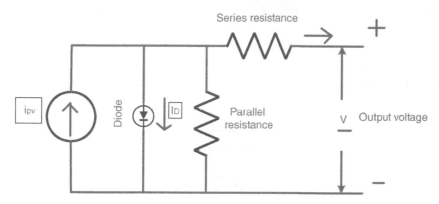

Figure 24.3 Equivalent circuit of single PV cell.

in series, it produces a high amount of output voltage; when they are connected in parallel, it produces a higher amount of output current. This type of group combination of PV cells is called a PV array. Figure 24.3 depicts the equivalent circuit of a single PV cell.

(i) **PV Module:** A single solar cell may be able to generate very little power, which is insufficient to meet any load on the application side. Furthermore, the efficiency of up to 25% is dependent on the type of cell used in manufacturing. So, to compensate for the high-power demand, we have to interconnect the groups of cells either in a series pattern or in parallel. Finally, the group of cells is nothing but a solar PV module, and the group of modules is nothing but a solar PV array. Figure 24.4 shows the pictorial representation of the PV cell, array, and module.

(ii) **PV System Integration:** Solar-grid integration is the most essential technology nowadays, which improves energy sharing, energy balance, and reduces the consumption of non-renewable energy. This integration technology should be used on both the generation and utility sides. PV system integration is mainly classi-fied into two sections. There are two types of systems: on-grid and off-grid. Off-grid and on-grid systems are also known as standalone and grid-connected PV systems, respectively. The off-grid system is best suited for use in remote areas, where a grid cannot be built. In this standalone system, an inverter is essential to convert DC–AC for residential applications.

To balance the energy demand on the power grid, a grid-connected PV system is preferably used in large solar power plants for excess power generation via PV cells or concentrated PV plants. This technology is not easy to integrate, and it should be installed with more care and caution. Power converters, such as the DC–DC boost converter and the DC–AC converter for AC applications, also play an important role in PV system integration [4]. Solar PV power is fed to the normal power grid effectively through the point of common coupling (PCC) for perfect

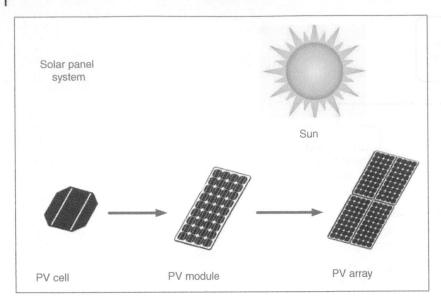

Figure 24.4   Solar PV cell, PV module, and PV array.

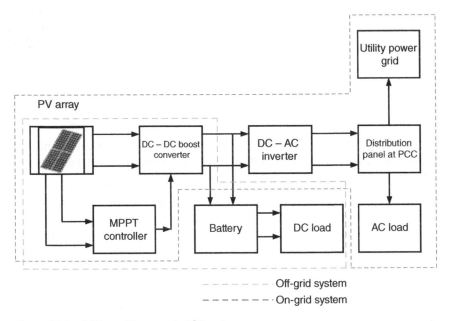

Figure 24.5   Off/on-grid-connected PV system.

synchronization of the electrical parameters such as voltage, current, and frequency. Figure 24.5 shows the standalone PV system and the grid-connected PV system with load and power conversion blocks.

24.1.2   Wind Energy

Wind energy plays an important role among the traditional renewable-energy sources, which limit environmental impacts such as the emission of carbon dioxide, sulfur dioxide, and some other toxic gases. Global warming and

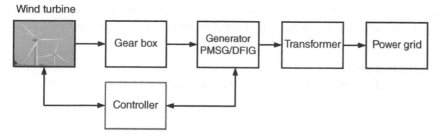

Figure 24.6 General block diagram of WECS.

nuclear power plant exhaust waste. Wind energy is pollution-free, very clean, eco-friendly, and seasonal [5]. In the past two decades, wind energy power generation has rapidly developed all over the world. In addition, wind power helps us to balance the power demand globally when fossil fuels tend to reduce production.

24.1.2.1 Wind Energy Conversion System (WECS)

The wind energy conversion system (WECS) principle is widely used in all windmills, which mainly consists of an interconnected aerodynamic turbine, gear setup, generator, controller section, transformer, and power grid. The general conceptual diagram of the WECS is depicted in Figure 24.6. An aerodynamic wind turbine (WT) is used to convert air to rotational mechanical energy based on the speed of the wind. A coupling connects the gearbox between the turbine and the generator to increase rotation. Hence, the rotating energy is transmitted to an electrical generator.

When compared to the doubly fed induction generator (DFIG), the permanent magnet synchronous generator (PMSG) is more commonly used in wind turbines due to the lower reliability of the gearing system in DFIG and less maintenance with higher efficiency achieved in PMSG [6]. The wind speed is not at the rated level (minimum speed). The aero turbine does not rotate, which is interconnected with the controller. The main purpose of the centralized controller is to sense the speed of the wind, the direction of the wind, and the torque. There are different control strategies with emerging techniques available to sense wind parameters for extracting maximum power. The fluctuations in generating output power are also controlled by supervised fuzzy control methods, as well as proportional integral (PI) and PID controllers.

24.1.2.2 Basic Equation of Wind Power

Generally, the wind speed is not constant, and it always changes in each area based on its geographic conditions and climatic changes [6]. As a result, the basic equation of wind power includes air, velocity, density, and air flux. It can be stated as kinetic energy given in Eq. (24.1).

$$K_e = \frac{1}{2}mv^2 \tag{24.1}$$

The mass of air is considered as a mass flow rate in mechanics,

m = air density × air volume

Hence, power is kinetic energy, so the equation (Eq. (24.2)) represents the fundamental equation of wind energy.

$$P = \frac{1}{2}\rho A v^3 \tag{24.2}$$

24.1.2.3 Wind Energy Site Selection

In wind energy conversion, systems continuously provide a high range of output power depending on the location preferred, where the wind availability is mostly based on the annual history of that particular area [7]. Therefore, site selection is critical for wind energy harvesting. The most suitable location could be selected in an area where the seasonal wind speed is almost constant or more than in the normal range. The anemometer is used to assess the potential of wind availability before site selection. Some important considerations to be followed in this WECS are as follows.

- The average wind speed is expected to be high.
- Historical wind data about that site should be satisfied based on the anemometer
- Wind curve at the selected site
- Wind structure at the selected site
- Local ecology – rough surfaces
- Low land prices

24.1.3   Hydropower

The electric energy which is generated from the flow of water is known as hydropower. The most cost-effective power generation among renewable-energy sources is hydropower. According to the international renewable energy agency (IRENA), India has 45 895 MW of installed hydropower capacity, which is more useful for meeting power demand. The contribution of hydropower plants is about 20–22% of overall electricity demand. Hydroelectricity has a relatively low investment cost and is a clean energy source [8].

24.1.3.1   Working Principle of Hydropower Plant

The main components of the hydropower station are the storage dam, control gate, penstock, water turbine, and generator. It works under the basic principle of water flow driving the water turbine. Hydroelectric power plants are classified into two types:

- Hydropower plant with dam and reservoir
- Hydropower plant without dam and reservoir

The dam/reservoir is necessary for hydropower plants at a suitable height at the level of high-water level, and the water has high-potential energy. When the control gate is opened, water with a high potential enters the turbine via the penstock. Due to the gravitational force of water, the high potential has to be converted into kinetic energy. Hence, the flow of water makes the turbines rotate continuously, which is interconnected with the generator through the water turbine shaft. This mechanism drives the generator and then electricity is generated. Figure 24.7 shows the general block diagram of the hydropower plant.

The hydropower plant is without any dam or reservoir. The normal river flow with a slight slope is identified and directs the water to the penstock. The efficiency of hydropower plants is almost 75–80% when compared to other renewable-energy sources, such as solar, wind, and biomass. The life cycle of a hydropower plant is almost 50 years.

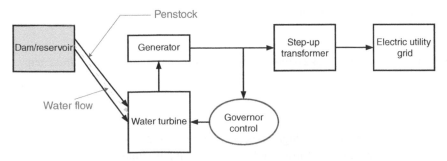

Figure 24.7   General block diagram of hydropower plant.

24.2   ML Algorithms for Green Energy

ML techniques have been widely used in a variety of industries, including data-driven challenges. Statistics, artificial neural networks (ANNs), optimization, mathematics, and data mining are all examples of ML methods. The ML approaches attempt to find the relationships between both input and output data without prescribed forms of problems. After implementing ML approaches, the decision makers can get satisfactory estimated output values by putting the predicting input data into the perfect trained models.

The ML approach primarily employs three types of learning: supervised, unsupervised, and reinforcement learning. Classified data is preferred in the training and development of supervised learning. Unsupervised learning, on the other hand, entails automatically inputting data into data based on specific standards. The number of clusters is determined by using clustering criteria. Learning through contact with the external surroundings to gather feedback in order to optimize the predicted advantages is known as reinforcement learning.

24.2.1   Forecasting Renewable-Energy Generation Using ML

The prediction of the power production of renewable energy is critical since it is dependent on a number of inhuman control elements, like environmental conditions. On the basis of energy resource used, the power plant exposes specific features that allow ML techniques to be used for prediction [9]. In this section, how the various ML methods can be used in solar, wind, and hydropower plants is discussed.

24.2.1.1   Solar-Energy Generation

Solar PV systems are used in a wide range of applications, from single-family homes to large PV power plants with a capacity of 1–100 MW. Because solar PV has long been used in small-scale household applications, and many research studies utilizing ML approaches to estimate PV performance have been conducted in recent years.

The ANN-based power prediction in solar PV systems is discussed in [10]. The accomplishment of the system was determined by the amount of usable energy recovered and the rise in the temperature of the water stored. The ANN is trained using productivity data from four different kinds of systems, all of which use the same collector panel and are in different weather states. The ANN outputs are the useable energy extracted from the system and the rise in water temperature. The network model of 7 input neurons, 2 output neurons, 24 hidden neurons, and a sigmoid transfer function is preferred.

The location forecasting models for solar PV systems based on meteorological factors using ML techniques are represented in Ref. [11]. Several regression approaches were utilized, including the least square support vector machines (LSSVMs) with kernel functions. The support vector machine (SVM) model performed well when compared to others, by up to 27%, in terms of accuracy. A quadratic least-square regression model was used for prediction with seven meteorological factors, and the results show that the root mean square error (RMSE) for the validating and estimation sets is 165 and 130 W/m^2, respectively.

The hybrid intelligent classifier for six-hour solar-energy forecasting is discussed in Ref. [12]. The system employed an ensemble approach that included seven commonly employed regression models: linear regression (LR), radial basis function (RBF), additive regression (AR), simple linear regression (SLR), locally weighted learning (LWR), SVM, and MLP. According to their findings, the three most effective regression models in terms of mean absolute percentage error (MAPE) and mean absolute error (MAE) are LMS, multi-layer perceptron (MLP), and SVM.

24.2.1.2   Wind Power Generation

There are many factors influencing wind power production, and the output power of the wind turbine could be estimated using Eq. (24.3). Where "*A*" denotes the area enclosed by the blades of the wind turbine (WT), "*V*" refers to the wind velocity, ρ denotes the density of the air, and "C_p" is the efficiency determinant that was set by the designer.

$$P = \frac{\rho A v^3 C_p}{2} \tag{24.3}$$

The power output is related to the wind speed, and normally, the wind speed is a key component in the above equation. It was also discovered that there is a cut-off speed after which the power output remains constant. Temperature and humidity have an effect on air density, which in turn has an effect on power production. As a result, it's critical to anticipate these variables and, eventually, a wind farm's total power output.

The two primary types of prediction models are physical and descriptive statistics models [13]. The long-term prediction is mainly handled with physical models, whereas the short- and medium-term forecasting is usually accomplished with statistical models. Statistical models are more directly connected with ML approaches. The auto-regressive integrated moving average (ARIMA) and auto-regressive moving average (ARMA) models are given for wind velocity prediction and then wind power prediction.

In Ref. [14], the Kalman filter approach is used to predict wind speed, with the state variables serving as the wind speed. The obtained model is preferable for the online prediction of wind velocity and power production. The forecasting of online power production is more significant because it gives the latest future forecasts, which can subsequently be utilized for power grid operation.

The comparative analysis of the ANN and ARIMA models for wind speed prediction is illustrated in Ref. [15]. Their findings revealed that the seasonal ARIMA model outscored the ANN model in terms of predicting accuracy, but that the performance of the ANN model might be increased by increasing the number of training vectors. The average hourly wind speed is calculated from the consecutive forecasts, and the electricity produced by the wind turbine is calculated.

The recurrent MLP model is a subtype of ANN that uses a Kalman filter based on the back-propagation network (BPN) [16]. This model is better at predicting long-term energy production than at predicting short-term power generation.

The ML can be used to forecast using fuzzy models [17]. The developed fuzzy model with a two-dimensional technique is used to forecast wind power generation. The presented model works well on flat terrain WT, but not so well on deteriorating terrain WT. This might be attributed to variations in wind speed as a function of tower height above ground level, as well as variances in air quality.

The numerical weather forecasting (NWF) models were utilized for wind power prediction [18]. The selection of an appropriate NWF model is critical in this technique since its accuracy is highly dependent on the initial NWF model. An aggregate technique was developed in Ref. [19] to minimize the influence of a single NWF model. The aggregate model enables the use of the original NWP with varied characteristics, such as initial states, physical characterization of the sub-grid system, or parameter estimation systems. It may also use a variety of NWF models to create the final ensemble learning.

The comprehensive overview of various ML approaches for predicting wind power production is represented in Ref. [18]. The random forest SVM, regression trees, ANN, and MLP are mentioned as the most preferred ML algorithms in the field of wind power.

24.2.1.3 Hydro Power Generation

The most commonly utilized and well-established renewable-energy source is hydropower. Several third-world nations rely on collecting energy from their accessible sources of water due to its features and economic feasibility. Hydropower employs moving water or stored water sources that are dependent on local rainfall. It is clearly influenced by non-human-controlled meteorological conditions that must be anticipated for effective scheduling and management.

The hybrid of recurrent SVM and ANN to anticipate the rainfall depth values were described in Ref. [20]. Furthermore, the variables for the SVM model were chosen using the chaotic particle swarm optimization (PSO) algorithm. For the validation, training, and testing sets, the model's performance was 0.40, 1.159, and 0.375, respectively, in terms of normalized mean square error (NMSE).

A periodic decomposition-based LSSVM method with a supervised learning model for hydroelectric energy utilization prediction is represented in Ref. [21]. The initial time series data was divided into geographical factors and irregular elements, and then the LSSVM analysis was performed. The periodic factor, trend cycle, and irregularity element were predicted using this developed LSSVM model, which was then inputted into another LSSVM to aggregate the prediction results.

The genetic algorithm (GA) for supporting optimum governor setting in hydropower plants is discussed in Ref. [22]. The GA was investigated as a mechanism for optimizing the gains of PI controllers in an adaptive manner. This approach is used for adjusting to changing plant factors, like load self-regulation.

An ANN-based analysis and control for both governor and exciter in a micro hydropower plant are represented in Ref. [23]. Their design, which is based on self-organization and the prediction estimating properties of the ANN, was implemented using the cluster-wise segment with the associative memory technique.

24.3 Managing Renewable-Energy Integration with Smart Grid

Users expect more effective and efficient processes and management of the grid as fast advancement in the power utility grid persists to create it smarter. As more people become involved in the electricity grid, the control of a large network becomes increasingly difficult. As a result, smarter approaches are needed to effectively control the smart grid. In this section, some of the challenges that electricity grids face, such as supply–demand balancing, grid management and operation, as well as a proposed ML solution to them.

24.3.1 Supply–Demand Balancing

When a power system is connected to renewable-energy resources, it is much more important to accurately forecast energy output and demand. With changes in the output of solar and wind power, as well as irregular and unpredictable power production, vulnerabilities are created in the power system, which weakens the grid. Hence, it is important to integrate the existing power generation on time, disconnect failing wind power plants, and employ smoothing methods for solar PV plants and grid connections to preserve the stability of the grid. The different ML approaches were used to detect the variables impacting the stability of the grid and to assure its proper maintenance.

The system generates a predictive model that focuses on the flexibility of energy supply and demand, assisting in the management of consumption and production on a smart grid powered by renewable-energy resources [24]. Additionally, the prediction model can rapidly examine new energy data in order to detect future changes in energy consumption. The model employs a mix of well-known techniques such as SVM and cluster learners. It utilizes various models at different time scales to achieve supply and demand in the time domain.

The smart grid will be made up of many separate autonomous components viz. smart homes, smart vehicles, and smart offices. Due to the fact that their requirements and lifestyles fluctuate over time, these demand-side users generate changing demand on the grid. The smart grid also prefers deregulated pricing operations at a variety of levels, and demand is driven by per-grid price rules that are specific to certain grids. The deregulated market provides users with flexibility, enabling them to negotiate for the energy that they require.

24.3.2 Grid Management and Operations

The summary of several ML approaches used for grid management and operations is described in Ref. [25]. The system included a variety of models for predicting at various levels throughout the whole power grid. Electricity providers may also prefer these models to assist with repair and maintenance work. A variety of specific variants of the suggested method are used to provide the feeder deterioration rankings, the feeder mean time between fault predictions, and the manhole events vulnerability rankings. In their model, the researchers used over 300 characteristics generated from time series analysis data and related factors. The models were built using classification and regression trees, SVM, and supervised learning approaches, including statistical methods and random forests.

24.3.3 Grid-Data Management

Smart grid implementations grow in the number of users, and the information must typically be shared across many stakeholders [26]. Many users created a lot of data that needed to be shared with interested parties in order to make better smart grid executive decisions. As a result, efficient and reliable ways of sharing smart grid data are necessary.

The model-based statistical modeling approach should be utilized for smart grid data, and it has the following benefits. The model-based approach has a small computation time while retaining all of the necessary data. It's also a useful tool for communicating data because it doesn't involve the sharing of raw data, but rather the representation of a model, which allows for faster computation when compared to retrieving information from raw data.

24.4 DL Models for Renewable Energy

A subset of Artificial Intelligence (AI) approaches is Deep Learning (DL). A computer can be used in the AI approach to perform difficult tasks like modeling, learning, or extracting features from experimental data. It is also known as the soft-computing approach. The subcategories of AI techniques are ANN, GA, fuzzy inference system (FIS), and analogous methods evolved from these methods [27]. AI approaches are used in all sectors of science to aid in the organization and investigation of datasets, particularly large datasets. These strategies are actually employed as a single or hybrid method. Each one has its own set of benefits and drawbacks. The processing, training, decision, and result parts are the four primary components of this procedure.

The primary goal of this section is to analyze DL architectures in the non-conventional energy sector (green/renewable/sustainable energy) and to assess DL performance in the solar and wind energy field. In addition, the performance of single and hybrid DL techniques on the same dataset is studied and compared. In the case of single techniques, AI techniques such as ANN, GA, P & O, BPN, FIS, PSO, or other AI technique is used simply for modeling, examining, or forecasting green-energy data [28]. One of the key benefits of AI approaches is that they produce results quickly and have an uncomplicated structure. The only drawback of these strategies is their limited accuracy when dealing with large datasets.

Hybrid techniques refer to the use of multiple AI techniques or the optimization of an AI technique using other learning models. This approach can aid in overcoming the drawbacks of employing single methods in large datasets and enhance the performance of the system. As a result, the key advantages of hybrid approaches to large green-energy datasets are their scalability, adaptability, and reliability. Pre-processing, training, and testing are three significant aspects of the AI approach's fundamental blocks. The AI technique procedure begins by importing datasets. Next, the data is pre-processed in order to be segregated into two major groups: training datasets and testing datasets, which are the pre-processing section's results.

The raw data is transformed into a comprehensible form in the pre-processing stage in order to continue the procedure. As a result, the model or network is built using the training data. The training data is then loaded into the training block of the program. The AI techniques learn the model using training data. The training outcomes are imported into the decision stage, in order to measure the correctness of the model generated by using comparison variables such as MAE, MSE, *R*, RMSE, or MAPE in the training phase. The next step is to assess the prediction performance using the testing data.

These variables compare the outputs of the composed model against actual values to assess how similar they are. The procedure proceeds to the pre-processing stage if the estimated error values are within the permissible range. In this step, the testing data is fed into the selected data and the results are used in the detection, diagnosis, or prediction of the output. If the estimated error exceeds the permitted range, the procedure returns to the training phase to select a new algorithm or training strategy to lower the anticipated error. Repeat this procedure until the appropriate error range between the actual and output values is attained.

24.4.1 Solar Energy

This is a substantial source of energy that is used in a wide range of applications, including fuel generation, heating, agriculture, transportation, and horticulture [29, 30]. This is a green-energy resource and one of the research fields with a vast dataset. The use of AI approaches to a range of problems is a prominent topic of research for scholars. Several issues affect the process of estimating, optimizing, making decisions, and policymaking systems, according to data collected on the large amount of big data used in the solar-energy sector. There is also the necessity for careful investigation in order to accomplish a green-energy system. As a result, a method with high accuracy is required. For these goals, there are a variety of AI techniques that can be used. To develop a procedure with greater accuracy, a trial and error procedure may be used. Large datasets in solar-energy systems lend themselves to the application of precise and long-term methodologies such as DL techniques.

24.4.2 Energy from the Wind

Wind energy is another green-energy source that may be used to meet load demand in remote places and cities [31]. This is a non-renewable, clean, and environmentally beneficial energy source that is widely available in most countries [32]. The main reason for establishing various methods in wind energy generation is similar to the reasoning for developing various methods in the solar-energy sector. Additionally, both the manufacturing and deployment of laboratory equipment in wind energy are costly. As a result, modeling approaches for estimating, optimization, and prediction can be a helpful step toward achieving greater goals, provided the method is accurate. In order to meet these targets, the development of DL technologies could be an important step.

24.4.3 Techniques of DL

The DL approach is a subset of ML approach. WNN, RNN, DBN, RWNN, SAE, LSTM, MRBM, RMT, DNN, and convolutional neural network (CNN) are some of the most common and widely used deep-learning algorithms, which are briefly discussed in this section. These techniques can be applied to image/audio/video processing, natural language processing, computer vision, game creation, machine translation, and social network filtering, as well as applications that generate or utilize a significant amount of data or depend on a big data aspect [33]. DL approaches have recently been used to solve traditional artificial-intelligence challenges. The ideas of deep learning approaches and their architectures are discussed in general terms in this section.

24.4.3.1 Convolutional Neural Network (CNN)

When learning on CNNs, many layers must be trained in order to obtain good performance. This approach is widely used in a variety of applications, including computer vision. In order to maximize the effectiveness of each CNN, a two-stage training process is used, which includes the back-propagation and feed-forward phases. In the initial step, the input data is fed into the network. The product of the input data and each neuron's parameters is then multiplied with a convolutional operation to compute the next layer's parameters. After computing the network output, the network performance is derived. The network outputs must be compared against the correct response using a loss function in order to calculate the error.

Based on the computed error, the back-propagation phase begins at the next stage. In this stage, the chain rule is used to compute the gradient of each parameter, and according to the effects of the error caused in the network, all parameters are modified [34]. The feed-forward is the following step after the parameters have been adjusted. The network training session comes to an end once the correct number of these steps has been completed.

CNN technique is proposed in Ref. [35] that used cloud movement data collected from four different cardinal directions in order to estimate four different wind fields. Based on the findings, this method of establishing grid sensor networks for wind field assessment was both accurate and cost-effective. Sun et al. [36] devised a method for estimating solar PV output using the CNN technique with sky images. On the basis of this assumption, images of the sky might be used to determine the amount of clouds covering over the sun's location. The datasets were divided into three categories: sunny, overcast, and partly cloudy. PV panels were installed on the roofs of two structures which were evaluated using the RMSE method.

To estimate probabilistic wind power, a hybrid WPF–CNN was proposed for wind farm data collected from China [37]. The SVM-QR and BP-QR algorithms were used as benchmarks. The constructed models were evaluated throughout four seasons. The CPRS value was used as a criterion for evaluation. The increase in power supplies improved CPRS in all strategies, with the exception of the proposed methodology. In general, the proposed approach exhibited the lowest CPRS when compared to other techniques. This demonstrated the suggested technique's significant potential for estimating wind power. This also lowered the method's inaccuracy and improved its accuracy.

24.4.3.2 Restricted Boltzmann Machine (RBM)

Restricted Boltzmann machine (RBM) is a Boltzmann machine with the drawback of producing a bipartite graph with hidden (H) and visible (V_1) units. As a result of this constraint, more efficient training methods, such as the gradient-based approach [36], have been developed. Due to the bipartite nature of this graph, the hidden units (H) and visible units (V_1) in this model are conditionally independent. Boltzmann's distribution can be found in Eq. (24.4) for both H and V_1.

$$P(HV_1) = P(H_1V_1)P(H_2V_2)\dots P(H_nV_1) \tag{24.4}$$

where $P(HV_1)$ can be used to get V_1. As a result, V_2 can be accessed via $P(H_2V_1)$. The difference between V_1 and V_2 can be minimized by specifying the parameters and the generated H as the best feature of V_1. RBMs can be used as learning modules to aid in the development of DBMs, DEMs, and DBNs. The lower levels of DBNs have directed interconnections, while the upper two layers of an RBM contain undirected connections. DBMs have connections between all network layers that are unmatched. Deterministic hidden units are found in lower layers of DEMs, while stochastic hidden units are found in the top layer.

The DBN model is a theoretical potential generator of a certain probability distribution that may be applicable to observable data. DBN implements an effective layer-by-layer learning strategy for parameter initialization, and then carefully adjusts all weights and expected outputs simultaneously. There are two benefits to the greedy learning technique [38]:

- Assures proper network initialization, since parameter selection can be problematic, resulting in an insufficient selection of local optima;
- It allows for the use of an unsupervised learning model without a class label because it eliminates labeled training data.

The DBN model's calculation process is costly as numerous RBM training is necessary and the possibility of learning to optimize the model is unclear [39]. DBNs have successfully steered researchers into DL, resulting in the creation of a diverse range of species [40].

DBM is one of the Boltzmann family's subsets. DBMs, in practice, have many layers, each with a hidden unit corresponding to the even-numbered individual layer number, and vice versa. It is impossible to estimate the posterior distribution of hidden units when entities are present, even if they are invisible. DBM teaches all the layers of a specified unsupervised model, including the probability boundaries, during network training. When employing the MCMC technique, this means that just one or more updates are made. To avoid local weak minima that leave many hidden units inactive, a greedy layer-based education strategy was utilized on layers in the pre-training DBM network, similar to what was done in DBN [39].

The integrated training has the potential to enhance feature-rich learners' capabilities as well as their proficiency. The complexity of the approximation inference in DBMs is substantially larger than in DBNs, which is a critical flaw. As a result, DBM settings for huge datasets have also been improved. A simplified solution has been introduced to improve DBMs' efficiency. To initialize the values of the latent variables, this approach employs a recognition model in all layers. In order to meet the requirements, updates can be performed in the pre-training stage or at the beginning of the training stage.

The LSTM approach is also one of the techniques for forecasting problems. Because most development studies combine or compare LSTM approaches. The RNN technique, which employs temporal information from the input data, is the foundation of the LSTM technique. Memory cells benefit from LSTM since it is a unique neuron structure that can store information for an extended period of time. The controlling of input and output values of a neuron's memory cell has been done with input, output, and forgetting gates. Activation functions are performed by each gate in response to the input neuron [41].

24.4.3.3 Auto-Encoder

Auto-encoder (AE) is a type of ANN that is used to increase learning encoding [42]. An AE will be learned in order to reconstruct one's original input X, as opposed to training the network to predict the target value of Y in exchange for the initial input X. As a result, the output vectors will have the same dimensions as the input vectors. By minimizing reconstruction errors, the AE is optimized.

A single layer, in general, is incapable of receiving various aspects of raw input. Researchers are currently working on the transfer of code from one AE to another in order to complete their tasks, and they are employing a deep AE to do this. A type of back-propagation operation is frequently used to train AE. Although the majority of this architecture is efficient, it can be exceedingly inefficient if the earliest levels contain errors. This difficulty can be solved by pre-training the network with initial weights that are close to the final answer [43].

In Ref. [44], the researchers used a novel hybrid SAEC Rough Regression approach to estimate the wind velocity. This framework aided in the development of a dependable DNN for generating output values that employ the RMSE as a performance metric. To demonstrate the suggested method's capabilities, its results were compared to that of FFNN, PR, TDNN, SAE, and NARNN, in five-time steps: 10-minutes, 30-minutes, 1-hour, 2-hour, and 3-hour. As a result of the increased number of datasets, raising the time step increased the RMSE value. In comparison to the other strategies, the proposed method exhibited the lowest RMSE value over all time steps. As a result of this accomplishment, efforts to develop a new hybrid strategy have been made. Additional findings were discovered as a result of this investigation. When comparing single techniques, SAE has the lowest RMSE among the other single processes (TDNN, PR, FFNN, TDNN, and NARNN). Despite this, the recommended technique was a combination of the two. The findings suggest that hybrid approaches are more accurate than single-mode methods at predicting data.

Hybrid SDAE approach is proposed in Ref. [45] to estimate the wind velocity. Four different wind farms located in Jilin, Ningxia, Gansu, and Inner Mongolia were each hooked up to four sets of sensors and collected 10-minute, 30-minute, 1-hour, and 2-hour historical wind speed data for a total of seven sets of data points (0.5, 1, 2, 3, 4, 5, and 6 months). Using MAPE and MAE, the suggested method's results were compared to single DNN, ELM, and SVR. In all situations, the suggested hybrid technique had the lowest MAPE and MAE, suggesting that it outperformed single techniques. Qureshi et al. [46] used AE based on transfer learning (TL) and the meta-regression

(MR) technique to estimate the wind velocity. MAE, SDE, and RMSE were the performance metrics. Power measurement data and Meteorological for five wind farms in Europe were collected over three years. For each farm, the suggested method's performance was compared to that of SVR (linear/RBF kernel) and GPeANN. The adopted DNN–MRT technique had the lowest SDE, RMSE, and MAE, followed by GPeANN, on average. GPeANN and DNN–MRT were both hybrid methods, whereas SVR was a single approach. The accuracy of hybrid procedures was much higher than that of single methods, according to the findings. As a result, their findings revealed two key points: the relevance of hybrid approaches and DL techniques' high accuracy in huge datasets when compared to other AI techniques.

24.5 Conclusion

This chapter presents details of a modern renewable-energy prediction model based on ML and DL techniques. The ML approach employs the load from the previous period of time as the training set, builds an appropriate network model, and then trains the network to satisfy the accuracy criteria using a specific training method. The DL has several hidden layers that are called MLP. It combines low-level characteristics to create more complex high-level features or characteristics to learn about the intrinsic nature of the input model. This chapter explains the various data pre-and post-processing approaches that can help to enhance the accuracy of ML and DL predictions. Then, some of the challenges and future research areas of ML and DL-based prediction models for green energy are discussed. This chapter fulfills the existing gaps in order to uncover the ML and DL applied to renewable-energy prediction.

References

1 Sampaio, P.G. and González, M.O. (2017). Photovoltaic solar energy: conceptual framework. *Renewable and Sustainable Energy Reviews* 74: 590–601.

2 Akrami, M. and Pourhossein, K. (2018). A novel reconfiguration procedure to extract maximum power from partially-shaded photovoltaic arrays. *Solar Energy* 173: 110–119.

3 Wu, Y., Zhang, B., Xu, C., and Li, L. (2018). Site selection decision framework using fuzzy ANP-VIKOR for large commercial rooftop PV system based on sustainability perspective. *Sustainable Cities and Society* 40: 454–470.

4 Senthil Kumar, R., Gerald Christopher Raj, I., Suresh, K.P. et al. (2021). A method for broken bar fault diagnosis in three phase induction motor drive system using artificial neural networks. *International Journal of Ambient Energy* 1–7.

5 Asadi, M. and PourHossein, K. (2019) Wind and solar farms site selection using geographical information system (GIS), based on multi criteria decision making (MCDM) methods: a case-study for East-Azerbaijan. *2019 Iranian Conference on Renewable Energy & Distributed Generation (ICREDG)*, Tehran, Iran, 11–12 June 2019.

6 Höfer, T., Sunak, Y., Siddique, H., and Madlener, R. (2016). Wind farm siting using a spatial analytic hierarchy process approach: a case study of the Städteregion Aachen. *Applied Energy* 163: 222–243.

7 Wu, Y., Zhang, T., Xu, C. et al. (2019). Optimal location selection for offshore wind-PV-seawater pumped storage power plant using a hybrid MCDM approach: a two-stage framework. *Energy Conversion and Management* 199: 112066.

8 Williamson, S.J., Stark, B.H., and Booker, J.D. (2014). Low head pico hydro turbine selection using a multi-criteria analysis. *Renewable Energy* 61: 43–50.

9 Perera, K.S., Aung, Z., and Woon, W.L. (2014). Machine learning techniques for supporting renewable energy generation and integration: a survey. In: *Data Analytics for Renewable Energy Integration* (ed. W.L. Woon, Z. Aung, S. Madnick), 81–96, Springer International Publishing.

10 Kalogirou, S.A., Panteliou, S., and Dentsoras, A. (1999). Artificial neural networks used for the performance prediction of a thermosiphon solar water heater. *Renewable Energy* 18 (1): 87–99.

11 Sharma, N., Sharma, P., Irwin, D. et al. 2011. Predicting solar generation from weather forecasts using machine learning. *2011 IEEE International Conference on Smart Grid Communications (SmartGridComm)*, Brussels, Belgium, 17–18 October 2011.

12 Hossain, M.R., Oo, A.M., and Ali, A.B. (2013). Hybrid prediction method for solar power using different computational intelligence algorithms. *Smart Grid and Renewable Energy* 4 (1): 76–87.

13 Lei, M., Shiyan, L., Chuanwen, J. et al. (2009). A review on the forecasting of wind speed and generated power. *Renewable and Sustainable Energy Reviews* 13 (4): 915–920.

14 Li, F., Wu, X., and Zhu, Y. (2014). Wind speed short-term forecast for wind farms based on ARIMA model. *Industrial Electronics and Engineering* 13: 151–162.

15 Sfetsos, A. (2000). A comparison of various forecasting techniques applied to mean hourly wind speed time series. *Renewable Energy* 21 (1): 23–35.

16 Li, S. (2003). Wind power prediction using recurrent multilayer perceptron neural networks. *2003 IEEE Power Engineering Society General Meeting (IEEE Cat. No.03CH37491)*, Toronto, Canada, 13–17 July 2003.

17 Damousis, G., Alexiadis, M.C., Theocharis, J.B., and Dokopoulos, P.S. (2004). A fuzzy model for wind speed prediction and power generation in wind parks using spatial correlation. *IEEE Transactions on Energy Conversion* 19 (2): 352–361.

18 Foley, A.M., Leahy, P.G., Marvuglia, A., and McKeogh, E.J. (2012). Current methods and advances in forecasting of wind power generation. *Renewable Energy* 37 (1): 1–8.

19 Al-Yahyai, S., Charabi, Y., Al-Badi, A., and Gastli, A. (2012). Nested ensemble NWP approach for wind energy assessment. *Renewable Energy* 37 (1): 150–160.

20 Hong, W.C. (2008). Rainfall forecasting by technological machine learning models. *Applied Mathematics and Computation* 200 (1): 41–57.

21 Wang, S., Tang, L., and Yu, L. (2011). SD-LSSVR-based decomposition-and-ensemble methodology with application to hydropower consumption forecasting. *2011 Fourth International Joint Conference on Computational Sciences and Optimization*, Kunming, Yunnan China, 15–19 April 2011.

22 Lansberry, E., Wozniak, L., and Goldberg, D.E. (1992). Optimal hydrogenerator governor tuning with a genetic algorithm. *IEEE Transactions on Energy Conversion* 7 (4): 623–630.

23 Djukanovic, M.B., Calovic, M.S., Vesovic, B.V., and Sobajic, D.J. (1997). Neuro-fuzzy controller of low head hydropower plants using adaptive-network based fuzzy inference system. *IEEE Transactions on Energy Conversion* 12 (4): 375–381.

24 Revankar, P.S., Thosar, A.G., and Gandhare, W.Z. (2010). Maximum power point tracking for PV systems using MATALAB/SIMULINK. *2010 Second International Conference on Machine Learning and Computing*, Bangalore, India, 09–11 February 2010.

25 Rudin, C., Waltz, D., Anderson, R.N. et al. (2012). Machine learning for the New York City power grid. *IEEE Transactions on Pattern Analysis and Machine Intelligence* 34 (2): 328–345.

26 Ramu, S.K., Irudayaraj, G.C., and Elango, R. (2021). An IoT-based smart monitoring scheme for solar PV applications. In: *Electrical and Electronic Devices, Circuits, and Materials: Technological Challenges and Solutions* (ed. S.L. Tripathi, P.A. Alvi, U. Subramaniam), 211–233. Scrivener Publishing LLC.

27 Shamshirband, S., Rabczuk, T., and Chau, K.W. (2019). A survey of deep learning techniques: application in wind and solar energy resources. *IEEE Access* 7: 164650–164666.

28 Kalantari, A., Kamsin, A., Shamshirband, S. et al. (2018). Computational intelligence approaches for classification of medical data: state-of-the-art, future challenges and research directions. *Neurocomputing* 276: 2–22.

29 Zou, L., Wang, L., Xia, A. et al. (2017). Prediction and comparison of solar radiation using improved empirical models and adaptive neuro-fuzzy inference systems. *Renewable Energy* 106: 343–353.

30 Yelmen, A. and Çakir, M.T. (2015). Estimation of average monthly total solar radiation on horizontal surface for Mediterranean region. *Energy Sources, Part A: Recovery, Utilization, and Environmental Effects* 37 (5): 543–551.

31 Alavi, O., Mohammadi, K., and Mostafaeipour, A. (2016). Evaluating the suitability of wind speed probability distribution models: a case of study of east and southeast parts of Iran. *Energy Conversion and Management* 119: 101–108.

32 Mohammadi, K. and Mostafaeipour, A. (2013). Using different methods for comprehensive study of wind turbine utilization in Zarrineh, Iran. *Energy Conversion and Management* 65: 463–470.

33 Zhang, Q., Yang, L.T., Chen, Z., and Li, P. (2018). A survey on deep learning for big data. *Information Fusion* 42: 146–157.

34 Krizhevsky, A., Sutskever, I., and Hinton, G.E. (2017). ImageNet classification with deep convolutional neural networks. *Communications of the ACM* 60 (6): 84–90.

35 Lee, D., Cisek, D., and Yoo, S. (2017). Sensor network-based wind field estimation using deep learning. *2017 New York Scientific Data Summit (NYSDS)*, New York, 06–09 August 2017.

36 Sun, Y., Szűcs, G., and Brandt, A.R. (2018). Solar PV output prediction from video streams using convolutional neural networks. *Energy & Environmental Science* 11 (7): 1811–1818.

37 Wang, H.Z., Li, G.Q., Wang, G.B. et al. (2017). Deep learning based ensemble approach for probabilistic wind power forecasting. *Applied Energy* 188: 56–70.

38 Arel, D.C.R. and Karnowski, T.P. (2010). Deep machine learning - a new frontier in artificial intelligence research [research frontier]. *IEEE Computational Intelligence Magazine* 5 (4): 13–18.

39 Ngiam, Z., Chen, P.W.K., and Ng, A.Y. (2011). Learning deep energy models. In: *Proceedings of the 28th International Conference on International Conference on Machine Learning*, Bellevue Washington USA, 28 June–02 July 2011, 1105–1112.

40 Lee, H., Grosse, R., Ranganath, R., and Ng, A.Y. (2011). Unsupervised learning of hierarchical representations with convolutional deep belief networks. *Communications of the ACM* 54 (10): 95–103.

41 Gensler, A., Henze, J., Sick, B. et al. (2016). Deep learning for solar power forecasting — an approach using autoencoder and LSTM neural networks. *2016 IEEE International Conference on Systems, Man, and Cybernetics (SMC)*, Budapest, 09–12 December 2016.

42 Liou, C.Y., Cheng, W.-C., Liou, J.W., and Liou, D.R. (2014). Autoencoder for words. *Neurocomputing* 139: 84–96.

43 Hinton, G.E. and Salakhutdinov, R.R. (2006). Reducing the dimensionality of data with neural networks. *Science* 313 (5786): 504–507.

44 Khodayar, M. and Teshnehlab, M. (2015). Robust deep neural network for wind speed prediction. *2015 4th Iranian Joint Congress on Fuzzy and Intelligent Systems (CFIS)*.

45 Hu, Q., Zhang, R., and Zhou, Y. (2016). Transfer learning for short-term wind speed prediction with deep neural networks. *Renewable Energy* 85: 83–95.

46 Qureshi, A.S., Khan, A., Zameer, A., and Usman, A. (2017). Wind power prediction using deep neural network based meta regression and transfer learning. *Applied Soft Computing* 58: 742–755.

25

Light Deep CNN Approach for Multi-Label Pathology Classification Using Frontal Chest X-Ray

Souid Abdelbaki[1], Soufiene B. Othman[2], Faris Almalki[3], and Hedi Sakli[1,4]

[1]*Department of Electrical Engineering, MACS Research Laboratory RL16ES22, National Engineering School of Gabes, Gabes University, Gabes, Tunisia*
[2]*Department of Telecom, PRINCE Laboratory Research, IsitCom, Hammam Sousse, Higher Institute of Computer Science and Communication Techniques, University of Sousse, Sousse, Tunisia*
[3]*Department of Computer Engineering, College of Computers and Information Technology, Taif University, Taif, Kingdom of Saudi Arabia*
[4]*EITA Consulting 5 Rue du Chant des oiseaux, Montesson, France*

25.1 Introduction

Today healthcare is one of the most promising sectors for the use of artificial intelligence and internet of things (IoTs)-based technologies, since patients may utilize wearable or implanted medical sensors to assess medical parameters they choose. Energy efficiency and aggregation of data are some of the critics of building an effective IoT–artificial-intelligence (AI) system [1–7]. Other research on blockchain is used to minimize security threats that may encounter data while being in transfer [8–13]. Our research focuses on the application of machine learning combined with computer vision to serve the medical field.

The chest X-ray scan is considered one of the most often utilized exams for screening and diagnosing lung disorders such as "infiltration, effusion, atelectasis, nodule, and mass" [14]. Because of increased health consciousness, the demand for chest X-ray scans examination is increasing around the world. Radiologists can benefit from advanced technology and automated algorithms by helping them efficiently diagnose illness with more precision. Trained tools can automatically predict abnormality from chest X-ray scans and categorize into groups, thus radiologists may pay closer attention to these aberrant images [15].

There are numerous classic categorization method as well improved versions. Because of its solid approach and incremental building, naive Bayes [16] is among the most effective inductive-learning algorithms for classification [17, 18]. In practice, these techniques cannot handle a huge number of chest X-ray pictures. Furthermore, most of them, such as SVM and naive Bayes, can only provide decent results on limited datasets. Before classification, researchers must manually extract image features such as local binary pattern and histogram of oriented gradient, which is time-consuming and challenging [19, 20]. Deep learning, particularly the architectural classification technique based on convolutional neural networks (CNNs), has grown in prominence in recent years due to its capacity to learn representative picture features via automated back propagation [21]. We did an extensive study on CNN referred to it in the related work section.

CNN is a strong method for training a robust image classifier based on a training set, on the other hand, is typically not available in the medical area due to patient privacy concerns. ImageNet has 1.3 million natural pictures for training large deep-CNN architectures, while most well-known medical image collections only have a few thousand images [22]. Reference [14] published Chest-X-ray14, a considerably larger collection with almost 30 000 patients and 112 120 annotated chest X-ray pictures. Despite being unbalanced and defective, Chest X-ray14 is considered a huge dataset to train on; therefore, we chose it as our dataset.

Machine Learning Algorithms for Signal and Image Processing, First Edition.
Edited by Deepika Ghai, Suman Lata Tripathi, Sobhit Saxena, Manash Chanda, and Mamoun Alazab.
© 2023 The Institute of Electrical and Electronics Engineers, Inc. Published 2023 by John Wiley & Sons, Inc.

Many publications have used transfer learning to analyze medical pictures, particularly chest X-ray scans, to excel in tasks like classification where the source and target domains should be separate yet linked [23], by fine-tuning state-of-the-art deep networks pretrained on a large dataset.

Under this study, we implemented a CNN to assess a range of chest diseases. Previous research used cutting-edge methods and produced highly accurate results, but they were time-consuming and inapplicable to a wide spectrum of diseases. The suggested methods' reliability and efficiency are slightly inferior due to a lack of fully standardized and inefficient data, to our knowledge; there is no practical technique to correctly diagnose numerous chest-related illnesses from X-ray pictures with trustworthy results. The objectives of this chapter are enlisted as follows:

- Using deep CNNs to categorize numerous chest-related illnesses and transfer learning to deal with inadequate and imbalanced data.
- To use deep CNNs for successful data augmentation in the categorization of 14 chest-related lung illnesses.
- To test MobileNet V2's performance on the task of classifying multiple classes from one input.

25.2 Related Work

A dozen studies in medicine have used artificial intelligence to help with medical diagnosis, and some of them have shown positive and accurate outcomes. This section examines approaches utilized by past researchers who used deep-neural networks and artificial intelligence to combat chest-related illnesses.

CXRs are used to identify the majority of chest-related illnesses. However, these CXRs are mostly exploited to identify extremely severe illnesses, which must be performed by a highly trained radiologist because it is a difficult process [24]. Sivasamy and Subashini [25] utilized a deep-learning approach to read CXRs and predict lung illnesses with 86.14% accuracy for the goal of using deep-neural networks, one-class detection of cardiothoracic illnesses is possible.

Zhang et al. [26] proposed building a supervised multi-organ segmentation model in X-ray images utilizing pixel-wise translated DRRs data. The X-ray image structure, as well as the gap in nodule annotations, is difficult and time-consuming to close. Rajpurkar et al. [27] demonstrated a 0.435 f1 binary CT scan pneumonia diagnosis classifier. Deep-neural networks are used for detecting multi-class pulmonary diseases. A few studies employed cutting-edge methods and had favorable outcomes for one or two pulmonary illnesses used meta-analysis, albeit they frequently had misclassification issues. Wang et al. [14] made one of the largest publicly available datasets of X-ray scans available to the scientific community, which they used for experiments and model training, and the majority of research has shown encouraging results implementing deep CNNs. They also stated that the dataset might be expanded to include more diseases in the future.

Rajpurkar et al. [28] presented an approach for classifying 14 underlying chest-related illnesses. To assess the likelihood of each of the 14 observations, the proposed model was trained using a single chest X-ray image. They trained many models to determine which one worked best, and in the end, DenseNet121 [29] had the highest accuracy and was picked for testing, despite the fact that it was limited only to the CheXpert [30] database and was prone to overfitting.

In addition, Sahlol et al. [31] proposed a hybrid technique for rapid identification and categorization of tuberculosis whether exists in the chest X-ray scan or not. MobileNet [32] is used in conjunction with transfer learning based on a CNN model previously trained on the ImageNet dataset to extract features from chest X-ray images. The AEO meta-heuristic method was employed as a component predictor to determine which of these qualities appeared to be the most relevant. The suggested method was evaluated on two publicly available datasets. It helps them to achieve substantial efficiency while lowering computational costs.

Souid, A. et al. [33] presented an approach to detect the 14 lung diseases classes provide by Wang et al. The proposed MobileNet V2 [34], in combination with transfer learning and classifier block, approach inspired us to

Table 25.1 Related work to chest X-ray classification.

References	Dataset	Method	Pros	Cons
Souid, A. et al. [33]	NIH Chest X-ray 14	MobileNet V2	Multi-class classification	Fine-tuning missing Data imbalance
Apostolopoulos et al. [35]	Customized	MobileNet V1	Implement feature extraction approaches	Targeting six classes base problem
Yao et al. [36]	NIH Chest X-ray 14	CNN and RCNN exploit the dependency on the labels	DenseNet [29] AsCNN skeleton	Low AUC
Li et al. [37]	NIH Chest x-ray 14	Detection + classification	ResNet50 [38]	Low AUC

improve his work by adding layers in the classifier block, also by fine-tuning the model and adjusting the dataset to prevent data imbalance.

Table 25.1 is a description of some relevant studies that highlight the benefits and drawbacks of existing techniques.

25.3 Materials and Method

In this section, we will go through the detail about our method in which we did use many techniques to achieve our proposed result. We will start by talking about the base model, and then we will go through our strategy of modifying data to build the model.

25.3.1 MobileNet V2

In our study, we used 88 layers architecture from the MobileNet family [32]. The depth-wise separable convolutions are used extensively in the MobileNet V1 convolution layer (Conv) presented in Figure 25.1. Each Conv layer is

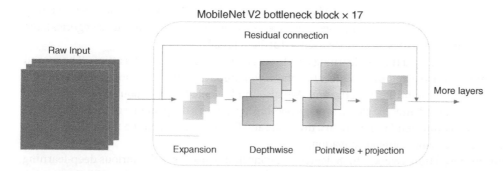

Figure 25.1 MobileNet V2 bottleneck: The base diagram of the bottleneck block, the idea is to add one expansion block before two depthwise blocks then reduces the computational cost we add the pointwise block to do a projection to the accumulated features.

Table 25.2 MobileNet V2 bottleneck block architecture.

Input	Operator	t	c	n	s
2242×3	conv2d	–	32	1	2
1122×32	Bottleneck	1	16	1	1
1122×16	Bottleneck	6	24	2	2
562×24	Bottleneck	6	32	3	2
282×32	Bottleneck	6	64	4	2
142×64	Bottleneck	6	96	3	1
142×96	Bottleneck	6	160	3	2
72×160	Bottleneck	6	320	1	1
72×320	conv2d 1x1	–	1280	1	1
72×1280	avgpool 7×7	–	–	1	–
$1 \times 1 \times 1280$	conv2d 1×1	–	k		–

followed by a batch normalization (BN) [34] and ReLU. In MobileNet V2, there are two major improvements from the first generation presented in Table 25.2.

The extension layer is the first new feature. The expansion layer is a 1×1 convolution whose function is to increase the number of channels in the picture data before proceeding to the depth convolution. As a result, the exit channel of this expansion layer is always bigger than the entry channel, which is the reverse of the projection layer. The second feature that was added to the MobileNet V2 building block is the residual connection inspired by the work presented in the ResNet [38] paper and assists in gradient transmission across the network. The feature channels are given an expansion factor t. For testing, we used MobileNet V2 with channel multipliers of 0.5× and 1× and input size of 224 224.

As illustrated in Table 25.2, the whole MobileNet V2 design consists of 17 of these construction blocks in a row, followed by a conventional 11 convolution, a global average pooling layer, and a classification layer (the very first block is slightly different; it uses a regular 33 convolution with 32 channels instead of the expansion layer).

25.3.2 Model Architecture

Transfer learning with extremely deep-neural networks is based on retraining a CNN model that was previously trained on ImageNet [39]. Because the dataset comprises a diverse range of objects (1000 unique categories), the model may learn a variety of characteristics that can be used to further classify tasks [40].

The proposed model is constructed of a transfer learning part to accelerate the extraction of features. The second part is the classification block constructed from a global average pooling layer with 1024 nodes followed by a fully connected layer with 512 nodes and ReLU activation. We also implemented two dropout layers to prevent overfitting. The first was implemented after the global average pooling layer with a dropout rate of 50%. The second layer was placed after the fully connected layer with 70% dropout rate. Finally, we rapped the model with a fully connected layer that has14 classes with sigmoid activation.

Figure 25.2 depicts our proposed framework, which depicts the conceptual process of the various deep-learning model architectures used in this study. TensorFlow, developed by Google, is the primary deep-learning framework used throughout the development process to build, train, validate, fine-tune, and test different architectures. Aside from TensorFlow, Pandas was utilized for data processing, while Matplotlib was used to create the visualization.

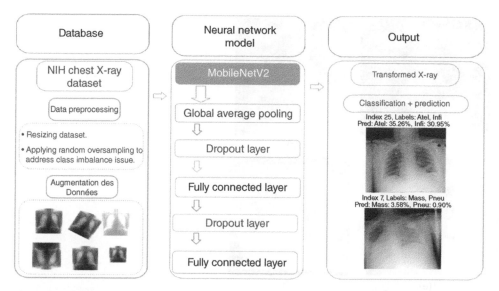

Figure 25.2 Proposed model architecture: The first block focuses on the database preprocessing that include data cleaning and image processing, our architecture is based on MobileNet V2 with and global averaging pooling layer presented in the second block and a fully connected layer with 14 outputs.

Raw data and labels are initially imported into the workspace using Panda's data frame. The whole dataset is based on 14 pathological illness designations in total.

25.4 Proposed Methodology

In this section, we present the study and analysis from our model based on existing works. We divide the section into two parts; the first part discusses the data utilized for the work, including data processing. The second part presents the followed method for training our proposed model.

25.4.1 Dataset Preparation and Preprocessing

We used the Chest X-ray14 database in this work, which is a comprehensive state-of-the-art dataset obtained by Wang et al. [14]. This resource is present as a highly contributed publicly available X-ray database.

It contains 112 120 frontal-view samples (CXR's) gathered from 30 805 different individual patients. All of the pictures were tagged with one of the 14 pathological illnesses and labeled as "No Findings" for the negative cases; the pathologies that existed in the NIH Chest X-ray14 are named as follows (Figure 25.3 shows some of the diseases: "Infiltration, Mass, Nodule, Pleural Thickening, Atelectasis, Cardiomegaly, Consolidation, Edema, Effusion, Emphysema, Fibrosis, Hernia, Pneumonia, and Pneumothorax").

The amount of data image samples were biased as we got the "In-filtration" as the most frequent with (9546) images sample (mentioned in Figure 25.4), and the "Hernia" as the least prevalent with only 227 samples, while "Emphysema, Fibrosis, Edema, and Pneumonia" are close but low, with a number of samples ranging from (322) to (727). The frequency of classes is more than (1093) and less than (4214) for the other diseases, as illustrated in Figure 25.4c. The issue of class disparity will have a big impact on the outcome. We have done data augmentation and split the dataset into training, validation, and testing. Figure 25.4a,b shows the distribution of the least common illness, "Hernia," by relating to meta-data (age, gender, negative case, or positive case).

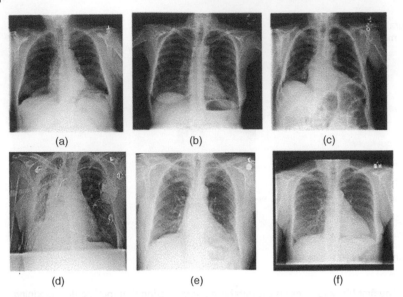

Figure 25.3 The Chest X-ray 14 dataset is represented by six samples. The utilized dataset comprises 112 120 frontal chest X-rays (Anterior Posterior–Posterior Anterior) from 30 805 patients. Each image is labeled with one of up to 14 pathologies or "No Finding." The dataset does include acute symptoms, such as the Nodule in (f), as well as those who were treated with a drain for "Pneumothorax" (d). (a) Emphysema, (b) Infiltration, (c) Atelectasis, (d)Pneumothorax, (e) No Findings, and (f) Nodule.

To address this class imbalance problem, many studies are being done in the field. We implemented the random over-sampling method. This method tries to resample a given data to minimize the impact of the data imbalance.

To get a more stable dataset, we remove the No finding class, as it presents about 35% from the total dataset by incorporating data augmentation. We were able to generalize the data.

Image augmentation generates training data by combining diverse processing or assembly methods used by various systems, such as random rotation, turns, shear, flips, and magnification, among others.

The original dataset image size is $(1024 \times 1024 \times 3)$. To get the dataset compatible with our model, we resize the data from $(1024 \times 1024 \times 3)$ to $(224 \times 224 \times 3)$.

(a) Training process
- First, as shown in Figure 25.2, we construct our neural network with a pre-trained weight set to the "ImageNet" dataset. Then, we included the [GAP-Drpt-FC (ReLU) Drpt-FC (Sigmoid)] classification block. The last FC layer corresponds to the 14 labels and has a sigmoid activation function.
- We split our dataset to 80% training and 10% validation and 10% testing, size of the input $(224 \times 224 \times 3)$, and 36 for the batch size.
- To build the model, 'binary-cross-entropy losses were utilized as the loss function and "Adam" as the optimizing algorithm, using "accuracy" as a measure.
- The training begins with a learning rate of 0.1e−2 but is gradually reduced to 0.01e−3 using the Reduce LR on Plateau scheduler.
- The model was initially trained by freezing all of the layers of the pre-trained model for 5 epochs with half the steps per epoch, followed by a second training phase with 40 epochs to fine-tune the model, the second training phase is started with 40 epochs, and training on all of the model's layers, including the pre-trained weights from "ImageNet," as well as a 0.5 threshold to improve the accuracy and allow the model to generalize better on the validation dataset.

(b) Evaluation

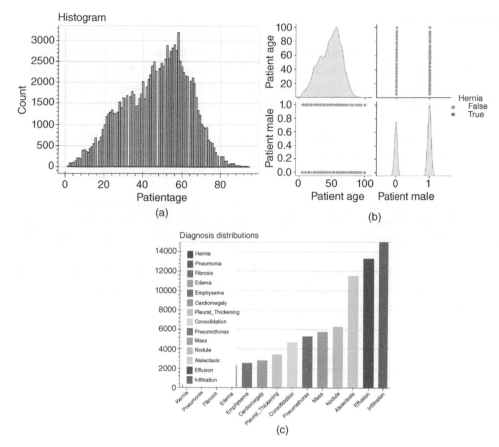

Figure 25.4 Class distribution inside the dataset. The distribution of patient gender. (a) Male and female age distribution in the dataset. (b) "Hernia" distribution with meta-information features, including age, gender, and true/false disease (true with dark gray, false with light gray). (c) 14 class frequency distribution the dataset. The least frequent class is hernia with 422 sample.

As for performance metrics, we utilized the following Eqs. (25.1)–(25.5) accuracy (Acc), sensitivity (Sens), specificity (Spec), AUC, F_1-score, and time consumption. The following are their definitions:

$$\text{Accuracy} = \frac{TP + TN}{TP + TN + FP + FN} \tag{25.1}$$

$$\text{Precision} = \frac{TP}{TP + FP} \tag{25.2}$$

$$F_1 = \frac{TP}{TP + \frac{1}{2}(TP + FN)} \tag{25.3}$$

$$\text{Sensitivity} = \frac{TP}{TP + FN} \tag{25.4}$$

$$\text{Specificity} = \frac{TN}{TN + FP} \tag{25.5}$$

TP (true positive) represents the number of disease samples that are properly classified, whereas TN (true negative) represents the number of miss-identified disease samples. The number of disease samples that were false identified as disease-free in the sample is denoted by "FP"(false positive), whereas the number of disease-free samples that were not classified as disease samples is denoted by "FN"(false negative).

25.5 Result and Discussions

In the presented study, we investigate the performance of MobileNet V2 architecture to design a solid predictive model. After applying the needed preprocessing on the dataset referred in section 4.1, we chose to work with the MobileNet V2 neural network based on the work of Souid, A. et al. [33]. They proposed to apply the MobileNet V2 with transfer learning to classify lung diseases, in comparison to the state-of-the-art CNN, they get extremely respectable results.

In the first iteration of training (five epochs with half the steps per epoch), the neural network did not surpass the 90% of accuracy. When we push the full steps per epochs with 40 epochs, we got a better result from the previous work of Souid, A. et al. [33], as illustrated in Figures 25.5 and 25.6. Our model accuracy did surpass the 95% of accuracy.

As we can see, the loss did not get a result as we would expect from our work, as the value of the minimal loss did not get lower than 0,156, thus we utilized the other metric provided in section 4.3. Table 25.3 illustrates the experimental result of our model, including the accuracy, sensitivity, specificity, and F_1 score from some of the diseases classes.

Based on these findings, we may conclude that specificity yielded some promising results. The sensitivity, on the other hand, did not provide a satisfactory outcome. The sensitivity of the courses was quite low, even in the other classes, and this poor sensitivity significantly reduced the F_1 score.

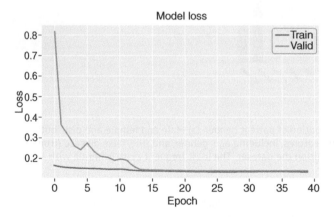

Figure 25.5   Model loss: The evolution of model Loss curve during training and validation phase.

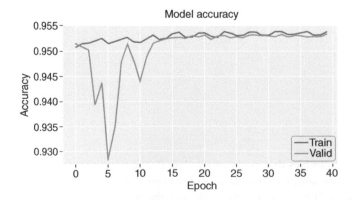

Figure 25.6   Model accuracy: The evolution of model Accuracy during training and validation phase.

Table 25.3 Our method's experimental results.

Pathology	Accuracy	Sensitivity	Specificity	F_1-score
Emphysema	0.953	0.115	0.998	0.2
Infiltration	0.616	0.015	0.989	0.560
Mass	0.885	0.22	0.964	0.5
Atelectasis	0.79	0.386	0.906	0.45
Cardiomegaly	0.826	0.717	0.833	0.326
Effusion	0.721	0.787	0.699	0.583

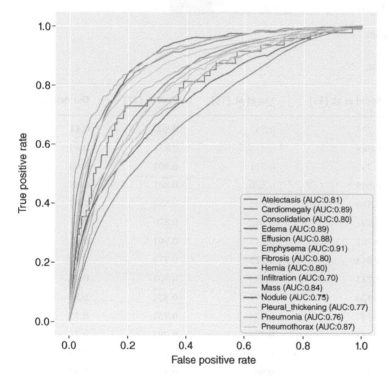

Figure 25.7 ROC curves for our model across 40 epochs: (AUC 0.91) for "Emphysema" and (AUC 0.70) for "Infiltration".

The AUC_ROC curve is used to understand more about our model's performance. Figure 25.7 depicts a ROC curve that compares the projected picture's false positive rate to its true positive rate. The AUC for "Emphysema" is (0.91), the AUC for "Effusion, Edema" are (0.88) and (0.89), the AUC for "Pneumothorax, Cardiomegaly" are (0.87) and (0.89), respectively, while the AUC for "Mass, Hernia, and Atelectasis are (0.84), (0.80), and (0.81). They had the lowest AUC for the remaining three diseases, with "Nodule" AUC of (0.75), "Pneumonia" AUC of (0.76), and "Infiltration" AUC of (0.70). Figure 25.8 shows that the ground truth label in the image with index 7 was Mass and Pneumonia, given that the model had (AUC 0.76). It's not unexpected that he got a 99% and 6% Mass forecast. In the second image, the model produced a solid result; the real illness in this image was Effusion, and the model predicted that it would be 64% Effusion.

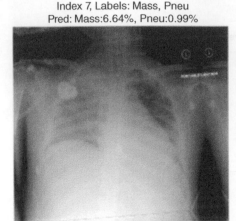
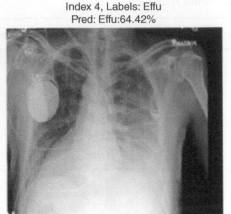

Figure 25.8 Model prediction based on the test set.

Table 25.4 AUC result overview for our experiment.

Pathology	Wang et al. [14]	Souid et al. [33]	Yao et al. [36]	Li et al. [37]	Our new model
Atelectasis	0.700	0.794	0.733	0.800	**0.81**
Cardiomegaly	0.810	0.885	0.856	0.871	**0.89**
Consolidation	0.703	0.790	0.711	**0.801**	**0.80**
Edema	0.805	**0.884**	0.806	0.881	**0.88**
Effusion	0.759	0.876	0.806	0.859	**0.89**
Emphysema	0.833	0.891	0.842	0.870	**0.91**
Fibrosis	0.786	0.762	0.743	**0.901**	0.80
Hernia	0.872	**0.811**	0.775	0.773	0.80
Infiltration	0.661	**0.711**	0.673	0.701	0.70
Mass	0.693	0.826	0.777	**0.831**	**0.84**
Nodule	0.669	0.743	0.724	**0.751**	**0.75**
Pleural Thickening	0.669	0.763	0.724	**0.791**	0.77
Pneumonia	0.658	0.733	0.684	0.671	**0.76**
Pneumothorax	0.799	**0.880**	0.805	0.871	0.87
Average	0.745	0.810	0.761	0.812	**0.819**

Text in bold present where our method exceeds other researches.

According to the findings, our model performed well in terms of AUC, with an average AUC score of 0.819. Table 25.4 compares our MobileNetV2 model to deep learning-based studies in the realm of chest pathology. It can be seen that our models had a significantly higher AUC in 9 of the 14 cardiothoracic diseases studied.

To begin, we compared the obtained results to those by Wang et al. [13] and Yao et al. [36]. While ResNet-38-large-meta and Yao et al. [36] have a higher overall AUC of 0.819 and a superior individual AUC in 9 out of 14 disease classes, MobileNet V2 has a 30% improvement in more than 3 of them. Li et al. [37] are among those who have contributed to this study. While our MobileNet V2 was trained with fewer pictures, it nevertheless produced good results for "Atelectasis, Cardiomegaly, Consolidation, Effusion, Edema, Infiltration, Pneumonia," and "Pneumothorax" as well, a slightly higher average AUC of 0.819.

25.6 Conclusion

When compared to Souid, A. et al. [33], In 9 of 14 classes, our improved MobileNet V2 architecture produces extremely respectable results (who had state-of-the-art results in 5 out of 14 classes on the official split). In the literature, other techniques have received higher evaluations.

In this study, our model lays the way for on-the-fly use of light neural network architecture in the medical sector (AUC was good, average 0.819, accuracy 95%), which will be built upon in future experiments.

Despite the good AUC, our model suffers from the sensitivity measure, which is driven by the forecast's increased false negative classification of data. The low sensitivity rate is caused by an imbalance in the data class distribution, which we addressed in the technique section by utilizing random average sampling; for better results, for better outcomes, we may use a more advanced methodology such as SEMOTE.

While the analysis suggests that deep neural network training in the medical profession is a viable alternative, the actual implementation of deep learning in clinical practice remains in progress until more public databases become more available. The extremely high tag noise of 10%, notably for the Chest X-ray14 datasets, makes assessing the real network difficult. As a result, a clean test set without a label is necessary for clinical efficiency evaluation.

References

1 Faris, A.A., Othman, S.B., Almalki, F.A., and Sakli, H. (2021). EERP-DPM: Energy efficient routing protocol using dual prediction model for healthcare using IoT. *Journal of Healthcare Engineering* 2021: Article No: 9988038.

2 Faris, A.A. and Othman, S.B. (2021). EPPDA: an efficient and privacy-preserving data aggregation scheme with authentication and authorization for IoT-based healthcare applications. *Wireless Communications and Mobile Computing* 2021: Article No: 5594159.

3 Bahhar, C., Baccouche, C., Othman, S.B., and Sakli, H. (2021). Real-time intelligent monitoring system based on IoT. In: *2021 18th International Multi-Conference on Systems, Signals & Devices (SSD)*, 93–96. https://doi.org/10.1109/SSD52085.2021.9429358.

4 Leila, E., Othman, S.B., and Sakli, H. (2020). An internet of robotic things system for combating coronavirus disease pandemic (COVID-19), *20th International Conference on Sciences and Techniques of Automatic Control & Computer Engineering*, Monastir, Tunisia (20–22 December).

5 Othman, S.B., Bahattab, A.A., Trad, A. and Youssef, H. (2020). PEERP: an priority-based energy-efficient routing protocol for reliable data transmission in healthcare using the IoT. *The 15th International Conference on Future Networks and Communications (FNC)*, Leuven, Belgium (9–12 August).

6 Othman, S.B., Bahattab, A.A., Trad, A. et al. (2020). LSDA: lightweight secure data aggregation scheme in healthcare using IoT. *ACM - 10th International Conference on Information Systems and Technologies*, Lecce, Italy (June).

7 Othman, S.B., Bahattab, A.A., Trad, A. et al. (2019). RESDA: robust and efficient secure data aggregation scheme in healthcare using the IoT. *The International Conference on Internet of Things, Embedded Systems and Communications (IINTEC 2019)*, Hammamet, Tunisia (20–22 December).

8 Saxena, S., Bhushan, B., and Ahad, M.A. (2021). Blockchain based solutions to secure IoT: background, integration trends and a way forward. *Journal of Network and Computer Applications* 181: 103050. https://doi.org/10.1016/j.jnca.2021.103050.

9 Haque, A.K., Bhushan, B., and Dhiman, G. (2021). Conceptualizing smart city applications: requirements, architecture, security issues, and emerging trends. *Expert Systems* https://doi.org/10.1111/exsy.12753.

10 Kumar, A., Abhishek, K., Bhushan, B., and Chakraborty, C. (2021). Secure access control for manufacturing sector with application of ethereum blockchain. *Peer-to-Peer Networking and Applications* https://doi.org/10 .1007/s12083-021-01108-3.

11 Saxena, S., Bhushan, B., and Ahad, M.A. (2021). Blockchain based solutions to secure Iot: background, integration trends and a way forward. *Journal of Network and Computer Applications* 103050: https://doi.org/10.1016/j .jnca.2021.10305.

12 Bhushan, B., Sahoo, C., Sinha, P., and Khamparia, A. (2020). Unification of blockchain and internet of things (BIoT): requirements, working model, challenges and future directions. *Wireless Networks* https://doi.org/10 .1007/s11276-020-02445-6.

13 Bhushan, B., Sinha, P., Sagayam, K.M., and J, A. (2021). Untangling blockchain technology: a survey on state of the art, security threats, privacy services, applications and future research directions. *Computers & Electrical Engineering* 90: 106897. https://doi.org/10.1016/j.compeleceng.2020.106897.

14 Wang, X., Peng, Y., Lu, L. et al. (2017). Chest X-ray8: hospital-scale chest x-ray database and benchmarks on weakly-supervised classification and localization of common thorax diseases. In: *2017 IEEE Conference on Computer Vision and Pattern Recognition (CVPR)*, 3462–3471. IEEE.

15 Kamel, S.I., Levin, D.C., Parker, L. et al. (2017). Utilization trends in noncardiac thoracic imaging, 2002–2014. *Journal of the American College of Radiology* 14 (3): 337–342.

16 Zhang, H. (2004). The optimality of naive Bayes. *The Association for the Advancement of Artificial Intelligence (AAAI) Conference* (25–29 July 2004). San Jose, California.

17 Amor, N.B., Benferhat, S., and Elouedi, Z. (2004). Naive Bayes vs decision trees in intrusion detection systems. In: *Proceedings of the 2004 ACM Symposium on Applied Computing*, 420–424. ACM.

18 Chen, W., Yan, X., Zhao, Z. et al. (2019). Spatial prediction of landslide susceptibility using data mining-based kernel logistic regression, naive Bayes and RBF network models for the Long County area (China). *Bulletin of Engineering Geology and the Environment* 78: 247–266.

19 Hu, Z., Tang, J., Zhang, P. et al. (2018). Identification of bruised apples using a 3-D multi-order local binary patterns based feature extraction algorithm. *IEEE Access* 6: 34846–34862.

20 Xiang, Z., Tan, H., and Ye, W. (2018). The excellent properties of a dense grid-based HOG feature on face recognition compared to Gabor and LBP. *IEEE Access* 6: 29306–29319.

21 Bar, Y., Diamant, I., Wolf, L. et al. (2015). Deep learning with non-medical training used for chest pathology identification. In: *Medical Imaging 2015: Computer-Aided Diagnosis*, vol. 9414, 94140V. International Society for Optics and Photonics.

22 Krizhevsky, A., Sutskever, I., and Hinton, G.E. (2012). Imagenet classification with deep convolutional neural networks. In: *Advances in Neural Information Processing Systems*, 1097–1105.

23 Sevakula, R.K., Singh, V., Verma, N.K. et al. (2018). Transfer learning for molecular cancer classification using deep neural networks. *IEEE/ACM Transactions on Computational Biology and Bioinformatics* 16 (6): 2089–2100.

24 Franquet, T. (2018). Imaging of community-acquired pneumonia. *Journal of Thoracic Imaging* 33 (5): 282–294. https://doi.org/10.1097/rti.0000000000000347.

25 Sivasamy, J. and Subashini, T. (2020). Classification and predictions of lung diseases from chest X-rays using MobileNet. *International Journal of Analytical and Experimental Modal Analysis* 12 (3): 665–672.

26 Zhang, Y., Miao, S., Mansi, T. et al. (2018). Task driven generative modeling for unsupervised domain adaptation: application to X-ray image segmentation. In: *Medical Image Computing and Computer Assisted Intervention MICCAI*, 599–607. https://doi.org/10.1007/978-3-030-00934-2_67.

27 Rajpurkar, P., Park, A., Irvin, J. et al. (2020). AppendiXNet: deep learning for diagnosis of appendicitis from a small dataset of CT exams using video pretraining. *Scientific Reports* 10 (1): 3958. https://doi.org/10.1038/ s41598-020-61055-6.

28 Rajpurkar, P., Irvin, J., Ball, R.L. et al. (2018). Deep learning for chest radiograph diagnosis: a retrospective comparison of the CheXNeXt algorithm to practicing radiologists. *PLoS Medicine* 15 (11): e1002686. https://doi .org/10.1371/journal.pmed.1002686.

29 Huang, G., Liu, Z., Maaten, L.V.D. et al. (2017). Densely connected convolutional networks. *IEEE Conference on Computer Vision and Pattern Recognition (CVPR)* (21–26 July). Honolulu, HI, USA. https://doi.org/10.1109/CVPR.2017.243.

30 Irvin, J., Rajpurkar, P., Ko, M. et al. (2019). CheXpert: a large chest radiograph dataset with uncertainty labels and expert comparison. arXiv:1901.07031 [cs, eess], janvier, 1704.04861.

31 Sahlol, A.T., AbdElaziz, M., Tariq, J.A. et al. (2020). A novel method for detection of tuberculosis in chest radiographs using artificial ecosystem-based optimization of deep neural network features. *Symmetry* 12 (7): 1146. https://doi.org/10.3390/sym12071146.

32 Howard, A.G., Zhu, M., Chen, B. et al. (2017). MobileNets: efficient convolutional neural networks for mobile vision applications. arXiv:1704.04861[cs], avril.

33 Souid, A., Sakli, N., and Sakli, H. (2021). Classification and predictions of lung diseases from chest X-rays using MobileNet V2. *Applied Sciences* 11: 2751. https://doi.org/10.3390/app11062751.

34 Sandler, M., Howard, A., Zhu, M. et al. (2018). MobileNet V2: inverted residuals and linear bottlenecks. *IEEE/CVF Conference on Computer Vision and Pattern Recognition*, Salt Lake City. 1801.04381.

35 Apostolopoulos, I.D., Aznaouridis, S.I., and Tzani, M.A. (2020). Extracting possibly representative COVID-19 biomarkers from X-ray images with deep learning approach and image data related to pulmonary diseases. *Journal of Medical and Biological Engineering* 40: 462–469. https://doi.org/10.1007/s40846-020-00529-4.

36 Yao, L., Poblenz, E., Dagunts, D. et al. (2018). Learning to diagnose from scratch by exploiting dependencies among labels. *International Conference on Learning Representations ICLR*, Vancouver. 1710.10501v2.

37 Li, Z., Wang, C., Han, Mei., et al. (2018). Thoracic disease identification and localization with limited supervision. *IEEE/CVF Conference on Computer Vision and Pattern Recognition*, Salt Lake City.

38 He, K., Zhang, X., Ren, S. et al. (2015). Deep residual learning for image recognition. arXiv preprint, 1512.03385.

39 Deng, J., Dong, W., Socher, R. et al. (2009). ImageNet: A large-scale hierarchical image database. *IEEE Conference on Computer Vision and Pattern Recognition* (20–25 June), 248–255. Miami, FL, USA. https://doi.org/10.1109/CVPR.2009.5206848.

40 Shukla, P.K., Sandhu, J.K., Ahirwar, A. et al. (2021). Multi-objective genetic algorithm and convolutional neural networks based COVID-19 identification in chest X-ray images. *Mathematical Problems in Engineering* 2021: 1–9. https://doi.org/10.1155/2021/7804540.

Index

Machine Learning Algorithms for Signal and Image Processing, First Edition.
Edited by Deepika Ghai, Suman Lata Tripathi, Sobhit Saxena, Manash Chanda, and Mamoun Alazab.
© 2023 The Institute of Electrical and Electronics Engineers, Inc. Published 2023 by John Wiley & Sons, Inc.